Student Study Guide and Problems Book
for

Principles of Biochemistry With a Human Focus

Garrett & Grisham

David K. Jemiolo
Vassar College

Steven M. Theg
University of California, Davis

Harcourt College Publishers
Fort Worth | Philadelphia | San Diego | New York | Orlando | Austin
San Antonio | Toronto | Montreal | London | Sydney | Tokyo

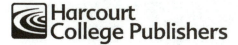

Printed in the United States of America

ISBN 0030973716

123 202 7654321

Preface

• •

The Course

In my early years at Vassar, the College had a senior-level, full-year course in biochemistry that required both organic chemistry and intermediate-level biology. The one disadvantage to this course was that students had to wait until their senior year to discover this fascinating and exciting subject. The course was largely taken by biochemistry majors but occasionally other majors would take the course and later lament about how they would have majored in the field had they discovered it earlier in their college tenure. To address this shortcoming we now teach a one-semester course. It still requires organic chemistry but we allow students to take it after completing one semester of organic chemistry.

Since initiating the curriculum, we have experienced a renewed interest in biochemistry. Student demand for the course is so high that we teach two sections typically at capacity. In addition to biochemistry majors, we now attract biology majors, other science majors, and students interested in completing premedical requirements. "*The Principles of Biochemistry with a Human Focus*" presents biochemistry at a pace and depth ideally suited for our course. In addition, it does a great job of showing biochemistry at work in human physiology, an approach always popular with students.

The Approach

In one scene in the movie Stripes (Columbia Picture Corporation 1981), privates John Winger and Russell Zissky (played by Bill Murray and Harold Ramis) attempt to persuade their platoon to an all night training session to prepare for the next day's final parade. The troops are skeptical of the plan; however, Zissky wins them over by his testimony of the importance of cramming. He proudly reports that he had, in fact, once learned two semesters of geology in a single three-hour all nighter.

It would seem unlikely that this approach would work well with biochemistry (or even geology). Rather a steady diet of reading, problem solving, and reviewing might be a better plan of attack. This study guide was written to accompany "*The Principles of Biochemistry with a Human Focus*" by Garrett and Grisham. It includes chapter outlines, guides to key points covered in the chapters, in-depth solutions to the problems presented in the textbook, additional problems, and detailed summaries of each chapter. In addition, there is a glossary of biochemical terms and key text figures.

Several years ago I spent part of a sabbatical in Italy and in preparation took a year-long course in elementary Italian. I had not been on the student-end of an academic interaction for several years and taking a language course was an excellent opportunity to be reminded of the difficulties of learning something for the first time. Memorization is part and parcel to the study of any language and so I found myself committing to memory nouns, verbs, adverbs, adjectives, and complex, irregular verb conjugations. The study of biochemistry has parallels to language studies in that memorization is necessary. What makes the study of biochemistry somewhat easier, however, are the common themes, the interconnections between various facets of biochemistry, and the biological and chemical principles at work. The authors have done a marvelous job in presenting these aspects of biochemistry and I have attempted to highlight them here. Biochemistry is a demanding discipline but one well worth the effort for any student of the sciences. *Buona fortuna.*

Acknowledgments

It is often stated that teaching a subject is the best way to learn it. In teaching my one-semester biochemistry course, because I never have sufficient time to cover all the topics, I used to worry about forgetting certain aspects of biochemistry. Thanks to Charles Grisham and Reginald Garrett this fear is no longer with me. I thank both authors for the marvelous text and the opportunity to relearn all of biochemistry. I also thank them for chapter summaries. I thank Sandi Kiselica of Saunders College Publishing for all her hard work and for the glossary of terms. I thank my co-author, Steven Theg. When you walk along through the dark you always hear noises but when you walk with someone else, it's either better or worse. With Steven, it was better. Finally, I thank my wife Kristen P. Jemiolo for carrying my share of the family responsibilities and Kaedin M. Jemiolo and Tristan M. Jemiolo for understanding why I couldn't always have a catch or shoot on goal.

Special Thanks

Several years ago, John Vondeling, my publisher from Saunders, asked me to review the first edition of a biochemistry book by Garrett and Grisham and to write its accompanying study guide. John passed away this year. Although I never met him, once or twice a year over the last eight or so years he and I would chat over the phone about projects. It was always exciting working with him and I will miss this. John knew what he wanted and knew what was fair.

David K. Jemiolo
Poughkeepsie, NY
May 2001

Acknowledgments

I am a teacher of introductory biochemistry, but also a researcher. In the latter role, my research group and I regularly make use of our collective biochemical expertise, and as with all who become familiar with a subject through daily use, we rarely revisit the fundamental concepts that led us to our working knowledge. In designing my biochemistry course, I was forced to think deeply about each topic presented and extract those parts that seemed particularly important and illuminating to me. I came away with a new, deeper understanding of my science. The work on this Study Guide took me down a similar, but much broader path. Certainly, I learned or relearned many facts, but what I really enjoyed was thinking again about many old ones and coming to new insights about them. I am grateful to Saunders College Publishing for making this possible, and to Charles Grisham and Reginald Garrett for providing such an insightful and well-written text with which to do it. It is a rare pleasure to have the opportunity to re-examine after many years the fundamentals of one's profession, and I recommend it to all.

I am grateful to my co-author, David Jemiolo, for his openness and encouragement, for his hard work, and for our many laughter-filled, transcontinental conversations. I couldn't have found a better colleague with whom to work. Finally, I thank Jill Theg and our sons Chris, Alex and Sam for their constant support. Without them little of what I do would seem much worth doing.

Steven M. Theg
Davis, CA
May 2001

Why study biochemistry?

This excerpt from *Poetry and Science* by the Scottish poet Hugh MacDiarmid (1892-1978), which first appeared in *Lucky Poet* (1943) might help with an answer.

Poetry and Science

...
Wherefore I seek a poetry of facts. Even as
The profound kinship of all living substance
Is made clear by the chemical route.
Without some chemistry one is bound to remain
Forever a dumbfounded savage
In the face of vital reactions.
The beautiful relations
Shown only by biochemistry
Replace a stupefied sense of wonder
With something more wonderful
Because natural and understandable.
Nature is more wonderful
When it is at least partly understood.
Such an understanding dawns
On the lay reader when he becomes
Acquainted with the biochemistry of the glands
In their relation to diseases such as goitre
And their effects on growth, sex, and reproduction.
He will begin to comprehend a little
The subtlety and beauty of the action
Of enzymes, viruses, and bacteriophages,
These substances which are on the borderland
Between the living and the non-living.
He will understand why the biochemist
Can speculate on the possibility
Of the synthesis of life without feeling
That thereby he is shallow or blasphemous.
He will understand that, on the contrary,
He finds all the more
Because he seeks for the endless
---'Even our deepest emotions
May be conditioned by traces
Of a derivative of phenanthrene!'

Science is the
Differential Calculus
of the mind,
Art is the Integral
Calculus; they may be
Beautiful apart, but
are great only when
combined.

Sir Ronald Ross

Comments, corrections, and suggestions may be directed to GGGuide@Vassar.edu.

Table of Contents

Chapter 1

Chemistry Is the Logic of Biological Phenomena

• •

Chapter Outline

- Properties of living systems
  - Highly organized: Cells > organelles > macromolecular complexes > macromolecules (proteins, nucleic acids, polysaccharides)
  - Structure/function correlation
  - Energy transduction
  - Steady state by virtue of energy flow
  - Self-replication and importance of structural complementarity
- Biomolecules
  - Elements: Hydrogen, oxygen, carbon, nitrogen (lightest elements of the periodic table capable of forming a variety of strong covalent bonds)
  - Compounds: Carbon-based compounds
- Biomolecular hierarchy
  - Simple compounds: H_2O, CO_2, NH_4^+, NO_3^-, N_2
  - Metabolites
  - Building blocks: Amino acids, nucleotides, monosaccharides
  - Macromolecules: Proteins, nucleic acids, polysaccharides
- Membranes: Lipid bilayers with membrane proteins
- Organelles: Mitochondria, chloroplasts, nuclei, etc.
- Properties of biomolecules
  - Directionality or structural polarity
    - Proteins: N-terminus and C-terminus
    - Nucleic acids: 5'- and 3'- ends
    - Polysaccharides: Reducing and nonreducing ends
  - Information content: Sequence of monomer building blocks and 3-dimensional architecture
- 3-Dimensional architecture and intermolecular interactions (via complementary surfaces) of macromolecular are based on weak forces
  - van der Waals interactions
  - Hydrogen bonding
  - Ionic interactions
  - Hydrophobic interactions
- Life restricted to narrow range of conditions (temperature, pH, salt concentration, etc.) because of dependence on weak forces
- Enzymes: Biological catalysts capable of being regulated
- Cell types
  - Prokaryotes: Eubacteria and archaea: Plasma membrane but no internal membrane-defined compartments

- Eukaryotes: Internal membrane-defined compartments: nuclei, endoplasmic recticulum, Golgi, mitochondria, chloroplasts, vacuoles

Chapter Objectives

Understand the basic chemistry of H, O, N and C.

H forms a single covalent bond. When bound to an electronegative element, like O or N, the electron pair forming the covalent bond is not equally shared, giving rise to a partial positive charge on the hydrogen (this is the basis of H bonds which will be covered in the next chapter). In extreme cases the H can be lost as a free proton.

O forms two covalent bonds and has two lone pairs of electrons. It is an electronegative element and when bound to hydrogen it will cause H to be partially positively charged. O is highly reactive due to its high electronegativity.

N forms up to three covalent bonds and has a single lone pair of electrons. It is an electronegative element and will create a partial positive charge on a hydrogen bonded to it.

C forms four covalent bonds. With four single bonds, tetrahedral geometry is predominant. With one double bond, carbon shows trigonal planar geometry, with an additional pair of electrons participating in a pi bond.

Macromolecules and subunits

Proteins are formed from amino acids composed of C, H, O, N and in some instances S.

Nucleic acids are formed from nucleotides, which are composed of phosphate, sugar and nitrogenous base components. (Nucleosides lack phosphate).

Polysaccharides are made of monosaccharides or sugar molecules (carbohydrates).

Lipids are a class of mostly nonpolar, mostly hydrocarbon molecules.

Macromolecular structures

Macromolecular structures are composed of complexes of macromolecules (i.e., proteins, nucleic acids, polysaccharides and lipids). The ribosome, made up of protein and ribonucleic acid, is a prime example.

Organelles

Organelles are subcellular compartments defined by lipid bilayer membranes.

Cell types

There are two fundamental cell types: eukaryotic, having organelles and a defined nuclear region, and prokaryotic, lacking organelles and a membrane-enclosed region of genetic material. The archaea and eubacteria comprise the prokaryotes.

Problems and Solutions

1. The nutritional requirements of* Escherichia coli *cells are far simpler than those of humans, yet the macromolecules found in bacteria are about as complex as those of animals. Since bacteria can make all their essential biomolecules while subsisting on a simpler diet, do you think bacteria may have more biosynthetic capacity and hence more metabolic complexity than animals? Organize your thoughts on this question, pro and con, into a rational argument.

Answer: Although it is true that *Escherichia coli* are capable of producing all of their essential biomolecules (there is no minimum daily requirement for vitamins in the world of wild-type *E. coli*), they are rather simple, single-cell organisms capable of a limited set of responses. They are self sufficient, yet they are incapable of interactions leading to levels of organization such as multicellular tissues. Multicellular organisms have the metabolic complexity to produce a number of specialized cell types and to coordinate interactions between them.

2. Without consulting chapter figures, sketch the characteristic prokaryotic and eukaryotic cell types and label their pertinent organelle and membrane systems.

Answer: Prokaryotic cells lack the compartmentation characteristic of eukaryotic cells and are devoid of membrane bound organelles such as mitochondria, chloroplasts, endoplasmic reticulum, Golgi apparatus, nuclei, and vacuoles. Both cell types are delimited by membranes and contain ribosomes.

3. Escherichia coli cells are about 2 μm (microns) long and 0.8 μm in diameter.
a. How many E. coli cells laid end to end would fit across the diameter of a pinhead? (Assume a pinhead diameter of 0.5 mm.)

Answer:

$$\text{E. coli/pinhead diameter} = \frac{0.5\ \text{mm/dia.}}{2\ \mu\text{m/E. coli}}$$
$$= \frac{0.5\times10^{-3}\text{m}}{2\times10^{-6}\text{m}}$$
$$= 250\ \text{E. coli/pinhead diameter}$$

b. What is the volume of an E. coli cell? (Assume it is a cylinder, with the volume of a cylinder given by $V=\pi r^2h$, where $\pi = 3.14$.)

Answer:

$$V = \pi r^2 h$$
$$= 3.14\times(\frac{0.8\ \mu\text{m}}{2})^2\times 2\ \mu\text{m}$$
$$= 3.14\times(0.4\times10^{-6}\text{m})^2\times(2\times10^{-6}\text{m})$$
$$= 1\times18^{-18}\text{m}^3$$
$$\text{But, } 1\ \text{m}^3 = (10^2\ \text{cm})^3 = 10^6\text{cm}^3 = 10^6\text{ml} = 10^3\text{L}$$
$$V = 1\times10^{-18}\text{m}^3 = 1\times10^{-15}\text{L}$$
$$= 1\ \text{fL (femtoliter)}$$

c. What is the surface area of an E. coli cell? What is the surface-to-volume ratio of an E. coli cell?

Answer:

$$SA = 2\pi r^2 + \pi dh$$
$$= 2\times3.14\times(0.4\times10^{-6}\text{m})^2 + 3.14\times(0.8\times10^{-6}\text{m})(2\times10^{-6}\text{m})$$
$$= 6.03\times10^{-12}\text{m}^2$$

$$SA/V = 6\times10^{-12}\text{m}^2 / 1\times10^{-18}\text{m}^2$$
$$= 6.03\times10^{6}\text{m}^{-1}$$

d. Glucose, a major energy-yielding nutrient, is present in bacterial cells at a concentration of about 1 mM. How many glucose molecules are contained in a typical E. coli cell? (Recall that Avogadro's number = 6.023×10^{23}.)

Answer:

$$\text{moles glucose} = \text{concentration} \times \text{volume}$$
$$= 1\text{ mM} \times (1 \times 10^{-15}\text{L})\text{ (from b.)}$$
$$= 1 \times 10^{-18}\text{moles}$$
$$\#\text{molecules} = 1 \times 10^{-18}\text{moles} \times (6.023 \times 10^{23})(\text{molecule/mol})$$
$$= 6 \times 10^{5}\text{ molecules}$$

e. A number of regulatory proteins are present in E. coli at only one or two molecules per cell. If we assume that an E. coli contains just one molecule of a particular protein, what is the molar concentration of this protein in the cell?

Answer:

$$\frac{1\text{ molecule}}{6.023 \times 10^{23}\text{ molecules/mole}} = 1.7 \times 10^{-24}\text{ mole}$$

$$\text{Molar concentration} = \frac{\text{moles}}{\text{volume(in liters)}}$$
$$= \frac{1.7 \times 10^{-24}\text{ mole}}{10^{-15}\text{ L (from b.)}}$$
$$= 1.7 \times 10^{-9}\text{ M} = 1.7\text{ nM}$$

f. An E. coli cell contains about 15,000 ribosomes, which carry out protein synthesis. Assuming ribosomes are spherical and have a diameter of 20 nm (nanometers), what fraction of the E. coli cell volume is occupied by ribosomes?

Answer:

$$\text{volume of one ribosome} = \frac{4}{3}\pi r^3$$
$$= \frac{4}{3} \times 3.14 \times (10 \times 10^{-9}\text{ m})^3$$
$$= 4.2 \times 10^{-24}\text{m}^3$$
$$\text{volume of 15,000 ribosomes} = 4.2 \times 10^{-24}\text{m}^3 \times 15{,}000$$
$$= 6.3 \times 10^{-20}\text{m}^3$$
$$\text{fractional volume} = \text{volume ribosome/volume } \textit{E. coli}$$
$$= \frac{6.3 \times 10^{-20}\text{m}^3}{1 \times 10^{-18}\text{m}^3\text{(from b.)}}$$
$$= 0.063 \text{ or } 6.3\%$$

g. The E. coli chromosome is a single DNA molecule whose mass is about 3 x 10⁹ daltons. This macromolecule is actually a circular array of nucleotide pairs. The average molecular weight of a nucleotide pair is 660 and each pair imparts 0.34 nm to the length of the DNA molecule. What is the total length of the E. coli chromosome? How does this length compare with the overall dimensions of an E. coli cell? How many nucleotide pairs does this DNA contain?

Answer: The number of moles of base pairs in 3.0×10^9 D dsDNA is given by:

$$= \frac{3.0 \times 10^{9}\text{ (gm/mole dsDNA)}}{660\text{(gm/mole bp)}}$$
$$= 4.55 \times 10^{6}\text{ mole bp/mole dsDNA}$$

$$1 \text{ molecule dsDNA} = 4.55 \times 10^6 \text{ bp}$$

$$\text{length} = 4.55 \times 10^6 \text{ bp} \times 0.34 \text{ (nm/bp)} = 4.55 \times 10^6 \text{ bp} \times 0.34 \times 10^{-9} \text{ (m/bp)}$$

$$= 1.55 \times 10^{-3} \text{ m} = 1.55 \text{ mm} = 1{,}550\ \mu\text{m}$$

$$\text{length } E. coli = 2\mu\text{m}$$

$$\frac{\text{length DNA}}{\text{length E. coli}} = \frac{1{,}550\ \mu\text{m}}{2\mu\text{m}} = 775$$

h. The average E. coli protein is a linear chain of 360 amino acids. If three nucleotide pairs in a gene encode one amino acid in a protein, how many different proteins can the E. coli chromosome encode? (The answer to this question is a reasonable approximation of the maximum number of different kinds of proteins that can be expected in bacteria.)

Answer: To calculate the number of different proteins:

$$360 \text{ aa/protein} \times 3 \text{ bp/aa} = 1080 \text{ bp/protein}$$

$$\# \text{ different proteins} = \frac{4.55 \times 10^6 \text{ bp}}{1080 \text{ bp/protein}} = 4{,}213 \text{ proteins}$$

4. Assume that mitochondria are cylinders 1.5 µm in length and 0.6 µm in diameter.
a. What is the volume of a single mitochondrion?

Answer :

$$V = \pi r^2 h$$
$$= 3.14 \times (3 \times 10^{-7} \text{ m})^2 \times 1.5 \times 10^{-6} \text{m}$$
$$= 4.24 \times 10^{-19} \text{m}^3$$
$$(1 \text{ m}^3 = 10^3 \text{ L})$$
$$= 4.24 \times 10^{-16} \text{ L} = 0.424 \text{ fL}$$

b. Oxaloacetate is an intermediate in the citric acid cycle, an important metabolic pathway localized in the mitochondria of eukaryotic cells. The concentration of oxaloacetate in mitochondria is about 0.03 µM. How many molecules of oxaloacetate are in a single mitochondrion?

Answer:

$$\# \text{ molecules} = \text{molar conc.} \times \text{volume} \times 6.023 \times 10^{23} \text{ molecules/mol}$$
$$= (0.03 \times 10^{-6})(4.24 \times 10^{-16})(6.023 \times 10^{23})$$
$$= 7.66 \text{ molecules (less than 8 molecules)}$$

5. Assume that liver cells are cuboidal in shape, 20 µm on a side.
a. How many liver cells laid end to end would fit across the diameter of a pinhead? (Assume a pinhead diameter of 0.5 mm.)

Answer:

$$\# \text{ liver cells} = \frac{0.5\,(\text{mm/pinhead})}{20(\mu\text{m/cell})}$$
$$= \frac{0.5 \times 10^{-3}(\text{m/pinhead})}{20 \times 10^{-6}(\text{m/cell})}$$
$$= 25 \text{ cells}$$

b. What is the volume of a liver cell? (Assume it is a cube.)

Answer:

$$\text{Volume of cube} = \text{length}^3 = (20 \times 10^{-6}\,\text{m})^3$$
$$= 8 \times 10^{-15}\,\text{m}^3 \times \left(\frac{100\text{cm}}{\text{m}}\right)^3 \times \left(\frac{1\text{L}}{1000\text{cm}}\right) = 8 \times 10^{-12}\,\text{L} = 8\text{pL}$$

c. What is the surface area of a liver cell? What is the surface-to-volume ratio of a liver cell? How does this compare to the surface-to-volume ratio of an E. coli cell? (Compare this answer to that of problem 3c.) What problems must cells with low surface-to-volume ratios confront that do not occur in cells with high surface-to-volume ratios?

Answer:

$$\text{Surface area} = 6 \times (20 \times 10^{-6}\,\text{m})(20 \times 10^{-6}\,\text{m}) = 2.4 \times 10^{-9}\,\text{m}^2$$
$$\frac{\text{Surface area}}{\text{Volume}} = \frac{2.4 \times 10^{-9}\,\text{m}^2}{8 \times 10^{-15}\,\text{m}^3}$$
$$= 3.0 \times 10^5\,\text{m}^{-1}$$

The surface-to-volume ratio of liver to that of *E. coli* is given by:

$$\frac{3.0 \times 10^5\text{m}^{-1}}{6 \times 10^6\text{m}^{-1}} = 0.05\ (1/20\text{th})$$

The volume of a cell sets or determines the cell's maximum metabolic activity while the surface area defines the surface across which nutrients and metabolic waste products must pass to meet the metabolic needs of the cell. Cells with a low surface-to-volume ratio have a high metabolic capacity relative to the surface area for exchange.

d. A human liver cell contains two sets of 23 chromosomes, each set being roughly equivalent in information content. The total mass of DNA contained in these 46 enormous DNA molecules is 4 x 10^{12} daltons. Since each nucleotide pair contributes 660 daltons to the mass of DNA and 0.34 nm to the length of DNA, what is the total number of nucleotide pairs and the complete length of the DNA in a liver cell? How does this length compare with the overall dimensions of a liver cell?

Answer:

$$\# \text{ base pairs} = \frac{4.0 \times 10^{12}\text{D}}{660\ \text{D/bp}}$$
$$= 6.1 \times 10^9\ \text{bp}$$

$$\text{length} = 0.34(\text{nm/bp}) \times (6.1 \times 10^9)\text{bp} = 0.34 \times 10^{-9}(\text{m/bp}) \times (6.1 \times 10^9)\text{bp}$$
$$= 2.06\ \text{m!}$$

$$\text{length relative to liver cell} = \frac{2.06\ \text{m}}{20\ \mu\text{m}}$$
$$= 1.03 \times 10^5$$

or about 100,000 times greater!

e. The maximal information in each set of liver cell chromosomes should be related to the number of nucleotide pairs in the chromosome set's DNA. This number can be obtained by

dividing the total number of nucleotide pairs calculated above by 2. What is this value? If this information is expressed in proteins that average 400 amino acids in length and three nucleotide pairs encode one amino acid in a protein, how many different kinds of proteins might a liver cell be able to produce? (In reality livers cells express at most about 30,000 different proteins. Thus, a large discrepancy exists between the theoretical information content of DNA in liver cells and the amount of information actually expressed.)

Answer:

$$\text{The information content} = 3.0 \times 10^9$$

$$\#\text{ different proteins} = 400\ (\text{aa/protein}) \times 3\ (\text{bp/aa}) = 1200\ \text{bp/protein}$$

$$\#\text{ different proteins} = \frac{3.0 \times 10^9}{1200\ (\text{bp/protein})} = 2.5 \times 10^6\ \text{proteins}$$

6. Biomolecules interact with one another through molecular surfaces that are structurally complementary. How can various proteins interact with molecules as different as simple ions, hydrophobic lipids, polar but uncharged carbohydrates, and even nucleic acids?

Answer: The amino acid side chains of proteins can participate in a number of interactions through hydrogen bonding, ionic bonding, hydrophobic interactions, and van der Waals interactions. For example, the polar amino acids, basic amino acids, and acidic amino acids and their amides all have groups that can participate in hydrogen bonding. Those amino acid side chains that are charged can form ionic bonds. The hydrophobic amino acids can interact with nonpolar, hydrophobic surfaces of molecules. Thus, amino acids are capable of participating in a variety of interactions. A protein can be folded in three dimensions to organize amino acids into surfaces with a range of properties.

7. What structural features allow biological polymers to be informational macromolecules? Is it possible for polysaccharides to be informational macromolecules?

Answer: Biopolymers like proteins and nucleic acids are informational molecules because they are vectorial molecules, composed of a variety of building blocks. For example, proteins are linear chains of some 20 amino acids joined head-to-tail to produce a polymer with distinct ends. The information content is the sequence of amino acids along the polymer. Nucleic acids (DNA and RNA) are also informational molecules for the same reason. Here, the biopolymer is made up of 4 kinds of nucleotides. Monosaccharides are capable of forming polymeric structures but typically with little information content. When a polymer is formed from only one kind of monosaccharide, as for example in glycogen, starch, and cellulose, even though the molecule is vectorial (i.e., it has distinct ends) there is little information content. (The polysaccharides have an advantage in chemistry over the nucleic acids and proteins in that they readily form branch structures. Branched polysaccharides composed of a number of different monosaccharides are rich in information.)

8. Why is it important that weak forces, not strong forces, mediate biomolecular recognition interactions?

Answer: Life is a dynamic process characterized by continually changing interactions. Complementary interactions based on covalent bonding would of necessity produce static structures, structures difficult to change and slow to respond to outside stimuli.

9. Why does the central role of weak forces in biomolecular interactions restrict living systems to a narrow range of environmental conditions?

Answer: The weak forces such as hydrogen bonds, ionic bonds, hydrophobic interactions, and van der Waals interactions can be easily overcome by low amounts of energy. Slightly elevated temperatures are sufficient to break hydrogen bonds. Changes in ionic strength, pH, concentration of particular ions, etc., all potentially have profound effects on macromolecular structures dependent on the weak forces.

10. Describe what is meant by the phrase "cells are steady-state systems".

Answer: Life is characterized as a system through which both energy and matter flow. The consequence of energy flow in this case is order, the order of monomeric units in biopolymers, which in turn produces macromolecular structures that function together as a living cell.

Additional Problems

1. Silicon is located below carbon in the periodic chart. It is capable of forming a wide range of bonds similar to carbon yet life is based on carbon chemistry. Why are biomolecules made of silicon unlikely?

2. Identify the following characters of the Greek alphabet:
α, β, γ, δ, Δ, ε, ζ, θ, κ, λ, μ, ν, π, ρ, σ, Σ, τ, χ, ϕ, ψ, and ω.

3. Give a common example of each of the weak forces at work.

4. On a hot dry day, leafy plants may begin to wilt. Why?

Abbreviated Answers

1. Covalent silicon bonds are not quite as strong as carbon covalent bonds because the bonding electrons of silicon are shielded from the nucleus by an additional layer of electrons. In addition, silicon is over twice the weight of carbon. Also, silicon oxides (rocks, glass) are extremely stable and not as reactive as carbon.

2. These Greek letters are commonly used in biochemistry but this set is not the complete Greek alphabet. alpha (α), beta (β), gamma (γ), delta (δ), capital delta (Δ), epsilon (ε), zeta (ζ), theta (θ), kappa (κ), lambda (λ), mu (μ), nu (ν), pi (π), rho (ρ), sigma (σ), capital sigma (Σ), tau (τ), chi (χ), phi (ϕ), psi (ψ), and omega (ω, the last letter of the Greek alphabet).

3. Ice is an example of a structure held together by hydrogen bonds. Sodium and chloride ions are joined by ionic bonds in table salt crystals. A stick of butter is a solid at room temperature because of van der Waals forces. Oil and water don't mix because hydrophobic interactions between oil molecules cause the oil to coalesce.

4. The tonoplast loses water and begins to shrink causing the plant cell membrane to exert less pressure on the cell wall.

Summary

The chapter begins with an outline of the fundamental properties of living systems: complexity and organization, biological structure and function, energy transduction, and self replication. What are the underlying chemical principles responsible for these properties? The elemental composition of biomolecules is dominated by hydrogen, carbon, nitrogen and oxygen. These are the lightest elements capable of forming strong covalent bonds. In particular, carbon plays a key role serving as the backbone element of all biomolecules. It can participate in as many as four covalent bonds arranged in tetrahedral geometry and can produce a variety of structures including linear, branched, and cyclic compounds.

The four elements are incorporated into biomolecules from precursor compounds: CO_2, NH_4^+, NO_3^- and N_2. These precursors are used to construct more complex compounds such as amino acids, sugars, and nucleotides, which serve as building blocks for the biopolymers; proteins, polysaccharides, and nucleic acids, as well as fatty acids and glycerol, which are the building blocks of lipids. These complex macromolecules are organized into supramolecular complexes such as membranes and ribosomes that are components of cells, the fundamental units of life.

Proteins, nucleic acids and polysaccharides are biopolymers with structural polarity due to head-to-tail arrangements of asymmetric building block molecules. In these biopolymers, the building blocks are held together by covalent bonds, but they assume an elaborate architecture, because of weak, noncovalent forces such as van der Waals interactions, hydrogen bonds, ionic bonds and hydrophobic interactions. The three-dimensional shape is important for biological function, especially for proteins. At extreme conditions such as high temperature, high pressure, high salt concentrations, extremes of pH, and so on, the weak forces may be disrupted, resulting in loss of both shape and function in a process known as denaturation. Thus, life is confined to a narrow range of conditions.

Life demands a flow of energy during which energy transductions occur in the organized, orderly, small, manageable steps of metabolism, each step catalyzed by enzymes.

The fundamental unit of life is the cell. There are two types: eukaryotic cells with a nucleus and prokaryotic cells without a nucleus. Prokaryotes are divided into two groups, eubacteria and archaea. All cells contain ribosomes, which are responsible for protein synthesis. Eukaryotic cells, found in plants, animals, fungi, and single cell organisms (protista), contain an array of membrane-bound compartments or organelles, including a nucleus, mitochondria, chloroplasts, endoplasmic reticulum, Golgi apparatus, vacuoles, lysosomes, and perixosomes. Organelles are internal compartments in which particular metabolic processes are carried out.

Chapter 2

Water: The Medium of Life

• •

Chapter Outline

- Properties of water: High boiling point, high melting point, high heat of vaporization, high surface tension, high dielectric constant, maximum density in liquid state: All due to ability of water to hydrogen bond
- Water structure
  - Electronegative oxygen, two hydrogens: Nonlinear arrangement: Dipole
  - Two lone pairs of electrons on oxygen: H-bond acceptors
  - Partially positively charged hydrogens: H-bond donors
- Ice
  - Lattice with each water interacting with 4 neighboring waters
  - H-bonds: Linear geometry and stable
- Liquid: H-bonds present but less than 4 per molecule and transient
- Solvent properties of water
  - High dielectric constant decreases strength of ionic interactions between other molecules
    - Force of ionic interactions, $F = e_1e_2/Dr^2$, inversely dependent on D.
    - Salts dissolve in water.
  - Interaction with polar solutes through H-bonds
  - Hydrophobic interactions: Entropy-driven process results in minimization of solvation cage
  - Amphiphilic (amphipathic) compounds have polar and non-polar groups. In solution will form micelles
- Colligative properties: Freezing point depression, boiling point elevation, lowering of vapor pressure osmotic pressure effects: Depend on number of solute molecules per unit volume
- Ionization of water
  - Ions: Hydrogen ion H^+ (protons), hydroxyl ion OH^-, hydronium ion H_3O^+
  - Ion product: $K_w = [H_2O]\times K_{eq} = 55.5\times K_{eq} = 10^{-14} = [H^+][OH^-]$
  - $pH = -\log_{10} [H^+]$, $pOH = -\log_{10} [OH^-]$, $pH + pOH = 14$
- Strong electrolytes: Completely dissociate: Salts, strong acids, strong bases
- Weak electrolytes: Do not fully dissociate: Hydrogen ion buffers
- Buffers
  - Henderson-Hasselbalch equation: $pH = pK_a + \log_{10} ([A^-]/[HA])$
  - Biological buffers: Phosphoric acid ($pK_1 = 2.15$, $pK_2 = 7.2$, $pK_3 = 12.4$); histidine ($pK_a = 6.04$) bicarbonate ($pK_{overall} = 6.1$)

Chapter Objectives

Water

Its properties arise because of the ability of water molecules to form H bonds and to dissociate to H^+ and OH^-. Thus, water is a good solvent, has high heat capacity and a high dielectric constant.

Acid-Base Problems

For acid-base problems the key points to remember are:

Henderson-Hasselbalch: $pH = pK_a + \log([A^-]/[HA])$

Conservation of acid and conjugate base:

$[A^-] + [HA]$ = Total amount of weak electrolyte added.

Conservation of charge: $\sum$ [cations] = $\sum$ [anions] i.e., the sum of the cations must equal the sum of the anions.

In many cases, simplifications can be made to these equations. For example, $[OH^-]$ or $[H^+]$ may be small relative to other terms and ignored in the equation. For strong acids, it can be assumed that the concentration of the conjugate base is equal to the total concentration of the acid. For example, an x M solution of HCl is x M in Cl^-. Likewise for x M NaOH, the $[Na^+]$ is x M.

In polyprotic buffers (e.g., phosphate, citrate, etc.), the group with the pK_a closest to the pH under study will have to be analyzed using the Henderson-Hasselbalch equation. For groups with pK_a's two or more pH units away from the pH, they are either completely protonated or unprotonated.

The solution to a quadratic equation of the form, $y = ax^2 + bx + c$ is:

$$x = \frac{-b \pm \sqrt{b^2 - 4ac}}{2a}$$

Problems and Solutions

1. Calculate the pH of the following.

a. 5×10^{-4} M HCl

Answer: HCl is a strong acid and fully dissociates into $[H^+]$ and $[Cl^-]$.
Thus, $[H^+] = [Cl^-] = [HCl]_{\text{total added}}$

$$pH = -\log_{10}[H^+] = -\log_{10}[HCl_{total}]$$
$$= -\log_{10}(5 \times 10^{-4}) = 3.3$$

b. 7×10^{-5} M NaOH

Answer: For strong bases like NaOH and KOH,

$$pH = 14 + \log_{10}[\text{Base}]$$
$$= 14 + \log_{10}(7 \times 10^{-5}) = 9.85$$

c. 2×10^{-6} M HCl

Answer:

$$pH = -\log_{10}[H^+] = -\log_{10}[HCl_{total}]$$
$$= -\log_{10}(2 \times 10^{-6}) = 5.70$$

d. 3×10^{-2} M KOH

Answer:

$$pH = 14 + \log_{10}(3 \times 10^{-2}) = 12.5$$

e. 4×10^{-5} M HCl

Answer:

$$pH = -\log_{10}[H^+] = -\log_{10}[HCl_{total}]$$
$$= -\log_{10}(0.04 \times 10^{-3}) = 4.4$$

f. 6 x 10⁻⁹ M HCl

Answer: Beware! Naively one might fall into the trap of simply treating this like another strong acid problem and solving it like so:

$$pH = -\log_{10}[H^+] = -\log_{10}[HCl_{total}]$$
$$= -\log_{10}(6 \times 10^{-9}) = 8.22$$

However, something is odd. This answer suggests that addition of a small amount of a strong acid to water will give rise to a basic pH! What we have ignored is the fact that water itself will contribute H^+ into solution so we must consider the ionization of water as well. There are two approaches we can take in solving this problem. As a close approximation we can assume that:

$$[H^+] = 10^{-7} + [HCl] \text{ or,}$$
$$[H^+] = 10^{-7} + 6\times 10^{-9} = 10^{-7} + 0.06 \times 10^{-7} = 1.06 \times 10^{-7}$$
$$pH = -\log_{10}(1.06 \times 10^{-7}) = 6.97$$

The exact solution uses the ion product of water.

$$[H^+][OH^-] = K_W = 10^{-14} \quad \text{(the ion product of water)} \quad (1)$$

In solution HC1 fully dissociates into $H^+ + Cl^-$. For any solution the sum of negative and positive charges must be equal. Since we are dealing with monovalent ions we can write:

$$[H^+] = [Cl^-] + [OH^-] \quad (2)$$

Now because HC1 is fully dissociated:

$$[Cl^-] = [HCl_{total}] = 6 \times 10^{-9} \quad (3)$$

Substituting this into equation (2), solving equation (2) for $[OH^-]$ and substituting in equation (1) we have a quadratic equation in H^+:

$$[H^+]^2 - 6\times 10^{-9}[H^+] - 10^{-14} = 0$$

whose general solution is given by

$$[H^+] = \frac{-b \pm \sqrt{b^2 - 4ac}}{2a} = \frac{6\times 10^{-9} \pm \sqrt{(6\times 10^{-9})^2 + 4\times 10^{-14}}}{2}$$

Before firing up the calculator a little reflection suggests that the argument under the square root is dominated by 4×10^{-14} whose root is 2×10^{-7}. Furthermore, of the two solutions (i.e., ±) it must be the + solution (- given rises to a negative $[H^+]$!)
Therefore,

$$[H^+] = \frac{6 \times 10^{-9} + 2\times 10^{-7}}{2}$$

and, $pH = -\log_{10}[H^+] = 6.99$

2. Calculate from the values in Table 2.3:
a. the [H⁺] in vinegar

Answer:

From Table 2.3 we find that pH = 2.9
since $pH = -\log_{10}[H^+]$
$[H^+] = 10^{-pH} = 10^{-2.9} = 1.26 \times 10^{-3} M = 1.26$ mM

b. the [H⁺] in saliva

Answer:

From Table 2.3 we find that in saliva pH = 6.6

since pH = $-\log_{10}[H^+]$

$[H^+] = 10^{-pH} = 10^{-6.6} = 2.5 \times 10^{-7}\ M = 0.25\ \mu M$

c. the [H⁺] in household ammonia

Answer:

The pH of ammonia is 11.4 thus

$[H^+] = 10^{-pH} = 10^{-11.4} = 4 \times 10^{-12}\ M = 4\ pM$

d. the [OH⁻] in milk of magnesia

Answer:

The pH of milk is 10.3.

From pH + pOH = 14,

pOH = 14 - 10.3 = 3.7

$[OH^-] = 10^{-pOH} = 10^{-3.7} = 2 \times 10^{-4}\ M = 0.2\ mM$

e. the [OH⁻] in beer

Answer:

The pH of beer is 4.5.

From pH + pOH = 14,

pOH = 14 - 4.5 = 9.5

$[OH^-] = 10^{-pOH} = 10^{-9.5} = 3.16 \times 10^{-10}\ M = 0.316\ nM$

f. the [H⁺] inside a liver cell

Answer:

The pH of a liver cell is 6.9 thus

$[H^+] = 10^{-pH} = 10^{-6.9} = 1.26 \times 10^{-7}\ M = 0.126\ \mu M$

3. Measurement of the pH of a 0.02 M solution of an acid gave a value of 4.6.
a. What is the [H⁺] in this solution?

Answer:

$[H^+] = 10^{-pH} = 10^{-4.6} = 2.5 \times 10^{-5}\ M = 25\ \mu M$

b. Calculate the acid dissociation constant K_a and pK_a for this acid.

Answer:

$$HA \leftrightarrows H^+ + A^-$$

$$K_a = \frac{[H^+][A^-]}{[HA]}$$

Assume that a small amount, *x*, of HA dissociates into equal molar amounts of H^+ and A^-. We then have:

$$HA \leftrightarrows H^+ + A^- \text{ or,}$$
$$0.02 - x \leftrightarrows x + x$$
$$\text{and, } K_a = \frac{x^2}{[HA]}$$

From (a) we know that $[H^+] = 25\ \mu M = x = [A^-]$

And, $[HA] = 0.02 - 25 \times 10^{-6} \approx 0.02$

$$\text{Thus, } K_a = \frac{(25 \times 10^{-6})^2}{0.02} = \frac{625 \times 10^{-12}}{0.02} = 3.13 \times 10^{-8}$$

$$pK_a = -\log_{10}(3.13 \times 10^{-8}) = 7.5$$

4. The K_a for formic acid is 1.78 x 10^{-4} M.

a. What is the pH of a 0.1 M solution of formic acid?

Answer:

$$pH = pK_a + \log_{10}\frac{[A^-]}{[HA]} \text{ or } [H^+] = K_a\frac{[HA]}{[A^-]}\ (1)$$

For formic acid, $[H^+] \approx [A^-]$ (2)

and $[HA] + [A^-] = 0.1$ M or

$[HA] = 0.1 - [A^-]$ (3)

and, using equation (2) we can write

$[HA] = 0.1 - [H^+]$

Substituting this equation and equation (2) into (1) we find :

$$[H^+] = K_a\frac{0.1 - [H^+]}{[H^+]} \text{ or}$$

$[H^+]^2 + K_a[H^+] - 0.1K_a = 0$, a quadratic whose solutions are

$$[H^+] = \frac{-K_a \pm \sqrt{K_a^2 + 0.4K_a}}{2}$$

The argument under the square root sign is greater than K_a. Therefore, the correct solution is the positive root. Furthermore, K_a^2 is small relative to $0.4K_a$ and can be ignored.

$$[H^+] = \frac{-K_a + \sqrt{0.4K_a}}{2} = \frac{1.78 \times 10^{-4} + \sqrt{0.4 \times 1.78 \times 10^{-4} K_a}}{2}$$

$[H^+] = 0.00431$ M

$pH = -\log_{10}[H^+] = 2.37$

b. 150 ml of 0.1 M NaOH is added to 200 ml of 0.1 M formic acid, and water is added to give a final volume of 1 L. What is the pH of the final solution?

Answer:

The total concentration of formic acid is :

$$[HA_{total}] = \frac{0.2\ L \times 0.1\ M}{1\ L} = 0.02\ M$$

Prior to addition of NaOH, essentially all of the formic acid is in the protonated, HA, form. When NaOH is added a stoichiometric amount of the conjugate base form, A^-, is produced. Thus, the concentration of conjugate base is equal to the concentration of NaOH added, which is given by:

$$[A^-] = [OH^-] = \frac{0.15\ L \times 0.1\ M}{1\ L} = 0.015\ M$$

And, $[HA] = [HA_{total}] - [A^-] = 0.02\ M - 0.015\ M = 0.005\ M$

From the Henderson-Hasselbalch we have

$$pH = pK_a + \log_{10}\frac{[A^-]}{[HA]} = 3.75 + \log_{10}\frac{0.015}{0.005}$$

$pH = 4.23$

5. *Given 0.1 M solutions of acetic acid and sodium acetate, describe the preparation of 1 L of 0.1 M acetate buffer at a pH of 5.4.*

Answer: From the Henderson-Hasselbalch equation, i.e.,

$$pH = pK_a + \log_{10}\frac{[A^-]}{[HA]}$$

$$\frac{[A^-]}{[HA]} = 10^{(pH-pK)} = 10^{(5.4-4.76)} = 10^{0.64} = 4.37$$

$$\text{or, } [A^-] = 4.37 \times [HA]\ (1)$$

Further, we want

$$[HA] + [A^-] = 0.1\ M \quad (2)$$

By combining equations (1) and (2) we find that

$$[HA] = 0.0187\ M \text{ and } [A^-] = 0.0813\ M$$

Therefore, combine 187 ml 0.1 M acetic acid (HA) with 813 ml 0.1 M sodium acetate (sodium salt of A^-).

6. *If the internal pH of muscle cells is 6.8, what is the $[HPO_4^{2-}]/[H_2PO_4^-]$ ratio in this cell?*

Answer: The dissociation of phosphoric acid proceeds as follows:

$$H_3PO_4 \underset{2.15}{\leftrightarrows} H^+ + H_2PO_4^- \underset{7.20}{\leftrightarrows} H^+ + HPO_4^{2-} \underset{12.40}{\leftrightarrows} H^+ + PO_4^{3-}$$

Now, at pH = 6.8, we expect the first equilibrium to be completely to the right and the last equilibrium to be to the left (i.e., we expect phosphoric acid to be in the doubly or singly protonated forms). From the Henderson Hasselbalch equation we find:

$$pH = pK_a + \log_{10}\frac{[A^-]}{[HA]} \text{ where}$$

$$[A^-] = [HPO_4^{2-}]; [HA] = [H_2PO_4^-]$$

$$\log_{10}\frac{[HPO_4^{2-}]}{[H_2PO_4^-]} = pH - pK_a$$

$$\frac{[HPO_4^{2-}]}{[H_2PO_4^-]} = 10^{(pH-pK_a)} = 10^{(6.8-7.2)} = 0.4$$

7. *What are the approximate fractional concentrations of the following phosphate species at pH values of 2, 3, 6, 8, 10, and 13?*

a. H_3PO_4 ***b.*** $H_2PO_4^-$ ***c.*** HPO_4^{2-} ***d.*** PO_4^{3-}

Answer: For phosphoric acid the following equilibria apply:

$$H_3PO_4 \underset{2.15}{\leftrightarrows} H^+ + H_2PO_4^- \underset{7.20}{\leftrightarrows} H^+ + HPO_4^{2-} \underset{12.40}{\leftrightarrows} H^+ + PO_4^{3-}$$

At pH = 2 we expect the first two species to be involved in the equilibrium, thus,

$$pH = pK_a + \log_{10}\frac{[H_2PO_4^-]}{[H_3PO_4]} \text{ or}$$

$$[H_2PO_4^-] = [H_3PO_4] \times 10^{(pH-pK_a)} = [H_3PO_4] \times 10^{(2.0-2.15)} = 0.708[H_3PO_4]$$

$$\text{Fraction of } [H_2PO_4^-] = f_{H_2PO_4^-} = \frac{[H_2PO_4^-]}{[H_2PO_4^-] + [H_3PO_4]} = \frac{0.708[H_3PO_4]}{0.708[H_3PO_4] + [H_3PO_4]} = 0.415$$

$$\text{Fraction of } [H_3PO_4] = f_{H_3PO_4} = 1 - f_{H_2PO_4^-} = 1 - 0.415 = 0.585$$

The same equations is used for pH = 3 where we find:

At pH = 3

$$f_{H_2PO_4^-} = 0.876, \text{ and } f_{H_3PO_4} = 0.124$$

For pH = 6 one must consider the equilibrium between $H_2PO_4^-$ and HPO_4^{2-} using pK_a = 7.20.

$$pH = pK_a + \log_{10}\frac{[HPO_4^{2-}]}{[H_2PO_4^-]} \text{ or}$$

$$[HPO_4^{2-}] = [H_2PO_4^-] \times 10^{(pH-pK_a)} = [H_2PO_4^-] \times 10^{(6-7.20)} = 0.0631[H_2PO_4^-]$$

$$\text{Fraction of } [HPO_4^{2-}] = f_{HPO_4^{2-}} = \frac{[HPO_4^{2-}]}{[HPO_4^{2-}] + [H_2PO_4^-]} = \frac{0.0631[H_2PO_4^-]}{0.0631[H_2PO_4^-] + [H_2PO_4^-]} = 0.06$$

$$\text{Fraction of } [H_2PO_4^-] = f_{H_2PO_4^-} = 1 - f_{HPO_4^{2-}} = 1 - 0.06 = 0.94$$

The same equations is used for pH = 8 where we find:

$$f_{HPO_4^{2-}} = 0.86, \text{ and } f_{H_2PO_4^-} = 0.14$$

The value pH = 10 is approximately halfway between pK = 7.2 and pK = 12.0 so either may be used. The result is that essentially all of the phosphate is in the HPO_4^{2-} form. Thus,

$$f_{H_2PO_4^-} = 1.00$$

Finally, at pH = 13 the last two species in the equilibrium are considered:

$$pH = pK_a + \log_{10}\frac{[PO_4^{3-}]}{[HPO_4^{2-}]} \text{ or}$$

$$[PO_4^{3-}] = [HPO_4^{2-}] \times 10^{(pH-pK_a)} = [HPO_4^{2-}] \times 10^{(13-12.4)} = 3.98[HPO_4^{2-}]$$

$$\text{Fraction of } [PO_4^{3-}] = f_{PO_4^{3-}} = \frac{[PO_4^{3-}]}{[PO_4^{3-}] + [HPO_4^{2-}]} = \frac{3.98[HPO_4^{2-}]}{3.98[HPO_4^{2-}] + [H_2PO_4^-]} = 0.8$$

$$\text{Fraction of } HPO_4^{2-} = f_{HPO_4^{2-}} = 1 - f_{PO_4^{3-}} = 1 - 0.8 = 0.2$$

8. If 50 ml of 0.01 M HCl is added to 100 ml of 0.05 M phosphate buffer at pH 7.2, what is the resultant pH? What are the concentrations of $H_2PO_4^-$ and HPO_4^{2-} in the final solution?

Answer: The relevant pK_a for phosphoric acid is 7.2 governing the following equilibrium:

$$H_2PO_4^- \leftrightarrows H^+ + HPO_4^{2-}$$

From the Henderson-Hasselbalch equation we find that

$$pH = pK_a + \log_{10}\frac{[HPO_4^{2-}]}{[H_2PO_4^-]} \text{ or}$$

$$[HPO_4^{2-}] = [H_2PO_4^-] \times 10^{(pH-pK_a)} = [H_2PO_4^-] \times 10^{(7.20-7.20)} = [H_2PO_4^-]$$

And, since $[HPO_4^{2-}] + [H_2PO_4^-] = 0.05M$

$$[HPO_4^{2-}] = [H_2PO_4^-] = 0.025M$$

In 100ml we have $100\text{ml} \times \frac{1\text{L}}{1000\text{ml}} \times 0.025\text{ M} = 0.0025$ mole of each.

Now, addition of 50 ml of 0.1M HCl accomplishes two things:

(1) It dilutes the solution; and,

(2) It introduces protons that will convert HPO_4^{2-} to $H_2PO_4^-$.

The moles of protons is given by:

$$50\text{ ml} \times \frac{1\text{L}}{1000\text{ml}} \times 0.01\text{M} = 0.0005 \text{ mole}$$

Thus, 0.0005 mole of HPO_4^{2-} will be converted to $H_2PO_4^-$ or

There will be 0.0025 - 0.0005 mole HPO_4^{2-} and 0.0025 + 0.0005 mole $H_2PO_4^-$

$$pH = pK_a + \log_{10}\frac{[HPO_4^{2-}]}{[H_2PO_4^-]} = 7.2 + \log_{10}\frac{0.0025 - 0.0005}{0.0025 + 0.0005} = 7.02\,(1)$$

And

$$[HPO_4^{2-}] = \frac{(0.0025 - 0.0005)\text{ mole}}{100\text{ ml} + 50\text{ ml}} \times \frac{1000\text{ ml}}{1\text{ L}} = 0.0133\text{ M}$$

$$[H_2PO_4^-] = \frac{(0.0025 + 0.0005)\text{ mole}}{100\text{ ml} + 50\text{ ml}} \times \frac{1000\text{ ml}}{1\text{ L}} = 0.0200\text{ M}$$

Note: This amount of acid added to 100 ml of water gives pH = 2.5.

b. If 50 ml of 0.01 M NaOH is added to 100 ml of 0.05 M phosphate buffer at pH 7.2, what is the resultant pH? What are the concentrations of $H_2PO_4^-$ and HPO_4^{2-} in this final solution?

Answer: For NaOH, the same equations apply as above with one important difference: HPO_4^{2-} is increased and $H_2PO_4^-$ is decreased by addition of 0.0005 moles of base (0.05L × 0.01M).

Thus, 0.0005 mole of $H_2PO_4^-$ will be converted to HPO_4^{2-} or

There will be 0.0025 - 0.0005 mole $H_2PO_4^-$ and 0.0025 + 0.0005 mole HPO_4^{2-}

$$pH = pK_a + \log_{10}\frac{[HPO_4^{2-}]}{[H_2PO_4^-]} = 7.2 + \log_{10}\frac{0.0025 + 0.0005}{0.0025 - 0.0005} = 7.38\,(1)$$

And

$$[H_2PO_4^-] = \frac{(0.0025 - 0.0005)\text{ mole}}{100\text{ ml} + 50\text{ ml}} \times \frac{1000\text{ ml}}{1\text{ L}} = 0.0133\text{ M}$$

$$[HPO_4^{2-}] = \frac{(0.0025 + 0.0005)\text{ mole}}{100\text{ ml} + 50\text{ ml}} \times \frac{1000\text{ ml}}{1\text{ L}} = 0.0200\text{ M}$$

Note: If added to water instead, this amount of NaOH would result in a solution with pH = 11.5.

Additional Problems

1. Tris (Tris[hydroxymethyl]aminomethane) is a commonly used buffer with a pK_a = 8.0. In making up a Tris solution, an appropriate amount of the free base is dissolved in water and the pH of the solution is adjusted, often with HCl. The M_r of Tris is 121.1.
   a. Describe the preparation of 1 L of 25 mM solution, pH = 8.4 using solid Tris base and 1 M HCl.
   b. Draw the structure of Tris in its protonated and unprotonated forms.

2. When working with biomolecules it is often important to control the concentration of divalent cations. This may be accomplished by using chelating agents that form complexes with divalent cations making them unavailable for binding to biomolecules such as proteins or nucleic acids. One common agent used to control the availability of magnesium and calcium (among other cations) is EDTA (Ethylenediaminetetraacetic acid).
   a. Draw the structure of EDTA
   b. EDTA contains four carboxyl groups; however, it binds only one divalent cation. Divalent cation binding is pH-dependent and so stock solutions of EDTA are often adjusted to pH = 8.0. The pK_a's of the four carboxyl groups on EDTA are 2.0, 2.67, 6.16 and 10.26. At pH = 8.0, what is the predominant ionic species of EDTA?

3. To make up 1 L of a 0.5 M solution of EDTA starting with the free acid, approximately how much 10 M NaOH will have to be added to adjust the pH to 7.0? Do you expect this solution to have a pH-buffering capacity? Explain.

4. Good's buffers were developed to provide buffers that would not interact strongly with divalent cations such as Mg^{2+} and Ca^{2+}. One such buffer is PIPES (Piperazine-N,N'-bis[2-ethanesulfonic

acid]), pKa = 6.8. You are asked to make 1 L of 25 mM solution of PIPES, pH = 7.0. In preparing the solution starting from the free acid, you notice that PIPES is not very soluble in water. You find that the solution is acidic (around 2), and so you begin to add NaOH. After addition of approximately 3.75 mL of 10 N NaOH, the solution clears. The pH is around 7. Explain.

5. (a). On the same graph, sketch the titration curves of acetic acid, pKa = 4.76; ethylamine, pKa = 10.63; and, glycine, pKa = 2.34 and 9.6.
(b). The titrations of acetic acid, ethylamine, and glycine involve a single carboxyl group, a single amino group, and a carboxyl and amino group respectively. Can you suggest why the carboxyl group and the amino groups on glycine have lower pK_a's than the same groups on acetic acid and ethylamine?

Abbreviated Answers

1a. Add 3.028 g Tris to approximately 900 mL of water. Carefully titrate the pH to 8.4 using approximately 7.1 mL of 1 M HCl. Finally, adjust the volume of the solution to 1 L.

b.

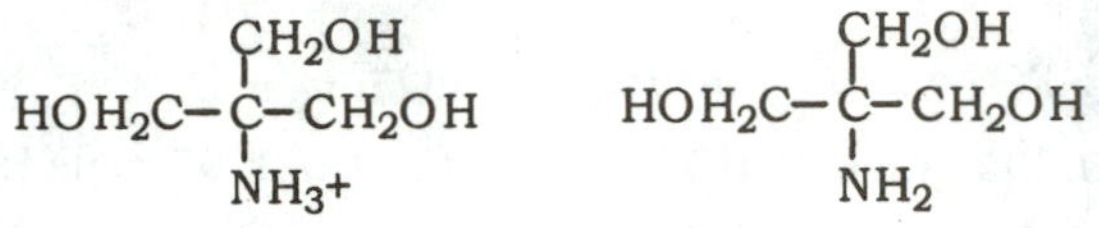

```
        CH2OH                    CH2OH
        |                        |
HOH2C—C—CH2OH            HOH2C—C—CH2OH
        |                        |
        NH3+                     NH2
```

2a.

```
HOOC—H2C                      CH2-COOH
        \                    /
         N—H2C—CH2—N
        /                    \
HOOC—H2C                      CH2—COOH
```

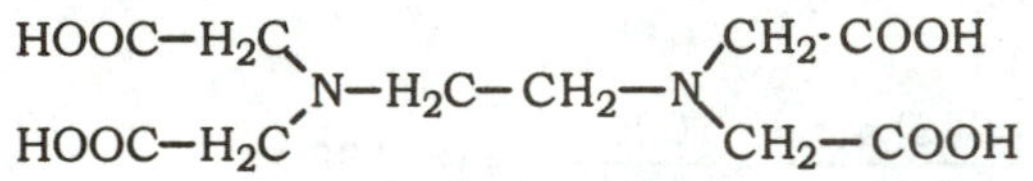

b. $EDTA^{3-}$ (i.e., the singly-protonated species).

3. To adjust the solution to pH = 7.0, sufficient NaOH must be added to titrate fully the two carboxyl groups with pK_a's = 2.0 and 2.67. This will require 100 mL of 10 N NaOH. Further, an additional 43.7 mL will be required to titrate the carboxyl group whose pK_a = 6.16. This is calculated using the Henderson-Hasselbalch equation. The total amount of NaOH is 143.7 mL. The solution will not be a good buffer at pH = 7.0 because the pH is 0.8 of a pH unit away from the nearest pK_a.

4. To bring the pH of the solution to its pK_a normally requires approximately one-half of an equivalent of (in this case) base. The 3.75 mL of NaOH represents 0.0375 equivalent of OH^- (=3.75 mL x 10 N). One liter of 25 mM Pipes contains 0.025 moles of Pipes. The 0.0375 equivalent of OH^- is 1.5 times the molar amount of Pipes. Therefore, there must be two titratable groups on PIPES.

5. The pK_a's of the carboxyl group and the amino group of glycine are shifted to lower pH values relative to the same groups on acetic acid and ethylamine. Clearly, the groups are influencing each other's pK_a's. The pK_a of the carboxyl group is lowered because the positively charged amino group influences it. The unprotonated carboxyl group is negatively charged. This state is favored by having a positively charged amino group nearby. Thus, the carboxyl group becomes a slightly stronger acid and will give up its proton at lower pH values. Similarly, the amino group becomes a slightly stronger acid because in its protonated state it is positively charged.

Summary

The most abundant molecule in living systems is H_20. What physical properties make water such an important component of life? The two H's in a water molecule make covalent bonds with a single O. However, there is something special about these covalent bonds. Oxygen is an electronegative element with a high affinity for electrons, surpassed only by F. (This high attraction for electrons is a consequence of a positively charged nucleus rather unshielded by electron clouds). In covalent bonds with H, the electrons will be attracted to O, giving rise to charge separation or a dipole moment. The hydrogens become partially positively charged whereas the oxygen becomes partially negatively charged. The four electron pairs in H_2O form a distorted tetrahedron, with two lone pairs separated by an angle greater than 109° and the two shared electron pairs (shared between O and H) at 104.5°. The partially positively charged hydrogens can interact with lone-pair electrons to form hydrogen bonds. We will see that hydrogen bonds, although weak compared to covalent bonds, are important interactions in life.

Water molecules can interact with each other to form hydrogen-bonded structures. Ice is a hydrogen-bonded structure, with each water molecule bonded to four other water molecules. In two of these bonds, the hydrogens interact with lone-pair electrons on two other water molecules. In these cases, water is serving as a hydrogen-bond donor. The two electron pairs make hydrogen bonds, participating as hydrogen-bond acceptors, with positively charged hydrogens on two other water molecules. The thermal properties of water are a consequence of hydrogen bonds. For example, the density of water increases with decreasing temperature down to 4°C; as water molecules lose kinetic energy, they can approach each other more closely. As the temperature moves to the freezing point of water, more and more hydrogen bonds form, causing water molecules to move apart and density to decrease. Thus, ice floats. The environmental consequence is that ponds and lakes freeze from the surface, allowing life to continue in the liquid interior. The transition from solid to liquid to gas is accompanied by disruption of hydrogen bonds. Many hydrogen bonds remain in liquid water, as evidenced by the difference between the heat of melting and the heat of sublimation. Water has a high heat capacity (the ability to absorb heat without a large increase in temperature) because energy in the form of heat is used to break hydrogen bonds rather than increase temperature (increase kinetic energy). Finally, the rather large amount of energy required to vaporize water means that water can absorb energy with little increase in temperature.

Water is a good solvent for ionic and polar substances, because it forms hydrogen bonds with these substances. Salts dissolve and dissociate because the ionic components become hydrated with water molecules. The high dielectric constant of water is responsible for decreasing the attractive force between two ionic components of a salt. (The force between two charges, e_1 and e_2, is given by $F = e_1e_2/Dr^2$; it is inversely dependent on the dielectric constant.)

The attraction between water molecules is strong and the tendency of water molecules to make these interactions is great. For example, at an air/water interface, interaction of water molecules on the surface is responsible for surface tension. But surface tension comes at a price. Water molecules at an interface are forced to make irregular hydrogen bonds, often with less than perfect geometry. In fact, this occurs at any interface, and when nonpolar molecules are put into water, each molecule in effect represents an interface. To minimize the interface area, there is a tendency, driven by hydrogen bonding of H_2O, for nonpolar substances to coalesce, to minimize their contact with water molecules. This tendency is termed hydrophobic interactions. We will see that biological membranes and proteins rely on hydrophobic interactions to maintain an ordered structure.

There are two other properties of water that are of extreme importance to biological systems: osmotic pressure and ionization of water. Osmotic pressure arises when a semipermeable membrane through which water can pass separates two aqueous compartments. If a solute is dissolved in one of the aqueous compartments (and the solute is not freely permeable to the membrane), water will move from the solute-free compartment into the compartment containing solute. The amount of pressure necessary to prevent this movement of water is the osmotic pressure. Osmotic pressure is a colligative property and as such its magnitude is directly proportional to concentration.

Water is capable of ionizing in solution to form hydrogen ions and hydroxyl ions. The pH is a measure of the hydrogen ion concentration; $pH = -\log[H^+]$. The pH scale in aqueous solution is set by the magnitude of the ion product. In aqueous solutions, pH + pOH = 14. Using this formula, the pH of solutions of strong acids or bases can be calculated. For weak acids and bases, the Henderson-Hasselbalch equation must be used: $pH = pK_a + \log([A^-]/[HA])$. Several of the problems illustrate the use of this equation.

Chapter 3

Thermodynamics of Biological Systems

Chapter Outline

- Thermodynamic concepts
  - Systems
    - Isolated systems cannot exchange matter or energy with surroundings
    - Closed systems exchange energy but not matter
    - Open systems exchange both energy and matter
  - First Law: $\Delta E = q + w$
    - ΔE = change in internal energy (state function), q = heat absorbed, and w = work done on the system
    - $H = E + PV$: H = enthalpy (energy transferred at constant P): $\Delta H = q$ when work limited to $P\Delta V$
    - $\Delta H = -Rd(\ln Keq)/d(1/T)$: van't Hoff plot RlnKeq vs. 1/T (R = gas constant = 8.314J/mol°K
  - Second Law: Disorder or randomness
    - $S = k \times \ln W$ where k = Boltzmann's constant = 1.38×10^{-23} J/°K, W = the number of ways of arranging the system without changing its internal energy
    - $dS = dq/T$ for reversible process
  - Third law: Entropy of a perfectly ordered, crystalline array at 0°K is exactly zero
    - $S = \int_0^T C_p \, d\ln T$, C_p = heat capacity = dH/dT for constant P process
  - Gibbs free energy: $\Delta G = \Delta H + T\Delta S$
    - $\Delta G = \Delta G° + RT\ln([P]/[R])$ and $\Delta G° = -RT\ln([P]eq/[R]eq)$
    - When protons involved in process: $\Delta G°' = \Delta G° \pm RT\ln[H^+]$
    - ("+" if reaction produces protons; "-" if protons consumed)
- Coupled processes: Enzymatic coupling of a thermodynamically unfavorable reaction with a thermodynamically favorable reaction to drive the unfavorable reaction. Thermodynamically favorable reaction is often hydrolysis of high-energy molecule
- Energy transduction: High-energy phosphate ATP and reduced cofactor NADPH
  - Phototrophs use light energy to produce ATP and NADPH
  - Chemotrophs use chemical energy to produce ATP and NADPH
- High-energy molecules
  - Phosphoric anhydrides (ATP, ADP, GTP, UTP, etc.)
  - Enol phosphates (phosphoenolpyruvate, [PEP])
  - Phosphoric-carboxylic anhydrides (1,3-bisphosphoglycerate)
  - Guanidino phosphates (creatine phosphate)
- Hydrolysis of high-energy bonds is favorable
  - Destabilization due to electrostatic repulsion
  - Product isomerization and resonance stabilization
  - Entropy factors
- Factors influencing thermodynamics of ATP hydrolysis: pH, cation concentration, reactant and product concentrations

Chapter Objectives

Laws of Thermodynamics

The first law of thermodynamics is simply a conservation of energy statement. The internal energy changes if work or heat is exchanged. In biological systems, we are usually dealing with constant-pressure processes and in this case the term enthalpy, **H**, is used. Enthalpy is the heat exchanged at constant pressure. Since enthalpy is heat, it is readily measured using a calorimeter or estimated from a plot of $R(\ln K_{eq})$ versus $1/T$, a van't Hoff plot. (To get ahead of the story, the point is that if ΔG and ΔH are known, ΔS can be calculated.)

The second law of thermodynamics introduces the term entropy, **S**, which is a measure of disorder or randomness in a system. A spontaneous reaction is accompanied by an increase in disorder. For a reversible reaction, $dS_{reversible} = dq/T$. Also, $S = k \ln W$ where k = Boltzmann's constant, and W = the number of ways to arrange the components of a system without changing the internal energy.

Gibbs Free Energy

The change in Gibbs free energy for a reaction is the amount of energy available to do work at constant pressure and constant volume. This is an important concept and should be understood. For a general reaction of the type

$$A + B \leftrightarrows C + D$$

$$\Delta G = \Delta G^\circ + \ln \frac{[C][D]}{[A][B]}$$

When a reaction is at equilibrium, it can do no work and $\Delta G = 0$. At equilibrium, $\Delta G^\circ = -RT \ln K_{eq}$. By measuring the equilibrium concentrations of reactants and products, ΔG° can be evaluated. Knowing the initial concentrations of reactants and products and ΔG° allows a calculation of ΔG. Why is this important? The sign on ΔG tell us in which direction the reaction will proceed. A negative ΔG indicates that the reaction has energy to do work and will be spontaneous in the direction written. A positive ΔG indicates work must be done on the reaction for it to proceed as written, otherwise it will run in reverse. The magnitude of ΔG is the amount of energy available to do work when the reaction goes to equilibrium. Finally, the relationship $\Delta G = \Delta H - T\Delta S$ can be used to evaluate ΔS, the change in disorder, if ΔG and ΔH are known.

High Energy Compounds

ATP is the energy currency of cells and its hydrolysis is used to drive a large number of reactions. For ATP and other high-energy biomolecules, you should understand the properties that make them energy-rich compounds. These include the following: destabilization due to electrostatic repulsion, stabilization of hydrolysis products by ionization and resonance, and entropy factors. Examples of high-energy compounds (from Table 3.3 in Garrett and Grisham) include: phosphoric acid anhydrides (e.g., ATP, ADP, GTP, UTP, CTP, and PP_i), phosphoric-carboxylic anhydrides (acetyl phosphate and 1,3-bisphosphoglycerate), enol phosphates (PEP), and guanidinium phosphates (creatine and arginine phosphate). ATP is the cardinal high-energy compound. Hydrolysis of GTP is important in signal transduction and protein synthesis. UTP and CTP are used in polysaccharide and phospholipid synthesis, respectively. We will encounter 1,3-bisphosphoglycerate and PEP in glycolysis. These high-energy compounds, along with creatine and arginine phosphates, are used to replenish ATP from ADP. Other important high-energy compounds include coenzyme A derivatives such as acetyl-CoA and succinyl-CoA, important in the citric acid cycle. Aminoacylated-tRNA's, the substrates used by the ribosome for protein synthesis, are also high-energy compounds.

Problems and Solutions

1. An enzymatic hydrolysis of fructose-1-P

$$\textit{Fructose} - 1 - P + H_2O \leftrightarrows \textit{fructose} + P_i$$

was allowed to proceed to equilibrium at 25°C. The original concentration of fructose-1-P was 0.2 M, but when the system had reached equilibrium the concentration of fructose 1-P was only 6.52×10^{-5} M. Calculate the equilibrium constant for this reaction and the standard free energy of hydrolysis of fructose 1-P.

Answer: For fructose – 1 – P + $H_2O \leftrightarrows$ fructose + P_i

The equilibrium constant, K_{eq}, is given by:

$$K_{eq} = \frac{[\text{fructose}]_{eq}[P_i]_{eq}}{[\text{fructose} - 1 - P]_{eq}}$$

At 25°C or 298 K°, $[\text{fructose-1-P}]_{eq} = 6.52 \times 10^{-5}$ M. Initially [fructose-1-P] = 0.2 M. The amount of the fructose produced is 0.2 M - 6.52×10^{-5} M.

And, since an equal amount of $[P_i]$ is produced, Keq may be written as follows:

$$K_{eq} = \frac{(0.2 - 6.52 \times 10^{-5})(0.2 - 6.52 \times 10^{-5})}{6.52 \times 10^{-5}}$$

$$K_{eq} = 613\text{ M}$$

$$\Delta G° = -RT\ln K_{eq}$$

$$\Delta G° = -(8.314\text{ J/mol K}) \times 298\text{ K} \times \ln 613$$

$$\Delta G° = -15.9\text{ kJ/mol}$$

2. *The equilibrium constant for some process A ⇆ B is 0.5 at 20°C and 10 at 30°C. Assuming that ΔH° is independent of temperature, calculate ΔH° for this reaction. Determine ΔG° and ΔS° at 20° and at 30°C. Why is it important in this problem to assume that ΔH° is independent of temperature?*

Answer:

At 20°C

$$\Delta G° = -RT\ln K_{eq}$$

$$\Delta G° = -(8.314\text{ J/mol K}) \times (273 + 20)\text{ K} \times \ln 0.5$$

$$= 1.69\text{ kJ/mol}$$

At 30°C

$$\Delta G° = -RT\ln K_{eq}$$

$$\Delta G° = -(8.314\text{ J/mol K}) \times (273 + 30)\text{ K} \times \ln 10$$

$$= -5.80\text{ kJ/mol}$$

From the equation ΔG° = ΔH° - TΔS°, we see that ΔG° is linearly related to T when ΔH° is independent of temperature. If this is the case, then dΔH°/dT = 0 (i.e., the heat capacity is zero). A plot of ΔG° versus T will be linear with a slope = -ΔS° and a y intercept = ΔH°.

$$-\Delta S° = \text{slope} = \frac{\Delta G°_{303\text{ K}} - \Delta G°_{293\text{ K}}}{T_{303\text{ K}} - T_{293\text{ K}}} = \frac{-5.8 - 1.69}{303 - 293}$$

$$\Delta S° = 0.75\text{kJ/mol K}$$

ΔH° can be calculated using ΔH° = ΔG° + TΔS°.

$$\text{For } 20°\text{C}, \Delta H° = 1.69\text{ kJ/mol} + 293\text{ K} \times 0.75\text{ kJ/mol K} = 221.5\text{ kJ/mol}$$

$$\text{For } 30°\text{C}, \Delta H° = -5.80\text{ kJ/mol} + 303\text{ K} \times 0.75\text{ kJ/mol K} = 221.5\text{ kJ/mol}$$

Therefore, ΔH° = 221.5 kJ/mol and ΔS° = 0.75 kJ/mol K at both temperatures.
And, ΔG° = 1.69 kJ/mol at 20° C and ΔG° = -5.80 kJ/mol at 30° C.

3. *The standard-state free energy of hydrolysis of acetyl phosphate is ΔG° = -42.3 kJ/mol.*

Acetyl-P + H_2O → acetate + P

Calculate the free energy change for the acetyl phosphate hydrolysis in a solution of 2 mM acetate, 2 mM phosphate and 3 nM acetyl phosphate.

Answer:

$$\Delta G = \Delta G° + RT\ln\frac{[\text{Product}]_{\text{initial}}}{[\text{Reactant}]_{\text{initial}}}$$

$$\Delta G = \Delta G° + 8.314\text{ J/mol K} \times (273 + 20)\text{ K} \times \ln\frac{[\text{acetate}][P_i]}{[\text{acetyl - P}]}$$

$$\Delta G = -42{,}300\ \text{J/mol} + 8.314\ \text{J/mol K} \times 293\ \text{K} \times \ln\frac{(2\times10^{-3})(2\times10^{-3})}{(3\times10^{-9})}$$

$$\Delta G = -24.8\ \text{kJ/mol}$$

4. Define a state function. Name three thermodynamic quantities that are state functions and three that are not.

Answer: State functions are quantities that depend on the state of a system not on the process or path taken to reach this state. For example, volume is a state function as is the change in volume ΔV. The quantity ΔV depends only on the final and initial value of V; ΔV is independent of the path taken from V_i to V_f. Pressure and temperature are also state functions. The total internal energy of a system is a state function. Recall from the first law of thermodynamics, the total internal energy changes by heat being absorbed or released and by work being done on or by the system. In going from one state to another, heat may or may not be exchanged and work may or may not be done. Energy expended to do work is not state function. Consider changing the gravitational potential of an object by doing work on the object. The gravitational potential energy change is a state function because it depends only on the distance separating two objects. The energy expended, however, depends on path. (The amount of energy expended pushing a stone uphill clearly depends on the route taken.)

5. ATP hydrolysis at pH 7 is accompanied by release of a hydrogen ion to the medium

$$ATP^{4-} + H_2O \leftrightarrows ADP^{3-} + HPO_4^{2-} + H^+$$

If the ΔG°' for this reaction is -30.5 kJ/mol, what is ΔG° (that is, the free energy change for the same reaction with all components, including H⁺, at a standard state of 1 M)?

Answer: The reaction produces H^+ and we can use the following equation to calculate ΔG°: ΔG° = ΔG°' - RT ln $[H^+]$ where ΔG°' = -30.5 kJ/mol, T = 298 K, R = 8.314 J/mol K and $[H^+] = 10^{-7}$. Thus,

$$\Delta G^{\circ\prime} = -30.5\ \text{kJ/mol}) - 8.314(\text{J/mol K} \times (273+25)\ \text{K} \times \ln 10^{-7}$$

$$\Delta G^{\circ\prime} = -30.5 + 39.9 = 9.4\ \text{kJ/mol}$$

6. For the process A ⇆ B, $K_{eq}(AB)$ is 0.02 at 37°C. For the process B ⇆ C, $K_{eq}(BC)$ = 1000 at 37°C.

a. Determine $K_{eq}(AC)$, the equilibrium constant for the overall process A ⇆ C, from $K_{eq}(AB)$ and $K_{eq}(BC)$.

b. Determine standard state free energy changes for all three processes, and use ΔG°(AC) to determine $K_{eq}(AC)$. Make sure that this value agrees with that determined in part a, above.

Answer:

For A ⇆ B and B ⇆ C,

$$K_{eq}(AB) = \frac{[B]_{eq}}{[A]_{eq}},\ \text{and}\ K_{eq}(BC) = \frac{[C]_{eq}}{[B]_{eq}}$$

By solving for $[B]_{eq}$ in the above two equations we find :

$$[B]_{eq} = K_{eq}(AB) \times [A]_{eq} = \frac{[C]_{eq}}{K_{eq}(BC)}$$

This equation can be rearranged to give:

$$\frac{[C]_{eq}}{[A]_{eq}} = K_{eq}(AC) = K_{eq}(AB) \times K_{eq}(BC) = 0.02 \times 1000\ \text{or,}$$

$$K_{eq}(AC) = 20$$

The standard free energy change is calculated as follows:

$$\Delta G°(AB) = -RT\ln K_{eq}(AB) = -8.314\ J/mol\ K \times 310\ K \times \ln 0.02$$
$$\Delta G°(AB) = 10.1\ kJ/mol$$

$$\Delta G°(BC) = -RT\ln K_{eq}(BC) = -8.314\ J/mol\ K \times 310\ K \times \ln 1000$$
$$\Delta G°(BC) = -17.8\ kJ/mol$$

$$\Delta G°(AC) = -RT\ln K_{eq}(AC) = -8.314\ J/mol\ K \times 310\ K \times \ln 20$$
$$\Delta G°(AC) = -7.72\ kJ/mol$$

or

$$\Delta G°(AC) = \Delta G°(AB) + \Delta G°(BC)$$
$$\Delta G°(AC) = 10.1\ kJ/mol - 17.8\ kJ/mol = -7.70\ kJ/mol$$

7. Draw all possible resonance structures for creatine phosphate and discuss their possible effects on resonance stabilization of the molecule.

Answer: Creatine phosphate

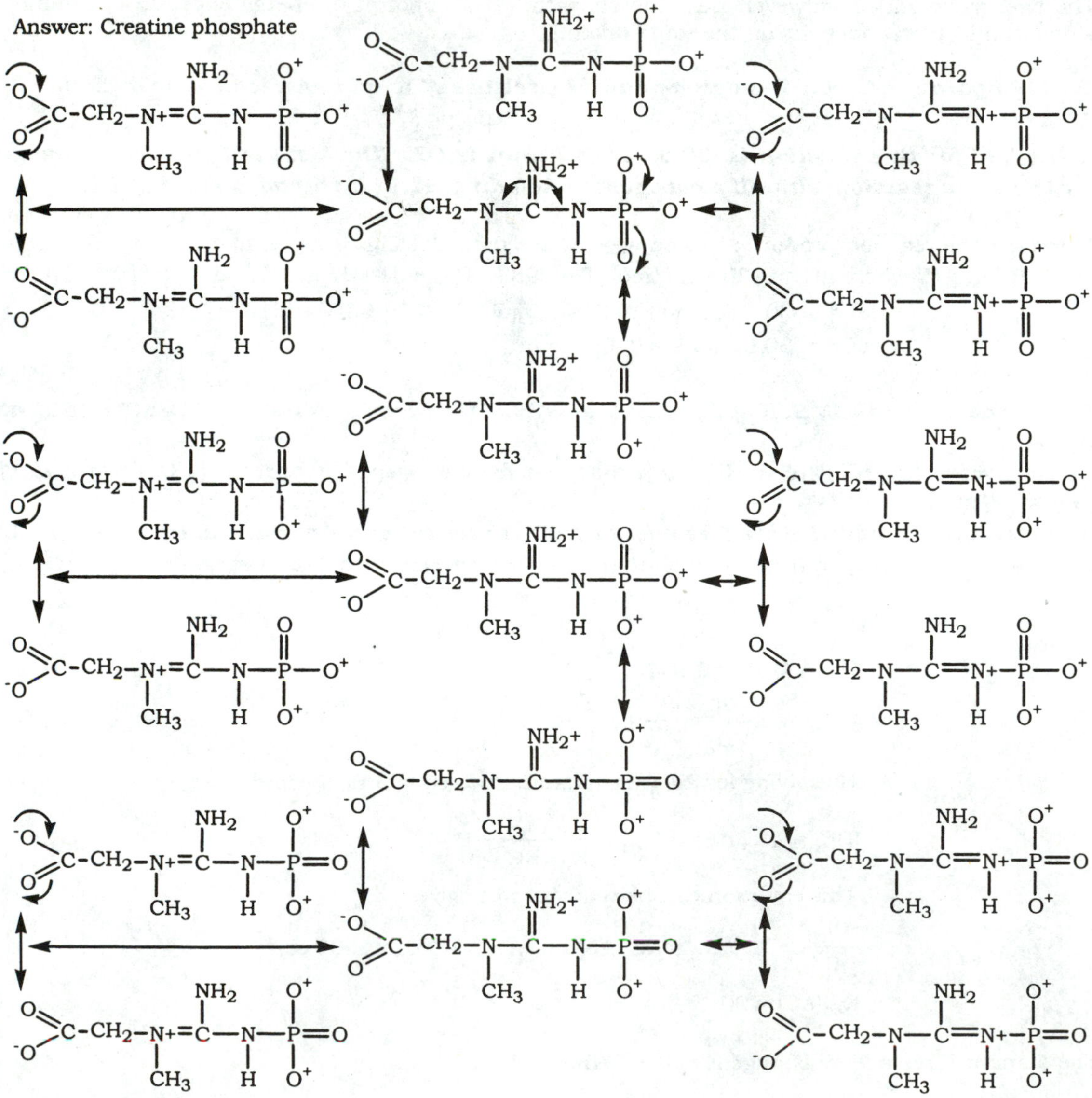

8. Write the equilibrium constant, K_{eq}, for the hydrolysis of creatine phosphate and calculate a value for K_{eq} at 25°C from the value of $\Delta G°'$ in Table 3.3.

Answer: For the reaction:

$$\text{creatine phosphate} + H_2O \leftrightarrows \text{creatine} + P_i,\ \Delta G°' = -43.3\ \text{kJ/mol}$$

$$K_{eq} = \frac{[\text{creatine}]_{eq}[P_i]_{eq}}{[\text{creatine phosphate}]_{eq}}$$

From $\Delta G°' = -RT\ln K_{eq}$, we can write:

$$K_{eq} = e^{\frac{-\Delta G°'}{RT}} = e^{\frac{-(-43.3\times 10^3)}{8.314\times 298}}$$

$$K_{eq} = 3.89\times 10^7$$

9. Imagine that creatine phosphate, rather than ATP, is the universal energy carrier molecule in the human body. Repeat the calculation presented in section 3.8, calculating the weight of creatine phosphate that would need to be consumed each day by a typical adult human if creatine phosphate could not be recycled. If recycling of creatine phosphate were possible, and if the typical adult human body contained 20 grams of creatine phosphate, how many times would each creatine phosphate molecule need to be turned over or recycled each day? Repeat the calculation assuming that glycerol-3-phosphate is the universal energy carrier, and that the body contains 20 grams of glycerol-3-phosphate.

Answer: The calculation presented in section 3.8 determined the number of moles of ATP that must be hydrolyzed under cellular conditions to provide 5,860 kJ of energy. Under standard conditions, ATP hydrolysis yields 35.7 kJ/mol (See Table 3.3) whereas under cellular conditions the value is approximately 50 kJ/mol. In order to repeat the calculation using creatine phosphate, we must estimate the free energy of hydrolysis under cellular conditions. Alternatively, we can assume that the same number of moles of creatine phosphate must be hydrolyzed as ATP. The energy of hydrolysis of creatine phosphate is larger (i.e., more negative) than that of ATP. So, hydrolysis of an equivalent molar amount of creatine phosphate will release considerably more energy. Let us try both solutions.

First, let us assume that an equivalent number of moles of creatine phosphate is hydrolyzed. From section 3.8 we see that 117 moles of ATP are required. An equal number of moles of creatine phosphate weighs:

$$117\ \text{moles} \times 180\ \text{g/mol} = 21{,}060\ \text{g}$$

And, the turnover of creatine phosphate is

$$\frac{21{,}060\ \text{g}}{20\ \text{g}} = 1{,}053\ \text{times}$$

To calculate the free energy of hydrolysis of creatine phosphate under cellular conditions, let us assume that in resting muscle, creatine phosphate is approximately 20 mM, the concentration of P_i is approximately 5 mM (See Problem 10), and approximately 10% of creatine phosphate or 2 mM is as creatine. Using these values and the standard free energy of hydrolysis of creatine phosphate (from Table 3.3) we calculate the energy of hydrolysis under cellular conditions as follows:

$$\Delta G = \Delta G° + RT\ln\frac{[\text{creatine}][P_i]}{[\text{creatine phosphate}]}$$

$$\Delta G = -43.3\ \text{kJ/mol} + 8.314\times 10^{-3}\text{kJ/mol K}\times(273+37)\text{K}\times\ln\frac{(2\times 10^{-3})(5\times 10^{-3})}{20\times 10^{-3}}$$

$$\Delta G = -62.9\ \text{kJ/mol}$$

The number of moles of creatine phosphate is given by

$$\frac{5860\ \text{kJ}}{62.9\ \text{kJ/mol}} = 93.2\ \text{moles}$$

The molecular weight of creatine phosphate is 180. Therefore

93.2 moles × 180 g/mole = 16,780 g is required.

$$\text{The turnover of creating phosphate is } \frac{16{,}780\ \text{g}}{20\ \text{g}} = 839\ \text{times.}$$

The solution to the problem using glycerol-3-phosphate is slightly more complicated. From Table 3.3 we see that the standard free energy of hydrolysis of the compound is only -9.2 kJ/mol, considerably lower than the -35.7 kJ/mol listed for ATP hydrolysis. So, we cannot simply assume that an equivalent number of moles of glycerol-3-phosphate will substitute for ATP hydrolysis as we did for creatine phosphate above because hydrolysis of an equivalent number of moles of glycerol-3-phosphate will not supply sufficient energy. To solve the problem, we must estimate the energy of hydrolysis of glycerol-3-phosphate under cellular conditions as we did for creatine phosphate hydrolysis above. Let us assume that [P_i] = 5 mM, and that the ratio of [glycerol]:[glycerol-3-phosphate] is 1:10. Under these conditions,

$$\Delta G = -9.2\ \text{kJ/mol} + 8.314 \times 10^{-3}\ \text{kJ/mol K} \times (273 + 37)\ \text{K} \times \ln\frac{5 \times 10^{-3}}{10}$$

$$\Delta G = -28.8\ \text{kJ/mol}$$

The number of moles of glycerol - 3 - phosphate is given by

$$\frac{5860\ \text{kJ}}{28.8\ \text{kJ/mol}} = 203\ \text{moles}$$

The molecular weight of glycerol - 3 - phosphate is 172.08 g/mole. Therefore

203 moles × 172.08 g/mole = 34,932 g is required.

$$\text{The turnover of glycerol - 3 - phosphate is } \frac{34{,}932\ \text{g}}{20\ \text{g}} = 1{,}747\ \text{times.}$$

If we use only the standard free energy, we require:

$$\frac{5860\ \text{kJ}}{9.2\ \text{kJ/mol}} = 637\ \text{moles or } 637\ \text{moles} \times 172.08\ \text{g/mole} = 109{,}615\ \text{g}$$

$$\text{The turnover of creating phosphate is } \frac{109{,}615\ \text{g}}{20\ \text{g}} = 5{,}480\ \text{times.}$$

10. Calculate the free energy of hydrolysis of ATP in a rat liver cell in which the ATP, ADP, and P_i concentrations are 3.4, 1.3, and 4.8 mM, respectively.

Answer: For [ATP] = 3.4 mM, [ADP] = 1.3 mM, and [P_i] = 4.8 mM, calculate the ΔG of hydrolysis of ATP.

$$\Delta G = \Delta G^{\circ\prime} + RT\ln\frac{[\text{ADP}][\text{P}_i]}{[\text{ATP}]}$$

$$\Delta G = -30.5\ \text{kJ/mol (From Table 3.3)} + 8.314 \times 10^{-3} \times (293 + 37)\ \text{K} \times \ln\frac{(1.3 \times 10^{-3})(4.8 \times 10^{-3})}{3.4 \times 10^{-3}}$$

$$\Delta G = -46.7\ \text{kJ/mol}$$

11. Hexokinase catalyzes the phosphorylation of glucose from ATP, yielding glucose-6-P and ADP. Using the values of Table 3.3, calculate the standard-state free energy change and equilibrium constant for the hexokinase reaction.

Answer: Hexokinase catalyzes the following reaction:

Glucose + ATP ⇆ glucose-6-P + ADP

This reaction may be broken down into the following two reactions:

Glucose + P_i ⇆ glucose-6-P + H_2O (1), and

ATP + H_2O ⇆ ADP + P_i (2)

From Table 3.3, we find that $\Delta G^{\circ\prime}$= -13.9 kJ/mol for glucose-6-P hydrolysis.
Thus, the reverse reaction, namely reaction (1), must have $\Delta G^{\circ\prime}$ = +13.9 kJ/mol.

From Table 3.3, we also find that ATP hydrolysis has $\Delta G^{\circ\prime}$ = - 30.5 kJ/mol.
The overall $\Delta G^{\circ\prime}$ for phosphoryl transfer from ATP to glucose is:

$$\Delta G^{\circ\prime} = +13.9 + (-30.5) = -16.6 \text{ kJ / mol and,}$$

$$K_{eq} = e^{-\frac{\Delta G^{\circ\prime}}{RT}} = e^{-\frac{(-16.6\times 10^{3})}{8.314\times 310}} = 626.9$$

12. Would you expect the free energy of hydrolysis of acetoacetyl-coenzyme A (see diagram) to be greater than, equal to, or less than that of acetyl-coenzyme A? Provide a chemical rationale for your answer.

$$H_3C-\overset{O}{\overset{\|}{C}}-CH_2-\overset{O}{\overset{\|}{C}}-S-CoA$$

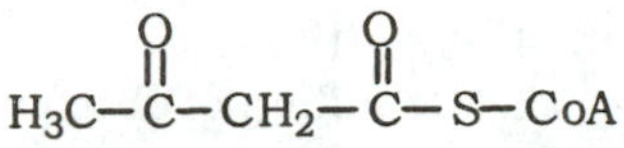

Answer: Hydrolysis of acetyl-coenzyme A produces free coenzyme A and acetate whereas hydrolysis of acetoacetyl-coenzyme A releases acetoacetate and coenzyme A. Acetate is relatively stable; however, acetoacetate is unstable and will break down to acetone and CO_2. Thus, the instability of one of the products of hydrolysis of acetoacetyl-coenzyme A, namely acetoacetate, will make the reverse reaction (i.e., production of acetoacetyl-coenzyme A) unlikely. Furthermore, the terminal acetyl group of acetoacetyl-coenzyme A is electron-withdrawing in nature and will destabilize the thiol ester bond of acetoacetyl-CoA. Thus, the free energy of hydrolysis of acetoacetyl-coenzyme A is expected to be greater than that of acetyl-coenzyme A and in fact, the free energy of hydrolysis is -43.9 kJ/mol for acetoacetyl-CoA and -31.5 kJ/mol for acetyl-CoA.

13. Consider carbamoyl-phosphate, a precursor in the biosynthesis of pyrimidines:

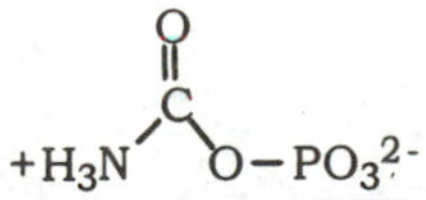

$$^{+}H_3N-\overset{O}{\overset{\|}{C}}-O-PO_3^{2-}$$

Based on the discussion of high-energy phosphates in this chapter, would you expect carbamoyl phosphate to possess a high free energy of hydrolysis? Provide a chemical rationale for your answer.

Answer: Is carbamoyl phosphate destabilized due to electrostatic repulsion? The carbonyl oxygen will develop a partial-negative charge causing charge-repulsion with the negatively charged phosphate. So, charge destabilization exists. Are the products stabilized by resonance or by ionization? Without regard to resonance states of the phosphate group, there are two possible resonance structures for carbamoyl phosphate:

$$H_2N-\overset{O}{\overset{\|}{C}}-O-PO_3^{2-} \rightleftharpoons {}^{+}H_2N=\overset{O^-}{\overset{|}{C}}-O-PO_3^{2-}$$

However, we expect that the carbonyl-carbon must pass through a positively charged intermediate and in doing so, affect phosphate resonance. Thus, the products are resonance stabilized. Finally, are there entropy factors? The products of hydrolysis are phosphate and carbamic acid. Carbamic acid is unstable and decomposes to CO_2 and NH_3 unless stabilized as a salt by interacting with a cation. With these considerations in mind, we expect carbamoyl phosphate to be unstable and therefore a high-energy compound. The free energy of hydrolysis of carbamoyl phosphate is -51.5 kJ/mol, whereas for acetyl phosphate it is only -43.3 kJ/mol.

Questions for Self Study

1. True of False.
   a. The internal energy of an isolated system is conserved. _____
   b. A closed system can exchange matter but not energy with the surroundings. ____
   c. An open system includes a system and its surroundings. _____

d. An open system can exchange matter with another open system. _____
e. The internal energy of an open system is always constant. _____

2. The first law of thermodynamics states that there are only two ways to change the internal energy of any system. What are they?

3. Enthalpy, H, is defined as $H = E + PV$ and $\Delta E = q + w$. Under what conditions is $\Delta H = q$?

4. Define the terms in the following expression: $S = k \ln W$.

5. Match the items in the two columns.

a. $\Delta G = \Delta H - T\Delta S$	1. Reaction spontaneous as written.
b. $\Delta G = \Delta G° + RT \ln([P]/[R])$	2. Used to determine standard ΔGibbs free energy.
c. $\Delta G° = 0$	3. Reaction unfavorable.
d. $\Delta G = 0$	4. $K_{eq} = 1$.
e. $\Delta G > 0$	5. Definition of change in Gibbs free energy.
f. $\Delta G < 0$	6. System at equilibrium.
g. $\Delta G° = - RT \ln K_{eq}$	7. Used to calculate amount of free energy released when reaction proceeds to equilibrium.

6. The compounds shown below include ATP, pyrophosphate, phosphoenolpyruvate, creatine phosphate, and 1,3-bisphosphoglycerate. They are all examples of high-energy compounds. Identify each, locate the high-energy bond and list the products of hydrolysis of this bond.

a.

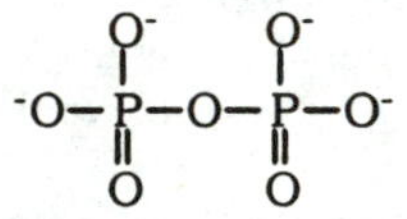

b.

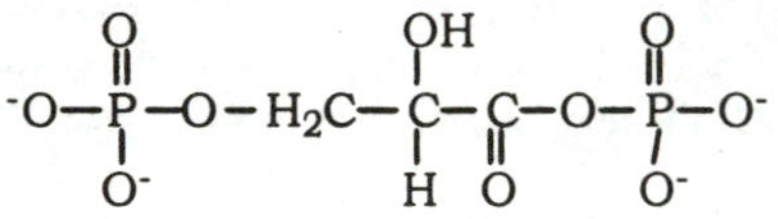

c.

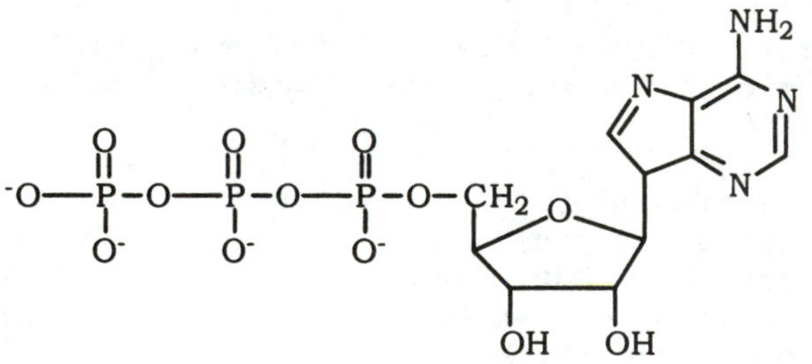

d.

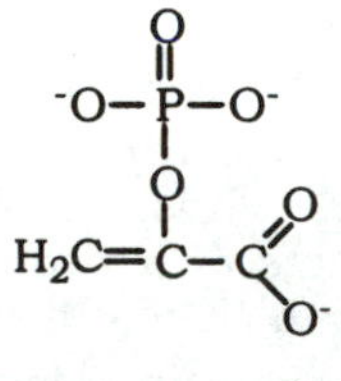

e.

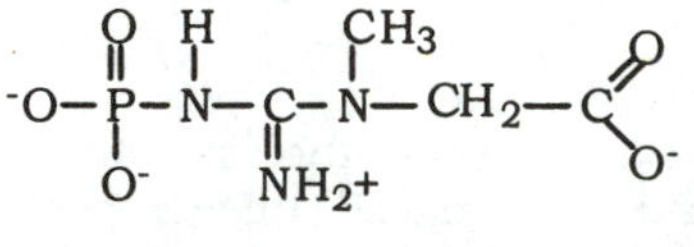

7. What are the three chemical reasons for the large negative energy of hydrolysis of phosphoric acid anhydride linkage for compounds such at ATP, ADP, and pyrophosphate?

8. Creatine phosphate is an example of a guanidinium phosphate, a high-energy phosphate compound. This compound is abundant in muscle tissue. What is its function?

9. During protein synthesis amino acids are joined in amide linkage by the ribosome. The substrates for this reaction are not free amino acids but rather amino acids attached via their

carboxyl groups to the 3' hydroxyl group of tRNA's. What kind of bond is formed between the amino acid and the tRNA? Is this a high-energy bond? Given the fact that aminoacyl-tRNA formation is accompanied by hydrolysis of ATP to AMP and PP_i and that PP_i is subsequently hydrolyzed to 2 P_i, how many high-energy phosphates are consumed to produce an aminoacyl-tRNA?

10. Although the standard free energy of hydrolysis of ATP is around -30 kJ/mol the cellular free energy change is even more negative. What factors contribute to this?

Answers

1. a. T; b. F; c. F; d. T; e. F.

2. Energy flow in the form of heat or work.

3. In general, $\Delta H = \Delta E + P\Delta V + V\Delta P = q + w + P\Delta V + V\Delta P$. When the pressure of the system remains constant (i.e., $\Delta P = 0$) and work is limited to only mechanical work (i.e., $w = -P\Delta V$) then $\Delta H = q$.

4. S is the entropy, k is Boltzmann's constant, ln W is the natural logarithm of the number of ways, W, of arranging the components of a system.

5. a. 5; b. 7; c. 4; d. 6; e. 3; f. 1; g. 2.

6a. pyrophosphate, hydrolysis products 2 P_i. b. 1,3-bisphosphoglycerate, hydrolysis products 3-phosphoglycerate and P_i. c. ATP, hydrolysis products either ADP and P_i or AMP and pyrophosphate. d. phosphoenolpyruvate, hydrolysis products pyruvate and P_i. e. creatine phosphate, hydrolysis products creatine and P_i. The location of high-energy bonds is shown below.

a.

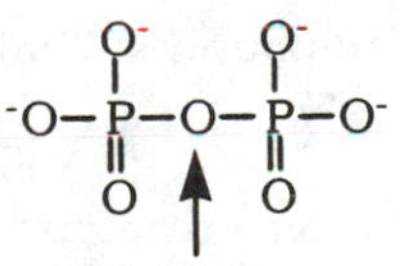

b.

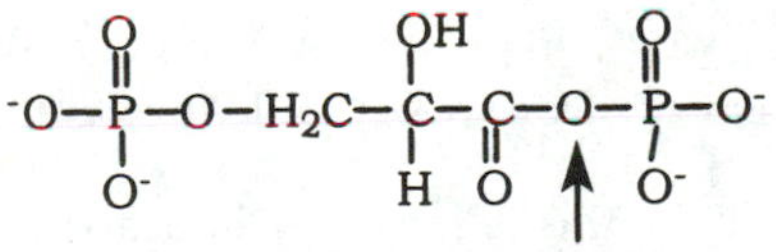

c.

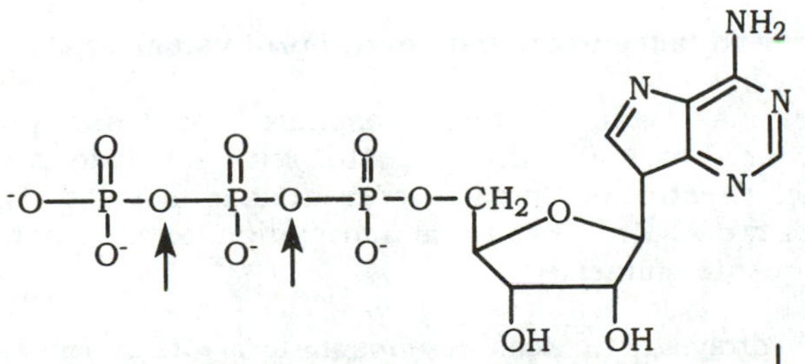

d.

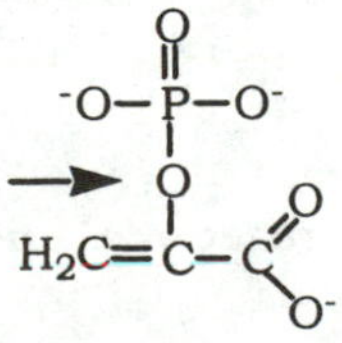

e.

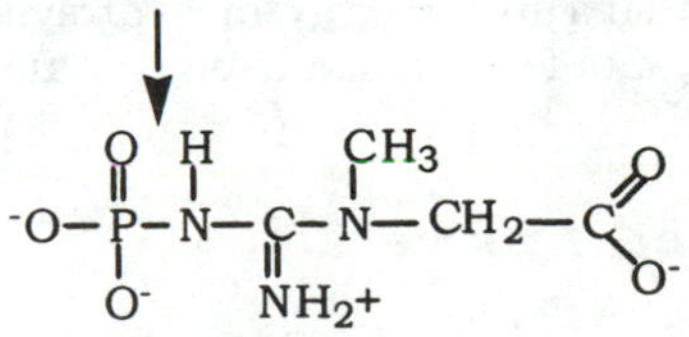

7. Bond strain due to electrostatic repulsion, stabilization of products by ionization and resonance, and entropy factors.

8. Creatine phosphate is used to replenish supplies of ATP by transferring phosphate to ADP in a reaction catalyzed by creatine kinase.

9. Amino acid ester bonds in aminoacyl-tRNA's are high-energy bonds produced at the expense of two high-energy phosphate bonds.

10. The presence of divalent and monovalent cations and the maintenance of low levels of ADP and P_i and high levels of ATP are responsible for the large negative free energy change of ATP under cellular conditions.

Additional Problems

1. Show that $\Delta G° = -RT \ln K_{eq}$.

2. The term ΔG° may be evaluated by measuring ΔG for a reaction in which reactants and products start out at 1 M concentration. The ΔG measured in this case is the standard-state free energy and is equal ΔG°. Prove this.

3. If a particular reaction is allowed to reach equilibrium, is the concentration of product ever dependent on the initial concentrations of reactant and product?

4. Confusion reigns supreme when work and energy expenditure are discussed. Define the term work. Are work and energy expenditure synonymous?

5. There are two statements about spontaneity of reactions: $\Delta S > 0$, and $\Delta G < 0$. Justify that these statements are in fact true descriptions of spontaneity.

6. DNA ligase catalyzes formation of a phosphodiester bond between a 5'-phosphate and a 3'-hydroxyl group on the ends of two DNA's to be joined. Many ligases use hydrolysis of ATP to drive phosphodiester bond synthesis; however, the *E. coli* ligase uses NAD^+ as a high-energy compound. Explain why NAD^+ is considered a high-energy compound. What type of reaction is required to release energy from NAD^+ and what are the products?

7. You find yourself in the laboratory, late at night, working on an important assay and you discover that the last of the ATP stock is used up. After frantically searching everywhere, you find a 10 mg bottle of 2'-deoxyadenosine 5'-triphosphate. Is this a high-energy compound? Will it substitute for ATP in your assay?

8. Stock solutions of ATP are usually adjusted to around neutral pH to stabilize them. Why?

9. Intense muscle activity depletes ATP and creatine phosphate stores and produces ADP, creatine, AMP, and Pi. ADP is produced by ATP hydrolysis catalyzed by myosin, a component of the contractile apparatus of muscle. Creatine is a product of creatine kinase activity, and AMP is produced by adenylate kinase from two ADP's. Given this information, explain how high-energy phosphate compounds in muscle are interconnected.

10. The standard free energy of hydrolysis of glucose-1-phosphate is -21 kJ/mol whereas it is only -13.9 kJ/mol for glucose-6-phosphate. Provide an explanation for this difference.

Abbreviated Answers

1. At equilibrium $\Delta G = 0$ and the concentration of reactants and products are at their equilibrium values. Using

$$\Delta G = \Delta G°'+RT\ \ln\frac{[C][D]}{[A][B]} \text{ we see that}$$

$$\Delta G = 0 = \Delta G°+RT\ \ln\frac{[C_{eq}][D_{eq}]}{[A_{eq}][B_{eq}]}$$

or, solving for $\Delta G°$ we find that :

$$\Delta G° = -RT\ \ln\frac{[C_{eq}][D_{eq}]}{[A_{eq}][B_{eq}]} = -RT \ln K_{eq}$$

2. Using

$$\Delta G = \Delta G°'+RT\ \ln\frac{[C][D]}{[A][B]} \text{ we see that}$$

$$\Delta G = \Delta G°+RT \ln\frac{1M\times 1M}{1M\times 1M} = \Delta G°+RT \ln 1$$

But, $\ln 1 = 0$, and

$$\Delta G = \Delta G°$$

3. The equilibrium constant is independent of the initial concentrations of reactants and products but the equilibrium constant is the ratio of the product of the concentration of products to the product of the concentration of reactants. This ratio is independent of initial concentrations. Clearly, the absolute amount of product depends on the initial concentration of reactant.

4. The first definition of work found in The Random House College Dictionary, Revised Edition is "exertion or effort directed to produce or accomplish something...." This is close to the thermodynamic definition of work but not quite right. We can modify the definition by replacing "exertion or effort" with "energy" an important modification because it introduces a term that can be quantitated. The rest of the definition is acceptable; however, an important addendum is necessary. Work is not all of the energy expended to produce or accomplish something. Rather, it is only that portion of expended energy that is equal to the amount of energy released when the something that was accomplished is allowed to return to its original state.

5. Entropy is a measure of disorder and for a reaction to be spontaneous $\Delta S > 0$ or since $S_f - S_i > 0$, $S_f > S_i$. A spontaneous reaction results in an increase in disorder.

Gibbs free energy is the amount of energy available to do work at constant pressure and temperature. Using G as a criterion for spontaneity requires that $\Delta G < 0$ or since $G_f - G_i < 0$, $G_f < G_i$. The amount of energy available to do work decreases for a spontaneous reaction.

6. NAD^+ contains a single high-energy phosphoric anhydride linkage between AMP and nicotinamide monophosphate nucleotide. Hydrolysis will release AMP and nicotinamide nucleoside monophosphate.

7. 2'-Deoxyadenosine 5'-triphosphate or dATP contains two high-energy phosphoric anhydride bonds. However, if the assays being performed are enzymatic assays (as opposed to some chemical assay), then it is out-of-the-question to even think about substituting dATP for ATP. ATP is a ribonucleotide; dATP is a deoxyribonucleotide. dATP is used for synthesis of DNA and little else.

8. At first thought it might seem reasonable to make the ATP solution slightly acidic. This will neutralize the negatively charged phosphates and should lead to stabilization of the phosphoric anhydride linkages. The problem with this idea is that the N-glycosidic linkage is acid labile. This bond is more stable in alkaline solution but the anhydride bonds are readily cleaved by hydroxide attack.

9. The key here is to recognize that kinases are enzymes that transfer the γ-phosphate of ATP to a target molecule. Thus, creatine kinase phosphorylates creatine and adenylate kinase

phosphorylates AMP. The following scheme outlines how ATP, ADP, AMP, creatine phosphate, creatine, and Pi are related.

ATP + AMP ← ADP + ADP (adenylate kinase)
↓
ATP + H_2O → ADP + P_i (myosin)
↑
ATP + creatine ← ADP + creatine phosphate (creatine kinase)

10. Phosphate is an electron-withdrawing group that is attached to quite different carbons in glucose-6-phosphate versus glucose-1-phosphate. In the latter, carbon-1 is already bonded to an electronegative oxygen in the pyranose form.

Summary

Thermodynamics - a collection of laws and principles describing the flows and interchanges of heat, energy and matter - can provide important insights into metabolism and bioenergetics. Thermodynamics distinguishes between isolated systems, which cannot exchange matter or energy with the surroundings; closed systems, which may exchange energy, but not matter, with the surroundings; and open systems, which may exchange matter, energy or both with the surroundings.

The first law of thermodynamics states that the total energy of an isolated system is conserved. E, the internal energy function, which is equal to the sum of heat absorbed and work done on the system, is a useful state function, which keeps track of energy transfers in such systems. In constant pressure processes, it is often more convenient to use the enthalpy function ($H = E + PV$) for analysis of heat and energy exchange. For biochemical processes, in which pressure is usually constant and volume changes are small, enthalpy (H) and internal energy (E) are often essentially equal. Enthalpy changes for biochemical processes can often be determined from a plot of R(ln Keq) versus 1/T (a van't Hoff plot).

Entropy is a measure of disorder or randomness in the system. The second law of thermodynamics states that systems tend to proceed from ordered (low entropy) states to disordered (high entropy) states. The entropy change for any reversible process is simply the heat transferred divided by the temperature at which the transfer occurs: $dS_{rev} = dq/T$.

The third law of thermodynamics states that the entropy of any crystalline, perfectly ordered substance must approach zero as the temperature approaches 0° K, and at 0° K, entropy is exactly zero. The concept of entropy thus invokes the notion of heat capacity - the amount of heat any substance can store as the temperature of that substance is raised by one degree K. Absolute entropies may be calculated for any substance if the heat capacity can be evaluated at temperatures from 0° K to the temperature of interest, but entropy changes are usually much more important in biochemical systems.

The free energy function is defined as $G = H - TS$ and, for any process at constant pressure and temperature (i.e., most biochemical processes), $\Delta G = \Delta H - T\Delta S$. If ΔG is negative, the process is exergonic and will proceed spontaneously in the forward direction. If ΔG is positive, the reaction or process is endergonic and will proceed spontaneously in the reverse direction. The sign and value of ΔG, however, do not allow one to determine how fast a process will proceed. It is convenient to define a standard state for processes of interest, so that the thermodynamic parameters of different processes may be compared. The standard state for reactions in solution is 1 M concentration for all reactants and products. The free energy change for a process $A + B \leftrightarrows C + D$ at concentrations other than standard state is given by:

$$\Delta G = \Delta G° + RT \ln([C][D]/[A][B])$$

and the standard state free energy change is related to the equilibrium constant for the reaction by $\Delta G° = -RT \ln K_{eq}$. This states that the equilibrium established for a reaction in solution is a function of the standard state free energy change for the process. In essence, $\Delta G°$ is another way of writing an equilibrium constant. Moreover, the entropy change for a process may be determined if the enthalpy change and free energy change calculated from knowledge of the equilibrium constant and its dependence on temperature are known.

The standard state convention of 1 M concentrations becomes awkward for reactions in which hydrogen ions are produced or consumed. Biochemists circumvent this problem by defining a modified standard state of 1 M concentration for all species except H^+, for which the standard state is 1×10^{-7} M or pH 7.

Thermodynamic parameters may provide insights about a process. As an example, consider heat capacity changes: Positive values indicate increase freedom of movement whereas negative values indicate less freedom of motion.

Many processes in living things must run against their thermodynamic potential, i.e., in the direction of positive ΔG. These processes are driven in the thermodynamically unfavorable direction via coupling with highly favorable processes. Coupled processes are vitally important in intermediary metabolism, oxidative phosphorylation, membrane transport and many other processes essential to life.

There is a hierarchy of energetics among organisms: certain organisms capture solar energy directly, whereas others derive their energy from this group in subsequent chemical processes. Once captured in chemical form, energy can be released in controlled exergonic reactions to drive life processes. High-energy phosphate anhydrides and reduced coenzymes mediate the flow of energy from exergonic reactions to the energy-requiring processes of life. These molecules are transient forms of stored energy, rapidly carrying energy from point to point in the organism. One of the most important high-energy phosphates is ATP, which acts as an intermediate energy shuttle molecule. The free energy of hydrolysis of ATP (for hydrolysis to ADP and phosphate) is less than that of PEP, cyclic-AMP, 1,3-BPG, creatine phosphate, acetyl phosphate and pyrophosphate, but is greater than that of the lower energy phosphate esters, such as glycerol-3-phosphate and the sugar phosphates. Group transfer potential - the free energy change that occurs upon hydrolysis (i.e., transfer of a chemical group to water) - is a convenient parameter for quantitating the energetics of such processes. The phosphoryl group transfer potentials for high-energy phosphates range from -31.9 kJ/mole for UDP-glucose to -35.7 kJ/mole for ATP to -62.2 kJ/mole for PEP.

The large negative $\Delta G°'$ values for the hydrolysis of phosphoric acid anhydrides (such as ATP, GTP, ADP, GDP, sugar nucleotides and pyrophosphate) may be ascribed to 1) destabilization of the reactant due to bond strain caused by electrostatic repulsion, 2) stabilization of the products by ionization and resonance, and 3) increases in entropy upon product formation. Mixed anhydrides of phosphoric and carboxylic acids - known as acyl phosphates - are also energy-rich. Other classes of high-energy species include enol phosphates, such as PEP, guanidinium phosphates, such as creatine phosphate, cyclic nucleotides, such as 3',5'-cyclic AMP, amino acid esters, such as aminoacyl-tRNA and thiol esters, such as coenzyme A. Other biochemically important high-energy molecules include the pyridine nucleotides (NADH, NADPH), sugar nucleotides (UDP-glucose) and S-adenosyl methionine, which is involved in the transfer of methyl groups in many metabolic processes.

Though the hydrolysis of ATP and other high-energy phosphates are often portrayed as simple processes, these reactions are far more complex in real biological systems. ATP, ADP and similar molecules can exist in several different ionization states, and each of these individual species can bind divalent and monovalent metal ions, so that metal ion complexes must also be considered. For example, ATP has five dissociable protons. The adenine ring amino group exhibits a pK_a of 4.06, whereas the last proton to dissociate from the triphosphate chain possesses a pK_a of 6.95. Thus, at pH 7, ATP is a mixture of ATP^{4-} and $HATP^{3-}$. If equilibrium constants for the hydrolysis reactions are re-defined in terms of total concentrations of the ionized species, then such equilibria can be considered quantitatively, and fractions of each ionic species in solution can be determined. The effects of metal ion binding equilibria on the hydrolysis equilibria can be quantitated in a similar fashion.

The free energy changes for hydrolysis of high-energy phosphates such as ATP are also functions of concentration. In the environment of the typical cell, where ATP, ADP and phosphate are usually 5 mM or less, the free energy change for ATP hydrolysis is substantially larger in magnitude than the standard state value of -30.5 kJoule/mole.

High-energy molecules such as ATP are rapidly recycled in most biological environments. The typical 70 kg human body contains only about 50 grams of ATP/ADP. Each of these ATP molecules must be recycled nearly 2,000 times each day to meet the energy needs of the body.

Chapter 4

Amino Acids and Polypeptides

• •

Chapter Outline

- Amino acids: Tetrahedral alpha (α) carbon with H, amino group, carboxyl group, and side chain
- Classification
  - Nonpolar: Ala, Val, Leu, Ile, Pro, Met, Phe, Trp
  - Polar uncharged: Gly, Ser, Thr, Tyr, Asn, Gln, Cys
  - Acidic: Asp, Glu
  - Basic: His, Lys, Arg
- Ionic properties of amino acids
  - α–Carboxyl group: pKa's range from 2.0 to 2.4
  - α–Amino group: pKa's range from 9.0 to 9.8
  - Side chains
    - Acidic pKa's for Asp, Glu
    - Basic pKa's for His, Arg, Lys
- Amino acid chemistry
  - Carboxyl chemistry
  - Amino chemistry
  - Side chain chemistry
- Chirality: Asymmetric carbon has four different groups attached
  - Optically active compounds
    - Dextrorotatory (+), clockwise rotation
    - Levorotatory (-), counterclockwise rotation
  - Asymmetric carbons
    - One asymmetric carbon: Mirror image isomers or enantiomers
    - More than one asymmetric carbon: enantiomers and diasteriomers
  - D,L System: L-Amino acids (naturally occurring): Related to L-glyceraldehyde
  - (R,S) System
- Ultraviolet spectral properties of amino acids: Phe, Try, Trp absorb UV light above 250 nm
- Separation of amino acids
  - Ion exchange chromatography: Based on electrical charge
    - Anion exchangers: Matrix is positively charged so anions bind
    - Cation exchangers: Matrix is negatively charged so cations bind
  - Partition chromatography: Based on solubility
- Proteins: Linear polymers of amino acids
  - Peptide bond
    - Carboxyl group joined to amino group with loss of water - condensation reaction incorporates amino acid residue
    - Double bond characteristics
      - π-Bond and peptide bond resonance
      - No free rotation about peptide bond
      - Cα, O, N, H all in peptide plane

- Size classification
  - Peptides: Dipeptide to dodecapeptide (2 to 12 amino acid residues, 1 to 11 peptide bonds)
  - Oligopeptides: 12 to 20 Residues
  - Polypeptides: >20 Amino acid residues
  - Proteins
    - Monomeric protein -single chain-polypeptide
    - Multimeric protein -multiple chains: Homomultimeric or heteromultimeric
- Protein purification
  - Size distinction: Size exclusion chromatography, ultracentrifugation, ultrafiltration
  - Charge distinction: Ionic exchange, electrophoresis
  - Solubility distinction: Ammonium sulfate (salts), organic solvent, pH (solubility minimum at isoelectric point)
  - Differential interactions: Hydrophobic interaction chromatography
- Amino acid composition
  - 6N HCl, 110°C, 24, 48, 72 hr
    - Ser/Thr extrapolate to zero time, Val/Ile extrapolate to infinity
    - Trp destroyed, Asn/Gln converted to Asp/Glu
  - N- and C-terminus identification
    - N-terminus chemistry: Edman degradation - phenylisothiocyanate to product PTH derivative of N-terminal amino acid
    - C-terminus: Carboxypeptidase C or Y cleaves any C-terminal amino acid, carboxypeptidase B cleaves only Arg or Lys, carboxypeptidase A does not cleave Arg, Lys, or Pro
- Protein Sequencing
  - Polypeptide chain fragmentation: Sets of short, overlapping peptides
    - Trypsin: Carbonyl side of Arg, Lys
    - Chymotrypsin: Carbonyl side of Phe, Tyr, Trp
    - Cyanogen bromide: Carbonyl side of Met
    - Others: Clostripain, pepsin, papain, subtilisin, elastase
  - Sequence determination of peptides from fragmented proteins
    - Edman degradation
    - Mass Spectrometry
  - Sequence reconstruction: Match sequence of overlapping peptides
- Amino acid sequence analysis
  - Homologous proteins: Proteins that perform the same function in different organisms
  - Homologous proteins: Proteins related evolutionarily and detected by sequence similarities
  - Proteins with related function may have common evolutionary origin
  - Proteins with different functions may have common evolutionary origin
- Protein complexity
  - Simple proteins: Composed only of polypeptide chain(s)
  - Conjugated proteins: Protein with prosthetic group: Glyco- sugar, lipo- lipids, nucleo- nucleic acid, phospho- phosphorylated amino acid, metallo- metal atom, hemo- protoporphyrin, flavo- FMN or FAD
- Biological function of proteins
  - Enzymes: Biological catalysts
  - Regulatory proteins: Hormones, DNA binding proteins
  - Transport proteins: Hemoglobin, serum albumin
  - Storage proteins: Zein, ovalbumin, phaseolin, ferritin
  - Contractile and motile proteins: Actin, myosin, dynein, kinesin, tubulin
  - Structural proteins: α-Keratins, collagens, elastin, β-keratin, proteoglycans
  - Protective proteins: Antibodies, blood clotting factors, antifreeze proteins, toxins
  - Exotic proteins: Monellin, glue protein

Chapter Objectives

It is imperative to learn the structures and one-letter codes for the amino acids. Amino acids are key molecules, both as components of proteins and as precursors of important biomolecules. The structures are shown in Figure 4.3 of *Biochemistry,* which is presented on the following this section.

Classification of Amino Acids
Here is an alternative classification scheme.
Acidic amino acids and their amides:
Aspartic acid, asparagine, glutamic acid, glutamine.
Basic amino acids:
Histidine, lysine, arginine.
Aromatic amino acids:
Phenylalanine, tyrosine, tryptophan.
Sulfur containing amino acids:
Cysteine, methionine.
Imido acid:
Proline.
Hydrophobic side chain amino acids:
Glycine, alanine, valine, leucine, isoleucine.
Hydroxylic amino acids:
Serine, threonine, (tyrosine).

To memorize the structures of the 20 amino acids, here are some aids:
The side chain of **glycine** is H; **alanine's** is a methyl group, CH_3.
The side chain of **valine** is the methyl side chain of alanine with two methyl groups substituting for two Hs. (In effect it is just dimethylalanine.)
The side chain of **leucine** is just valine with a $-CH_2-$ group between the valine side chain and the α–carbon (or, leucine is alanine with the valine side chain attached).
Isoleucine is an isomer of leucine with one of the methyl groups exchanged with a H on the carbon attached to the α–carbon.
Several of the amino acids have alanine as a base.
Phenylalanine has a phenyl group (i.e., benzene ring) attached to alanine.
Tyrosine is hydroxylated phenylalanine.
Serine has a hydroxyl group substituting for a H on the methyl group of alanine.
Cysteine can be thought of as the sulfur variety of serine i.e., with a sulfhydryl group in place of the hydroxyl group of serine.
Threonine, a hydroxylic amino acid, can be thought of as methylated serine.

The acidic amino acids and their amides are quite easy to remember. **Aspartic acid** and **glutamic acid** have carboxylic acid groups attached to $-CH_2-$ and $(-CH_2-)_2$, respectively. The amides of these amino acids, namely **asparagine** and **glutamine**, have ammonia attached to the side chain carboxyl groups in amide linkage.

The basic amino acids all have six carbons, counting the carboxyl carbon and the α–carbon. **Lysine's** side chain is a linear chain of four $-CH_2-$ groups with a terminal amino group. **Arginine** has a linear side chain of three $-CH_2-$ groups to which a guanidinium group is attached. (The sixth carbon is the central C of the guanidinium group.) Guanidinium is like urea but with a nitrogen group replacing the carbonyl oxygen. **Histidine** has an imidazole ring attached to alanine. The ring contains two nitrogens and three carbons, a five-membered ring.

An easy way to remember the structure of **proline** is to know that it is biosynthesized from glutamic acid. The carboxyl side chain forms a C-N bond to the amino group to form this imido acid. (In the process, the carbon is reduced so the side chain carboxyl oxygens do not show up in proline.)

Methionine is a sulfur-containing amino acid. Biologically, it is also used as a methyl group donor in the molecule S-adenosylmethionine (SAM). These two facts indicate that the side chain ends in a $S-CH_3$, which is attached to $(-CH_2-)_2$.

The remaining amino acid is **tryptophan**. Its side chain is an indole ring attached to a $-CH_2-$ group. In effect it can be thought of as alanine with an indole ring. Substituted indole rings are widely used in biochemistry. The basic structure is a six membered benzene ring fused to a five-membered ring containing a single N.

Nomenclature

The one-letter codes for the amino acids must be memorized. In journals, the one-letter codes are used to show amino acid sequences. See the answer to Problem 2 for a discussion of how the one-letter codes are organized.

Chirality

The α–carbons of amino acids excepting Gly, have four different groups attached to them, hence they are chiral carbons. Two different arrangements of the four groups about the chiral carbon can be made, giving rise to two structural isomers. Structural isomers that are mirror images of each other are called enantiomers. When two or more chiral centers exist on a single molecule, 2^n structural isomers are possible, where n = the number of chiral centers. The various structural isomers include enantiomers (mirror-image-related isomers) and non mirror-image-related isomers known as diastereomers.

The structural isomers can be distinguished from each other using the *R* and *S* prefixes assigned to each chiral center of a molecule. The *R* and *S* prefixes are assigned according to a set of sequence rules. The four groups about a chiral center are assigned priority numbers with the highest priority given to the atom with the highest atomic number, thus S > O > N > C > H (for elements found in amino acids). When two or more groups have the same primary element, then elements attached to the primary element are considered. So for example, SH > OR > OH > NH_2 > COOH > CHO > CH_2OH > CH_3. Once priority assignments are made, the molecule is viewed with the group having the lowest priority away from the reader. Moving from highest to lowest priority, for the three remaining groups, clockwise movement indicates ***R*** and counterclockwise indicates ***S*** configuration.

UV-absorbing Properties

Only three amino acids, phenylalanine, tyrosine, and tryptophan, absorb light in the near UV range (i.e., 230 nm to 310 nm). These amino acids dominate the UV absorption spectra of proteins. The wavelength maxima for tyrosine and tryptophan are around 280 nm. In addition, these two amino acids have extinction coefficients quite a bit larger than the extinction coefficient of phenylalanine, whose λ max = 260 nm (See Figure 4.15). Thus, very many proteins have maximum absorbance around 280 nm.

Peptide Bond

Know the bonding orbitals responsible for the peptide bond. The carbonyl oxygen and hydrogen attached to the nitrogen are usually trans to each other in the peptide plane, and free rotation about the C-N bond is restricted due to π-bonds. In considering the structure of a protein, without regard to the side chains of the amino acids, a protein is composed of successive peptide planes connected by α–carbons. The carbonyl oxygen has two sets of lone-pair electrons that can participate as hydrogen-bond acceptors. The hydrogen attached to the nitrogen in a peptide bond is a good hydrogen bond donor.

Protein Terminology

Short chains of amino acids held together by peptide bonds are referred to as polypeptides. Polypeptides contain amino acid residues and the number of residues may be indicated by a prefix. For example, dipeptides, tripeptides, and tetrapeptides contain two, three, and four amino acid residues respectively. An oligopeptide may contain in excess of 12 residues whereas a polypeptide may contain several dozen residues. The term protein typically refers to a unit exhibiting biological activity. Proteins may be single polypeptide chains referred to as monomeric proteins or they may be homomultimeric or heteromultimeric complexes.

Biological Functions

Proteins are a diverse category of biomolecules responsible for an array of biological functions. Protein functions include enzyme catalysis, regulation, transportation, storage, contractility and motility, structure, and as natural defense mechanisms.

Protein Sequencing

The basic strategy of protein sequencing is to first purify the protein and separate it into its component polypeptide chains. End group analysis identifies the N- and C-terminal residues. The chains are cleaved to produce smaller more manageable pieces whose amino acid sequence is determined using Edman degradation or mass spectrometry. Understand the basic idea behind Edman chemistry as applied to amino acid sequencing. It is often easier to sequence a protein by cloning its gene. However, knowing the sequence of even short oligopeptides from the protein gives information to search for the protein's gene.

In fragmenting a protein, remember: trypsin cuts after lysine and arginine; clostripain cuts after arginine; chymotrypsin cleaves after the aromatic amino acids, phenylalanine, tyrosine and tryptophan; and, cyanogen bromide cleaves after methionine.

Nonpolar (hydrophobic)

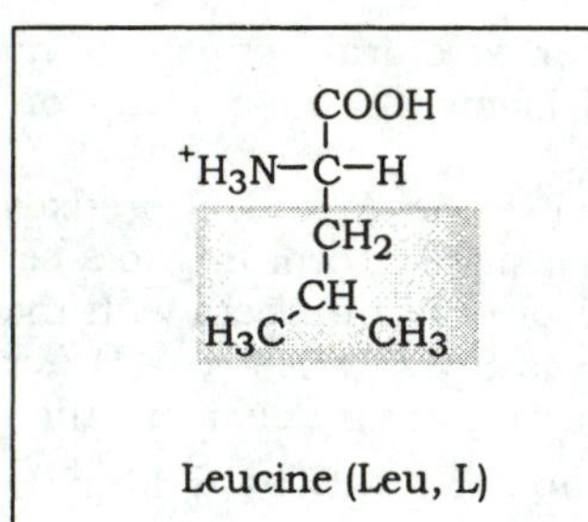

Leucine (Leu, L)

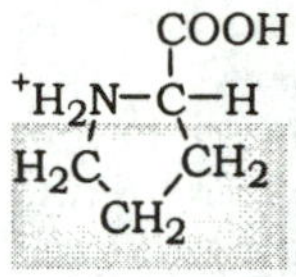

Proline (Pro,P)

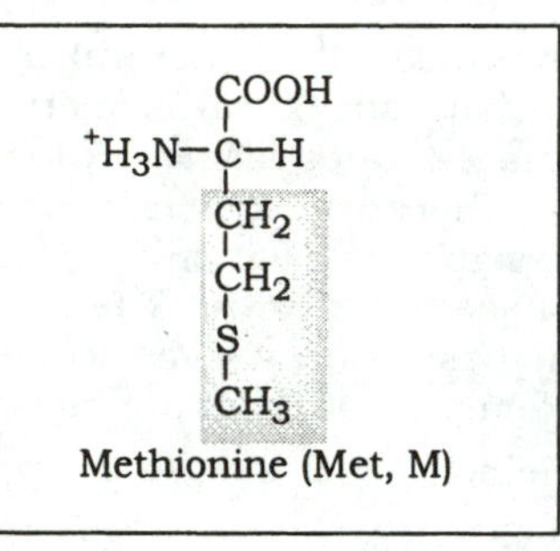

Methionine (Met, M)

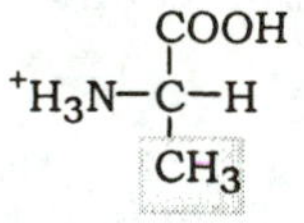

Alanine (Ala, A)

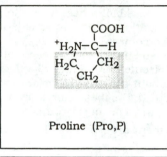

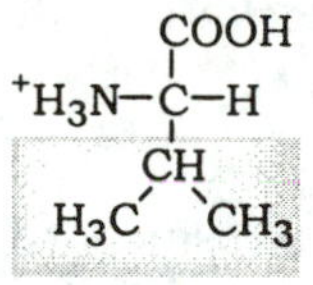

Valine (Val,V)

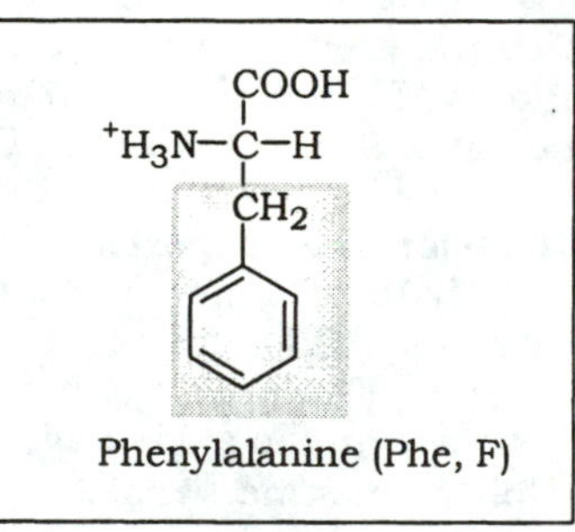

Phenylalanine (Phe, F)

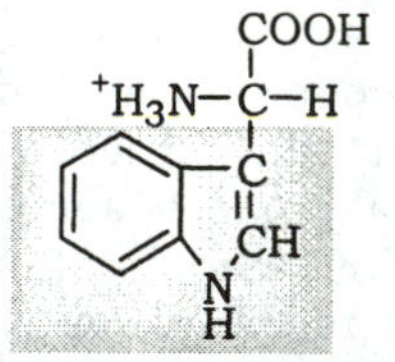

Tryptophan (Trp, W)

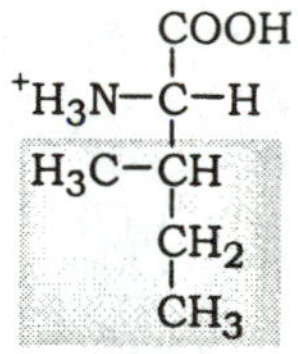

Isoleucine (Ile, I)

Acidic

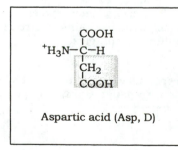

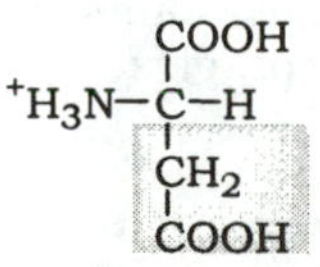

Aspartic acid (Asp, D)

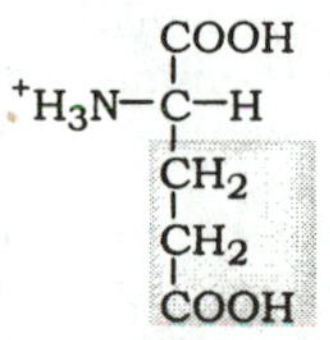

Glutamic acid (Glu, E)

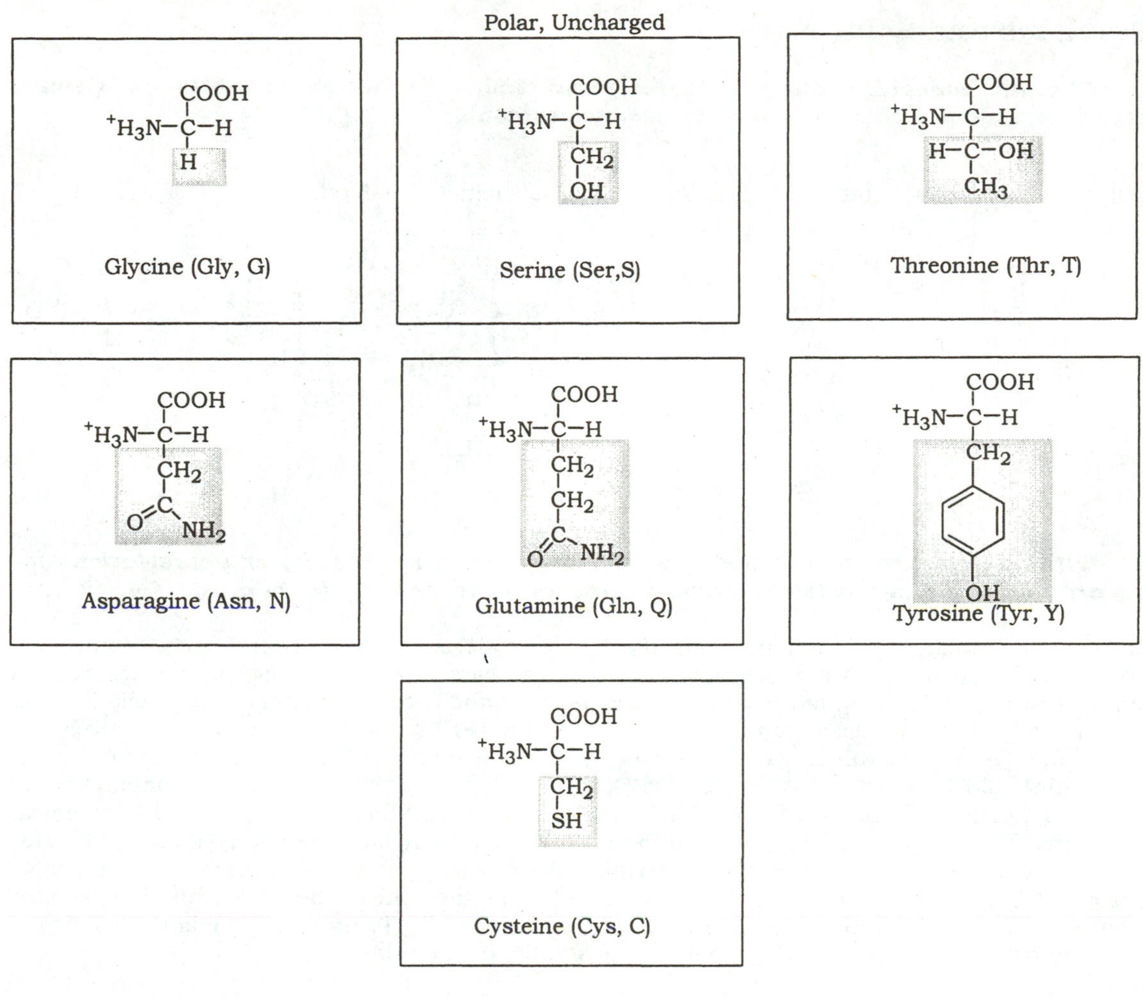

Polar, Uncharged

Histidine (His, H)

Lysine (Lys, K)

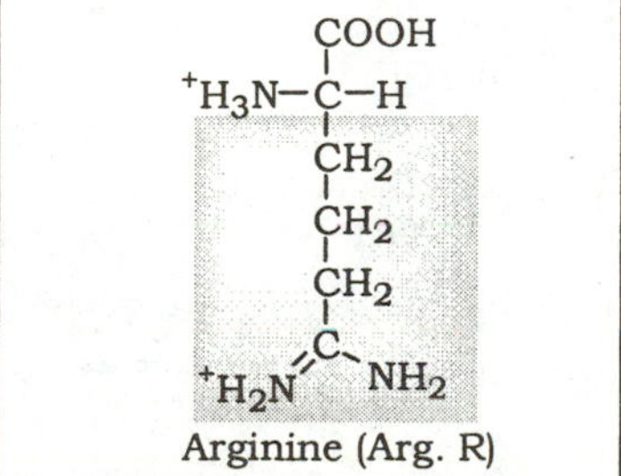

Figure 4.3 The amino acids that are the building blocks of most proteins can be classified as nonpolar (hydrophobic), acidic, polar -uncharged, and basic. Also shown are the one-letter and three-letter codes used to denote amino acids. For each amino acid, the side chains are highlighted.

Problems and Solutions

1. Without consulting chapter figures, draw Fisher projection formulas for glycine, aspartate, leucine, isoleucine, methionine, and threonine.

Answer:

Glycine: $^{+}H_3N-C(H)(H)-C(=O)-O^-$

Aspartate: $^{+}H_3N-C(H)(CH_2-C(=O)-O^-)-C(=O)-O^-$

Leucine: $^{+}H_3N-C(H)(CH_2-CH(CH_3)-CH_3)-C(=O)-O^-$

Isoleucine: $^{+}H_3N-C(H)(CH(CH_3)-CH_2-CH_3)-C(=O)-O^-$

Methionine: $^{+}H_3N-C(H)(CH_2-CH_2-S-CH_3)-C(=O)-O^-$

Threonine: $^{+}H_3N-C(H)(CHOH-CH_3)-C(=O)-O^-$

2. Without reference to the text, give the one-letter and three-letter abbreviations for asparagine, arginine, cysteine, lysine, proline, tyrosine, and tryptophan.

Answer: For many of the amino acids the 1 letter code is the first letter of the amino acid. Problems arise when two amino acids start with the same letter. Whenever this occurs, the amino acid with the lowest molecular weight is assigned the letter. The other amino acid is given a phonetic letter or the closest unused letter in the alphabet. The amino acids assigned their first letters are Glycine, Alanine, Valine, Leucine, Isoleucine, Serine, Threonine, Proline, Cysteine, Methionine and Histidine. Aspartic acid is D, an easy way to remember it is to pronounce it as "asparDic" acid. Glutamic acid is E, E follows D, glutamic acid follows aspartic acid in chemical structure. E is also the closest unused letter to G. F is for phenylalanine, (**F**enylalanine). R is for arginine, remembered by pronouncing arginine (**R**ginine). N is for asparagine; remember to stress the N sound. Glutamine follows asparagine; its 1 letter code is the next unused consonant after N, namely Q. Lysine is K, the closest unused letter to L. Tyrosine is Y; Y is the dominant sound in tyrosine. Tryptophan is W (remember "double ring, double you").

3. Write equations for the ionic dissociations of alanine, glutamate, histidine, lysine, and phenylalanine.

Answer: Alanine dissociation:

$$^{+}H_3N-C(H)(COOH)(CH_3) \rightleftharpoons {}^{+}H_3N-C(H)(COO^-)(CH_3) \rightleftharpoons H_2N-C(H)(COO^-)(CH_3)$$

Glutamate Dissociation:

$$^{+}H_3N-C(H)(COOH)(CH_2CH_2COOH) \rightleftharpoons {}^{+}H_3N-C(H)(COO^-)(CH_2CH_2COOH) \rightleftharpoons {}^{+}H_3N-C(H)(COO^-)(CH_2CH_2COO^-) \rightleftharpoons H_2N-C(H)(COO^-)(CH_2CH_2COO^-)$$

Histidine dissociation:

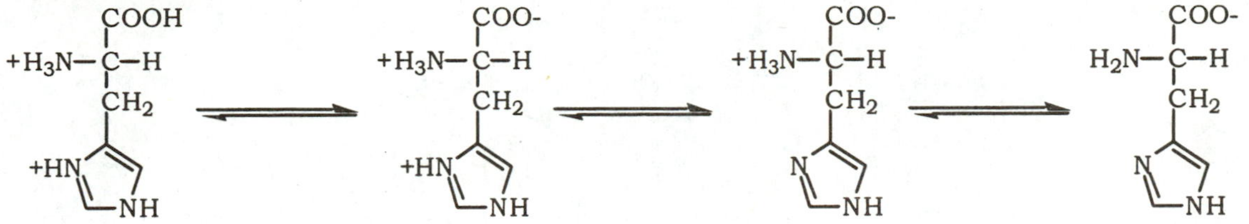

COOH
^+H_3N—C—H
CH_2
^+HN (imidazole ring) NH ⇌ COO-
^+H_3N—C—H
CH_2
^+HN (imidazole ring) NH ⇌ COO-
^+H_3N—C—H
CH_2
N (imidazole ring) NH ⇌ COO-
H_2N—C—H
CH_2
N (imidazole ring) NH

Lysine dissociation:

COOH
^+H_3N—C—H
CH_2
CH_2
CH_2
CH_2
NH_3+ ⇌ COO-
^+H_3N—C—H
CH_2
CH_2
COOH
CH_2
NH_3+ ⇌ COO-
H_2N—C—H
CH_2
CH_2
COO-
CH
NH_3+ ⇌ COO-
H_2N—C—H
CH_2
CH_2
COO-
CH_2
NH_2

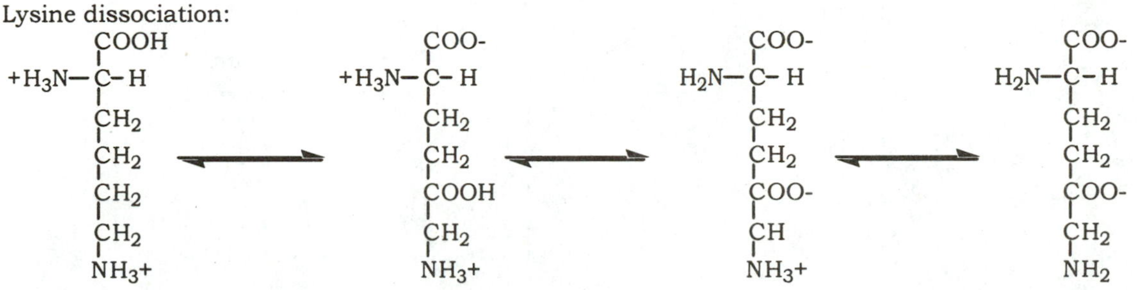

Phenylalanine dissociation:

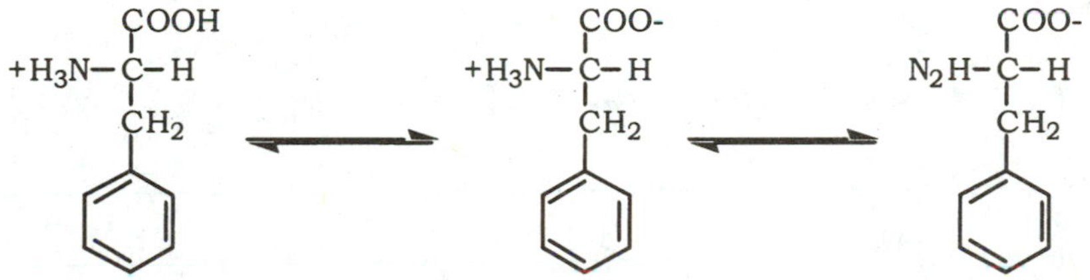

COOH
^+H_3N—C—H
CH_2
(phenyl ring) ⇌ COO-
^+H_3N—C—H
CH_2
(phenyl ring) ⇌ COO-
N_2H—C—H
CH_2
(phenyl ring)

4. How is the pK$_a$ of the a-NH$_3^+$ group affected by the presence on an amino acid of the a-COOH?

Answer: The pK$_a$ on an isolated amino group is around 10 (as for example, the side chain amino group of lysine whose pK$_a$ = 10.5 (see Table 4.1). In general, the pK$_a$'s of α–amino groups in the amino acids are around 9.5. Thus, the proximity of the α–carboxyl group lowers the pK$_a$ of the α–amino group.

5. Draw an appropriate titration curve for aspartic acid, labeling the axis and indicating the equivalence points and the pK$_a$ values.

Answer: For a titration curve the independent variable is the amount of acid or base used to titrate the amino acid. The dependent variable is pH. The pK$_a$'s for aspartic acid are as follows: 2.1 for the α-carboxylic acid group; 3.9 for the β-carboxylic group (side chain); 9.8 for the α - amino group.

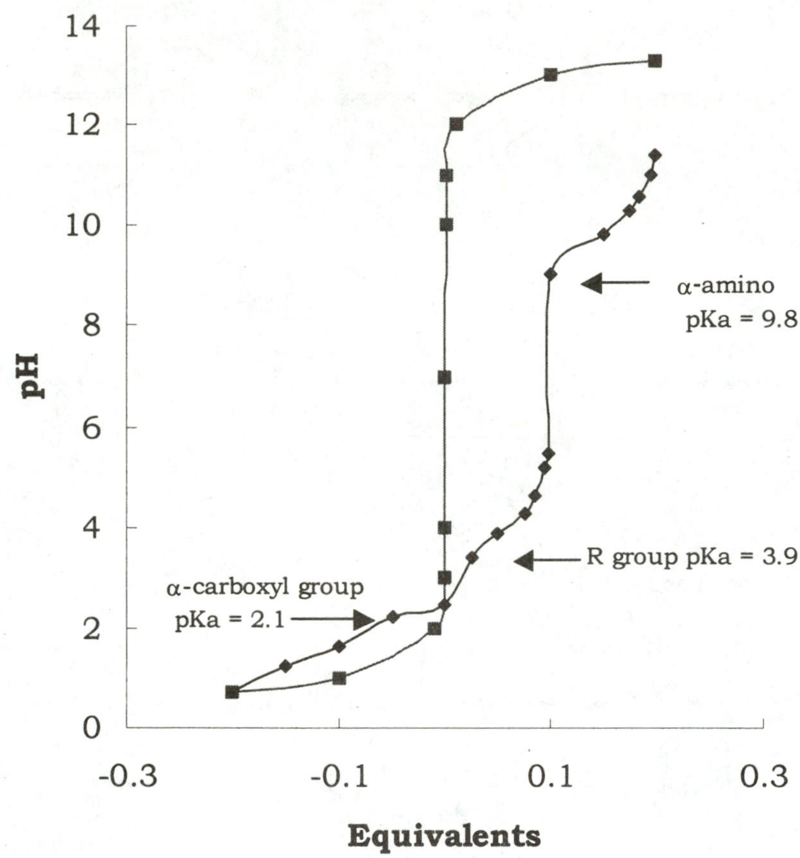

The data show the titration curve for aspartic acid (diamonds) and water (squares). The experiment is conducted by dissolving aspartic acid in water and reading its pH. The initial reading corresponds to the pH at zero equivalents. As you can see the aspartic acid solution is acidic and this is because its side chain is acidic. (Solutions of amino acids without ionizable side chains are near neutral pH.) In contrast, pure water's pH is 7.0. (This is actually a difficult measurement to make because care must be taken to insure that no contaminant like carbon dioxide is present and pH electrodes function poorly in low ionic strength solutions.) Addition of acid to the amino acid solution is depicted as negative equivalents and base addition positive equivalents. When water is challenged with acid or base the pH changes greatly. In contrast, the amino acid solution resists changes in pH when challenged with acid.

equivalence point is a point on the titration curve reached after a molar equivalent of acid or base is added to the amino acid. For aspartic acid there are actually three equivalence points at around 4.2, 5.9 and 11.8. These values are two pH units above the pK_a's and represent pH's at which the groups are approximately 99% saturated. The equivalence point at 4.2 is too close to the pK_a at 3.9 to be clearly recognized. On either side of pH = 5.9 addition of acid or base causes drastic changes in pH. This pH coincides with the addition of two molar equivalents of base representing the titration of the two carboxyl groups of aspartic acid. An additional equivalent of acid is required to reach pH = 11.8 and this corresponds to titration of the amino group of the amino acid.

6. Calculate the pH at which the γ–carboxyl group of glutamic acid is two-thirds dissociated.

Answer: The pK_a of γ–carboxyl group of glutamic acid is 4.3.

Using the Henderson - Hasselbalch equation

$$pH = pK_{COOH} + \log\frac{[C_\gamma OO^-]}{[C_\gamma OOH]}$$

When the γ - carboxyl group is two - thirds dissociated :

$$\frac{[C_\gamma OO^-]}{[C_\gamma OO^-]+[C_\gamma OOH]} = \frac{2}{3} \text{ or, } [C_\gamma OO^-] = \frac{2}{3}\left([C_\gamma OO^-]+[C_\gamma OOH]\right)$$

$$\text{Thus, } [C_\gamma OOH] = \frac{1}{2}[C_\gamma OO^-]$$

Upon substituting into the Henderson - Hasselbalch equation we find :

$$pH = 4.3 + \log\frac{[C_\gamma OO^-]}{[C_\gamma OOH]} = 4.3 + \log\frac{[C_\gamma OO^-]}{\frac{1}{2}[C_\gamma OO^-]} = 4.3 + \log 2$$

$$pH = 4.6$$

7. *Calculate the pH at which the ε-amino group of lysine is 20% dissociated.*

Answer: For the lysine side chain, pK_a = 10.5.

Using the Henderson - Hasselbalch equation

$$pH = pK_{N_\varepsilon H_3^+} + \log\frac{[N_\varepsilon H_2]}{[N_\varepsilon H_3^+]}$$

When the ε - amino group is 20% dissociated :

$$\frac{[N_\varepsilon H_2]}{[N_\varepsilon H_2]+[N_\varepsilon H_3^+]} = .2\ (20\%) \text{ or, } [N_\varepsilon H_2] = .2\left([N_\varepsilon H_2]+[N_\varepsilon H_3^+]\right)$$

$$\text{Thus, } [N_\varepsilon H_2] = \frac{1}{4}[N_\varepsilon H_3^+]$$

Upon substituting into the Henderson - Hasselbalch equation we find :

$$pH = 10.5 + \log\frac{[N_\varepsilon H_2]}{[N_\varepsilon H_3^+]} = 10.5 + \log\frac{\frac{1}{4}[N_\varepsilon H_3^+]}{[N_\varepsilon H_3^+]} = 10.5 - \log 4$$

$$pH = 9.9$$

8. *Calculate the pH of a 0.3 M solution of (a) leucine hydrochloride, (b) sodium leucinate, and (c) isoelectric leucine.*

Answer: The solution 0.3 M leucine HCl is composed of 0.3 M leucine and 0.3 M HCl. Thus, we expect it to have an acidic pH and only the α–carboxyl group of leucine is involved in proton equilibrium.

The pK_a of the α–carboxyl of leucine is 2.4. Thus,

$$pH = pK_a + \log\frac{[A^-]}{[HA]} = 2.4 + \log\frac{[COO^-]}{[COOH]} \text{ (1) and,}$$

$$[COO^-] + [COOH] = 0.3M \text{ (2)}$$

For charge neutrality :

$$[Cl^-] + [COO^-] = [H^+] + [NH_3^+]$$

If we recognize that $[Cl^-] = [NH_3^+]$ (because Cl^- and the amino acid are in equimolar amounts)
The charge neutrality equation may be simplified to:

$[COO^-] = [H^+]$ and combined with (1) and (2) we have:

$$pH = 2.4 + \log\frac{[H^+]}{0.3 - [H^+]}$$

If we make the assumption that $[H^+] << 0.3$

$$pH = 2.4 + \log\frac{[H^+]}{0.3} = 2.4 + \log[H^+] - \log 0.3 = 2.4 - \log 0.3 + \log[H^+]$$

Recognizing that $\log[H^+] = -pH$ we see that

$pH = 2.4 - \log 0.3 - pH$ or

$$pH = \frac{2.4 - \log 0.3}{2} = 1.46$$

Note: If we did not make the simplification that $[H^+]<0.3$ we would have had to solve a quadratic and would have found that pH = 1.49.

(b) sodium leucinate.
Here we assume the solution is basic; therefore, only the α–amino group is involved. The mathematical solution is similar to that given for (a) with $pK_a = 9.6$.

$$pH = pK_a + \log\frac{[NH_2]}{[NH_3^+]} = 9.6 + \log\frac{[NH_2]}{[NH_3^+]} \quad (1) \text{ and,}$$

$[NH_2] + [NH_3^+] = 0.3M$ (2)

For charge neutrality:

$[OH^-] + [COO^-] = [NH_3^+] + [Na^+]$

If we recognize that $[COO^-] = [Na^+]$ (because Na^+ and the amino acid are in equimolar amounts)
The charge neutrality equation may be simplified to:

$[OH^-] = [NH_3^+]$ and combined with (1) and (2) we have:

$$pH = 9.6 + \log\frac{0.3 - [OH^-]}{[OH^-]}$$

If we make the assumption that $[OH^-] << 0.3$

$$pH = 9.6 + \log\frac{0.3}{[OH^-]} = 9.6 + \log 0.3 - \log[OH^-]$$

Recognizing that $pOH \equiv -\log[OH^-]$ and that $pH + pOH = 14$ we have

$pH = 9.6 + \log 0.3 + pOH = 9.3 + \log 0.3 + 14 - pH$ or

$$pH = \frac{9.6 + 14 + \log 0.3}{2} = 11.5$$

(c) isoelectric leucine
Isoelectric leucine is a leucine solution at a pH at which there is no net charge on the molecule. For this to occur the carboxyl group must be unprotonated to the same extent as the amino group is protonated. This must occur at a pH half way between the pK_a's of the two groups

At the pI, $[COO^-] = [NH_3^+]$

Using the Henderson - Hasselbalch equations

$$pH = pK_{COOH} + \log\frac{[COO^-]}{[COOH]} \text{ and } pH = pK_{NH_3^+} + \log\frac{[NH_2]}{[NH_3^+]}$$

and recognizing that $[COOH] = [\text{Amino acid}]_{total} - [COO^-]$

and that $[NH_2] = [\text{Amino acid}]_{total} - [NH_3^+]$

upon substituting, solving for $[COO^-]$ and $[NH_3^+]$, and setting these terms equal at pH = pI we find that

$$pI = \frac{pK_{COOH} + pK_{NH_3^+}}{2} \text{ or}$$

$$pI = \frac{2.4 + 9.6}{2} = 6.0$$

9. Describe the stereochemical aspects of the structure of cystine, the structure that is a disulfide-linked pair of cysteines.

Answer: Cystine (dicysteine) has two chiral carbons, the two α-carbons of the cysteine moieties. Since each chiral center can exist in two forms there are 4 stereoisomers of cystine. However, it is not possible to distinguish the difference between L-cysteine/D-cysteine and D-cysteine/L-cysteine dimers. So three distinct isomers are formed.

L-cysteine-L-cysteine	L-cysteine-D-cysteine	D-cysteine-L-cysteine	D-cysteine-D-cysteine
H	H	H	H
$^-OOC-C-NH_3^+$	$^+H_3N-C-COO^-$	$^-OOC-C-NH_3^+$	$^+H_3N-C-COO^-$
CH_2	CH_2	CH_2	CH_2
S	S	S	S
S	S	S	S
CH_2	CH_2	CH_2	CH_2
$^+H_3N-C-COO^-$	$^+H_3N-C-COO^-$	$^-OOC-C-NH_3^+$	$^-OOC-C-NH_3^+$
H	H	H	H

L-cysteine-L-cysteine L-cysteine-D-cysteine D-cysteine-L-cysteine D-cysteine-D-cysteine

10. Describe the expected elution pattern for a mixture of aspartate, histidine, isoleucine, valine, and arginine on a column of Dowex-50.

Answer: Dowex-50 is a cation exchanger and binds most tightly to cations. The relative strength of the positive charges on the amino acids can be determined from their respective isoelectric points. The elution can be performed by running an increasing concentration of salt over the column (a salt gradient), or by running an increasing pH gradient. Then the amino acids should elute in the order of their increasing isoelectric points. (The isoelectric point, the pH at which a molecule has a net charge of zero, is simply the average value of the pK_as (for amino acids lacking ionizable R groups). Accordingly, we expect aspartate to elute first because it is an acidic, negatively-charged amino acid. Next will be valine followed by isoleucine. The charge on these amino acids will be close to (but not) zero; however, isoleucine will be slightly less negatively charged than valine. Thus, isoleucine's isoelectric point is (2.4 + 9.7)/2 = 6.05 whereas valine's isoelectric point is (2.3 + 9.6)/2 = 5.95. At neutral pH we expect isoleucine to be slightly less negatively charged than valine. Following isoleucine, we expect expect histidine to elute first followed by arginine. Of these two basic amino acids, arginine will have a +1 charge at neutral pH whereas histidine's charge will be slightly less than +1.

11. Assign (R,S) nomenclature to the threonine isomers of Figure 4.14.

Answer:

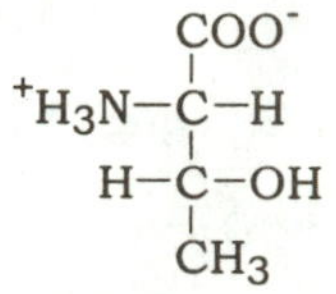

COO^-
$^+H_3N—C—H$
$H—C—OH$
CH_3

The groups about the α-carbon of L-threonine are assigned the following priority: NH_3^+ > COO^- > $C_β$> H. In Fisher projections, two groups joined to the α-carbon (i.e., NH_3^+ and H) are coming out of the plane while the other two groups are into the plane. By rotating the molecule such that the group with the lowest priority, namely H, is away from the reader, the NH_3^+ group will be located on the right side of the α-carbon, the carboxyl group above and to the left, and the remaining β-carbon to the left and below the α-carbon:

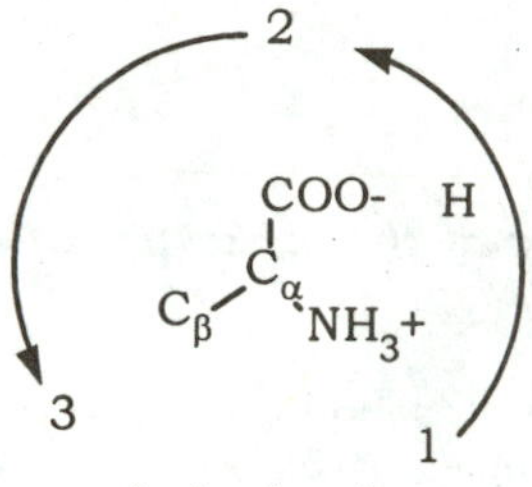

2
COO- H
$C_α$
$C_β$ NH_3^+
3
1

Since moving from highest to lowest priority involves counterclockwise movement the α-carbon is in the *S* configuration.

For the β-carbon, the priority is OH > $C_α$ > CH_3 > H. (The orientation of the $C_α$-$C_β$ bond was already defined for the α-carbon as being into the plane therefore it must be out of the plane from the perspective of the β-carbon, as must be the bond to -CH_3. Thus, bonds to OH and H are into

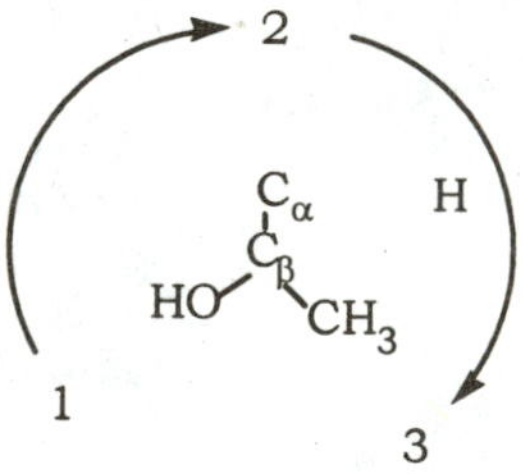

2
$C_α$ H
$C_β$
HO CH_3
1
3

the plane. By rotating the molecule such that H is behind the β-carbon, we find the OH to the left, CH_3 below and $C_α$ above:

Since moving from highest to lowest priority involves clockwise movement the α-carbon is in the *R* configuration. The β-carbon is in the *R* configuration.

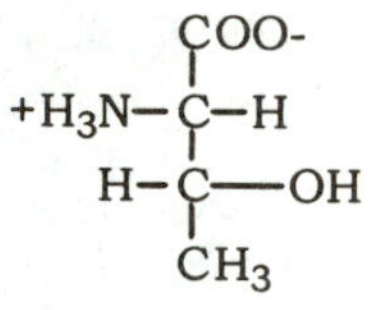

L-threonine	D-threonine	L-allothreonine	D-Allothreonine
COO-	COO-	COO-	COO-
$+H_3N—C—H$	$H—C—NH_3+$	$+H_3N—C—H$	$H—C—NH_3+$
H—C—OH	HO—C—H	HO—C—H	H—C—OH
CH_3	CH_3	CH_3	CH_3

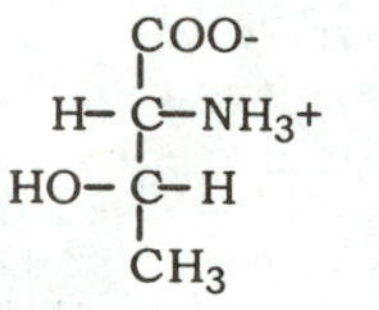

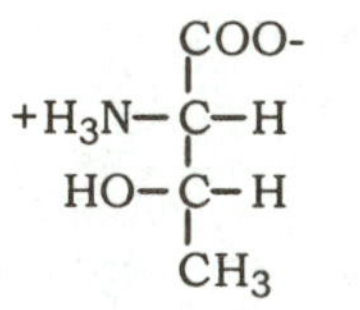

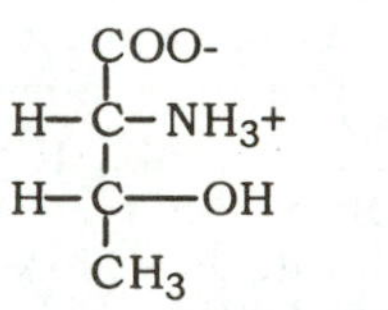

By inspection we see that,

L-threonine is (2*S*, 3*R*) threonine.
D-threonine is (2*R*,3*S*) threonine
L -allothreonine is (2*S*,3*S*) threonine
D-allothreonine is (2*R*,3*R*) threonine

12. The element molybdenum (atomic weight 95.95) constitutes 0.08% of the weight of nitrate reductase. If the molecular weight of nitrate reductase is 240,000, what is its likely quaternary structure?

Answer: If molybdenum (95.95 g/mole) is 0.08% of the weight of nitrate reductase, and nitrate reductase contains at a minimum 1 atom of Mo per molecule, then

$$95.95 = 0.08\% \times M_R$$

$$M_R = \frac{95.95}{0.0008} = 119{,}937 \approx 120{,}000$$

Since the actual molecular weight is 240,000, the quaternary structure must be a dimer.

13. Amino acid analysis of an oligopeptide seven residues long gave

Asp Leu Lys Met Phe Tyr

The following facts were observed:

a. Trypsin treatment had no apparent effect.

b. The phenylthiohydantoin released by Edman degradation was

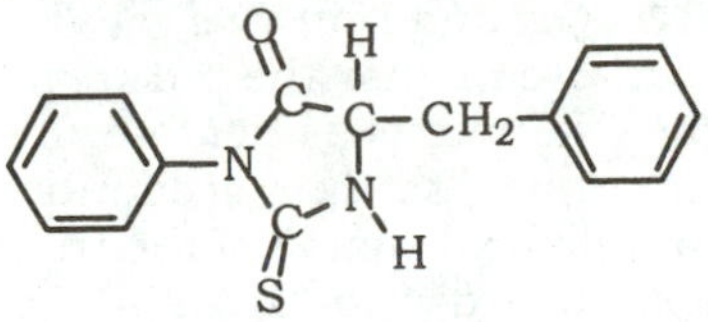

c. Brief chymotrypsin treatment yielded several products including a dipeptide and a tetrapeptide. The amino acid composition of the tetrapeptide was Leu, Lys, and Met.

d. Cyanogen bromide treatment yielded a dipeptide, a tetrapeptide, and free Lys.

What is the amino acid sequence of this heptapeptide?

Answer: From (a) we suspect that the C terminal amino acid is lysine (Trypsin cleaves on the carboxyl side of lysine or arginine) and that there is only one lysine in the peptide. (The amino acid composition lists only six amino acids. So, one amino acid must be present twice. It can't be lysine because trypsin would have cleaved the peptide.

The heptapeptide has phenylalanine as an N terminal amino acid because the PTH derivative shown in (b) is a modified phenylalanine.

Phe _ _ _ _ _ Lys

Chymotrypsin cleaves strongly at phenylalanine, tyrosine (and tryptophan). It must have released the N-terminal Phe in addition to a dipeptide and tetrapeptide. Since the tetrapeptide contains Lys and does not contain Phe or Tyr it must have derived from the C terminus by cleavage at position 3.

Phe _ Tyr _ _ _ Lys

From (d) we realize that there must be two methionines, which occupy two of the four unassigned positions. One Met must be located at position 6 to account for Lys production upon cyanogen bromide cleavage.

Phe _ Tyr _ _ Met Lys

We still have to account for one Met, one Leu, and one Asp. The Asp must be at position 2 because it is not found in the tetrapeptide generated by chymotrypsin cleavage. The only arrangement of Leu and Met that is consistent with the data is:

Phe Asp Tyr Met Leu Met Lys

14. Amino acid analysis of another heptapeptide gave

Asp Glu Leu Lys

Met Tyr Trp NH_4^+

The following facts were observed:

a. Trypsin had no effect.

b. The phenylthiohydantoin released by Edman degradation was

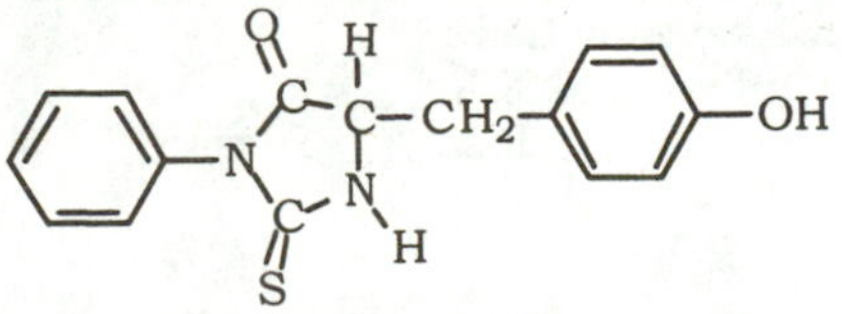

c. Brief chymotrypsin treatment yielded several products, including a dipeptide and a tetrapeptide. The amino acid composition of the tetrapeptide was Glx, Leu, Lys, and Met
d. Cyanogen bromide treatment yielded a tetrapeptide that had a net positive charge at pH 7 and a tripeptide that had a zero net charge at pH 7.
What is the amino acid sequence of this heptapeptide?

Answer: Trypsin cleaves after Lys and Arg. Since amino acid analysis revealed the presence of lysine, it must be on the C-terminus.
The PTH derivative amino acid is identified as tyrosine. So far we have:

Tyr _ _ _ _ _ Lys

From information in (d) we may place Met at positions 3 or 4 but from (c) we recognize that Met must be closer to Lys and so it must be at position 4.

Tyr _ _ Met _ _ Lys

From (c) we may conclude that Glx and Leu are located between Met and Lys. For chymotrypsin to produce a tetrapeptide and a dipeptide, Trp must be at position 3. Asx (Asp or Asn) is at position 2. Thus,

Tyr Asx Trp Met (Glx and Leu) Lys

Analysis of cyanogen bromide cleavage products reveals that the tetrapeptide has a net positive charge at neutral pH. Since cleavage leaves the C-terminus as homoserine lactone (and not as a negatively charged carboxyl group) then Asx must be Asn thus accounting for the ammonium. The Glx must be Glu and not Gln to account for a neutral charge of the tripeptide at neutral pH. From the information given we cannot assign an order to Glu and Leu, thus two possible heptapeptides are consistent with the information.

Tyr Asn Trp Met Glu Leu Lys
Tyr Asn Trp Met Leu Glu Lys

15. Amino acid analysis of a decapeptide revealed the presence of the following products:

NH_4^+ Asp Glu Tyr Arg
Met Pro Lys Ser Phe

The following facts were observed:
a. Neither carboxypeptidase A nor B treatment of the decapeptide had any effect.
b. Trypsin treatment yielded two tetrapeptides and free Lys.
c. Clostripain treatment yielded a tetrapeptide and a hexapeptide.
d. Cyanogen bromide treatment yielded an octapeptide and a dipeptide of sequence NP (using the one-letter codes).
e. Chymotrypsin treatment yielded two tripeptides and a tetrapeptide. The N-terminal chymotryptic peptide had a net charge of -1 at neutral pH and a net charge of -3 at pH 12.
f. One cycle of Edman degradation gave the PTH derivative:

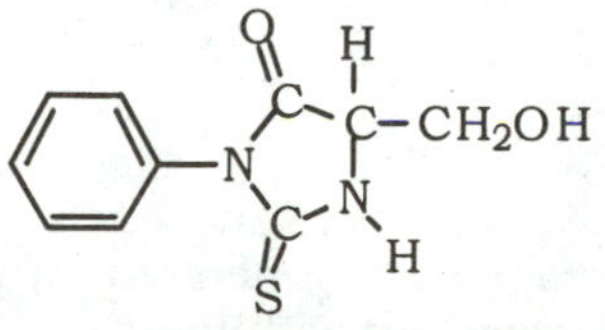

What is the amino acid sequence of this decapeptide?

Answer: Carboxypeptidase A will cleave at the C-terminus, except for Pro, Lys, and Arg whereas carboxypeptidase B cleavage occurs at C-terminal Arg and Lys. No action by either indicates either a circular peptide or proline as C-terminus.
Trypsin cleavage yielded two tetrapeptides and free lysine. Since we are working with a decapeptide, two moles of lysine must have been released. Further, the lysines must be located such that trypsin cleavage releases them both. We can further assume that one of the tetrapeptides ends in R. There are two possibilities:

(1) _ _ _ R K K _ _ _ P
(2) K _ _ _ R K _ _ _ P

From (c) we recognize that R could be located at positions 4 or 6 but only option (1) is consistent with this observation.
From (d) we learn that M is located at position 8 and that the last two amino acids are N P in this order. And, from (f) we identify the N terminal amino acid as serine. Thus,

(1) S _ _ R K K _ M N P

F or Y must be located at position 7 to insure that chymotrypsin cleavage produces a tripeptide from the C-terminus and nothing longer. In addition an aromatic must also be located at position 3.
The only amino acid unaccounted for is either Glu (E) or Gln (Q), which is located at position 2. This gives:

(1) S (E/Q) (F/Y) R K K (F/Y) M N P

For N-terminal chymotryptic peptide to have a net charge of -1 at neutral pH and -3 at pH = 12 it must contain glutamic acid (E) and tyrosine (Y) to account for the charge.

S E Y R K K F M N P

Questions for Self Study

1. Fill in the blanks: A _____ bond is an amide bond between two _____. The bond is formed by a reaction between the _____ of one amino acid and the ____ of a second amino acid with the elimination of the elements of _____. The result is a linear chain with an _____ end and a _____ end. The chain can be extended into a polymer with additional amino acids. The polymer formed is called either a _____ or a _____ depending on the number of amino acids joined. The amide bond can be broken by the addition of the elements of water across it in a ____ reaction.

2. List the four classes of amino acids.

3. Indicate which class of amino acids best represents the statements presented below:
   a. Their side chain may contain a hydroxyl group. _____
   b. Negatively charged at neutral pH._____
   c. Relatively poorly soluble in aqueous solution._____
   d. Relatively soluble in aqueous solution._____
   e. Positively charged at neutral pH._____
   f. Side chains are capable of hydrogen bonding._____
   g. Includes the amino acids arginine, histidine and lysine._____
   h. Side chains contain predominantly hydrocarbons._____
   i. Side chains are responsible for the hydrophobic cores of globular proteins._____
   j. The amides of glutamic and aspartic acid belong to this group._____

4. From the list of amino acids presented indicate which one best fits the description:
   a. The smallest amino acid (It also lacks a stereoisomer).
      alanine, glycine, phenylalanine, glutamine
   b. An aromatic amino acid.
      glutamic acid, tyrosine, isoleucine, proline
   c. A sulfur-containing amino acid.
      methionine, aspartic acid, arginine, leucine
   d. An amino acid capable of forming sulfur-sulfur bonds.
      methionine, proline, cysteine, tryptophan
   e. A branched chain amino acid.
      methionine, leucine, aspartic acid, asparagine

5. Fill in the blanks. Most of the amino acids are chiral compounds because they have at least one carbon atom with ___ different groups attached. Thus, there are ____ ways of arranging the four groups about the carbon atom. For the case of an amino acid with a single asymmetric carbon, the isomer pairs that can be formed are called ____; they are nonsuperimposable, ____ images of each other. Using the D,L system of nomenclature, the commonly occurring amino acids are all ____ amino acids.

6. What three amino acids absorb light in the ultraviolet region of the spectrum above 250 nm?

7. Of the following techniques which are commonly used to separate amino acids: ion exchange chromatography, NMR, electroporation, HPLC, UV spectroscopy, ninhydrin reaction?

8. Although there are twenty common amino acids, there are hundreds of different amino acids found in nature, some of which derive from the common amino acids by simple chemical (biochemical) modification. Name one common modification of amino acids found in nature.

9a. Of the twenty common amino acids, what amino acid has a pK_a around neutrality.
 b. What amino acid lacks an amino group?

10. Fill in the blanks. Proteins are linear chains of _____ _____ held together by covalent _____ bonds. Although these bonds are typically drawn as C-N single bonds, they in fact have partial double-bond characteristics because of delocalization of _____ _____. The result of this delocalization is to constrain four atoms in a single plane termed the _____ plane. The four atoms are _____, _____, _____, and _____. Thus, linear chains of amino acids can be considered as strings of _____ carbons joined together by amide or peptide planes.

11. Match the terms in the first column with terms in the second column.

a. Dipeptide	1. An amino acid minus the elements of water.
b. Protein	2. The bond joining amino acids in a protein.
c. Polypeptide	3. Two amino acids joined in amide linkage.
d. Amino acid residue	4. Subunit of a multimeric protein.
e. Peptide	5. A molecule composed of polypeptide chains.

12. Proteins represent an extremely diverse set of biomolecules. Match the functional class of proteins with their appropriate function.

Class	Function
a. Enzymes	1. Source of amino acids for developing organisms
b. Regulatory proteins	2. Move glucose across the cell membrane.
c. Transport proteins	3. Various venoms and toxins.
d. Storage proteins	4. Components of motile cellular appendages.
e. Contractile proteins	5. Biological catalysts.
f. Structural proteins	6. Certain extracellular, fibrous elements.
g. Protective proteins	7. Certain hormones.

13. Trypsin, chymotrypsin, clostripain, staphylococcal proteins, and cyanogen bromide are agents used in amino acid sequence determination. What is their purpose?

14. Short polypeptides are subjected to Edman degradation for what purpose?

15. What is the difference between amino acid composition and amino acid sequence?

16. Answer True or False
   a. Proteins sharing a significant degree of sequence similarity are homologous. _____
   b. Proteins that perform the same function in different organisms are referred to as homologous. _____
   c. Invariant residues usually are a result of random mutations. _____
   d. Evolutionarily related proteins always have similar biological activities. _____

Answers

1. peptide; amino acids; amino group; carboxyl group; water; N-terminal (or amino); C-terminal (or carboxyl); polypeptide; protein; hydrolysis.

2. Nonpolar or hydrophobic; neutral polar; acidic; basic.

3. a. neutral polar; b acidic; c. nonpolar; d. neutral polar, acidic, and basic; e. basic; f. neutral polar, acidic, and basic; g. basic; h. nonpolar; I. nonpolar; j. neutral polar.

4. a. glycine; b. tyrosine; c. methionine; d. cysteine; e. leucine.

5. Four; two; enantiomers (or mirror image isomers); mirror; L.

6. Phenylalanine, tyrosine, and tryptophan.

7. Ion exchange chromatography, HPLC

8. Hydroxylation, methylation, carboxylation, phosphorylation.

9a. histidine; b. proline.

10. Amino; acids; amide (or peptide); pi; electrons; amide (or peptide); the carbonyl carbon; the carbonyl oxygen; the amide nitrogen; amide hydrogen; alpha.

11. a. 3; b. 5; c. 4; d. 1; e. 2.

12. a. 5; b. 7; c. 2; d. 1; e. 4; f. 6; g. 3.

13. These agents are used to cleave a polypeptide into short oligopeptides. Each oligopeptide is then sequenced. At least two different agents are used separately to cleave the polypeptide in order to produce overlapping oligopeptides.

14. Edman degradation is used to determine the amino acid sequence of an oligopeptide from the N-terminus to the C-terminus.

15. The amino acid sequence is the primary structure of a polypeptide and refers to the sequence of amino acids, N-terminus to C-terminus. The amino acid composition is the total number of each amino acid in a polypeptide.

16. a. T; b. T; c. F; d. F.

Additional Problems

1. Sketch the titration curve of histidine.

2. In the protein hemoglobin, a number of salt bridges (ionic bonds) are broken in going from oxygenated to deoxygenated hemoglobin. One in particular involves an aspartic acid: histidine ionic interaction. (a). At pH = 7.0, would you expect this salt bridge to form? Explain. (b). Under certain conditions (low oxygen tension) the salt bridge does form. Explain how the influence of aspartic acid on histidine is similar to the influence of the α-carboxyl group on the pKa of the α-amino group (and vice versa).

3. How might the presence of calcium ions affect the apparent pKa's of the side chains of aspartic acid and glutamic acid?

4. Protein purification schemes typically involve several techniques. For the techniques listed below provide a brief explanation for each. Of the following properties: size, electric charge, partition properties, state which property is being exploited for each technique.

(a) Molecular Sieve Chromatography (b) Ultracentrifugation
(c) Ammonium Sulfate Fractionation (d) Affinity Chromatography
(e) Polyacrylamide Gel Electrophoresis (f) Reverse Phase HPLC

5. How are protein purification procedures monitored?

6. The apparent molecular weights of globular proteins are often determined by SDS polyacrylamide gel electrophoresis. Explain.

7. The ultraviolet absorption spectra of proteins rarely looks like the sum of the absorption spectra of the component amino acids but the situation improves when the protein is treated with high concentrations of urea or guanidinium hydrochloride. Explain.

8. Determining the amino acid sequence of even a small portion of a protein is extremely useful. Often this information can be used to identify the protein. Calculate the total number of different amino acid sequences for polypeptide chains of length 5, 10, and 15 amino acids.

Abbreviated Answers

1. Histidine is the only amino acid with a pKa around neutrality. The pKa's of histidine are 1.8, 6.0, and 9.2. The titration curve will be similar to the one shown in the answer to problem 5 but with an additional plateau around pH = 6.0.

2. (a) The pKa's of the side chains of aspartic acid and histidine are 3.9 and 6.0 respectively. At pH = 7.0, the side chain of aspartic acid is unprotonated and therefore negatively charged, whereas the side chain of histidine is also unprotonated and therefore uncharged. We might not expect this interaction to occur.

(b) The fact that the salt bridge forms indicates that the two amino acid side chains are close together in the three dimensional structure of the protein. The proximity of the negatively charged aspartic acid must therefore raise the pKa of the histidine side group.

3. Calcium (Ca^{2+}) may bind to carboxylate ions. Therefore, the presence of calcium may lower the apparent pKa's of aspartic acid and glutamic acid.

4. (a) Molecular sieve chromatography or molecular exclusion chromatography separates proteins on the basis of size. A porous matrix is employed such as beads of dextran, agarose, or polyacrylamide. As proteins pass through a column of this material, depending on the protein's size relative to the pore size, they either enter the porous beads or are excluded from them.

(b). In ultracentrifugation, very high centrifugal forces are employed to separate proteins. Depending on the technique employed, proteins may be separated according to size and shape, or according to density.

(c). Ammonium sulfate fractionation operates on the principle that under certain ionic conditions protein charge may be shielded by ionic interactions with solvent components allowing proteins to aggregate and precipitate Variations of this basic technique include the use of other salts as precipitating agents or changing pH. Proteins are least soluble at their isoelectric point, the pH at which a protein's net charge is zero.

(d). Affinity chromatography exploits the fact that proteins have unique surfaces. These surfaces might be a substrate binding site, a cofactor binding site, or a protein-protein interaction surface as examples. The principle is to attach a molecule, which interacts specifically with a particular protein, to an inert, insoluble matrix. A mixture containing the protein of interest is added to the affinity matrix under conditions favoring interactions. The matrix is washed to remove unbound protein and then subjected to conditions that disrupt the complementary interactions. These conditions may be high salt, or high ionic strength, a change in pH, or high concentration of cofactor or substrate.

(e). Polyacrylamide gel electrophoresis separates proteins according to their charge to mass ratio and their size. The driving force is a voltage gradient. Charged proteins are forced to move through a porous matrix (typically polyacrylamide). Separation occurs because the degree to which movement of a protein is impeded is proportional to the size of the protein. Under so-called native conditions, protein mixtures may be separated into component proteins retaining biological activity. In the presence of denaturing agents such as urea or SDS, individual polypeptide chains may be separated.

(f). HPLC refers to high performance (or high pressure) liquid chromatography. In general, the efficiency of chromatographic procedures depends on the number of interactions or discriminations made per volume. In high performance chromatography, a large number of interactions are made possible by using chromatography matrix material of very small size (resulting in an increase in the number of theoretical plates.) The small size greatly increases resistance to flow and so chromatography must be carried out using high pressure. Reverse phase HPLC is a particular application of HPLC that employs chromatography matrices with hydrophobic surfaces. Initial applications of chromatography employed cellulose, in the form of paper, as a stationary phase. Molecules are moved over this stationary phase in solutions containing organic solvents. Thus, the mobile phase favors hydrophobic compounds whereas the stationary phase favors hydrophilic compounds (because cellulose contains numerous polar groups). In reverse phase, the stationary phase is hydrophobic whereas the mobile phase, at least initially, is an aqueous solution. Proteins that adsorb to the column are removed by addition of increasing amounts of organic solvent (e.g., methanol) or by altering the ionic strength.

5. After each stage in a protein purification scheme, the total amount of protein is measured and the amount of the protein of interest is quantitated. Quantitation may include measuring the biological activity of the protein by performing an assay. The specific activity is the activity normalized to total protein concentration. SDS polyacrylamide gel electrophoresis (SDS PAGE) is routinely employed to monitor purification. Using very thin slab gels, large numbers of samples can be examined in only a few hours. SDS PAGE allows a determination of the complexity of samples with respect to the number and kinds of polypeptides present.

6. A globular protein contains a core of hydrophobic amino acids that is largely responsible for its globular shape. This core is formed by hydrophobic interactions of amino acids distributed throughout the polypeptide chain. For typical globular proteins, the number of hydrophobic amino acids is approximately a fixed percentage of the total number of amino acids. In solution, SDS forms micelles around each hydrophobic amino acid. This fact is exploited in SDS polyacrylamide gel electrophoresis to produce a protein/SDS complex with a charge dependent on the number of hydrophobic amino acids. The intrinsic charge of a protein contributes negligibly to the total charge of a protein/SDS complex. Because the total charge is proportional to the number of hydrophobic amino acids, and, the number of hydrophobic amino acids is a fixed percentage of the total number of amino acids, the charge to mass ratio of protein/SDS complexes for globular proteins is constant. Thus, protein/SDS complexes will migrate at a constant rate in an electric field, a rate completely independent of protein size. Separation of proteins is accomplished by forcing protein/SDS complexes through a polyacrylamide matrix. Here, migration is inversely dependent on size. By comparing the mobility of a protein of unknown size to the mobility of proteins of known size, an estimate of the molecular weight can be made.

7. The amino acids phenylalanine, tryptophan, and tyrosine absorb ultraviolet light and these amino acids are largely responsible for the UV-light absorption properties of most proteins. However, the physical dimensions of a protein may be comparable to the wavelength of ultraviolet light. In this case the protein will exhibit light scattering. Urea and guanidinium HCl are both powerful denaturing agents that will disrupt protein/protein interactions and unfold proteins, thus minimizing light scattering. Also, there are effects due to polarity of the environment in which the aromatic ring is located.

8. The number of arrangements of 20 different amino acids taken n at a time is $(20)^n$.

Polypeptide Length	Number of Polypeptides
5	3.2×10^6
10	1.0×10^{13}
15	3.3×10^{19}

Summary

Amino acids are the building blocks of proteins. There are 20 amino acids commonly found in proteins. The general structure of an amino acid is based on a central α-carbon, to which is attached a carboxyl group, a hydrogen, an amino group, and a variable side chain or R group. Amino acids can be classified according to the nature of their R groups into three classes. The nonpolar, hydrophobic amino acids include **alanine**, **valine**, **leucine** and **isoleucine** in addition to **proline**, **methionine**, **phenylalanine** and **tryptophan**. Polar, uncharged amino acids include

glycine, **serine**, **threonine**, **tyrosine**, **asparagine**, **glutamine,** and **cysteine**. The acidic amino acids are **aspartic acid** and **glutamic acid** and the basic amino acids include **histidine, lysine,** and **arginine**. The amino acids are weak polyprotic acids. The carboxyl group is a rather strong carboxylic acid with pK_a around 2.0, whereas the amino group is a base with pK_a around 9.5. At neutral pH, the carboxyl group is unprotonated and negatively charged whereas the amino group is protonated and positively charged. Considering only these two groups at pH = 7.0, amino acids are zwitterionic. Several of the side chains have titratable groups including aspartic acid, glutamic acid, cysteine, tyrosine, arginine, histidine, and lysine.

The amino acids can participate in a number of useful chemical reactions, some specific to either the carboxyl group or the amino group common to all amino acids. Other chemistries are specific for side groups. The α-carbons of all of the amino acids except glycine are chiral. Thus amino acids can exist as stereoisomers. In nature, the L amino acids predominate.

There are two general methods for separating and analyzing amino acid mixtures. In one method, the ionic properties of the amino acids are exploited in ion exchange chromatography. Ion exchange involves a stationary charged species in equilibrium with a counter ion. Conditions are established such that the amino acids can displace the counter ion and bind to the stationary phase. The degree to which amino acids bind to the stationary phase depends on their intrinsic charge and on the concentration of counter ion with which they compete. Removal of amino acids from the stationary phase can be accomplished either by changing the pH of the solution (and thus changing the charge on the amino acids) or increasing the concentration of counter ion. The conditions under which a particular amino acid is displaced from the stationary phase are characteristic of the amino acid.

The other technique for separating amino acids is to react the amino terminus with a hydrophobic group and to separate the derivatized amino acids by reverse phase chromatography. In reverse phase chromatography, derivatized amino acids are bound to a stationary, hydrophobic phase and subsequently eluted with increasing concentrations of a polar substance (e.g., methanol).

In proteins, amino acids are joined by peptide (or amide) bonds. The carboxyl group of one amino acid is joined to the amino group of another amino acid, with elimination of the elements of water. The chain thus formed is composed of amino acid residues. The bond formed, an amide or peptide bond, joins a carboxyl carbon with an amino nitrogen in a linkage with partial double bond character. These double bond characteristics arise due to formation of π-bonds with lone-pair electrons from N and C (which is a carbonyl carbon in the peptide bond). As a result, rotation about the C-N bond is restricted and the carbonyl group (i.e., the remainder of the carboxyl group), nitrogen and its attached hydrogen all lie within a single plane, the amide plane. Thus, the atoms of the ensemble define a plane. Proteins may be composed of a single polypeptide chain (monomeric protein), or more than one polypeptide chain of either the same type (homomultimeric protein), or different types (heteromultimeric). Single polypeptide chains with a small number of amino acid residues are dipeptides, tripeptides, tetrapeptides, or oligopeptides. The prefix refers to the number of amino acids, not the number of peptide bonds (which in general equals n-1, where n is the number of amino acid residues).

Protein characterization begins with purification of the protein. Protein purification is accomplished using a number of techniques including differential solubility, differential centrifugation and gradient centrifugation; ion exchange, molecular sieve, or affinity column chromatography.

Next, the purified protein is separated into its individual polypeptide chains. The nature of the N-terminal and C-terminal amino acids of the polypeptide components is determined using techniques specific for each group. For example, phenylisothiocyanate is used during Edman degradation to convert N-terminal amino acids into PTH (phenylthiohydantoin) derivatives. C-terminal residues may be identified after enzymatic cleavage with carboxypeptidase A, B, C or Y (A is blocked by Pro, Arg and Lys; B only cleaves at Arg or Lys; C and Y cleave at any amino acid).

The amino acid composition is determined by first hydrolyzing the polypeptide chains into amino acids using 6 N HCl at 110°C for 24, 48 and 72 hr to insure that total hydrolysis has occurred. A time course is performed because some amino acids (e.g., Ser and Thr) are unstable under these conditions and slowly degrade whereas others (e.g., hydrophobic amino acids) are only slowly released. The time course can be extrapolated to zero time to determine the initial level of unstable amino acids and to infinite time to predict when total hydrolysis will occur. The mixture of amino acids produced by acid hydrolysis is analyzed on an amino acid analyzer to determine the relative amount of each amino acid in the mixture. An amino acid analyzer separates and quantitates amino acids or amino acid derivatives. The separation may be based on the ionic properties of the amino acids (i.e., ion exchange chromatography), or it may be based on the degree of hydrophobicity of derivatized amino acids (i.e., reverse phase HPLC).

The primary structure or amino acid sequence is established by fragmenting polypeptides into short oligopeptides suitable for Edman degradation (see below). Fragmentation is accomplished by either 1) enzymatic cleavage at specific residues in the polypeptide using trypsin (hydrolyzes peptide bonds whose carbonyl group is donated by Arg or Lys), chymotrypsin (Phe, Tyr or Trp), clostripain (Arg), staphylococcal protease (Asp, Glu), or 2) using chemical means such as cyanogen bromide to cleave after Met.

The amino acid sequences of the short oligopeptides are established using Edman degradation. The basic idea is to subject an oligopeptide to multiple rounds of modification and cleavage. In Edman chemistry, the amino-terminal amino acid is modified using phenylisothiocyanate and specifically cleaved to release the derivatized N-terminal residue while producing a new terminus. The amino acid sequences of all of the oligopeptides produced by fragmentation using at least two different and overlapping fragmentation techniques are compared, to establish how the individual fragments are represented in the protein. Mass spectrometry is also being used to establish the sequence of short oligonucleotides.

As a crude classification, proteins are cataloged as fibrous, globular, or membrane. Fibrous proteins are elongated in shape; globular proteins are spherical in shape and water soluble; membrane proteins associate with biological membranes. Of the biopolymers, proteins are by far the most diverse group serving a number of biological purposes. Proteins composed only of peptide chains are termed simple proteins. Conjugated proteins contain additional chemical constituents as part of their structure. Examples of conjugated proteins include glycoproteins (containing carbohydrates), lipoproteins (lipids), nucleoproteins (RNA), phosphoproteins (phosphates on serine, threonine, or tyrosine), metalloproteins (metal ions), hemoprotein (heme groups) and flavoproteins (riboflavin derivatives).

The vast majority of biological catalysts are enzymes, proteins capable of increasing the rate of specific reactions. Regulatory proteins modulate physiological responses of cells by binding to specific cell components. Transport proteins facilitate the movement of substances. Storage proteins provide a source of amino acids. Contractile and motile proteins are proteinaceous elements involved in force generation and motility in cells. Structural proteins provide mechanical support. Protective proteins serve as a defense barrier.

Chapter 5

Proteins: Secondary, Tertiary, and Quaternary Structure

• •

Chapter Outline

- Protein conformation: Three-dimensional shape
  - Amino acid sequence contains information to fold protein into conformation
  - Weak forces act on amino acids to stabilize conformation: Hydrogen bonding, hydrophobic interactions, electrostatic interactions, van der Waals forces
- Secondary Structure: Structural elements formed by peptide plane interactions
  - ϕ and ψ Angles: Describe orientation of peptide planes about alpha carbons
  - Peptide Plane
    - Carbonyl oxygen: H-bond acceptor
    - Amide hydrogen: H-bond donor
  - Alpha Helix (Pauling and Corey)
    - Peptide carbonyl of ith residue hydrogen bonded to peptide H of i+4th residue
    - 3.6_{13}: 3.6 Residues per turn, 13 atoms between carbonyl oxygen and amide hydrogen
    - 0.54 nm per turn, 0.15 nm per residue along helix (Z) axis
    - H bonds parallel to helix axis
  - Other helices: 3_{10}: 3 residues per turn 10 atoms between carbonyl oxygen and amide hydrogen
  - Beta pleated sheets
    - Polypeptide chain fully extended
    - Chains or strands joined by H bonds
    - Parallel sheets
    - Antiparallel sheets
  - Beta turn
    - Carbonyl of ith residue H bonded to amide nitrogen of i+3
    - Peptide plane in *cis*
    - Gly or Pro common
- Tertiary structure principles
  - Secondary structures formed when possible
  - Secondary structures associate and pack to form layers
  - Elements between secondary structures are short and direct
  - Stability of final fold
    - Large number of intramolecular H bonds
    - Reduction is surface area
- Fibrous Proteins
  - α–Keratin: α Helical coiled-coils
    - Subunit: 311-314 Residue-long α helix-rich region flanked by nonhelical regions
    - Two-stranded coiled coils with helix repeat of 0.51 nm
    - 7-Residue quasi repeat with 1st and 4th residues nonpolar
    - Disulfide bonds: Form rigid, inextensible, and insoluble structure
  - Fibroin and β-keratin: Stacked β–sheets

- Collagen: Connective tissue protein in animals
  - Tropocollagen: three intertwined chains
  - Amino acid composition and amino acid sequence
    - One-third Gly and one-third Pro or Hyp
    - Modified amino acids: Hydroxyproline and hydroxylysine (vitamin C-dependent reaction)
    - Gly-Pro-Hyp repeat with G placed in center of three-stranded coil
- Globular proteins
  - Core composed of hydrophobic amino acids
  - Peptide groups often H bonded as helices or sheets
    - Surface helices amphiphilic
    - Internal helices or sheets hydrophobic
    - Solvent exposed helices hydrophilic
  - Packing of elements results in formation of small cavities imparting flexibility to structure
  - Ordered regions with well-defined, nonrepetitive structure
  - Disordered, flexible regions
- Molecular motion: Proteins are dynamic structures
  - Atomic fluctuations: Movement over small distances
  - Collective motions: Movement of groups of atoms
  - Conformational changes: Movement of large domains
- Chain folding
  - Structural stability
  - Right-handed structures
    - Right-handed twists in sheets
    - Right-handed crossovers joining secondary structural elements
  - β-strand connections
    - Antiparallel strands connected by hairpins
    - Parallel strands connected by crossovers forming βαβ loop
  - Hydrophobic amino acids sequestered in interior: Layers
    - Two layers of backbone form a single hydrophobic core
    - Three layers of backbone form two hydrophobic cores
    - Four-layered and five-layered structures
- Globular protein classification based on secondary structure
  - Antiparallel α helix
  - Parallel or mixed β-sheet
  - Antiparallel β-sheet
  - Metal- and Disulfide-rich
- Molecular chaperones: Protein complexes that catalyze process of protein folding
- Protein design: Domains: Regions 40-100 amino acids long that form stable tertiary structures
- Quaternary structure
  - Subunits typically fold into independent globular structures
  - Subunit association forces
    - Hydrophobic interactions between faces
    - Disulfide bond stabilized
  - Polymers: Open quaternary structures
- Structural and functional advantages to quaternary associations
  - Protein stabilized by reduction in surface-to-volume ratio
  - Genetic economy: Encode self-assembling subunits rather than single complex protein
  - Catalytic sites brought together
- Cooperativity: Influence on catalytic sites by neighboring sites

Chapter Objectives

The key to understanding protein structure is to understand the weak forces responsible for maintaining a protein in the folded state: hydrogen bonds, hydrophobic interactions, electrostatic interactions, and van der Waals forces. It will be helpful to review the structures of the amino acids and to recall the kinds of weak interactions each amino acid side chain is capable of making.

Peptide Bond

Know the characteristics of the peptide bond: the fact that it is planar because of π-bonding and resonance stabilization and the fact that it has a hydrogen bond donor and an acceptor.

Secondary Structure

Protein secondary structures are regular structures formed by hydrogen bonds between amide planes. Understand the structure of the α-helix. Sheet structures form between fully extended peptide chains. Beta-turns are tight turns stabilized by a hydrogen bond and requiring *cis* orientation of amide planes. The α-helix and β-sheet are shown in Figures 5.10 and 5.13.

Tertiary Structure

The tertiary structure of a protein is its three-dimensional structure. Fibrous proteins have secondary structure elements arranged parallel to a single axis. Examples include α-keratin and collagen. Be familiar with the structures of these two proteins. Globular proteins are by far the most abundant class of proteins. As additional protein structures are solved by x-ray crystallography and nuclear magnetic resonance, a greater understanding of the rules governing protein folding and structure stabilization will unfold. Currently a majority of globular proteins are classified into four broad groups: antiparallel α-helix, parallel or mixed β-sheet, antiparallel β-sheet, and the small metal- and disulfide-rich proteins.

Quaternary Structure

Quaternary structure is reserved for proteins composed of multiple subunits. The subunits represent distinct structural domains that interact in specific ways to form the native protein.

Protein Denaturation

The three-dimensional shape of a functional protein is called the native conformation. A conformational change to a state or states lacking activity is known as denaturation. Protein denaturation may occur when the weak forces, responsible for the native conformation, are disrupted. Denaturing agents or conditions include: heat, high concentrations of urea, guanidine HCl, high salt concentrations, low ionic strength solutions, SDS, organic solvents, or extremes in pH.

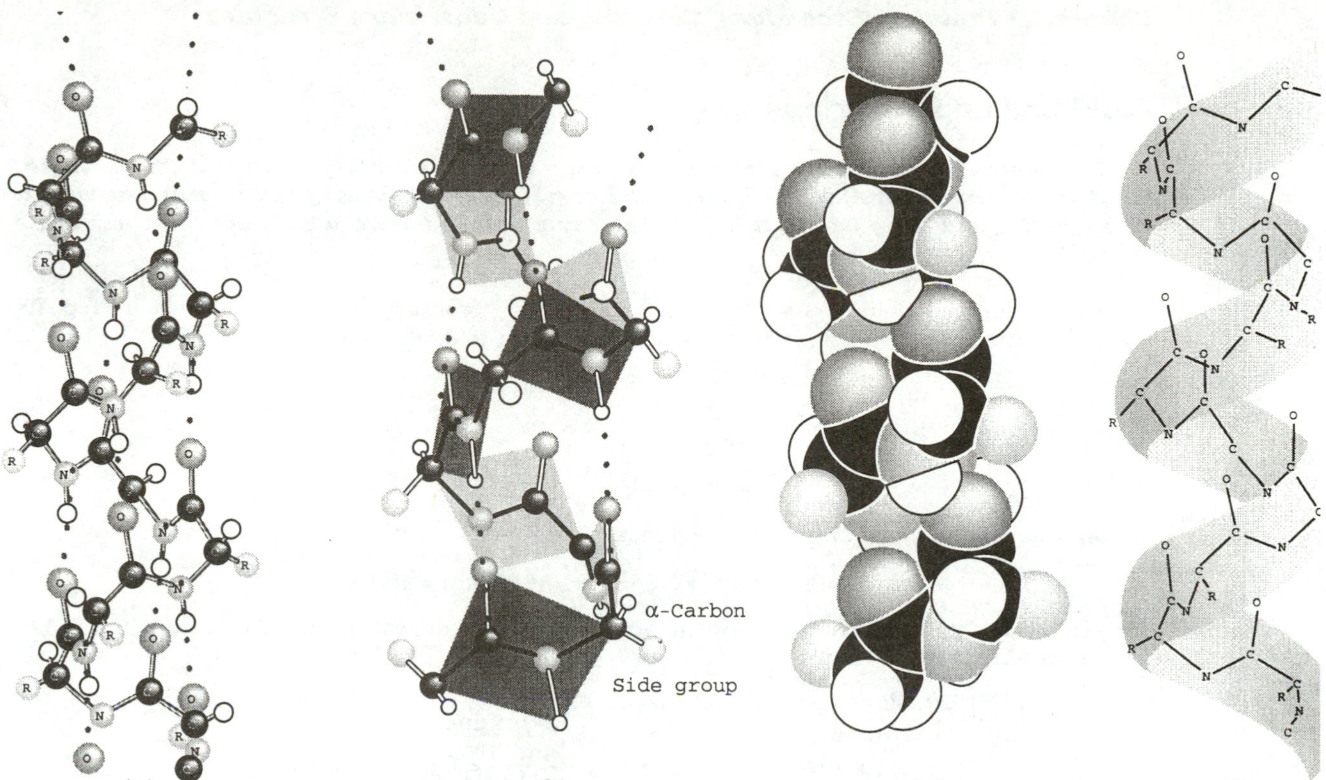

(a)
Hydrogen bonds stabilize the helix structure.

α-Carbon

Side group

(b)
The helix can be viewed as a stacked array of peptide planes hinged at the α-carbons and approximately parallel to the helix.

(c)

(d)

Figure 5.10 Four different graphic representations of the α-helix. (a) As it originally appeared in Pauling's 1960 *The Nature of the Chemical Bond*. (b) Showing the arrangement of peptide planes in the helix. (c) A space-filling computer graphic presentation. (d) A "ribbon structure" with an inlaid stick figure, showing how the ribbon indicates the path of the polypeptide backbone (*Irving Geis*).

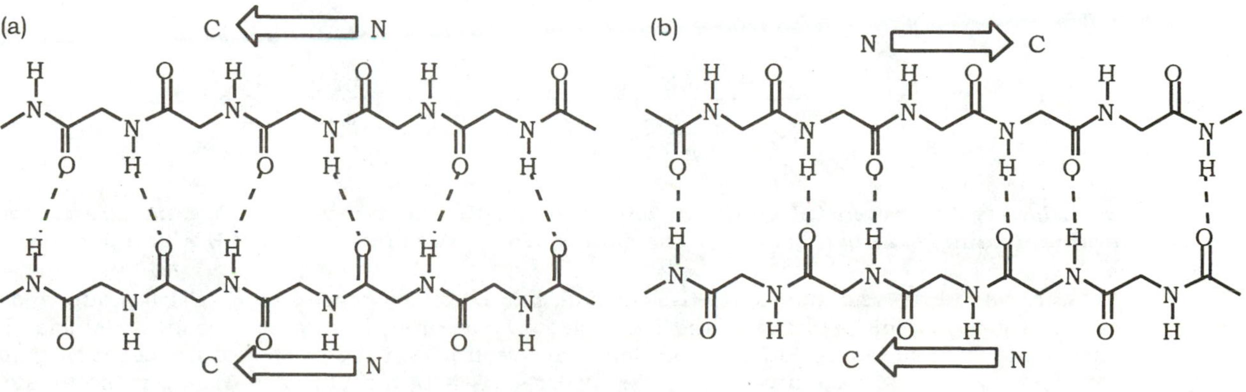

Figure 5.13 The arrangement of hydrogen bonds in (a) parallel and (b) antiparallel β–pleated sheet.

Problems and Solutions

1. The central rod domain of a keratin protein is approximately 312 residues in length. What is the length (in Å) of the keratin rod domain? If this same peptide segment were a true α-helix, how long would it be? If the same segment were a β-sheet, what would its length be?

Answer: α-Keratin has extensively distorted α-helical secondary structure. The helix repeats every 3.6 residues but has a pitch of 0.51 nm (compared to 0.54 nm for a true α-helix).

$$\frac{312\text{ residues}}{3.6\dfrac{\text{residues}}{\text{turn}}} = 86.7\text{ turns}$$

$$86.7\text{ turns} \times 0.51\frac{\text{nm}}{\text{turn}} = 44.2\text{ nm} = 422\text{ Å}$$

For an α-helix of the same number of residues:

$$86.7\text{ turns} \times 0.54\frac{\text{nm}}{\text{turn}} = 46.8\text{ nm} = 468\text{ Å}$$

For pleated sheets, the distance between residues is 0.347 nm for antiparallel sheets and 0.325 nm for parallel sheets. Thus,

$$312\text{ residues antiparallel sheet} \times 0.347\frac{\text{nm}}{\text{residue}} = 108.3\text{ nm} = 1083\text{ Å}$$

$$312\text{ residues parallel sheet} \times 0.325\frac{\text{nm}}{\text{residue}} = 101.4\text{ nm} = 1014\text{ Å}$$

2. A teenager can grow 4 in. in a year during a "growth spurt". Assuming that the increase in height is due to vertical growth of collagen fibers (in bone), calculate the number of collagen helix turns synthesized per strand per minute.

Answer: Four inches of growth corresponds to

$$4\text{ in} \times 2.54\frac{\text{cm}}{\text{in}} = 10.16\text{ cm}$$

How many collagen helix turns does 10.16 cm represent? The collagen helix has the following parameters: 0.29 nm per residue; 3.3 residues per turn; and 0.96 nm/turn. How many turns in 10.16 cm. (Note: 1cm = 10^{-2}m, 1nm = 10^{-9}m ∴ 1cm = 10^{7}nm)

$$\frac{10.16\times 10^{7}\text{ nm}}{0.96\dfrac{\text{nm}}{\text{turn}}} = 1.06\times 10^{8}\text{turns}$$

$$\frac{1.06\times 10^{8}\text{ turns}}{1\text{yr}\times\dfrac{365\text{days}}{\text{yr}}\times\dfrac{24\text{hr}}{\text{day}}\times\dfrac{60\text{min}}{\text{hr}}} = 201\frac{\text{turns}}{\text{min}}$$

3. Discuss the potential contributions to hydrophobic and van der Waals interactions and ionic and hydrogen bonds for the side chains of Asp, Leu, Tyr and His in a protein.

Answer: Aspartic acid has a relatively small side chain composed of a $-CH_2-$ group and a carboxyl group. The presence of the ionizable carboxyl group indicates that aspartic acid can participate in ionic bonds. In addition, lone-pair electrons on the oxygen of the carboxyl group can participate in H bonds, as can the hydrogen when the carboxyl group is protonated. Hydrophobic interactions and van der Waals interactions are negligible.

The leucine side chain is an alkane and as such will not participate in hydrogen bonds or ionic bonds. The side chain is hydrophobic and relatively bulky, indicating that it will participate in hydrophobic interactions and is capable of entering numerous van der Waals interactions.

Tyrosine has a phenolic group attached to the α carbon by a methylene bridge (i.e., CH_2). The phenolic group is weakly ionizing with a pK_a of 10. Thus, only under special conditions or environments would it be expected to participate in ionic bonds. The hydroxyl group can both donate and accept H bonds. When protonated, tyrosine is capable of hydrophobic interactions, and, because it is a bulky amino acid, it is expected to participate in numerous van der Waals interactions.

The imidazole side chain of histidine has an ionizable nitrogen and, when protonated, allows histidine to participate in ionic bonds. In addition, hydrogen bond donor and acceptor groups support participation in hydrogen bonds. Thus, the ability to form both hydrogen bonds and ionic bonds precludes hydrophobic bonding. The large side chain is expected to participate in van der Waals interaction.

4. Figure 5.38 shows that Pro is the amino acid least commonly found in α-helices but most commonly found in β-turns. Discuss the reasons for this behavior.

Answer: Proline is an imido acid; it has a secondary nitrogen with only one hydrogen. When participating in a peptide bond, the nitrogen no longer has hydrogen bound to it. Thus, peptide bonds in which the nitrogen is supplied by proline are incapable of functioning as hydrogen-bond donors. A prolyl residue will interrupt α-helices when located on the C-terminal end of a helix. If located within 3 residues of the N terminus, proline is capable of hydrogen bonding through its carbonyl oxygen. The β-bend requires a *cis* peptide bond and proline stabilizes the *cis* conformation. Recall a β-bend involves hydrogen bonding of the carbonyl oxygen of a peptide bond with an amide hydrogen three residues away. Usually, adjacent peptide bonds are *trans* but for β-bend formation a *cis* conformation is required.

5. For the flavodoxin in Figure 5.30, identify the right-handed crossovers and the left-handed crossovers in the parallel β-sheet.

Answer For a right-handed crossover, moving in the N-terminal to C-terminal direction, the crossover moves in a clockwise direction. The reverse is true for a left-handed crossover, namely, movement from N-terminus to C-terminus is accompanied by counterclockwise rotation. Although not necessary, it is helpful to identify the N- and C-terminus of the protein. For the N-terminus this is accomplished by starting on any strand of β-sheet and moving in the opposite direction of the arrow until an end is encountered. This locates the N-terminus at the bottom of the second strand of β-sheet. Starting on the N-terminus the chain connects to the first strand making a left-handed crossover. The first strand is connected to the third, the third to the fourth, and the fourth to the fifth all with right-handed crossovers. The diagram below illustrates these points.

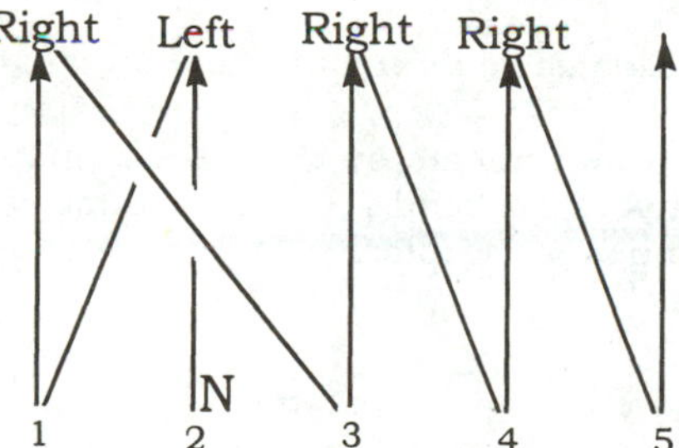

6. Choose any three regions in the Ramachandran plot and discuss the likelihood of observing that combination of ϕ and φ in a peptide or protein. Defend your answer using suitable molecular models of a peptide.

Answer: The Ramachandran plot reveals allowable values of ϕ and φ. The plots consider steric hindrance and will be somewhat specific for individual amino acids. For example, glycine has more allowable ϕ and φ angles in α-helical conformations than a bulky amino acid like phenylalanine.

7. Two polypeptides, A and B, have similar tertiary structures, but A normally exists as a monomer, whereas B exists as a tetramer, B_4. What differences might be expected in the amino acid composition of A versus B?

Answer: Oligomeric proteins are held together by a number of forces including hydrogen bonds, ionic bonds, hydrophobic interactions, van der Waals interactions, and covalent, disulfide bonds. Of these interactions, we might expect hydrophobic interactions to play a major part in subunit associations (subunit associations involve interactions of two or more surfaces). In comparing a monomeric protein with a homologous, polymeric protein, interacting regions in the polymer may reveal themselves as sequences of hydrophobic amino acid residues.

8. The hemagglutinin protein in influenza virus contains a remarkably long α-helix with 53 residues.
a. How long is this α-helix (in nm)?
b. How many turns does this helix have?
c. Each residue in an α-helix is involved in two H bonds. How many H bonds are present in this helix?

Answer: The α–helix repeats every 0.54 nm (3.6 amino acid residues) and the distance along the helical axis between residues is 0.15 nm. An α-helix of 53 residues is:

$$\frac{53 \text{ residues}}{3.6\frac{\text{residues}}{\text{turn}}} \times 0.54\frac{\text{nm}}{\text{turn}} = 7.95 \text{ nm or,}$$

$$53 \text{ residues} \times 0.15\frac{\text{nm}}{\text{residue}} = 7.95\text{nm}$$

The number of turns is:

$$\frac{53 \text{ residues}}{3.6\frac{\text{residues}}{\text{turn}}} = 14.7 \text{ turns}$$

To calculate the number of H bonds, the ith residue is H-bonded to the (i + 4)th residue to form a helix. For a 53-residue helix, the carbonyl groups of the first 49 residues are involved in H bonding (i + 4 = 53, i = 49). The amide H for the four residues on the N terminal side are not hydrogen-bonded nor are the carbonyl groups of the last 4 residues, on the C-terminal side. Thus, 49 hydrogen bonds are made.

Questions for Self Study

1. List the weak interactions responsible for maintaining protein conformation.

2. For the structures shown below indicate by D, A, D&A, or B if the structure contains only a hydrogen bond donor, only a hydrogen bond acceptor, a donor and separately an acceptor, or a group that functions both as donor or acceptor.

a. >C=O b. $-NH_2$ c. —OH d. —C(=O)—N(H)— e. =N—

3. Fill in the blanks. The α-helix is an example of a _____ structural element. It is a helical structure formed by _____ _____ between groups in peptide bonds. In particular, the _____ _____ of the ith residue of an α–helix is bonded to the _____ _____ of the (i+4)th residue. The result is a helical structure that repeats every _____ residues. The _____ of the helix, or distance along the z-axis per turn, is 0.54 nm. This arrangement orients the amide planes _____ to the helix axis and the side chains ____ to the helix axis.

4. Proteins composed predominantly of β-pleated sheets may be expected to form structures that are flexible yet not extensible. Please explain.

5. What two amino acids are most suited to beta-turns?

6. What posttranslational modification is necessary to produce collagen and what water-soluble vitamin is this modification dependent on?

7. Give two examples of fibrous proteins.

8. What are four common globular protein groups found in nature?

9. The agents listed below are all expected to affect protein tertiary structure. For each give a brief explanation of how they affect protein structure.
   a. Urea
   b. SDS
   c. High temperature
   d. β-mercaptoethanol
   e. Distilled water
   f. Organic solvents

10. What are the advantages to quaternary associations?

Answers

1. Hydrogen bonds, hydrophobic interactions, electrostatic interactions, van der Waals interactions.

2. a. A; b. D; c. B; d. D&A; e. A

3. Secondary; hydrogen; bonding; carbonyl oxygen; amide nitrogen; 3.6; pitch; parallel; perpendicular.

4. The polypeptide chain in a β-pleated sheet is already fully extended; however, the chains are free to bend.

5. Proline and glycine.

6. Hydroxylation; vitamin C.

7. α-Keratin, silk fibroin proteins, collagen.

8. Antiparallel α-helix; parallel or mixed β-sheet; antiparallel β-sheet; small metal- and disulfide-rich proteins.

9. Urea will disrupt hydrogen bonds. SDS is a detergent that will disrupt the hydrophobic cores of proteins. β–mercaptoethanol will reduce disulfide bonds. Distilled water will perturb ionic interactions. Organic solvents will interact with hydrophobic amino acid side chains.

10. Genetic economy and efficiency; structural stability due to reduction of surface-to-volume ratio; formation of catalytic centers; cooperativity.

Additional Problems

1. For the following conditions or agents, explain how they may denature proteins: heat, urea, guanidine HCl, distilled water, SDS, liquid phenol.

2. An often-cited example of protein denaturation is the change accompanying the heating of egg whites. Clearly, this process is irreversible yet the classic experiments of Anfinsen on RNase A showed that denaturation is reversible. Rectify these two seemingly conflicting facts.

3. In the movie Papillon, a criminal, played by the late Steve MacQueen, is imprisoned on a remote tropical island for many years. Needless to say, conditions in prison are not idyllic; the prisoners are served only enough gruel to keep them alive. In one scene, the main character reaches into his mouth and pulls out a tooth. From what disease might the prisoner have suffered?

4. Gelatin desserts are prepared by adding hot water to a dry powder. This popular dessert is made of collagen from pigskins. Given your understanding of the structure of collagen explain the gel-like consistency of gelatin.

5. Myoglobin was the first globular protein whose structure was solved by x-ray crystallography. The protein contains a hydrophobic core formed by the sides of several α-helices. For human myoglobin, the sequence of helix E is given below. For this helix, label the hydrophobic residues likely to contribute to the hydrophobic core.

E Helix Sequence: S E D L K K H G A T V L T A L G G I L

Abbreviated Answers

1. Gentle heating will disrupt hydrogen bonds. The structures of urea and guanidine HCl are shown below:

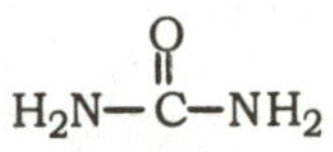

$$H_2N-\overset{\overset{\large O}{\|}}{C}-NH_2 \qquad\qquad H_2N-\overset{\overset{\large NH_2^+\ \ Cl^-}{\|}}{C}-NH_2$$

Urea 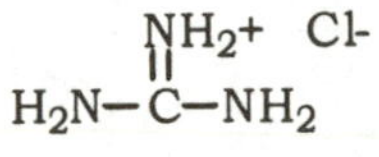 — Guanidinium HCl

Urea and guanidine HCl are both very soluble in water and may be used to form highly concentrated solutions. These agents denature proteins by disrupting hydrogen bonds. The low ionic strength and high dielectric constant of distilled water cause changes in ionic bonds that may lead to protein denaturation. SDS is an ionic detergent that disrupts hydrophobic interactions. In solution, SDS forms micelle structures by association of hydrocarbon tails. Micelles form around hydrophobic amino acid side chains, disrupting the normal hydrophobic interactions these side chains make. Liquid phenol will disrupt hydrophobic interactions.

2. Egg whites contain about 10% by weight of ovalbumin and, while it is true the heat of a hot frying pan is sufficient to denature the protein, the real problem comes subsequently. The denatured protein readily aggregates in this highly concentrated protein solution and the aggregation is for all intents and purposes irreversible. Anfinsen's experiments were done at much lower protein concentrations (and with a considerably smaller protein).

3. The disease was probably scurvy caused by a deficiency of vitamin C. Hydroxylation of proline in collagen fibers is catalyzed by prolyl hydroxylase in a vitamin C-dependent reaction. A deficiency of vitamin C results in improper assembly of collagen fibers. Collagen fibers are found in connective tissues, which are thus weakened in scurvy.

4. Collagen fibers hydrate and interact to form a tangled mass of crisscrossed fibers.

5. The residues in helix E are plotted in a helical wheel plot to identify the location of amino acids on the helix surface. To construct a helical wheel plot, a projection of the residues in a helix is made along the helical axis (Z-axis) onto the X-Y plane. In an α-helix, the helix repeats every 3.6 residues. Thus, each residue is 100° apart. The helical wheel plot is shown below with hydrophobic amino acids highlighted.

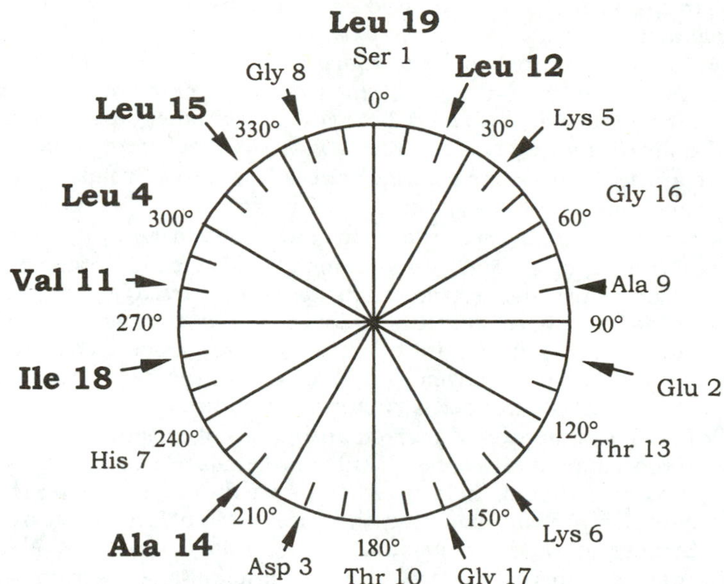

The plot clearly shows that hydrophobic amino acids (in bold) are concentrated along one side of the helix. This is the side of the helix that faces the hydrophobic core.

Summary

Proteins are composed of linear chains of amino acids held together by peptide bonds. The sequence of amino acids is the primary structure of the protein. It contains information necessary to fold up the chains into a functional protein. This folding involves weak forces such as hydrogen bonds, electrostatic interactions, hydrophobic interactions and van der Waals interactions. The peptide bond itself is capable of participating in hydrogen bonds with both hydrogen-bond donor and acceptor groups. For example, the carbonyl group has two pairs of lone-pair electrons capable of accepting hydrogen bonds. Also, the electronegative nitrogen induces a partial positive charge on its attached hydrogen, allowing the hydrogen to function as a hydrogen-bond donor. Hydrogen bonding between peptide bond groups is the basis of protein secondary structure elements, namely, helices, pleated sheets and turns.

The side chains of the amino acids are capable of a variety of interactions including, hydrogen bonds, ionic bonds, hydrophobic interactions, and van der Waals interactions. These interactions along with secondary structural elements are responsible for the tertiary structure of proteins, their actual three-dimensional shape.

The experiments of Anfinsen and White showed that the information to fold a peptide chain into a protein can be contained in the primary structure. Working with ribonuclease, a small protein capable of hydrolyzing phosphodiester bonds in RNA, they showed that enzymatic activity is dependent on the native conformation of the protein. The protein contains numerous disulfide bridges that stabilize the native state. By reducing these bonds and treating the protein with a denaturing agent, activity is lost. However, the activity is regained if the denaturant is first removed, allowing the chain to fold into its native conformation before disulfide bonds are allowed to reform.

Secondary structure elements include helices, pleated sheets and β-turns. Stable helices are a result of hydrogen bond formation between donor and acceptor groups in peptide bonds. The α-helix is a right-handed helix (in moving from the N terminus to the C terminus, rotation is clockwise) formed by the carbonyl oxygen of the ith residue hydrogen bonded to the amide hydrogen of the i + 4th residue. The helix makes one turn every 0.54 nm and has 3.6 amino acid residues per turn. Amino acids are spaced 0.15 nm along the helix axis The hydrogen bonds are parallel to the helical axis, with the peptide-bond planes and the amino acid side chains perpendicular. The α-helix is also referred to as a 3.6_{13} helix; it has 3.6 residues per turn with the hydrogen-bonded groups separated by 13 atoms. When the polypeptide chain is in an extended state, β-pleated sheets may form by hydrogen bonding between parallel chains (parallel β-pleated sheets) or antiparallel chains (β-pleated sheets). The disposition of α-carbons in these structures forms a pleated pattern.

The peptide chain is often required to make sharp turns (as for example when antiparallel β-pleated sheets are formed) and the β-turn is often used to accomplish this. In this secondary structure element, a sharp turn is stabilized by hydrogen bonding between the carbonyl of a peptide bond with the amide hydrogen 3 residues away. In a polypeptide chain, peptide bonds are usually *trans* but in the β-turn a *cis* arrangement is required. Proline and glycine are often found in β-turns because they can assume the *cis* configuration.

Some proteins are dominated by secondary structure. In α-keratin, found in hair, wool, claws, fingernails, and horns of animals, two-stranded coiled coils composed of distorted α-helices predominate. Connective tissues, including tendons, cartilage, bones, teeth, skin, and blood vessels, contain collagen, whose structure is dominated by a triple helix. The collagen helix has 3.3 residues per turn, a fact reflected in its amino acid sequence. The primary structure is dominated by repeats of Gly-Pro-Hyp. Glycines are located on the helix interior, making contact with each other to form the three-stranded structure. In contrast to these helical proteins, the fibroin protein of silk fibers is composed of stacked antiparallel β-sheets.

The largest class of proteins is that of the globular proteins. Here, helices, sheets, and turns are used to form a globular structure held together by weak forces. Hydrophobic interactions play a major role in globular protein structure; the core is often formed by hydrophobic amino acid side chains. Structural motifs consisting of secondary structure elements are often components contributing to protein structure, as for example β-sheet arrays with right-handed twists and βαβ-loops. The globular proteins can be classified based on the type and arrangement of secondary structure. In antiparallel helix proteins, the core structure is a bundle composed of antiparallel α–helices. Parallel or mixed β-sheet proteins have an extended sheet of β-structure. The sheet may be internal, in which case hydrophobic amino acids are distributed on both sides of the sheet. Alternatively, the sheet may be arranged in a barrel shape. Antiparallel β-sheet proteins have two sheets, each composed of antiparallel β-sheets. Finally, metal-rich proteins and disulfide-rich proteins rely on either metal binding sites or numerous disulfide bonds to maintain a stable globular structure.

Some proteins appear to be composed of more than one structural domain and each of the domains serves a particular role in protein function. Often, a distinct structural domain is a protein module dedicated to a particular function.

Quaternary structure is reserved for oligomeric proteins, proteins composed of distinct subunits. When the subunits are identical, the protein is a homomultimer; when the subunits are different, the protein is a heteromultimer. In multimeric proteins, the subunits typically fold into independent structures and make isologous (same surface) or heterologous (different surface) interactions with other subunits in the protein. These interactions are typically hydrophobic in nature and may be stabilized by disulfide bonds.

Chapter 6

Lipids, Membranes, and Transport

• •

Chapter Outline

- Lipids: Compounds composed largely of reduced carbons and exhibiting low water solubility. Lipids may be completely hydrophobic or, if they contain polar groups, amphipathic
- Fatty acids: Carboxyl head group and hydrocarbon tail
  - Typically even number of carbons (14 to 24)
  - Saturated fatty acids lack carbon-carbon double bonds
  - Unsaturated fatty acids contain one or more double bonds in cis configuration
    - Monosaturated: Single carbon-carbon double bond
    - Polyunsaturated: *cis* Double bonds separated by methylene carbon ($-CH_2-$)
  - Nomenclature
    - Systematic: e.g., Octadecanoic acid: 18-carbons long
    - Common names: Stearic acid
    - Short hand notation:
      - 18:0 -18 carbons long with no double bonds
      - 18:1(9) -18 carbons long with one double bond starting at carbon 9
  - Essential fatty acids: Linoleic and γ-linolenic acid: Plant fatty acids required in diets of animals: used to synthesize arachidonic acid -precursor to eicosanoids (prostaglandins)
- Triacylglycerols (triacylglycerides): Fats and oils
  - Glycerol esterified with three fatty acids
  - Metabolic oxidation of large number of reduced carbons provides large amount of energy
  - Anhydrous storage allows for efficient energy storage
- Glycerophospholipids: A class of phospholipids
  - 1,2-Diacylglycerol with phosphate ester at carbon 3
  - Stereospecific numbering system: Number glycerol carbons based on (R,S) system: C1 is carbon that would lead to S-configuration if its priority were increased
  - Phosphatidic acid: sn-Glycerol-3-phospate with fatty acids esterified to C1 and C2
    - Phosphatidic acid precursor of glycerophospholipids
    - C1 fatty acid typically saturated
    - C2 fatty aid typically unsaturated
  - Glycerophospholipids: Polar group esterified to phosphate of phosphatidic acid
    - Ethanolamine: Phosphatidylethanolamine
    - Choline: Phosphatidylcholine
    - Glycerol
    - Serine
    - Inositol
    - Diphosphatidyl glycerol (cardiolipin)
  - Ether glycerophospholipids
    - Ether linkage at carbon 1
    - Alkyl group esterified to carbon 2
    - Platelet activating factor has acetate unit as alkyl making molecule water soluble
  - Plasmalogens: Ether glycerophospholipids with *cis*-α,β-unsaturated alkyl group
- Sphingolipids: Backbone of sphingosine: 18-Carbon amino alcohol with C-C trans double bond

- Ceramide: Sphingosine with fatty acid in amide linkage
- Sphingomyelins: Alcohol esterified to phosphoceramide
  - Choline
  - Ethanolamine
- Glycosphingolipids: Ceramide with sugars in β-glycosidic linkage
  - Cerebroside: Sugar is glucose or galactose
  - Sulfatide: Sugar is galactose with sulfate esterified at carbon 3 of galactose
  - Gangliosides: Ceramide with three or more sugars including sialic acid
  - Important in cell recognition

- Waxes: Esters of long-chain alcohol and long-chain fatty acid
- Terpenes: Class of lipids formed from 2-methyl-1,3-butadiene (isoprene)
  - Monoterpene: Two isoprene units: 10 Carbons
  - Sesquiterpenes: Three isoprene units: 15 Carbons
  - Diterpenes: Four isoprenes: two monoterpenes: 20 Carbons
  - Triterpenes: Six isoprenes: 30 carbons: Cholesterol precursors squalene and lanosterol
  - Tetraterpenes: Eight isoprenes: 40 Carbons: Carotenoids
  - Polyprenols: Long-chain polyisoprenoid alcohols
- Steroids
  - Cholesterol
  - Steroid hormones
    - Androgens: Testosterone
    - Estrogens: Estradiol
    - Progestins: Progesterone
    - Glucocorticoids: Cortisol
    - Mineralocorticoids
  - Bile salts: Cholic acid and deoxycholic acid
- Membrane functions
  - Boundary for cell and organelles
  - Surface on which reactions can occur
  - Regulation of material flux through membrane proteins
  - Signal transduction interface
  - Specialized properties: Photosynthesis, electron transport, electrical activity
- Plasma membrane
  - Delimits cell
  - Excludes and retains certain ions and molecules
  - Major role in energy transduction
  - Cell locomotion
  - Reproduction
  - Signal transduction
  - Interactions with other cells or extracellular matrix
- Membrane components
  - Lipids: Amphipathic molecules arranged as bilayers with polar groups out and nonpolar groups in
  - Proteins: Surface associated or embedded in bilayer
- Lipid interactions
  - Monolayers: Formation of single-molecule-thick layer at air/water interface with polar groups in contact with water
  - Micelles: Lipid spheres with polar groups out and hydrophobic tails in the center: Critical micelle concentration is the concentration of amphiphilic compound at which micelles form
  - Lipid bilayer: Two lipid monolayers with hydrophobic surfaces face to face
- Fluid Mosaic Model
  - Singer and Nicholson, 1972
  - Phospholipid bilayer forming fluid matrix
  - Two classes of membrane proteins
    - Peripheral (extrinsic) proteins
      - Associated with bilayer surface via ionic interactions and H bonds
      - Extractable with high salt or agents that disrupt H bonds (urea)
    - Integral (intrinsic) proteins

- Associate with hydrophobic bilayer interior via hydrophobic interactions
- Extractable with detergents

❖ Membrane mobility
  - ⅄ Protein
    - ♦ Frye and Edidin, 1970: Lateral movement of membrane proteins following fusion of mouse and human cells
    - ♦ Lateral movement may be impeded by interactions with cytoskeleton
  - ⅄ Lipids
    - ♦ Rapid lateral movement
    - ♦ Slow transverse movement

❖ Membrane asymmetry
  - ⅄ Lateral asymmetry arises from clustering of membrane components within the plane
    - ♦ Lipid clustering: Phase separation induced by divalent cations and influenced by lipid type
    - ♦ Protein clustering: Self-associating membrane proteins e.g., bacteriorhodopsin
  - ⅄ Transverse asymmetry
    - ♦ Lipids: Lipid asymmetry due to two processes
      - Asymmetric synthesis
      - Energy-dependent transport: Flippases
    - ♦ Proteins: Asymmetric molecules
    - ♦ Carbohydrates: Glycoproteins and glycolipids on outer surface

❖ Membrane phase transitions: Radical change in physical state occurring within narrow range of temperature. Midpoint is T_m.
  - ⅄ Lipids close-pack: Lose lateral mobility and rotational mobility of fatty acid chains
  - ⅄ Consequences: Membrane thickens and decreases surface area
  - ⅄ Characteristics
    - ♦ T_m increases with chain length and degree of saturation and is influenced by nature of head group
    - ♦ Pure phospholipid bilayers show narrow temperature range
    - ♦ Native membranes show broad transition influenced by protein and lipid composition

❖ Membrane Proteins
  - ⅄ Functions: Transport, receptors, enzymes
  - ⅄ Two types: Peripheral and integral
    - ♦ Integral membrane proteins: Two classes
      - Single transmembrane segment proteins: Hydrophobic alpha helix that spans the lipid bilayer: Glycophorin: 19-Amino acid long alpha helix that spans the membrane with extracellular domain decorated with oligosaccharides that are ABO and MN blood group antigens
      - Multi-transmembrane segment proteins: Essentially globular proteins embedded in membrane
        - – Bacteriorhodopsin: Seven alpha helical segments embedded in bilayer: Segments organized into a channel
        - – Porins: Beta sheet barrel forms pore lined with polar residues

❖ Lipid-anchored membrane proteins: Four types
  - ♦ Amide-linked myristoylated proteins
    - Myristic acid (14:0 fatty acid)
    - Amide linkage to amino group of N-terminal glycine
  - ♦ Thioester-linked: Fatty acid attached to cysteine as thioester (or to Ser or Thr as ester)
  - ♦ Thioether-linked prenylated proteins
    - Prenyl: Long-chain isoprene polymers
    - Attachment as thioether to cysteine of CAAX (A= Aliphatic)
    - After attachment AAX cleaved: Carboxyl terminus methylated
  - ♦ Glycosylphosphotidylinositol (GPI) anchors
    - Lipid: Oligosaccharide-modified phosphoinositol
    - Linkage: Carboxy terminus attached via phosphoethanolamine to mannose residue of oligosaccharide

- Membrane transport: Three types
  - Passive diffusion
    - Entropically driven process: Molecules move down a concentration gradient
      - $\Delta G = RT\ln([C2]/[C1])$ for uncharged molecule
      - $\Delta G = RT\ln([C2]/[C1]) + Z\mathcal{F}\Delta\Psi$ for charged molecules
      - $R = 8.3145$ J/K·mol, T = K, Z = charge, $\mathcal{F} = 96485$ J/V·mol, $\Delta\Psi$ = electrical potential
    - Rate depends on concentration gradient and lipid solubility
  - Facilitated diffusion
    - Entropically driven process like passive diffusion
    - Involves integral membrane protein
      - Rate depends on concentration but rate is saturatable
      - Specificity and affinity due to protein/transported molecule interaction
    - Examples:
      - Glucose transporter: RBC band 4.5: 55 kD Protein functional as trimer
      - Anion transport system: RBC band 3: 95 kD Protein: Cl^-, HCO_3^+ exchange
  - Active transport: Energy driven process
    - Primary active transport
      - Energy sources
        - ATP hydrolysis (most common)
        - Light energy
      - Electrogenic transport: Net transport of charge transport
      - Examples of primary active transport
        - Na^+,K^+-ATPase (sodium pump)
          - 120 kD α-subunit, 35 kD β-subunit
          - 3 Na^+ out, 2 K^+ in per ATP hydrolyzed: Electrogenic
          - Ouabain: Cardiac glycoside that inhibits sodium pump
        - Calcium ATPase
          - 2 Ca^{2+} out of cytoplasm per ATP hydrolyzed
          - Restores/maintains low cytoplasmic calcium
        - H^+,K^+-ATPase
          - 1 H^+ out, K^+ in per ATP hydrolyzed
          - Gastric enzyme: ΔpH largest gradient known
        - Vacuolar ATPases: Pump H^+ in number of vacuoles and cells
        - Multidrug resistance in malignant cells and transport of yeast "a factor" peptide transported by ATPases
    - Secondary active transport (energy is ion gradient formed by some other process)
      - Na^+ or H^+ coupled movement of amino acids or sugars
      - Symport: Ion and substance move in same direction
      - Antiport: Ion and substance move in opposite directions
- Transmembrane ion channels
  - Melittin: Bee venom toxin: Alpha helical ion pore
  - Cecropin: Silkworm antibiotic: Alpha helical ion pore
- Ionophores: Facilitate movement of molecules across lipid bilayer
  - Mobile carrier
    - Sensitive to membrane phase transition
    - Forms complex with ion, complex diffuses across membrane
    - Example: Valinomycin: Potassium ionophore
  - Channel-forming ionophore
    - Insensitive to membrane phase transition
    - Forms ion-specific channel that spans the membrane
    - Example: Gramicidin

Chapter Objectives

Lipids are amphipathic molecules with both polar and nonpolar groups. Understand why this is the case for both simple lipids (like cholesterol) and complex lipids (like phospholipids).

Fatty acids are important components of membrane lipids and triacylglycerols (fats and oils). Fatty acids are most commonly composed of an even number of carbons. Saturated fatty acids have only carbon-carbon single bonds (their carbons are saturated with respect to hydrogens), whereas unsaturated fatty acids have one or more (polyunsaturated) carbon-carbon double bonds in *cis* configuration. When more than one double bond is present, the bonds are not conjugated but rather separated by a $-CH_2-$ group.

Complex Lipids

Complex lipids include triacylglycerols, glycerophospholipids and sphingolipids. They are classified as complex because their carbons derive from different sources. For example, triacylglycerols have a glycerol backbone to which three fatty acids are esterified. Glycerophospholipids again have a glycerol backbone but with only two fatty acids (in ester linkage to carbons 1 and 2) and, attached to the third carbon of glycerol, a phosphate group to which a polar alcohol is linked (like ethanolamine, choline, serine or inositol). Sphingolipids are composed of sphingosine, a fatty acid, and a polar head group (like phosphocholine or one or more sugar moieties).

Simple Lipids

The carbons of simple lipids (terpenes or cholesterol and its derivatives) all derive from isoprene. Know the general structure of isoprene and cholesterol and appreciate the fact that important biomolecules such as steroid hormones and bile salts are derivatives of cholesterol. Lipids associate to form two- and three-dimensional structures. Understand the forces responsible for this behavior, including hydrophobic interactions and van der Waals forces. Know why monolayers of lipids form at an air/water interface, and what a micelle is and how it forms. Lipids are also capable of forming bilayers, an important structural component of biological membranes.

Membranes

Biological membranes are composed of various lipids arranged in a bilayer and, embedded in the bilayer, integral (or intrinsic) proteins. The fluid mosaic model of membranes suggests that both lipids and proteins are free to move within a bilayer. The two surfaces of bilayers of biological membranes are asymmetric with respect to protein, lipid, and carbohydrate composition.

Membrane phase transitions occur when membrane components, in particular lipids, interact in a manner causing loss of fluidity. The temperature of this transition, a transition from solid to liquid, is known as the melting temperature (T_m). What are the effects of degree of saturation, of chain length, and of cholesterol on T_m?

Two types of membrane proteins are peripheral and integral proteins. Peripheral proteins interact through electrostatic bonds and hydrogen bonds with the surfaces of bilayers. Integral proteins are strongly associated with the bilayer. There are three kinds of protein motifs responsible for anchoring integral proteins to membranes. Certain integral proteins have a single transmembrane segment, in the form of an α-helix composed of hydrophobic amino acid residues, anchoring the protein to the lipid bilayer. Another structural motif found is the 7-helix, transmembrane segment used by integral proteins involved in transport and signaling activities. Certain proteins have covalently linked lipid molecules that serve as anchors. You should understand the four kinds of anchors.

Passive Diffusion

Passive diffusion proceeds down a concentration gradient. The driving force is a change in free energy given by $\Delta G = RT\ln([C_2]/[C]_1)$ where $[C_2] < [C]_1$ and the substance moves from side 1 to side 2. For a charged species, the driving force is an electrochemical potential given by $\Delta G = RT\ln([C_2]/[C_1]) + Z\mathcal{F}\Delta\Psi$ where Z is the charge, $\mathcal{F}$ is Faraday's constant, and $\Delta\Psi$ is the membrane potential.

Facilitated Diffusion

Facilitated diffusion is reminiscent of enzyme kinetics because it is a carrier-mediated process and as such involves an interaction between a carrier and a transported molecule. The flux is still dependent on a difference in concentration and it occurs from high concentration to low concentration but the dependence is no longer linear. The flux shows saturation at high concentrations and is critically dependent on stereochemistry of the compound. The glucose transporter and the anion transporter (both in erythrocytes) are examples of facilitated diffusion carriers.

Active Transport

Unlike passive and facilitated diffusion, active transport can move a substance against a concentration gradient. However the overall ΔG of the reaction must be favorable and this is achieved by coupling transport to some other energy-yielding process like ATP hydrolysis, capture of light energy, and coupled to other gradients. The sodium pump or Na^+,K^+-ATPase is a well characterized active transporter for movement of 3 Na^+ out of the cell and 2 K^+ into the cell coupled to hydrolysis of ATP. The enzyme is an intrinsic membrane protein that exists in two conformational states, which differ in ion- and ATP-binding properties. Understand how transient phosphorylation leads to conformational changes and movement of ions. The cardiac glycosides are important inhibitors of the sodium pump. Finally, because of the difference in charge transported, (a difference of one positive charge) the sodium pump is electrogenic leading to formation of a membrane potential. The calcium transporter of sarcoplasmic reticulum is also an ATP-dependent transporter, but of calcium. It has a similar mechanism of action to the sodium pump, shuffling between two conformational states with ATP hydrolysis driving calcium uptake. This transporter is a key player in relaxation of muscle and is also electrogenic. The H^+,K^+-ATPase moves protons out of the cell and potassium back into the cell with ATP hydrolysis. This nonelectrogenic pump is capable of producing extremely high concentration gradients of protons.

Bacteriorhodopsin, an H^+-pump, is a light energy-driven active transport systems. There are many important examples of transport systems driven by ion gradients. Proton gradients produced by electron-transport driven proton pumping or by proton-ATPases, sodium gradients produced by the sodium pump, and other cation and anion gradients are used to move a range of molecules such as sugars and amino acids. You should know the terms symport and antiport.

Specialized Pores

Ionophores are compounds that allow passage of specific ions across a membrane. There are two general classes of ionophores, mobile carriers and channel formers. Mobile carriers form a complex with the ion to be transported and this complex diffuses across the membrane. In effect, a mobile ionophore makes the ion more lipid soluble. Channel formers bridge the membrane and provide a hole or channel through which ions pass.

Problems and Solutions

1. Draw the structures of (a) all the possible triacylglycerols that can be formed from glycerol with stearic and arachidonic acid and (b) all the phosphatidylserine isomers that can be formed from palmitic and linolenic acids. Which of the PS isomers are not likely to be found in biological membranes?

Answer: Triacylglycerols have a glycerol backbone to which three fatty acids are esterified. With nonidentical fatty acids at carbons 1 and 3, carbon 2 is chiral and so two stereoisomers are possible. Biological triacylglycerols have the L-configuration. Stearic acid is an 18-carbon saturated fatty acid. Arachidonic acid is a 20-carbon fatty acid with four *cis* double bonds at carbons 5, 8, 11 and 14. With either stearic acid alone or arachidonic acid alone occupying carbons 1 and 3, carbon 2 is not chiral. Therefore, only two triacylglycerols are possible with two stearic acid groups or two arachidonic acid groups occupying carbons 1 and 3.

$_1CH_2OH$
HO▸$_2$C◂H
$_3CH_2OH$
glycerol

stearate
$^-O-C(=O)-$

arachidonate
$^-O-C(=O)-$

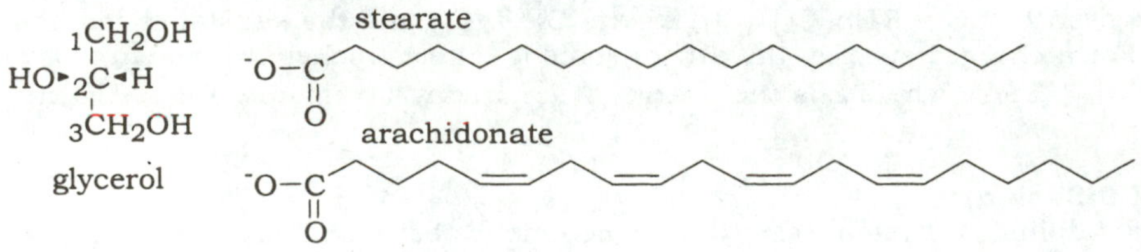

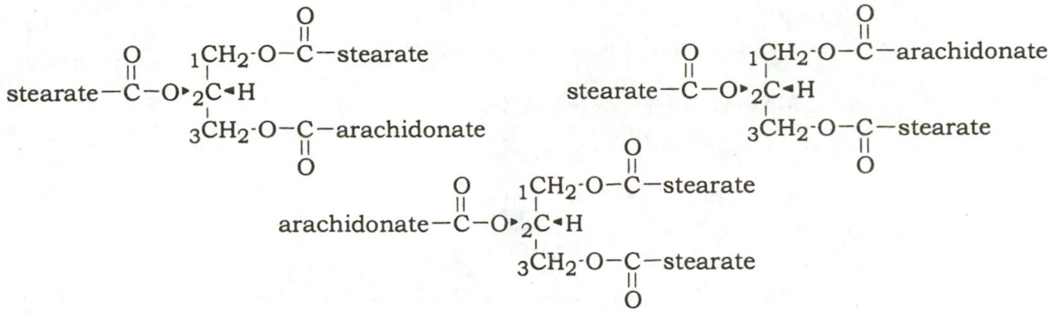

```
                       O                                                    O
         O   1CH2-O-C—stearate                           O   1CH2-O-C—arachidonate
stearate—C—O▸2C◂H                              stearate—C—O▸2C◂H
             3CH2-O-C—arachidonate                           3CH2-O-C—stearate
                    O                                             O
                            O                   O
               O   1CH2-O-C—stearate
arachidonate—C—O▸2C◂H
                   3CH2-O-C—stearate
                          O
```

For the top two structures, switching the positions of the fatty acids and the hydrogen about carbon 2 will generate two more stereoisomers. The same is true for the two structures shown below.

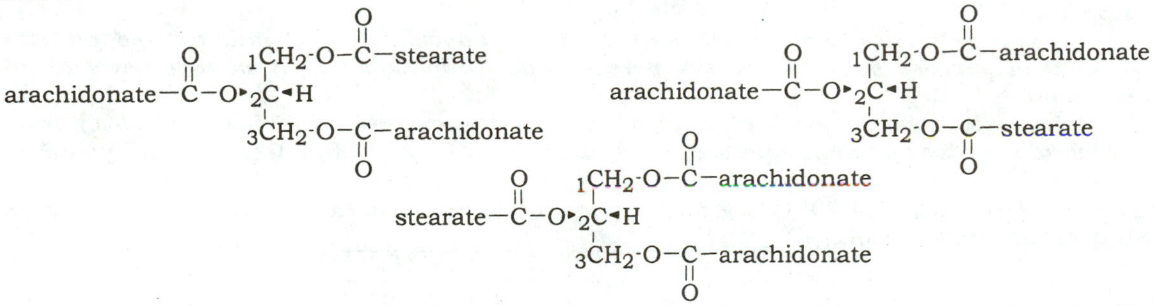

```
                           O                                                    O
             O   1CH2-O-C—stearate                           O   1CH2-O-C—arachidonate
arachidonate—C—O▸2C◂H                          arachidonate—C—O▸2C◂H
                 3CH2-O-C—arachidonate                           3CH2-O-C—stearate
                        O                          O                   O
           O   1CH2-O-C—arachidonate
stearate—C—O▸2C◂H
               3CH2-O-C—arachidonate
                      O
```

b. Palmitic acid is 16:0; linolenic acid is 18:3($\Delta^{9,12,15}$). The backbone structure of phosphatidylserine is 3-phosphoglycerol with L-serine in phosphate ester linkage. Fatty acids are esterified at carbons 1 and 2. There is a preference for unsaturated fatty acids at carbon 2. The alpha carbon of serine is chiral; L and D serine are possible; however, the L isomer occurs in phosphatidylserine. The central carbon in the glycerol backbone of phosphatidylserine is prochiral and only one isomer is used, the one based on *sn*-glycerol-3-phosphate.

```
    1CH2OH                   palmitate                                   COO-
HO▸2C◂H   O           -O-C                                         +H3N▸C◂H
   3CH2O—P-O-             O                                              CH2
          O-                 linolenate                                  OH
sn-glycerol-3-phosphate -O-C                                          L-serine
                            O
```

```
                                        O
                          O   1CH2-O-C—linolenate or palmitate
linolenate or palmitate—C—O▸2C◂H    OH          NH3+
                              3CH2-O-P—CH2—C—COO-
                                     O             H
```

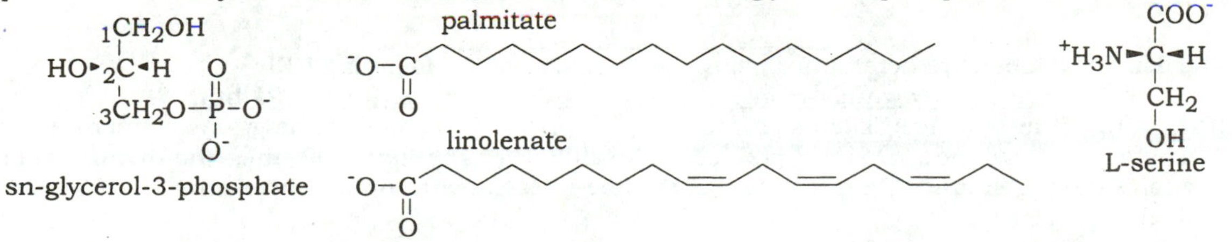

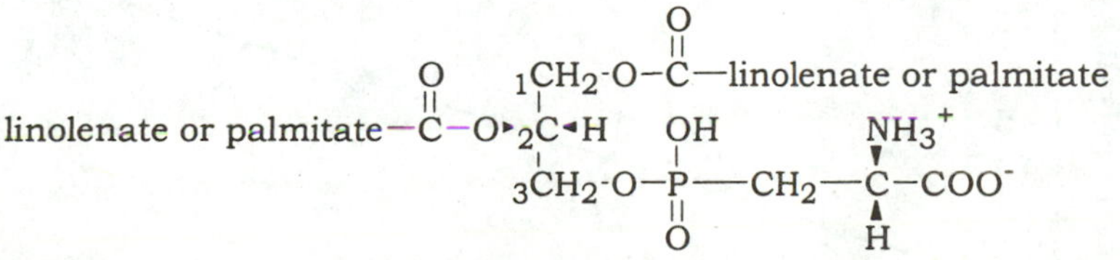

Note: Phosphatidylserine with unsaturated lipids at position 1 are very rare. Unsaturated fatty acids are usually found at position 2.

2. The purple patches of the* Halobacterium halobium *membrane, which contain the protein bacteriorhodopsin, are approximately 75% protein and 25% lipid. If the protein molecular weight is 26,000 and an average phospholipid has a molecular weight of 800, calculate the phospholipid to protein mole ratio.

Answer:

Let x = the weight of bacteriorhodopsin - lipid complex.
Weight of lipid in the complex = 0.25x
Weight of protein in the complex = 0.75x

$$\text{Moles lipid} = \frac{0.25x}{800\frac{g}{mole}} = 3.13 \times 10^{-4}x, \text{ and}$$

$$\text{Moles protein} = \frac{0.75x}{26{,}000\frac{g}{mole}} = 2.88 \times 10^{-5}x$$

$$\text{Molar ratio (lipid : protein)} = \frac{3.13 \times 10^{-4}x}{2.88 \times 10^{-5}x} = 10.8$$

3. Sucrose gradients for separation of membrane proteins must be able to separate proteins and protein-lipid complexes having a wide range of densities, typically 1.00 to 1.35 g/mL.
a. Consult reference books (such as the CRC Handbook of Biochemistry) and plot the density of sucrose solutions versus percent sucrose by volume (g sucrose per 100 mL solution).
b. What would be a suitable range of sucrose concentrations for separation of three membrane-derived protein-lipid complexes with densities of 1.03, 1.07, and 1.08 g/mL?

Answer: The density, at 20° C (ρ, g/mL), of sucrose solutions and their percent by volume (g per 100 mL) are shown below.

ρ (g /mL)	% Sucrose (g per 100mL)
0.9982	0
1.0382	10
1.0777	20
1.1168	30
1.1554	40
1.1935	50
1.2312	60
1.2683	70

The density values were determined using the calculator at the following URL:
http://www.univ-reims.fr/Externes/AVH/MementoSugar/001.htm
(One word of caution: The "sucrose content (%)" at this URL is equal to the grams of sucrose per 100 grams of solution. Since problem 3 asks for grams per 100 mL, the "Gram Liter Concentration (g sucrose / L solution)" should be selected with input values equal to 100, 200, 300, etc.)

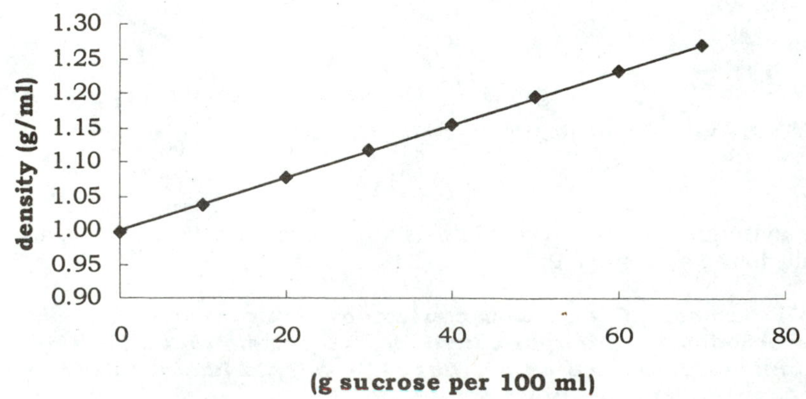

4. Phospholipid lateral motion in membranes is characterized by a diffusion coefficient of about 1 x 10^{-8} cm^2/sec. The distance traveled in two dimensions (in the membrane) in a given time is $r=(4Dt)^{1/2}$, where r is the distance traveled in centimeters, D is the diffusion coefficient, and t is the time during which diffusion occurs. Calculate the distance traveled by a phospholipid across a bilayer in 10 msec (milliseconds).

Answer:

$$\text{For } D = 1\times 10^{-8}\frac{cm^2}{sec}, t = 10 \text{ msec} = 10\times 10^{-3}\text{ sec}$$

$$r = \sqrt{4\times D\times t}$$

$$= \sqrt{4\times (1\times 10^{-8}\frac{cm^2}{sec})\times (10\times 10^{-3}\text{sec})}$$

$$= 2.0\times 10^{-5}\text{cm} = 2.0\times 10^{-7}\text{m} = 0.2\ \mu\text{m}$$

5. Protein lateral motion is much slower than that of lipids because proteins are larger than lipids. Also, some membrane proteins can diffuse freely through the membrane, whereas others are bound or anchored to other protein structures in the membrane. The diffusion constant for the membrane protein fibronectin is approximately 0.7 x 10^{-12} cm^2/sec, whereas that for rhodopsin is about 3 x 10^{-9} cm^2/sec.
a. Calculate the distance traversed by each of these proteins in 10 msec.
b. What could you surmise about the interactions of these proteins with other membrane components?

Answer:

$$\text{For fibronectin, } D = 0.7\times 10^{-12}\frac{cm^2}{sec}$$

$$\text{For rhodopsin, } D = 3.0\times 10^{-9}\frac{cm^2}{sec}$$

$$t = 10 \text{ msec} = 10\times 10^{-3}\text{ sec}$$

$$r = \sqrt{4\times D\times t}$$

$$\text{For fibronectin } r = \sqrt{4\times (0.7\times 10^{-12}\frac{cm^2}{sec})\times (10\times 10^{-3}\text{ sec})} = 1.67\times 10^{-7}\text{cm} = 1.67\text{ nm}$$

$$\text{For rhodopsin } r = \sqrt{4\times (3.0\times 10^{-9}\frac{cm^2}{sec})\times (10\times 10^{-3}\text{ sec})} = 1.10\times 10^{-5}\text{cm} = 110\text{ nm}$$

b. The diffusion coefficient is inversely dependent on size and unless we know the size of each protein we can surmise very little. The M_r of rhodopsin and fibronectin are 40,000 and 460,000 respectively. For spherical particles D is roughly proportioned to $[M_r]^{-1/3}$. We might expect the ratio of diffusion coefficients (rhodopsin/fibronectin) to be

$$\frac{(40,000)^{-\frac{1}{3}}}{(460,000)^{-\frac{1}{3}}} = 2.3$$

The measured ratio is 4286! Clearly the size difference does not explain this large difference in diffusion coefficients. Fibronectin is a peripheral membrane protein that anchors membrane proteins to the cytoskeleton. Its movement is severely restricted.

6. Discuss the effects on the lipid phase transition of pure dimyristoyl phosphatidylcholine vesicles of added (a) divalent cations, (b) cholesterol, (c) distearoyl phosphatidylserine, (d) dioleoyl phosphatidylcholine, and (e) integral membrane proteins.

Answer: Myristic acid is a 14 carbon saturated fatty acid and as a component of dimyristoyl phosphatidylcholine is expected to participate in hydrophobic interactions and van der Waals interactions. At a particular temperature, T_m, these forces are strong enough to produce local order in a bilayer of this phospholipid.

a. Divalent cations (e.g., Mg^{2+}, Ca^{2+}) interact with the negatively charged phosphate group and thus stabilize bilayers and increase the T_m.

b. Cholesterol does not change the T_m; however, it broadens the phase transition. As a lipid, it can participate in hydrophobic and van der Waals interactions. Above the T_m of dimyristoyl phosphatidylcholine, cholesterol stabilizes interactions; however, below the T_m, it interferes with the packing of dimyristoyl phosphatidylcholine.

c. Distearoyl phosphatidylserine contains stearic acid, an 18-carbon, fully saturated fatty acid, which should participate favorably in van der Waals interactions and hydrophobic bonds. Its slightly longer chain length may perturb the geometry of vesicles. Also, the longer chain and negatively-charged headgroup should raise T_m.

d. Oleic acid is a 18-carbon fatty acid with a single double bond in *cis* configuration between carbons 9 and 10. Although capable of hydrophobic interactions, the unsaturated fatty acids are expected to interfere with van der Waals interactions. The T_m will be decreased.

e. Integral proteins will broaden the phase transition and could either raise or lower the T_m depending on the nature of the protein.

7. Calculate the free energy difference at 25°C due to a galactose gradient across a membrane, if the concentration on side 1 is 2 mM and the concentration on side 2 is 10 mM.

Answer:

$$\Delta G = RT\ln\frac{[C_2]}{[C_1]},$$

where $[C_1]$ and $[C_2]$ are the concentrations of C on opposites of the membrane.

$$\Delta G = 8.314\times10^{-3}\frac{kJ}{K\cdot mol}\times298\ K\times\ln\frac{10\ mM}{2\ mM} = 4.0\frac{kJ}{mol}$$

8. Consider a phospholipid vesicle containing 10 mM Na⁺ ions. The vesicle is bathed in a solution that contains 52 mM Na⁺ ions, and the electrical potential difference across the vesicle membrane $\Delta\psi = \psi_{outside} - \psi_{inside}$ = -30 mV. What is the electrochemical potential at 25°C for Na⁺ ions?

Answer: The electrical potential is given by the following formula:

$$\Delta G = RT\ln\frac{[C_2]}{[C_1]} + Z\mathcal{F}\Delta\Psi$$

where R is the gas constant, T the temperature in degrees Kelvin, $\mathcal{F}$ is Faraday's constant (96.49 kJ/K·mol) and Z is the charge on the ion: +1 in this case.

$$\Delta G = 8.314\times10^{-3}\frac{kJ}{K\cdot mol}\times298°K\times\ln\frac{52\ mM}{10\ mM}+(+1)\times96.49\frac{kJ}{V\cdot mol}\times(-30\times10^{-3}\,V)$$

$$\Delta G = 1.19\frac{kJ}{mol}$$

9. Transport of histidine across a cell membrane was measured at several histidine concentrations:

[Histidine],μM	*Transport, μmol/min*
2.5	42.5
7	119
16	272
31	527

72	1220

Does this transport operate by passive diffusion or by facilitated diffusion?

Answer: A characteristic of transport by passive diffusion is that the rate of transport is linearly dependent on concentration and so a plot of transport rate versus concentration will be linear. For facilitated diffusion, the transported molecule interacts with a carrier protein in the membrane. The rate of transport will be dependent on concentration but the dependence is not linear. Rather the dependence is reminiscent of Michaelis-Menten enzyme kinetics in that it shows saturation. A plot of rate versus concentration is presented below. The data show a linear relationship indicating that transport is by passive diffusion. However, a facilitated transport system with a high K_m relative to the concentrations of histidine tested here will also be approximately linear. The concentrations tested here are high relative to physiologically reasonable concentrations of histidine and if facilitated diffusion is at work it may not be of physiological importance. One way to confirm that the transport is passive is to retest transport using D-histidine. Using a different stereoisomer of histidine will have no affect on passive diffusion. Facilitated diffusion will show specificity for one of the stereoisomers.

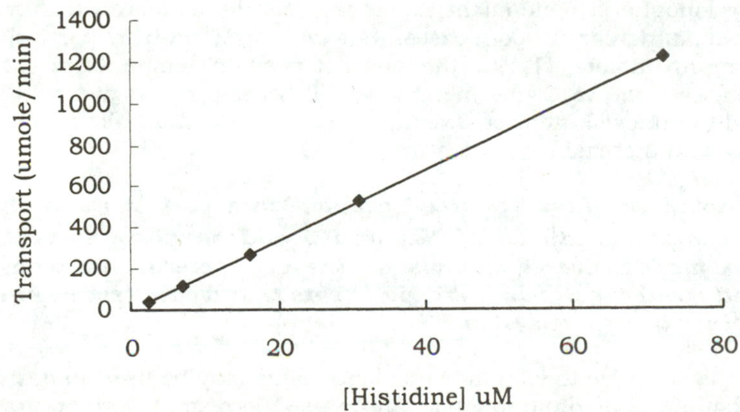

10. Fructose is present outside a cell at 1 μM concentration. An active transport system in the plasma membrane transports fructose into this cell, using the free energy of ATP hydrolysis to drive fructose uptake. What is the highest intracellular concentration of fructose that this transport system can generate? Assume that one fructose is transported per ATP hydrolyzed, that ATP is hydrolyzed on the intracellular surface of the membrane, and that the concentrations of ATP, ADP, and Pi are 3 mM, 1 mM, and 0.5 mM, respectively. T=298°K (Hint: Refer to Chapter 3 to read the effects of concentration on free energy of ATP hydrolysis.)

Answer: The free energy of hydrolysis of ATP is given by

$$\Delta G = \Delta G°' + RT \ln \frac{[ADP][P_i]}{[ATP]}$$

$$\Delta G = -30\frac{kJ}{mol} + 8.314 \times 10^{-3}\frac{kJ}{K \cdot mol} \times 298°K \times \ln\frac{1\ mM \times 0.5\ mM}{3\ mM}$$

$$= -51.6\frac{kJ}{mol}$$

The free energy of a gradient of a substance across a membrane is given by

$$\Delta G = RT \ln\frac{[C_2]}{[C_1]}$$

We can set this equal to the free energy of hydrolysis calculated above but with the opposite sign and solve for C_2 given that C_1 is equal to 1 mM.

$$\Delta G = RT\ln\frac{[C_2]}{[C_1]} = 51.6\frac{kJ}{mol} \text{ and}$$

$$[C_2] = [C_1] \times e^{\frac{51.6\frac{kJ}{mol}}{RT}} = 1.0 \times 10^{-6} M \times e^{\frac{51.6\frac{kJ}{mol}}{8.314\times10^{-3}\frac{kJ}{K\cdot mol}\times 298\,K}}$$

$$[C_2] = 1{,}109\,M\,!$$

11. The rate of K^+ transport across bilayer membranes reconstituted from dipalmitoylphosphatidylcholine (DPPC) and monensin is approximately the same as that observed across membranes reconstituted from DPPC and cecropin a at 35°C. Would you expect the transport rates across these two membranes to also be similar at 50°C? Explain.

Answer: The two ionophores have quite different modes of action. Monensin is a mobile ion carrier that forms a complex with K^+, diffuses across the membrane, and releases K^+. Cecropin is a channel-forming ionophore. One might expect that, as the temperature increases, the rate of movement of ion would increase in both cases. However, nigericin transport will show a dramatic change around approximately 41.4°C, the phase transition temperature for DPPC. As the temperature increases from 35°C, the membrane will become much more fluid and the rate of nigericin-mediated uptake will increase dramatically. The rate of uptake for cecropin is expected to increase slightly with increasing temperature.

12. In this chapter we have examined coupled transport systems that rely on ATP hydrolysis, on primary gradients of Na^+ or H^+, and on phosphotransferase systems. Suppose you have just discovered an unusual strain of bacteria that transports rhamnose across its plasma membrane. Suggest experiments that would test whether it was linked to any of these other transport systems.

Answer: If uptake is sensitive to ion gradients, ionophores may be used to destroy the gradients. Uncouplers like dicumarol or dinitrophenol can be used to degrade proton gradients. Ouabain can be used to inhibit the sodium pump. Dependence on ATP hydrolysis may be determined by using nonhydrolyzable ATP analogs. PEP dependent mechanisms similar to PTS in *E. coli* are sensitive to fluoride.

Questions for Self Study

1. Fill in the blanks. Important biomolecules composed of a long hydrocarbon chain (or tail) and a carboxyl group are known as _____ _____. When all of the carbon-carbon bonds are single bonds the compound is said to be _____. This term also indicates that the carbons in the tail are associated with a maximum number of _____ atoms. Compounds of this type with one carbon-carbon double bond are _____ whereas those with multiple carbon-carbon double bonds are _____. Usually there are an _____ number of carbons atoms. These compounds are components of fats and oils in which they are joined to a _____ backbone in _____ linkage.

2. True of False
   a. 2-methyl-1,3-butadiene is also known as isoprene. _____
   b. Cholesterol is a phospholipid. _____
   c. The androgens are a class of terpene-based lipids involved in absorption of dietary lipids in the intestine. _____
   d. Cholesterol is an example of a complex lipid. _____
   e. Cholesterol is a hydrocarbon composed of three six-membered rings and one five-membered ring in addition to a hydrocarbon tail. _____

3. Identify the following from the structures shown below: phosphatidic acid, phosphatidylcholine, phosphatidylserine, phosphatidylinositol, ceramide, phosphatidyl-ethanolamine.

a.
C(=O)–O–CH$_2$
C(=O)–O–CH
H$_2$C–O–P(=O)(O-)–O–CH$_2$–CH$_2$–N$^+$(CH$_3$)$_3$

b.
C(=O)–O–CH$_2$
C(=O)–O–CH
H$_2$C–O–P(=O)(O-)–O–CH$_2$–CH$_2$—NH$_3$+

c.
C(=O)–O–CH$_2$
C(=O)–O–CH
H$_2$C–O–P(=O)(O-)–O–CH$_2$–CH(COO-)(NH$_3$+)

d.
C(=O)–O–CH$_2$
C(=O)–O–CH
H$_2$C–O–P(=O)(O-)–O-

e.
C(=O)–O–CH$_2$
C(=O)–O–CH
H$_2$C–O–P(=O)(O-)–O–(inositol: OH, OH, OH, OH, OH)

f.
C(H)=C(H)
H–C—OH
C(=O)–N(H)–CH
H$_2$C–OH

4. Based on your knowledge of lipid and carbohydrate biochemistry identify components of the following compound and state how this compound is chemically similar in structure to triacylglycerols? How does it biochemically different?

5. Very often grocery stores sell produce with a waxy coating applied to their outside (cucumbers and turnips are often treated this way). What is the general structure of a wax? For what purpose is the layer of wax applied? Would something like a fatty acid or a triacylglycerol be a good substitute?

6. Match the items in the two columns

a. Singer and Nicolson	1. Peripheral protein.
b. Extrinsic protein	2. Lipid bilayer structure.
c. Integral protein	3. Lipid transfer from outside to inside.
d. Liposome	4. Phase change.
e. Micelle	5. Intrinsic protein.
f. Flippase	6. Lipid monolayer structure.
g. Transition temperature	7. Fluid mosaic model.

7. For proteins with a single transmembrane segment, explain why the segment is often helical and why it is often composed of hydrophobic residues.

8. Give three examples of lipid anchoring motifs.

9. Explain the term critical micelle concentration.

10. For each of the statements below indicate which of the following, passive diffusion (P), facilitated diffusion (F), and active transport (A), applies.
   a. Can only move down a concentration gradient. _____
   b. Is expected to transport L-amino acid and D-amino acid at the same rate. _____.
   c. Can be saturated. _____
   d. Can occur in both directions across a biological membrane. _____
   e. Can be used to concentrate substances. _____
   f. Movement is coupled to exergonic process. _____
   g. Glucose transporter in erythrocytes. _____
   h. Rate is linearly proportional to concentration difference. _____
   i. Movement across biological membrane dependent on lipid solubility. _____
   j. Sodium pump. _____

11. Match the active transport system with an appropriate function.

a. Na^+,K^+-ATPase	1. Acidifies membrane bound compartments.
b. Ca^{2+}-ATPase	2. Transports a host of cytotoxic drugs.
c. H^+,K^+-ATPase	3. Resets levels of important second message after stimulation.
d. Vacuolar ATPase	4. Electrogenic pump inhibited by cardiac glycosides.
e. MDR ATPase	5. Responsible for production of the largest concentration gradient known in eukaryotic cells.

12. What is a symport? Antiport? How can they be used to move a substance against its concentration gradient?

13. Bacteriorhodopsin and halorhodopsin are active transport proteins for the movement of protons and chloride ions respectively. What energy source do they use to support ion pumping?

14. What is the difference between a carrier ionophore and a channel-forming ionophore.

Answers

1. Fatty; acids; saturated; hydrogen; monounsaturated; polyunsaturated; even; glycerol; ester.

2. a.T; b.F; c.F; d.F; e.T.

3. d.; a.; c.; e.; f.; b.

4. You should readily identify the two rings as substituted sugars. The six-membered ring is glucose and the five-membered ring is fructose. The disaccharide they form is sucrose. Each of the hydroxyl groups of sucrose has a fatty acid attached by ester bonds. The compound is sucrose polyester or more commonly known as olestra (Trade name: Olean). Olestra is currently being used as a fat substitute because it has properties identical to fats and oils but in not digested.

Triacylglycerols contain fatty acids esterified to glycerol, a three-carbon alcohol. Both triacylglycerol and olestra are amphiphilic molecules with uncharged, weakly polar head groups and hydrocarbon tails.

5. Waxes are composed of a long-chain alcohol and a long-chain fatty acid joined in ester linkage. Waxes are often used to make surfaces water impermeable thus a waxy coating will prevent water loss and prolong shelf life. A layer of triacylglycerol might accomplish the same results; however, typical fats and oils have lower melting temperatures and would not be expected to form as stable a layer as wax.

6. a.7; b.1; c.5; d.2; e.6; f.3; g.4.

7. The hydrogen bonding groups in the peptide bond are all involved in hydrogen bonds in a helix. Hydrophobic residues can interact with the hydrophobic interior of membranes through hydrophobic interactions.

8. Any three of: Amide-linked myristoyl anchors, thioester-linked fatty acyl anchors, thioether-linked prenyl anchors, and amide-linked glycosyl phosphatidylinositol anchors.

9. The critical micelle concentration is that concentration of lipid at which micelle formation is supported. Concentrations of lipid below the critical micelle concentration do not form micelles. Lipid solutions whose concentration is greater than the critical micelle concentration contain micelles in equilibrium with free lipid molecules. The concentration of the free lipid is equal to the critical micelle concentration.

10. a. P, F, A; b. P; c. F and A; d. P and F; e. A; f. A; g. F; h. P; i. P; j. A.

11. a. 4; b. 3; c. 5; d. 1; e. 2.

12. A symport is a transport system that couples movement of two substances in the same direction. An antiport couples the movement of two substances in opposite directions. They can be used to move substances against a concentration gradient if transport of the coupled substance is down a concentration gradient. In this case the energy of the concentration gradient of the co-transported substance is used to drive uptake.

13. Light.

14. Ionophores are substances that can transport ions across a biological membrane. Carrier ionophores form a lipid-soluble complex with the ion and the complex diffuses from one side of a membrane to the other. A channel-forming ionophore is a lipid soluble compound that can dissolve in biological membranes and spans the membrane with a channel or pore through which an ion diffuses.

Additional Problems

1. At a romantic candlelight dinner, the conversation turns to properties of waxes and what exactly happens when a candle burns. Contribute to the conversation.

2a. Margarine is made from vegetable oil by a process called hydrogenation in which the oil is reacted with hydrogen gas in the presence of a small amount of nickel, which functions as a catalyst. Hydrogenation saturates double bonds. Explain why hydrogenated vegetable oil is a solid.
b. Margarine may be purchased in stick-form or in small tubs. What is the important chemical difference between these two kinds of margarine?

3. The transport properties of two potassium ionophores were being studied in synthetic lipid bilayers with a phase transition temperature of 50°C. Ionophore X transports potassium at a rate proportional to temperature from 20°C to 70°C. In contrast, ionophore Y transports potassium very well above 60°C; however, from 60°C to 40°C the rate of transport falls off precipitously to very low values below 40°C. Based on this information, can you suggest modes of action for these two ionophores?

4. How might you expect helical wheel plots of α-helical segments from the following proteins to differ: (a) a typical globular protein, (b) an integral membrane protein with a single transmembrane segment, and (c) an integral membrane protein with several α-helices forming a channel through which a water-soluble compound is transported?

5. Explain how lipid membrane asymmetry might arise in a natural membrane through action of a flippase that does not couple lipid movement to another thermodynamically favorable process (like ATP hydrolysis).

6. Activity of the sodium pump results in the net movement of a positive charge across the membrane. How does this lead to a change in the electrical potential of the membrane?

7. Would you expect proton pumps to be capable of creating large proton gradients if the pumps operated by an electrogenic mechanism? Explain.

8. Construct a helical wheel plot of melittin, whose amino acid sequence is: Gly-Ile-Gly-Ala-Val-Leu-Lys-Val-Leu-Thr-Thr-Gly-Leu-Pro-Ala-Leu-Ile-Ser-Trp-Ile-Lys-Arg-Lys-Arg-Gln-Gln. Assume that this peptide forms an α-helix and comment on the structure.

Abbreviated Answers

1. Waxes are esters of long-chain alcohols and fatty acids. For example, in beeswax, straight-chain alcohols 24 to 36 carbons in length are esterified to long, straight-chain fatty acids up to 36 carbons in length. The melting temperature of beeswax is around 63°C. When a candle burns, the lit wick produces heat that melts the wax. The liquid wax is drawn up the wick to be consumed in the flame. A good candle will produce very little dripping wax because the wax is all consumed in the flame.

Wax is a rich source of oxidizable hydrocarbons and serves the same purpose as oil does in a lamp, or gasoline does in an internal combustion engine. However, waxes and oils do not explode because they have very low vapor pressures.

2. Margarines are typically made from vegetable oils such as corn oil and soybean oil. They are oils, -liquids at room temperature- because their composition includes greater than 50% unsaturated fatty acids. Hydrogenation is the addition of hydrogen to double-bonds producing a saturated hydrocarbon. For a given chain length, saturated hydrocarbons have a higher melting temperature than do unsaturated hydrocarbons. Thus, saturation of the double-bonds in the fatty acids in corn and soy oil reduces the level of unsaturated fatty acids and as a consequence the melting temperature is increased.

The difference between tub-margarine and stick-margarine is the degree of saturation. Tub-margarine is distributed in a container because it is softer than stick-margarine due to a lower degree of hydrogenation.

3. Ionophore X may be a channel-forming ionophore, perhaps like the antibiotic gramicidin. Channel-forming ionophores span the membrane forming a channel through which an ion can diffuse across the membrane. Ionophore Y may be a carrier-ionophore. Carrier-ionophores form a complex with the ion to be transported. This complex diffuses across the membrane and dissociates, releasing the ion on the opposite side of the membrane. Valinomycin is an example of a potassium ionophore of this type.

4. In globular proteins containing α-helices, the helices often contribute to the hydrophobic core of the protein with hydrophobic amino acid residues located along one face of the helix. Integral membrane proteins, with a single stretch of α-helix responsible for anchoring the protein into the membrane, have a helix composed of hydrophobic amino acids. For integral proteins anchored by several helices, the helices are often amphipathic with both a hydrophobic surface and a hydrophilic surface. The hydrophobic surfaces contact the fatty acid side chains of the membrane lipid component, whereas the hydrophilic surfaces face inward and may form a pore though which hydrophilic substances diffuse.

5. Energy must be expended to produce an asymmetric distribution of lipids. In the case of a flippase, which simply equilibrates lipids in response to a concentration gradient of free lipids, lipid asymmetry might arise if lipids preferentially interact with a membrane protein.

6. There are two ways of looking at this. The net movement of a positive charge gives rise to an imbalance of charge across the membrane, resulting in an electrical potential. Alternatively, the movement of charge across the membrane represents a current. Current flowing across the

resistant of the membrane will produce a voltage change.

7. No. If the pumps were electrogenic an electrochemical gradient would be formed with an electrical potential component and a chemical gradient and this would require considerably more energy than just a chemical gradient.

8. The helical wheel plot of melittin in an α-helical conformation is shown below. One face of the helix is lined with hydrophobic amino acids, and a cluster of basic amino acids is found at the C-terminus. A proline residue is positioned approximately in the middle of the helix.

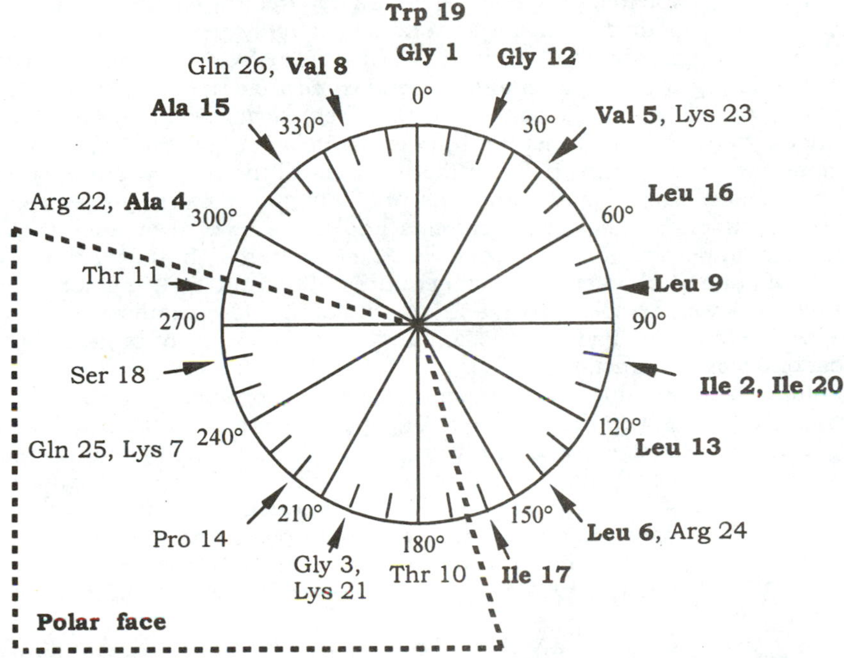

Summary

Lipids are a large and diverse class of cellular compounds defined by their insolubility in water and solubility in organic solvents. Lipids serve several biological functions. As highly reduced forms of carbon, lipids yield large amounts of energy in the oxidative reactions of metabolism. As hydrophobic molecules, lipids allow membranes to act as effective barriers to polar molecules. The unique bilayer structure of membranes derives mainly from the amphipathic nature of membrane lipids. Certain lipids also play roles as cell-surface components involved in immunity, cell recognition and species specificity. Other lipids act as intracellular messengers to regulate a variety of processes.

Most fatty acids found in nature have an even number of carbon atoms. Fatty acids may either be saturated or unsaturated, and double bonds are normally of the *cis* configuration. "Essential" fatty acids, including linoleic and linolenic acids, are not synthesized by mammals, but are required for growth and life. Triacylglycerols consist of a glycerol molecule with three fatty acids esterified. Triacylglycerols in animals are found primarily in adipose tissue, and serve as a major metabolic reserve for the organism.

Glycerophospholipids, a major class of lipids, are composed of sn-glycerol-3-phosphate with fatty acids esterified at the 1- and 2- positions. Many different "head groups" can be esterified to the phosphate, including choline, ethanolamine, serine, glycerol and inositol. Sphingolipids are lipids based on sphingosine, a long chain fatty alcohol, to which is often attached another fatty acid in an amide linkage to form a ceramide. Sphingomyelin is a phosphate-containing sphingolipid. Glycosphingolipids consist of a ceramide backbone with one or more sugars. Cerebroside is a glycosphingolipid containing either glucose or galactose. Gangliosides have three or more sugars esterified, one of which must be a sialic acid. Glycosphingolipids are present in only small amounts, but serve numerous important cell functions. Terpenes are a class of lipids derived from isoprene units. The steroids, including cholesterol, are an important class of terpene-based lipids. Other steroids in animals, including the androgens and estrogens (male and female hormones, respectively) and the bile acids (used in digestion) are derived from cholesterol.

Waxes are esters of long-chain fatty acids and long-chain alcohols. Because of their low water solubility and ability to aggregate, waxes are used to form water-impermeable surfaces.

Lipids form a variety of structures spontaneously in solution, including monolayers, micelles and bilayer structures. Lipid bilayers have a polar surface, composed of charged or neutral lipid headgroups, and a nonpolar interior, composed of hydrophobic lipid chains. The fluid mosaic model of membrane structure, proposed by Singer and Nicholson, pictures the lipid bilayer as a fluid, dynamic matrix, with lipids and proteins able to undergo free, rapid lateral motion.

The fatty acid chains in membrane lipids are oriented roughly perpendicular to the bilayer plane, and this ordering is more pronounced near the bilayer surface. As one proceeds into the bilayer interior, the ordering of the lipid chains decreases, so that the interior is a highly fluid environment. Transverse motion of lipids and proteins is very slow. Different lipid classes show different distributions between the inner and outer monolayers of the membrane bilayer. Proteins are asymmetrically distributed between the two faces of the bilayer, allowing a variety of vectorial (i.e., directionally dependent) functions including transport processes. Lateral asymmetries also exist in membranes, with proteins and lipids able to arrange themselves in clusters or aggregates important to cell function. Lipids in membranes exhibit dramatic, cooperative changes of state at characteristic temperatures. Such phase transitions between the solid, gel-like state at lower temperatures and the fluid, liquid-crystalline state at higher temperatures, are sensitive to the lipid composition and to the presence of proteins, and may be important in a host of biological functions.

Membrane proteins are of three fundamental types. Peripheral proteins form ionic interactions or hydrogen bonds with the surface of the lipid bilayer. Integral proteins intercalate into the lipid bilayer and are strongly associated with the membrane. The lipid anchored proteins attach to membranes via covalently linked lipid moieties. Peripheral proteins can be extracted with high salt, EDTA or urea, while integral proteins can only be removed with organic solvents or detergents. Detergents are amphipathic molecules, with both polar and nonpolar moieties, and function by intercalating into the membrane and solubilizing lipids and proteins. At the critical micelle concentration (CMC), detergents spontaneously form micelles and become much more effective solubilizing agents.

Integral membrane proteins take on a variety of conformations in membranes. Proteins such as glycophorin of the erythrocyte membrane have a single hydrophobic α-helix extending across the bilayer, with hydrophilic segments extending on either side of the lipid bilayer. Other proteins, such as bacteriorhodopsin of the purple patches of *Halobacterium halobium*, traverse the bilayer several times, with six or more hydrophobic alpha helices spanning the bilayer. These latter proteins are often involved in membrane transport activities and other processes that require a substantial portion of the peptide to be imbedded in the membrane.

Four different types of lipid anchoring motifs for membrane proteins have been found to date, including amide-linked myristic acid anchors, thioester-linked fatty acyl anchors, thioether-linked prenyl anchors, and amide-linked glycosyl-phosphatidylinositol anchors. With amide-linked myristic acid (14:0) proteins, the fatty acid is attached in amide linkage to the α-amino group of N-terminal glycine. In thioester- and thioether-linked proteins, lipids are attached to cysteine residues. Ether-linked lipids are long-chain polyisoprenoids. In GPI (glycosylphosphatidylinositol) anchors the C-terminus of the target protein is linked via phosphoethanolamine to a mannose residue on an oligosaccharide attached to phosphatidylinositol.

Transport processes are important to all life forms. The acquisition of nutrients, the elimination of waste materials and the generation of concentration gradients vital to nerve impulse transmission and the normal function of brain, heart, kidneys and other organs all depend on membrane transport systems. All transport processes are mediated by transport proteins, which may function either as channels or carriers. The three classes of transport are passive diffusion, facilitated diffusion and active transport. In passive diffusion, the transported species move across the membrane in the thermodynamically favored direction without the assistance of a specific transport system. Analogous to Brownian motion, passive diffusion is in essence an entropic process. The rate of flow of an uncharged molecule depends upon concentration difference and the permeability coefficient of the molecule. In facilitated diffusion, the transported species moves according to its thermodynamic potential, but with the help of a specific transport system. Facilitated diffusion systems display saturation behavior. The glucose transporter and the anion transporter of erythrocytes are both facilitated diffusion systems. For charged species, the charge of the molecule and the electrical potential difference also affect transport.

Active transport systems use energy input to drive a transported species against its thermodynamic potential. The most common energy input is ATP hydrolysis, but light energy and the energy stored in ion gradients may also be used. All active transport systems are energy-coupling devices. Na,K-ATPase, which transport Na^+ ions out of cells and transports K^+ ions into the cells, is an active transport system. Na,K-ATPase consists of a 120 kD α subunit and a 35 kD β subunit. The enzyme mechanism involves an aspartyl phosphate intermediate. Na,K-ATPase is strongly and specifically inhibited by cardiac glycosides such as ouabain. Calcium transport across the sarcoplasmic reticulum membrane is mediated by Ca-ATPase. The gastric H,K-ATPase transports protons across the membrane of stomach mucosal cells, generating the high concentrations of acid in the stomach that are essential to digestion of food. Proton pumps in osteoclasts enable these cells to degrade the mineral matrix of bone during the remodeling and reconstruction of bone tissue. ATPases also transport peptides and drugs. Yeast α-factor is transported out of yeast cells by a 1290-residue transport protein that consists of two identical halves formed from a gene duplication. An analogous transport protein known as the multidrug resistance (MDR) ATPase actively transports a wide spectrum of drugs out of human cells. This transport system is induced by the chronic administration of drugs (in cancer chemotherapy, for example). The yeast α-factor transporter and the MDR ATPase are two members of a superfamily of prokaryotic and eukaryotic transport proteins.

Bacteriorodopsin (bR) is a light-driven proton transport system from *Halobacterium halobium.*

Secondary active transport systems use the ion and proton gradients established by primary active transport systems to transport amino acids, sugars and other species in certain cells. Most of these operate as symport systems, with the ion or proton and the transported amino acid or sugar moving in the same direction.

Several small molecules produced by microorganisms and referred to as ionophore antibiotics facilitate ion transport across membranes. These ionophores may act either as channels or as mobile carriers. Valinomycin from *Streptomyces* is a mobile carrier. It is a cyclic structure containing 12 units from four different residues. In the valinomycin-K^+ complex, polar groups of the valinomycin structure face the center of the ring structure, coordinating K^+, and the nonpolar side chains are directed outward from the ring, where they interact favorably with the nonpolar interior of the membrane bilayer. Gramicidin from *Bacillus brevis* is a channel-forming ionophore. It is a linear peptide formed from both L- and D-amino acids, and it forms a head-to-head helical dimer in lipid membranes. The helix creates an ion channel through the bilayer membrane. Many other peptides form transmembrane channels, including melittin from bee venom and the cecropins from *Hyalophora cecropia*, the cecropia moth. These transmembrane helical channels are amphipathic, with polar residues clustered on one face of the helix and nonpolar residues elsewhere. In the membrane, the polar residues cluster to form an ion channel, leaving the nonpolar residues to interact with the hydrophobic interior of the bilayer membrane.

Chapter 7

Carbohydrates and Cell Surfaces

• •

Chapter Outline

- Carbohydrates $(CH_2O)_n$ $n \geq 3$
- Nomenclature
  - Monosaccharides (simple sugars)
  - Oligo- and polysaccharides: Polymers of simple sugars
- Classification
  - Aldose (aldehyde) and ketose (ketone)
  - Triose, tetrose, pentose, hexose, etc.
- Stereochemistry
  - Aldose $n \geq 3$, ketose $n \geq 4$ have asymmetric carbons (chiral centers)
  - D- and L- Configuration: Refer to configuration of highest numbered asymmetric carbon
  - D- and L- forms: Mirror images: Enantiomers
  - With >1 asymmetric carbon: Diastereomers: Configurations that differ at one or more chiral carbons but not mirror image molecules
  - Epimers: Two molecules that differ in configuration about one asymmetric carbon
- Ring structures
  - Pyranose: Six-membered, oxygen-containing ring
  - Furanose: Five-membered, oxygen-containing ring
  - Anomeric carbon: Ketone or aldehyde carbon that becomes chiral upon ring formation
  - Anomers: α, β Differ in configuration about anomeric carbon
    - α-Configuration: In Fisher projection, OH of anomeric carbon on same side as OH of highest numbered asymmetric carbon
    - β-Configuration: In Fisher projection, OH of anomeric carbon on opposite side as OH of highest numbered asymmetric carbon
  - Haworth projections: Three-dimensional representation: Groups to right in Fisher projection drawn down
  - Conformations
    - Chair and boat conformations due to ring pucker
    - Axial and equatorial orientation of groups attached to ring
- Monosaccharide derivatives
  - Free anomeric carbons can be reduced: Used to quantify amount of sugar in solution
  - Mild reduction produces sugar alcohol: Sorbitol, mannitol, xylitol
  - Deoxy sugars: 2-Deoxy-D-ribose in DNA
  - Phosphate esters of glucose, fructose -important metabolic intermediates- ribose phosphates in ATP and GTP
  - Amino sugars: Glucosamine and galactosamine
  - Muramic acid and neuraminic acid: Glucosamine with three-carbon acids linked at C-1 or C-3
  - Glycosides: Reaction of hemiacetal or hemiketal with alcohol with loss of water: Sugar polymers
- Oligosaccharides
  - Important disaccharides: Sucrose (non-reducing), maltose (diglucose), lactose (milk sugar)
  - Higher oligosaccharides: Stachyose, amygdalin, bleomycin
- Polysaccharides (glycans)

- Storage polysaccharides
  - Starch: α–Amylose and amylopectin
    - α–Amylose: Linear chains of α(1→4)D-glucose
    - Amylopectin: Linear chains of α(1→4)D-glucose with α(1→6)D-glucose branches every 12 to 30 residues
  - Glycogen: α(1→4) D-glucose chains with α(1→6) D-glucose branches every 8 to 12 residues
- Structural polymers
  - Cellulose: Plant cell wall: Linear polymer of β(1→4) D-glucose
  - Chitin: Exoskeleton and fungi cell wall: β(1→4) N-acetyl-D-glucosamine
    - α Chitin: Chains parallel
    - β Chitin: Chains antiparallel
    - δ Chitin: Sheets of parallel chains separated by sheets of antiparallel chains
  - Glycosaminoglycans: Polymers with disaccharide repeat: Negatively charged
  - Bacterial cell walls: Peptidoglycan: Murein
    - Oligosaccharide of N-acetylglucosamine-β(1→4)-N-acetylmuramic acid repeats
    - Peptide: L-Ala·D-Glu·L-Lys·D-Ala with Lys linked to side chain carboxyl group of Glu and peptide attached to lactate moiety of muramic acid via amide bond to N-terminal Ala
      - Gram-negative bacteria: Lys side chain linked to C-terminal D-Ala
        - Hydrophobic proteins attach peptidoglycan to outer membrane
        - C-terminal of protein: Lys side chain of protein in amide linkage to diaminopimelic acid groups (replace D-Ala) on about 10% of peptides
        - N-terminal of protein: Ser side chain linked to lipid
        - Lipopolysaccharides in outer membrane: O antigens
      - Gram-positive: Lys side chain linked to C-terminal D-Ala by pentaglycine chain
        - Techoic acid covalently attached to peptidoglycan: Techoic acid: Polymer of ribitol phosphate or glycerol phosphate linked by phosphodiester bonds: Hydroxyl groups substituted with glucose, N-acetylglucosamine, or D-Ala
  - Cell surface polysaccharides involved in modulating response of cells to other cells or to extracellular matrix: Interactions involve glycoproteins and proteoglycans

- Glycoproteins
  - Carbohydrate: Oligo- or polysaccharide
  - O-linked: N-acetylgalactosamine linked to Ser, Thr, or Hydroxylysine
    - Cell surface glycoproteins: Two structural motifs
      - Extended conformation: Leukosialin
      - Stem glycosylation: Separates functional extracellular domain from membrane surface
    - Mucins: Protective surfaces
    - Antifreeze glycoproteins: $[\text{Ala·Ala·Thr}]_n$·Ala·Ala where n ranges from 4 to 50 and with β-galactosyl-1,3 α-N-acetylgalactosamine attached to Thr
  - N-linked: Core of two N-acetylglucosamines linked to branched mannose triad: N-acetylglucosamine attached to amide nitrogen of Asn
    - N-linked triantennary oligosaccharides on serum glycoproteins used at timing device for protein degradation
  - Proteoglycans: O-linked glycosaminoglycans to Ser of Ser-Gly
    - Two types
      - Soluble -extracellular matrix proteins
      - Integral transmembrane proteins
    - Function: Binding specific proteins to glycosaminoglycan portion of proteoglycan

Chapter Objectives

It is important to know basic carbohydrate nomenclature, stereochemistry, and chemistry. Monosaccharides, with the general formula, $(CH_2O)_n$, are either aldoses or ketoses. The more important monosaccharides range from n = 3 to n = 7. Because carbohydrates contain a number of chiral centers (carbons with four different groups attached), L- and D- isomers can be formed. Figures 7.2 and 7.3 show D-aldoses and D-ketoses having from three to six carbons.

Cyclic carbohydrates are produced by an intramolecular reaction between an aldehyde or ketone group and a hydroxyl group. In this reaction, the carbonyl carbon of the aldehyde or ketose is converted to a chiral carbon, the anomeric carbon. As a consequence, two structural isomers, α or β anomers, can be formed. Stable five- and six-membered rings, furanoses and pyranoses, are very common. Sugars with free anomeric carbons are reducing sugars. Oligosaccharides and polysaccharides are composed of monosaccharides held together by glycosidic bonds. If a linear polymer contains a free anomeric carbon at one end, that end is called the reducing end and the other end the nonreducing end.

The following carbohydrates should be known:

trioses:	glyceraldehyde and dihydroxyacetone
pentoses:	ribose, ribulose, and deoxyribose
hexoses:	fructose, galactose, glucose, and mannose
heptoses:	sedulose
disaccharides:	lactose, maltose, and sucrose
polysaccharides:	amylose, amylopectin, glycogen, cellulose, and chitin

Important modified sugars include sugar acids, sugar phosphate esters, deoxy sugars, and amino sugars. Many of these modifications are found in the sugar moieties of important biological compounds. Monosaccharides do not display as wide a range of chemical characteristics as for example the amino acids. Nonetheless, polysaccharides are a diverse group of biomolecules because of the ability to join monosaccharides at a number of different positions to form branch polymers and the availability of numerous modified monosaccharides for polymer construction.

Understand bacterial cell wall structure and how peptides and saccharides are arranged in peptidoglycan. Peptide/saccharide molecules are found in glycoproteins. Know how saccharides are linked to proteins: O-linkage and N-linkage.

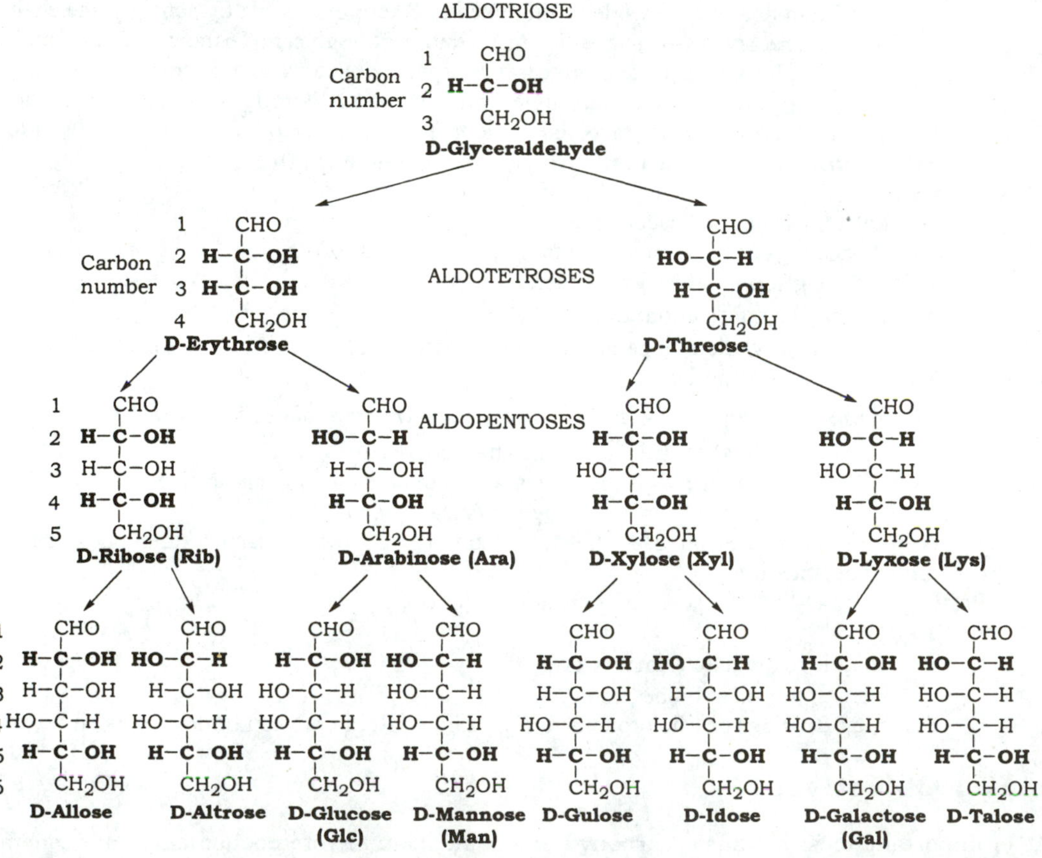

Figure 7.2 The structure and stereochemical relationships of D-aldoses having three to six carbons. The configuration in each case is determined by the highest numbered asymmetric carbon.

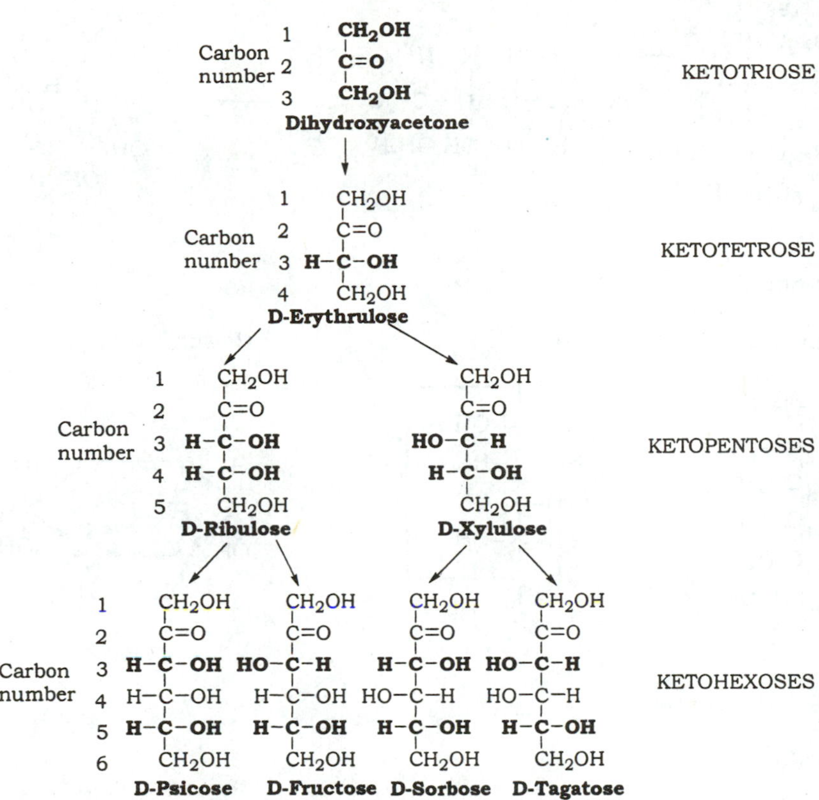

Figure 7.3 The structure and stereochemical relationships of D-ketoses having three to six carbons. The configuration in each case is determined by the highest numbered asymmetric carbon.

Problems and Solutions

1. Draw Haworth structures for the two possible isomers of D-altrose (Figure 7.2) and D-psicose (Figure 7.3).

Answer: To draw Haworth projections:

1. Rotate the Fisher projection by 90° as shown below.
2. Groups that are to the right in Fisher projections are down in Haworth projections.
3. To form the ring, the hydroxyl at C5 must be reoriented. This reorientation brings C6 above the plane of the ring.
4. For C1, the orientation of H and OH depends on which side of the carbonyl plane the OH attacks, β is up, α is down.

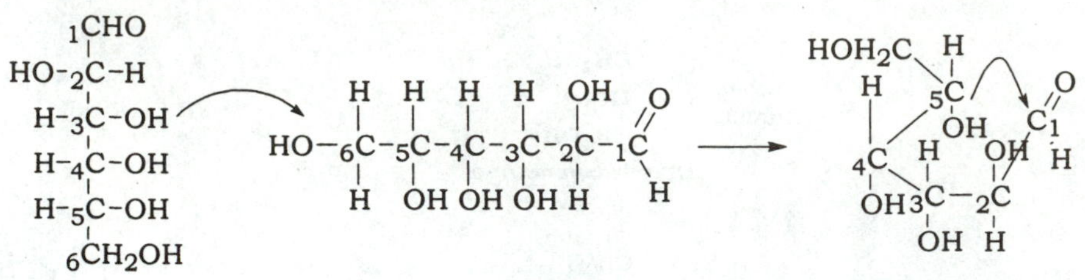

α-D-altrose

open form | ring | Haworth projection

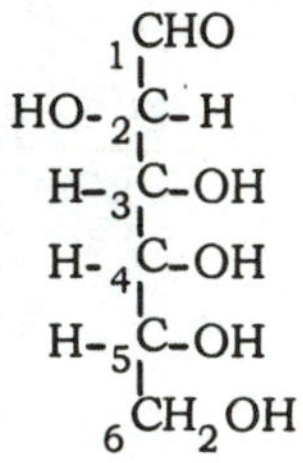

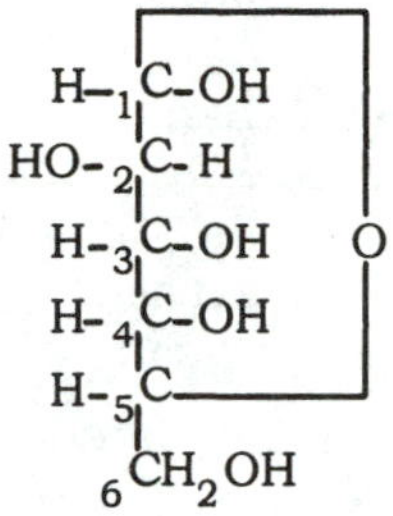

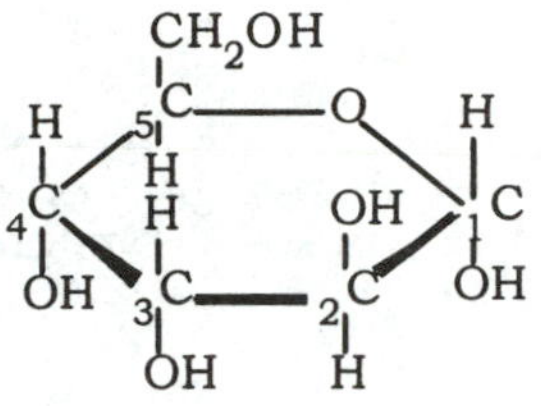

β-D-altrose

open form | ring | Haworth projection

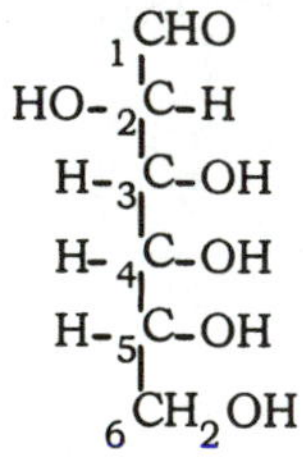

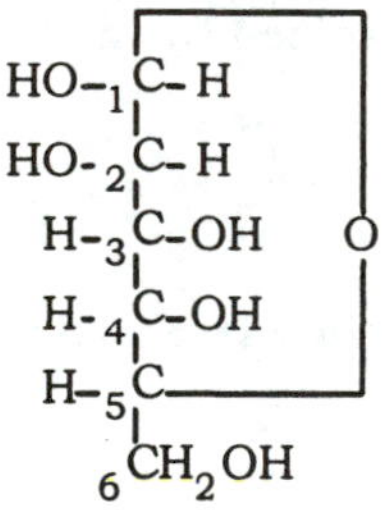

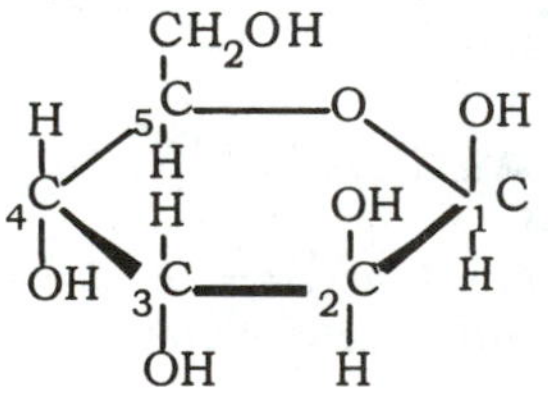

For psicose, a ketose, orientation about C2 depends on which side of the ketone group the hydroxyl of C5 attacks.

2. *Give the systematic name for stachyose (Figure 7.19).*

Answer:
β-D-Fructofuranosyl-O-α-D-galactopyranosyl-(1→6)-O-α-D-galactopyranosyl-(1→6)-O-α-D-glucopyranoside

3. *Trehalose, a disaccharide produced in fungi, has the following structure:*

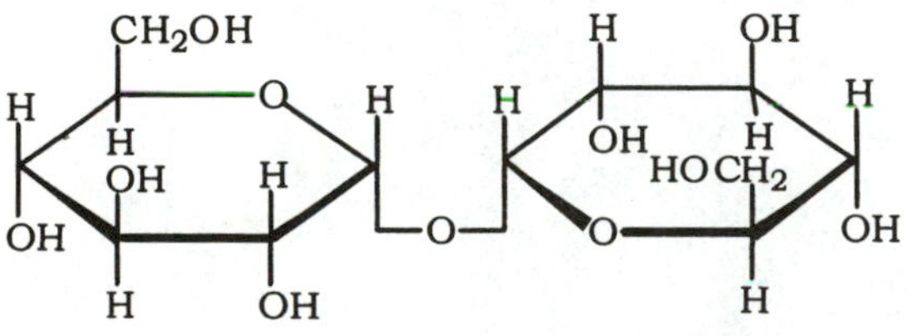

a) What is the systematic name for this disaccharide?

b) Is trehalose a reducing sugar? Explain.

Answer: Trehalose is a disaccharide of α–D-glucose and α–D-glucose, both in the pyranose form. Thus, the systematic name is:

O-α-D-glucopyranosyl-(1→1)-α-D-glucopyranoside.

A reducing sugar contains a free aldehyde or ketone group. In glucose, carbon 1 is an aldehyde. In trehalose, a glycosidic bond connects carbons 1 for both glucosyl moieties. Thus, trehalose is not a reducing sugar.

4. Draw Fischer projection structures for L-sorbose (D-sorbose is shown in Figure 7.3).

Answer: L-sorbose

open form

CH_2OH
$_2C{=}O$
$HO{-}_3C{-}H$
$H{-}_4C{-}OH$
$HO{-}_5C{-}H$
CH_2OH

ring

CH_2OH
$_2C{-}OH$
$HO{-}_3C{-}H$
$O\ \ H{-}_4C{-}OH$
$_5C{-}H$
CH_2OH

Haworth projections

α-L-sorbose

CH_2OH, O, OH, $_5C$, H, $OH\ _2C$, H, $_4C$, $_3C$, CH_2OH, OH, H

β-L-sorbose

CH_2OH, O, CH_2OH, $_5C$, H, $OH\ _2C$, H, $_4C$, $_3C$, OH, OH, H

5. α-D-glucose has a specific rotation, $[\alpha]_D^{20}$, of +112.2°, whereas β-D-glucose has a specific rotation of +18.7°. What is the composition of a mixture of α-D- and β-D-glucose, which has a specific rotation of 83.0°?

Answer: The specific rotation is the number of degrees through which plane polarized light is rotated in traveling 1-decimeter through a sample of 1 g/mL, or symbolically:

$$[\alpha]_D^{20} = \frac{\text{rotation (degrees)}}{\text{path length (dm)} \times \text{conc (g/mL)}}$$

A mixture of α-D- and β-D-glucose at 1 g/mL, with a specific rotation of 83°, contains x g/mL of α-D-glucose and y g/mL of β-D-glucose or:

$$(1)\quad x + y = 1\ \text{g/mL}$$

where x = concentration of α-D-glucose and y = concentration of β-D-glucose. The rotation of the mixture is equal to the sum of the rotations due to the two components. Thus,

rotation (mixture) = rotation(α-D-glucose) + rotation(β-D-glucose)

and,

$$\text{rotation} = [\alpha]_D^{20} \times 1\text{dm} \times \text{conc}$$

$$(2)\quad 83° = 112.2x + 18.7y$$

Solving equations (1) and (2) for x and y, we have:

x = 0.69 g/mL α-D-Glucose
y = 0.31 g/mL β-D-Glucose

6. A 0.2-g sample of amylopectin was analyzed to determine the fraction of the total glucose residues that are branch points in the structure. The sample was exhaustively methylated and then digested, yielding 50 μmol of 2,3-dimethylglucose and 0.4 μmol of 1,2,3,6-tetramethyl-glucose.

a. What fraction of the total residues are branch points?
b. How many reducing ends does this amylopectin have?

Answer: Methylation reactions are expected to modify hydroxyl groups. In amylopectin, 1→4 linkage ties up the hydroxyls on carbon 1 and 4. Of the remaining hydroxyls, on carbons 2, 3, 5 and 6, C-5 is involved in pyranose ring formation leaving only carbons 2, 3 and 6 to be modified. The observation that 2,3-dimethylglucose is produced implies that C-6 is unreactive, indicating that these residues are branch points. Thus, the 50 μmol of 2,3-dimethylglucose derives from branch points. The total number of moles of glucose residues in 0.2g amylopectin is:

$$\frac{0.2\text{g}}{162\frac{\text{g}}{\text{mole}}} = 1.23 \times 10^{-3}\text{ mole glucose residues}$$

(The molecular weight of a glucose residue is $C_6H_{12}O_6$ minus H_2O = 162 g/mol.)

$$\text{Fraction of residues at branches} = \frac{50 \times 10^{-6}\text{mol}}{1.23 \times 10^{-3}\text{ mol}} = 0.0405 \text{ or } 4\%$$

This analysis is expected to yield predominantly 2,3,6-trimethylglucose, a small amount of 2,3,4,6-tetramethylglucose from non-reducing ends of chains and 1,2,3,6-tetramethylglucose from reducing ends, in addition to 2,3-dimethylglucose from branch points. To determine the number of reducing ends in amylopectin, the concentration of 1,2,3,6-tetramethylglucose is used. The number of reducing ends is calculated as follows:

$$0.4 \times 10^{-6}\text{ mol} \times 6.022 \times 10^{23}\text{ molecules/mol} = 2.4 \times 10^{17}\text{ reducing ends}$$

Questions for Self Study

1. For the compounds shown below identify the following: aldose, ketose, chiral center, potential pyranose, potential furanose, anomeric carbon, enantiomers, epimers, diastereomers.

a.	b.	c.	d.
CH_2OH	O=C–H	CH_2OH	O=C–H
C=O	H–C–OH	C=O	H–C–OH
H–C–OH	HO–C–H	HO–C–H	HO–C–H
HO–C–H	H–C–OH	H–C–OH	HO–C–H
HO–C–H	H–C–OH	H–C–OH	H–C–OH
CH_2OH	CH_2OH	CH_2OH	CH_2OH

2. For each of the following disaccharides what are their monosaccharide components? Which are reducing sugars?
   a. lactose, b. sucrose, c. maltose.

3. Amylopectin (P), α-amylose (A) and glycogen (G) are storage polysaccharides found in plants and animals. Which applies the best to the following statements?
   a. Many α(1→6) branches
   b. Straight-chain polymer
   c. Reacts with iodine to give blue color
   d. Is cleaved by phosphorylation
   e. Single reducing end.

4. Amylose and cellulose are both plant polysaccharides composed of glucose in 1→4 linkage. However, amylose is a storage polysaccharide that is readily digested by animals whereas cellulose is a structural polysaccharide that is not digested by animals. What difference in the two polymers accounts for this?

5. Another structural polysaccharide, found in cell walls of fungi and the exoskeletons of crustaceans, insects, and spiders, is chitin. Chitin is similar to cellulose in that the repeat units are held together by β(1∏4) linkage. However, the repeat units are different for the two polymers. What are they?

6. One method of lysing bacteria involves the use of the enzyme lysozyme, which hydrolyzes the glycosidic bond between N-acetylmuramic acid and N-acetylglucosamine. Despite the fact that bacterial cell walls contain protein, why isn't a protease a good alternate choice to disrupt bacterial cell walls?

Answers

1. Aldoses: b and d; Ketoses: a and c; Chiral center: carbons 3,4,5 of a and c, and carbons 2,3,4,5 of b and d; Potential pyranose: b or d; Potential furanose: a and c; Anomeric carbon: carbon 2 of a and c, carbon 1 of b and d; Enantiomers: a and c; Epimers: b and d; Diastereomers: b and d.

2. For lactose: galactose and glucose. For sucrose: glucose and fructose. For maltose: glucose and glucose. Lactose and maltose are reducing sugars.

3. a. G and P; b. A; c. A; d. G; e. P,A,G.

4. Amylose is α-(1→4)-linked D-glucose whereas cellulose is β-(1∏4)-linked D-glucose. Animals have enzymes that cleave amylose linkage but not cellulose linkage.

5. Cellulose is a polymer of D-glucose; chitin is a polymer of N-acetyl-D-glucosamine.

6. The presence of D-amino acids, the unique linkage of amino acid groups, and the defined amino acid composition all make the peptide portion of bacterial cell walls difficult to hydrolyze by proteases.

Additional Problems

1. Protocols for ethanol precipitation of small quantities of DNA often include the addition of glycogen to act as a carrier. Typically, ethanol precipitation is carried out by adding two volumes of 95% ethanol to a solution of salty DNA at 4°C. Explain why glycogen will precipitate under these conditions. What properties of the two polymers, DNA and glycogen, make them behave in a similar manner under these conditions?

2. In one orientation, newsprint can be torn to produce a smooth, straight tear. However, in a perpendicular orientation, the tear line is jagged. Given the fact that newsprint is made from cellulose, explain this behavior.

3. The specific rotations of two freshly prepared solutions of α-D-glucose and β-D-glucose, each at 1 g/mL, were measured and found to be +112.2° and +18.7° respectively. After some time the specific rotations were remeasured and they were both found to be +52.7°. Why?

4a. Honeybees collect nectar, a dilute solution of approximately 10% sucrose, and convert it into honey, a concentrated solution of about 40% each of glucose and fructose. They accomplish this conversion by mixing the nectar with a salivary enzyme, then busily aerating and fanning the solution to drive off water. Given the specific rotations, $[\alpha]_D^{20}$, of sucrose (+66.5°), glucose (+52.7°), and fructose (-92°), calculate the rotation that accompanies honey production.
b. The enzyme is known as invertase. Why?

5a. Pectins are highly branched polysaccharides that are an important component of plant cell walls. The predominant fibers in plant cell walls are cellulose microfibrils composed of cellulose molecules in parallel arrays. In the plant cell wall, the cellulose microfibrils are embedded in a network that is composed of, among other things, pectin. Pectin is highly soluble and readily extracted from plant tissue with hot water whereas cellulose is not. Given this information, pectin is responsible for what property of plant cell walls?
b. Pectins are used to make jams and jellies. Typically, an acidic fruit is boiled with sugar and pectin and the mixture allowed to cool. With luck, a jam or jelly of the proper consistency is produced. The principal monosaccharide in pectin is galacturonic acid. Why is the combination of acid, sucrose, and pectin important for gelation?

Abbreviated Answers

1. Glycogen is a polymer of glucose that is soluble in aqueous solution because of the presence of

numerous hydroxyl groups capable of hydrogen bonding with water. It is insoluble in alcohol because this uncharged polymer readily aggregates. DNA is a sugar/phosphate polymer that is also quite soluble in aqueous solution. To precipitate DNA with ethanol, a counter ion must be present in order to form a DNA salt. In the absence of a counter ion, DNA is negatively charged and will not readily precipitate from dilute solutions with ethanol.

2. In newsprint, cellulose fibers are somewhat aligned along one axis of the print. This axis can be determined by ripping the paper in two orientations. A rip along the axis in which the cellulose fibers are aligned will produce a smooth tear whereas a rip perpendicular to this axis will produce a jagged tear.

3. Initially the two solutions start out with only one D-glucose anomer but with time they both equilibrate to a mixture of the two anomeric forms. The addition of alkali will speed up this conversion. Can you calculate the concentration of α-D-glucose and β-D-glucose in the mixture? See the answer to question 5 from Garrett and Grisham for an example of how to solve the problem.

4a. Assuming a 1 dm path length, the rotation of a 10% solution of sucrose (10 g/100 mL) is given by

$$\text{rotation (degrees)} = [\alpha]_D^{20} \times 1\ \text{dm} \times 0.1\ \text{g/mL}$$
$$= +66.5^\circ \times 1 \times 0.1 = +6.65^\circ$$

Similarly, the rotation of a mixture of 40% glucose and 40% fructose is given by

$$\text{rotation (degrees)} = (+52.7^\circ \times 1 \times 0.4) + (-92^\circ) \times 1 \times 0.4 = -15.7^\circ$$

b. The conversion of sucrose to a mixture of glucose and fructose is accompanied by inversion of rotation (the sign changes from positive to negative). A mixture of equal parts glucose and fructose is called invert sugar. Bees make invert sugar during honey production. Cane sugar is converted to invert sugar by catalysis using invertase in dilute HCl.

5a. Anyone who has eaten overcooked green vegetables should know that pectins are responsible for the rigidity of plant cell walls. As vegetables are cooked, pectins are slowly leached out of the plant cell walls and, as a result, the material becomes tender.

b. Because galacturonic acid is a sugar acid, pectin is a charged molecule at neutral pH. Pectin will form a gel-like mass when the acidic groups on galacturonic acid residues are neutralized at slightly acidic pH. Gel formation is also dependent on the presence of sucrose that apparently hydrogen bonds with neutralized pectin.

Summary

Carbohydrates are the most abundant class of organic molecules found in nature. As their name implies, they are hydrates of carbon, with the formula $(CH_2O)_n$. The versatile nature of carbohydrates arises from their chemical features, including: 1) the existence of one or more asymmetric centers, 2) the ability to adopt linear or ring structures, 3) the formation of polymer structures through glycosidic bonds, and 4) the ability to form multiple hydrogen bonds.

Carbohydrates include monosaccharides, oligosaccharides and polysaccharides. Monosaccharides include aldoses and ketoses, and may also be named for the number of carbons they contain (i.e., trioses, tetroses, etc.). Aldoses with at least three carbons and ketoses with at least four carbons contain chiral centers - carbon atoms with four different substituent groups. Monosaccharides are usually named using the Fischer nomenclature convention, which designates the chiral center farthest from the carbonyl carbon as either D- or L-configuration. The D-forms of monosaccharides predominate in nature, but L-forms are found in specialized roles. D- and L- forms of monosaccharides are mirror images of each other and are designated as enantiomers.

Monosaccharides spontaneously cyclize, forming cyclic hemiacetals with an additional chiral center, and these cyclic furanose and pyranose forms are the preferred structure for monosaccharides in solution. The α- and β-anomers of cyclic monosaccharides may undergo interconversion with intermediate formation of the linear aldehyde or ketone in a process called mutarotation. Pyranose forms are usually favored over furanose forms for aldohexose sugars, and furanose forms are usually more stable for ketohexoses. Furanose and pyranose rings are puckered, not planar, and the most stable conformations place as many bulky groups as possible in equatorial orientations. β-D-glucose, which can adopt a conformation with all its bulky groups in equatorial positions, is the most widely occurring organic group in nature and the central hexose in carbohydrate metabolism. Derivatives of monosaccharides include sugar acids, such as gluconic acid, sugar alcohols, such as sorbitol (used in sugarless gums), deoxy sugars found in DNA, sugar esters, such as ATP and GTP, amino sugars, such as muramic and neuraminic acids, and acetals, ketals and glycosides.

Oligosaccharides, the simplest of which are the disaccharides, consist of monosaccharides linked by glycosidic bonds. Oligosaccharides possessing a free, unsubstituted anomeric carbon are referred to as reducing sugars. Sucrose is not a reducing sugar, but maltose and lactose are. Lactose is the principal carbohydrate in milk. Some individuals do not produce the lactase, which breaks lactose down to galactose and glucose and cannot tolerate lactose in their diet. Higher oligosaccharides include stachyose (whose metabolism by intestinal bacteria causes flatulence) and a variety of antibiotic substances such as streptomycin and bleomycin.

Polysaccharides function as storage materials, as structural components of organisms, as protective substances, and as mediators of cellular recognition and communication. Starch and glycogen are the principal storage polysaccharides in plants and animals, respectively. Both are α(1∏4)-linked chains of glucose units. α-Amylose is a linear starch molecule and amylopectin is a highly branched structure. Glycogen is more highly branched than amylopectin. Structural polysaccharides of note include cellulose, a β(1∏4)-linked glucose polymer that is the most abundant carbohydrate polymer in nature. The β-linkage makes cellulose difficult to digest for most animals, but bacteria living symbiotically in the digestive tracts of termites, shipworms, and ruminant animals secrete cellulase, which effect the breakdown of cellulose. Chitin, a β(1∏4)-linked N-acetyl-D-glucosamine polymer, is the principal skeletal material in crustaceans, insects and spiders and is also present in the cell walls of fungi. It can occur in sheets composed of parallel or anti-parallel chains or in structures with mixed parallel and antiparallel sheets. Other structural polysaccharides include glycosaminglycans, such as heparin, hyaluronates, chondroitins, and keratin sulfate; peptidoglycan and lipopolysaccharide in bacterial cell walls; and glycoproteins and proteoglycans.

Bacterial cell walls are composed of a peptidoglycan layer consisting of a sugar polymer of N-acetylglucosamine-N-acetylmuramic acid repeats with a tetrapeptide attached to the lactic acid moiety of muramic acid. In Gram-negative bacterial cell walls the tetrapeptides are directly crosslinked to each other whereas in Gram-positive bacterial cell walls the tetrapeptides are crosslinked via pentaglycine bridges.

Glycoproteins contain O-linked and N-linked oligosaccharide structures. O-linked oligosaccharides are found in cell surface glycoproteins and also in mucins coating the mucous membranes of respiratory and gastrointestinal tracts. O-linked oligosaccharides occur frequently in the stem regions of membrane proteins. The O-glycosylated stem serves to raise the functional domains of these proteins far enough above the surface to make them accessible to extracellular macromolecules with which they interact. The antifreeze glycoproteins permit fish to live in the icy waters of Arctic and Antarctic Seas. N-linked oligosaccharides are found on many different proteins, and they function in some cases to stabilize protein conformations, to protect against proteolysis, to direct proteins to intracellular organelles, and to target proteins for degradation.

Proteoglycans are a family of glycoproteins whose carbohydrate moieties are predominantly glycosaminoglycans. These proteins may be soluble and located in the extracellular matrix, like serglycin, versican and the cartilage matrix proteoglycan, or they may be integral, transmembrane proteins, like syndecan. Both types function by interacting with other molecules through glycosaminoglycan components and through specific receptor domains in the polypeptide itself. Proteoglycans appear to be involved in the modulation or regulation of cell growth. Cartilage matrix proteoglycan is responsible for the flexibility and resilience of cartilage tissue in the body. In cartilage, the proteoglycan-hyaluronate aggregates are highly hydrated, by virtue of strong interactions between water and the polyanionic glycoprotein complex. When cartilage is compressed during running or walking, for example, water is briefly squeezed out of the cartilage tissue and is reabsorbed when the stress is diminished. This reversible hydration gives cartilage its flexible, shock-absorbing qualities and cushions the joints during stressful, potentially injurious activity.

Chapter 8

Nucleotides and Nucleic Acids

• •

Chapter Outline

- ❖ Nucleotide functions
  - ⅄ Essential metabolic intermediates
  - ⅄ Components of Nucleic acids: DNA and RNA: Heredity and gene expression
- ❖ Nucleotide structure
  - ⅄ Heterocyclic nitrogenous aromatic bases
    - ♦ Pyrimidines: Six-member heterocyclic aromatic ring with 2 nitrogens
      - • Cytosine: DNA and RNA
      - • Uracil: RNA
      - • Thymine: DNA
    - ♦ Purines: Five-member imidazole ring fused to six-member pyrimidine
      - • Adenine: 6-Amino purine: DNA and RNA
      - • Guanine: 2-Amino-6-oxy purine: DNA and RNA
    - ♦ Tautomeric forms
      - • Keto called lactam
      - • Enol called lactim
    - ♦ H-bond properties
    - ♦ UV-light absorbing properties
  - ⅄ Sugars: 5-Carbon pentose: Furanose rings: Numbering system primed
    - ♦ D-Ribose: RNA
    - ♦ 2'-Deoxy-D-ribose: DNA
  - ⅄ Nucleoside: N-glycosidic linkage of base to anomeric carbon of sugar
    - ♦ Anomeric carbon in β configuration
  - ⅄ Nomenclature
    - ♦ Pyrimidines: + idine: Cytidine, uridine, thymidine
    - ♦ Purines: + osine: Adenosine, guanosine
  - ⅄ Two conformations
    - ♦ Syn: Base over furanose ring: Purines
    - ♦ Anti: Base not over furanose ring: Pyrimidines and purines
- ❖ Nucleotides: Typically 5' nucleoside phosphate
  - ⅄ Phosphate esters
    - ♦ AMP, GMP, CMP, UMP: 5'-Ribonucleoside monophosphates
    - ♦ cAMP, cGMP: 3',5'-Cyclic ribonucleoside monophosphates: Regulation of cellular metabolism
  - ⅄ Phosphate esters and phosphoanhydrides
    - ♦ ADP, GDP, CDP, UDP: 5'-Ribonucleoside diphosphates
    - ♦ ATP (energy currency), GTP (protein synthesis and signal transduction), CTP (lipid synthesis), UTP (carbohydrate and polysaccharide synthesis): 5'-Ribonucleoside triphosphates
    - ♦ dATP, dGTP, dCTP, TTP (dTTP): DNA synthesis: 5'-Deoxyribonucleoside triphosphates
- ❖ Nucleic acids: Nucleoside monophosphates in phosphodiester linkage 5' to 3'
  - ⅄ Linear polymers: Directional: Base sequence always written 5' to 3'
  - ⅄ DNA: Genetic material: Typically double stranded

- dsDNA: Strands antiparallel
- Interchain H bonds form base pairs
- Chargaff's rules: A=T, G=C, Purines = Pyrimidines
- Structures
  - B-form DNA: Right-handed helix: X-ray diffraction of Franklin and Wilkins and model building of Watson and Crick
    - Parameters: 0.34 nm/base pair, 10 bp/turn, P = 3.4 nm
    - H-bonded base pairs AT and GC inside
    - Base pairs perpendicular and stacked
    - Negative charged phosphate and sugar outside
    - Major and minor groove
  - A-form DNA: Right-handed helix
    - Parameter: 0.23 nm/base pair, 11 bp/turn, P = 2.46 nm
    - Base pairs tilted
    - dsRNA and DNA/RNA hybrids
  - Z-form DNA: Left-handed helix
    - Dinucleotide repeat: Pyrimidine-purine
    - Conformation: Purine -syn, pyrimidine -anti
    - Zigzag pattern of sugar-phosphate backbone

- RNA: Typically single stranded: Produced by transcription
  - mRNA: Carries information encoded in genes to direct protein synthesis on ribosomes
    - Prokaryotic mRNA: Often codes for >1 protein: Polycistronic
    - Eukaryotic mRNA: Codes for 1 protein
      - Derive from heterogeneous nuclear RNA (hnRNA)
      - RNA processed
        - Splicing: Removal of introns (noncoding) and joining of exons(coding)
        - Capping (5' end), and polyA tail addition (3' end)
  - rRNA: Components of ribosome: Protein synthesis
  - tRNA: Carriers of activated amino acids used by ribosome for protein synthesis
  - snRNA: Small nuclear RNA's
- DNA vs. RNA
  - Composition
    - T in DNA not U to distinguish from T formed by deamination of C
    - 2' OH in RNA accounts for instability of RNA phosphodiester bond
  - Hydrolysis
    - RNA: Sensitive to base hydrolysis: Resistant to acid hydrolysis
    - DNA: Resistant to base hydrolysis: Depurinates by acid hydrolysis (apurinic base)

❖ Nucleases: Enzymes that hydrolyze nucleic acids: Phosphodiesterases: Specificity
- Ester bond attack
  - a-Side cleavage: Attacks on 3' side produces 5' phosphorylated product
  - b-Side cleavage: Attacks on 5' side produces 3' phosphorylated product
- Ester bond location
  - Exo-: Attack from ends
  - Endo-: Attack internally
- Nucleic acid specificity
  - RNase's: RNA specific
  - DNase's: DNA specific
  - Nucleases: DNA or RNA
  - Type II restriction endonucleases: Cleavage site 4, 6, or 8 base sequence with two-fold symmetry
  - Single strand or double strand

❖ DNA sequencing
- Chain termination (dideoxynucleotide): Sanger
  - Enzymatic synthesis of DNA using DNA polymerase
  - Primer extended using dNTP's but terminated using specific ddNTP's (lack 3' hydroxyl)

- Four reactions produce four sets of products
- Analyze products by electrophoresis

- DNA denaturation: Strand separation and loss of base stacking
  - Temperature, pH, ionic strength: Disrupt H bonds
  - Hyperchromic shift of UV absorbance upon denaturation: Absorbance increases 40%
  - T_m: Midpoint of transition
- Renaturation: Reannealing
  - Nucleic acid hybridization
    - Measure evolutionary relationships
    - Identify specific gene
- Supercoils: Release of tortional stress due to underwinding or overwinding of DNA
  - Negative supercoils: Underwound DNA
  - Positive supercoils: Overwound DNA
  - Natural DNA negatively supercoiled as a result of topoisomerase activity
- Chromosome structure: DNA and protein: Nucleoprotein complex called chromatin
  - Nucleosomes: Histone H2A, H2B, H3, H4 octamers with 146 bp of ds B-DNA wrapped in left-handed coil making about 1.65 turns
  - Histones: Basic: Arginine and lysine rich proteins
  - Histone H1: Link consecutive nucleosomes together
  - Solenoid: 30 nm Filaments: 6 Nucleosomes/turn
- Chemical synthesis of DNA
  - Solid phase synthesis using "gene machine"
  - Phosphoramidite chemistry
  - 3' to 5' Direction
- RNA structure: Typically single-stranded molecules: Intrastrand H bonding produces secondary structure
- RNA Classes
  - tRNA: Small (73 to 94 nucleotides) RNA charged with amino acids
    - Aminoacylated tRNA: Substrate for protein synthesis and decoding function
    - Cloverleaf secondary structure
      - Acceptor stem: NCCA 3': End at which amino acid is attached
      - D-loop: Modified base
      - Anticodon stem and loop: Three-base anticodon: Decodes mRNA
      - TψG loop: Modified base
  - rRNA: Component of ribosome: Protein synthetic machinery
    - Secondary structure used in phylogenetic comparisons: All ribosomes have common design and function
    - Tertiary structure: RNA structures: Coaxial stacking, pseudoknots, ribose zippers, tetraloops, internal loops

Chapter Objectives

Nucleotides

Understand the difference between a nucleoside and a nucleotide. The nucleotides are phosphorylated derivatives of nucleosides: compounds made up of a sugar and a nitrogenous base. (See Figure 8.11)

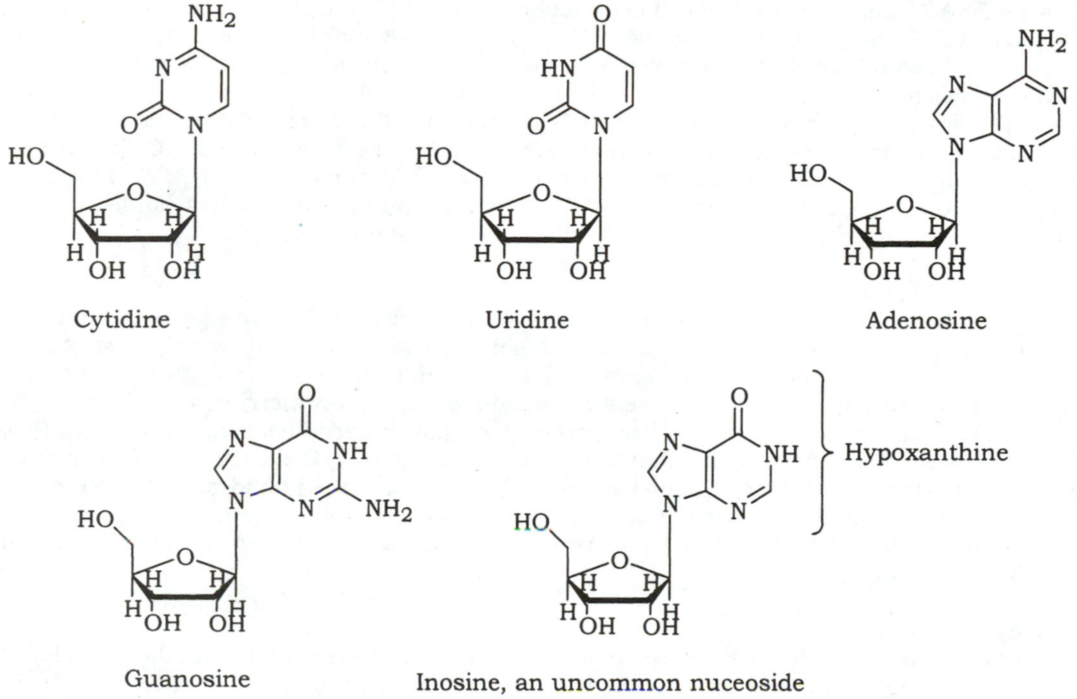

Cytidine

Uridine

Adenosine

Hypoxanthine

Guanosine

Inosine, an uncommon nuceoside

Figure 8.11 The common ribonucleosides - cytidine, uridine, adenosine, and guanosine- and inosine, an uncommon purine ribonucleotide.

It is important to know the structures of the pyrimidines and purines. Pyrimidines are six-membered heterocyclic, conjugated rings, whereas purines have a five-membered imidazole ring fused to a six-membered ring. (Remember: The small pyrimidine has a large name whereas the large purine has a small name.) The commonly occurring pyrimidines have two nitrogens separated by a carbonyl group. Uracil and thymine have an additional carbonyl group. Thymine is 5-methyluracil. Cytosine has an amino group but otherwise looks like uracil. Deamination of cytosine produces uracil. In the purine double ring, the carbon separating the two nitrogens of the six-membered ring section has a hydrogen in adenine and an amino group in guanine. Guanosine (2-amino-6-hydroxypurine) has a carbonyl group in the six-membered ring, which is substituted for an amino group in adenine (6-aminopurine).

You should be able to draw the structures of the bases and be able to identify hydrogen-bond donor and acceptor groups. Also, know the groups involved in base pairing in double-stranded DNA.

The compounds like ADP, ATP, GTP, UTP, CTP, NAD^+, FAD, and Coenzyme A are all either nucleotides or nucleotide derivatives of ribonucleotides. The phosphate groups are usually attached to the 5' carbon of the ribose sugar. The dNTP's are used as precursors for DNA.

Nucleic Acids

The convention for writing a nucleic acid sequence is to write, from left to right, the one-letter abbreviations of the bases from the 5'-end to the 3'-end.

The abbreviations most commonly found are A, C, G, T, U, N (any nucleotide, base unspecified), R (purine), and Y (pyrimidine).[1]

---

1 There is an extended code, which covers other combinations of bases. B specifies C, G, or T (i.e., anything but A). D indicates A, G, or T (i.e., not C). H is anything but G. V refers to A, C, or G. W signifies either A or T, the weak base-pair formers. S refers to G or C, the strong base-pair formers. A or C is denoted by M, for amino, and G or T is specified by K for keto.

Know what phosphodiester bonds are and that they join the 3'-hydroxyl group of a nucleotide to the 5' phosphate of an adjacent nucleotide in DNA and RNA. The single-stranded polymers thus formed have a 5'-end and a 3'-end. DNA is predominantly double-stranded, a fact reflected in Chargaff's rules.

The total DNA content of a haploid cell is its genome size. Diploid cells have two copies of DNA while haploid cells have a single copy of DNA. The discrete molecules of cellular DNA are chromosomes. Cells may have a single chromosome or several chromosomes. Chromosomes are circular DNA molecules in some organisms and linear DNA molecules in others.

RNA molecules are much less stable than DNA molecules. RNA is sensitive to base-catalyzed cleavage because of the presence of a hydroxyl group on the 2'-ribose carbon. Cells have an abundance of RNase's, which are often extremely stable, active enzymes. Finally, RNA is usually found as a single-stranded molecule whereas DNA is usually found in a double-stranded form in which bases are protected by being sequestered in the interior of the structure.

Special Endonucleases

The type II restriction endonucleases are DNA-specific nucleases that recognize palindromic sequences of DNA and cut within the sequence. Understand what is meant by palindromic DNA and that after restriction endonuclease cleavage, the resulting ends may be blunt-ended, or have either 5'- or 3'- overhangs. It is a rather easy task to identify an uninterrupted palindrome by inspection. Scan the sequence of bases for adjacent, base pair nucleotides (i.e., AT, TA, CG, GC). Once a pair is located, inspect the flanking two nucleotides; if they are also base-paired nucleotides, then they are part of a four-base palindrome. Six-base palindromes are common recognition sequences for a large number of restriction endonucleases.

The relationship between DNA, RNA, and protein is described in the central dogma outlined in Figure 8.1 shown below.

Sequencing Nucleic Acids

It is important to know how DNA is sequenced. For the chain termination method developed by F. Sanger, DNA is produced enzymatically using a template and a primer that defines the 5'-end of the DNA that is produced. As the primer is being extended, base specific chain-termination events produce a nested set of extension products. The sequence of the synthesized strand is determined by separating the products according to size using electrophoresis. The sequence of the synthesized strand is read from the gel or from an autoradiograph of the gel starting from short products and moving toward long products, reading 5' to 3'.

DNA Secondary Structure

Understand the structure of B-form double-stranded DNA. The two strands are antiparallel and are held together by complementary base pairs. The strands wrap around each other forming a right-handed helix with 10 base pairs per helix turn, 3.4 nm per turn, and 0.34 nm between base pairs along the helix axis. The helix axis runs perpendicular to the planar base pairs. Because the helix axis is centered within the base pairs, the helix has a major and a minor groove. Compared to B-form DNA, A-form DNA has 11 base pairs per helical turn and as a consequence the helix is broader, the base pairs are tilted with respect to the helix axis, and they are closer together. Double-stranded RNA and DNA-RNA hybrids assume the A-form conformation. Z-DNA is a left-handed structure requiring regions of alternating pyrimidine-purine sequence to form.

Solution Properties of DNA

Many techniques in genetic engineering and molecular biology exploit the properties of double-stranded DNA. For example, DNA solutions can be freed of protein by extraction with organic solvents such as liquid phenol or chloroform because DNA is a highly charged, polar molecule. It is not soluble and will therefore not partition into these immiscible organic solvents. To precipitate DNA, an alcohol like ethanol or propanol is added to an aqueous DNA solution. However, a counter ion (Na^+ or NH_4^+) must be present in order for the negatively-charged DNA to form a salt and precipitate. Many techniques require single-stranded DNA. DNA can be denatured by high temperatures or by treatment with urea or formamide to disrupt hydrogen bonds. Additionally, pH values in excess of 11.5 will disrupt hydrogen bonds.

Intracellular DNA

In general, cellular DNA's are extremely large molecules and, in order to fit inside a cell, DNA must be compacted. Prokaryotic organisms accomplish this, in part, by having circular DNA's that are supercoiled. Eukaryotic organisms take a different approach: DNA is associated with specific proteins, termed histones, into complexes known as nucleosomes. Nucleosomes are octamers of two each of histones H2A, H2B, H3, and H4 around which double-stranded DNA is wrapped to give the appearance of "beads-on-a-string". The nucleosomes are further organized into solenoids having six nucleosomes per turn to produce 30-nm filaments. The filaments are further organized to form miniband units of the chromosome. The structures actually seen by light microscopy during mitosis or meiosis are even more highly condensed.

RNA

RNA's are typically single-stranded molecules. However, they may have complementary regions that interact by base-pairing to form short stretches of A-form helix. An example of this is the cloverleaf secondary structure model of tRNA's. Know the general structure of a tRNA and be able to identify important regions such as the D-loop, anticodon, variable loop, TψC loop, and acceptor stem. Like proteins, RNA's have tertiary structure, and in tRNA, the interactions responsible for tertiary structure are made between residues far removed from each other in the primary sequence. In phylogenetic sequence comparisons, these residues have been identified as highly-conserved positions that are virtually identical in all tRNA's. Phylogenetic comparisons have been made with small subunit-ribosomal RNA's and with large-subunit RNA's. These large, stable RNA molecules play a key role in protein synthesis, a fundamental cellular reaction.

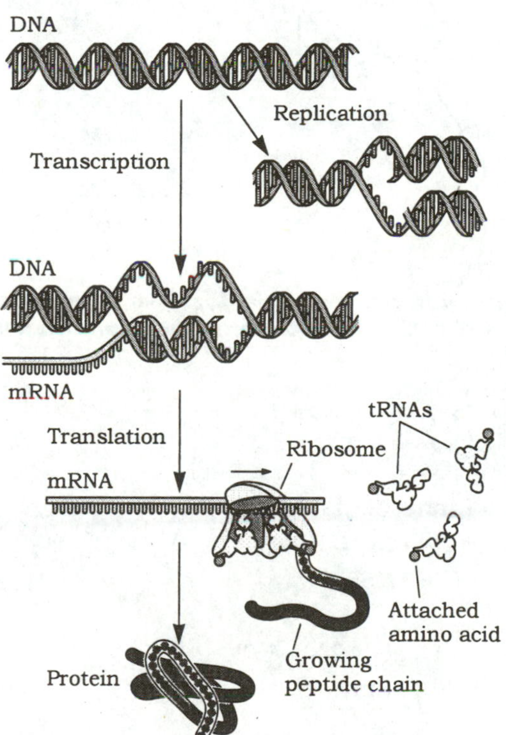

Replication
DNA replication yields two DNA molecules identical to the original one, ensuring transmission of genetic information to daughter cells with exceptional fidelity.

Transcription
The sequence of bases in DNA is recorded as a sequence of complementary bases in a single-stranded mRNA molecule.

Translation
Three-base codons on the mRNA corresponding to specific amino acids direct the sequence of building a protein. These codons are recognized by tRNAs (transfer RNAs) carrying the appropriate amino acids. Ribosomes are the "machinery" for protein synthesis.

Figure 8.1 The fundamental process of information transfer in cells. Information encoded in the nucleotide sequence of DNA is transcribed through synthesis of an RNA molecule whose sequence is dictated by the DNA sequence. As the sequence of this RNA is read (as groups of three consecutive nucleotides) by the protein synthesis machinery, it is translated into the sequence of amino acids in a protein. This information transfer system is encapsulated in the dogma: DNA→RNA→protein.

Problems and Solutions

1. Draw the chemical structure of pACG.

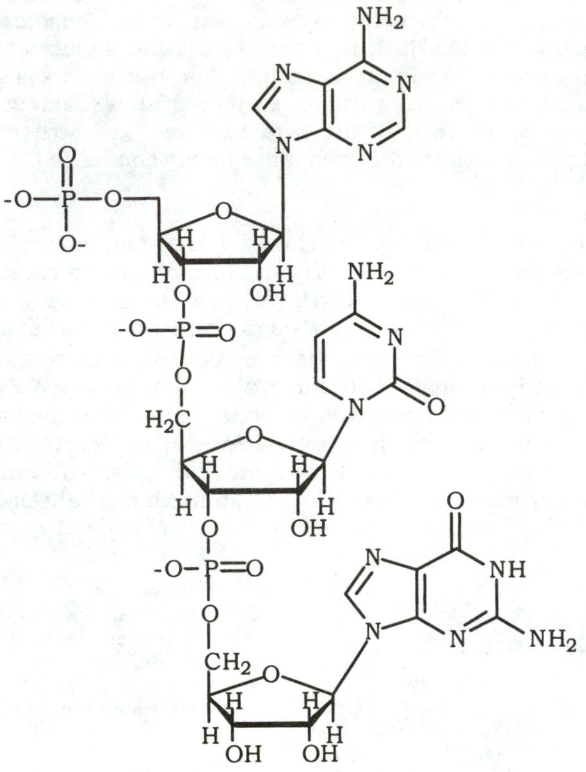

2. *Chargaff's results (Table 8.2) yielded a molar ratio of 1.56 for A to G in human DNA, 1.75 for T to C, 1.00 for A to T, and 1.00 for G to C. Given these values, what are the mole fractions of A, C, G, and T in human DNA?*

Answer: For human DNA: A/G = 1.56; T/C = 1.75; A/T = G/C = 1.

$$f_A + f_G + f_C + f_T = 1.0$$

where the terms are the mole fraction of each nucleotide.

In double - stranded DNA, $f_A = f_T$ and $f_G = f_C$ thus,

$2f_A + 2f_G = 1.0$ (or $2f_C + 2f_T = 1.0$)

By substituting $f_A = 1.56f_G$ (or $f_T = 1.75f_C$)

We find :

$2 \times 1.56f_G + 2f_G = 1.0$ ($2 \times 1.75f_C + 2f_C = 1.0$)

$f_G = 0.195$ ($f_C = 0.182$)

And,

$f_A = 1.56 \times f_G = 1.56 \times 0.195$ (or $f_T = 1.75 \times f_C = 1.75 \times 0.182$)

$f_A = 0.304$ (or $f_T = 0.318$)

Note : $\frac{f_G}{f_C} = 1.07$ and $\frac{f_A}{f_T} = 0.96$

3. *Adhering to the convention of writing nucleotide sequences in the 5' → 3' direction, what is the nucleotide sequence of the DNA strand that is complementary to d-ATCGCAACTGTCACTA?*

Answer:

5'-TAGTGACAGTTGCGAT-3'

Note: The complementary sequence written ‘5' to 3' is sometimes referred to as the reverse complementary sequence. (As an example, consider the sequence GAGGCTT. Its reverse is TTCGGAG i.e., the sequence literally written in reverse. For the reverse sequence by exchanging each base for its complementary base (i.e., A for T and G for C) we have AAGCCTC, which is the strand that is complementary to GAGGCTT but written 5' to 3'. When the reverse complementary sequence is so defined the complementary sequence is defined literally. Thus, the complementary sequence of GAGGCTT is CTCCGAA.)

4. Messenger RNA's are synthesized by RNA polymerases that read along a DNA template strand in the 3'→5' direction, polymerizing ribonucleotides in the 5'→3' direction (see Figure 8.22). Give the nucleotide sequence (5'→3') of the DNA template strand from which the following mRNA segment was transcribed:

5'-UAGUGACAGUUGCGAU-3'.

Answer: The DNA template strand is complementary to the mRNA but with T's in place of U's. For a mRNA of sequence

5'-UAGUGACAGUUGCGAU-3'

the complementary DNA sequence is

5'-ATCGCAACTGTCACTA-3'

5. The DNA strand that is complementary to the template strand copied by RNA polymerase during transcription has a nucleotide sequence identical to that of the RNA being synthesized (except T residues are found in the DNA strand at sites where U residues occur in the RNA). An RNA transcribed from this nontemplate DNA stand would be complementary to the mRNA synthesized by RNA polymerase. Such an RNA is called antisense RNA. A promising strategy to thwart the deleterious effects of genes activated in disease states (such as cancer) is to generate antisense RNA's in affected cells. These antisense RNA's would form double-stranded hybrids with minas transcribed from the activated genes and prevent their translation into protein. Suppose transcription of a cancer-activated gene yielded an mRNA whose sequence included the segment 5'-UACGGUCUAAGCUGA. What is the corresponding nucleotide sequence (5'→3') of the template strand in a DNA duplex that might be introduced into these cells so that an antisense RNA could be transcribed from it?

Answer: For a mRNA 5'-UACGGUCUAAGCUGA-3', what is the corresponding sequence of a template strand in DNA that might make antisense RNA? Antisense RNA has a sequence complementary to the mRNA shown above. It would have to be transcribed then from a DNA sequence complementary to itself. Since the antisense RNA has a sequence complementary to the mRNA, its DNA template strand must have a sequence identical (except T for U) to the original mRNA:

5'-TACGGTCTAAGCTGA-3'

In the spring of 1998 Phase III clinical trials for an antisense drug for the treatment of cytomegalovirus retinitis in AIDS patients were completed and a new drug application for this antisense drug was filed with the FDA.

6. A 10-kb DNA fragment digested with restriction endonuclease EcoRI yielded fragments 4 kb and 6 kb in size. When digested with BamHI, fragments 1, 3.5, and 5.5 kb were generated. Concomitant digestion with both EcoRI and BamHI yielded fragments 0.5, 1, 3, and 5.5 kb in size. Give a possible restriction map for the original fragment.

Answer:

EcoRI produces 4 kb and 6 kb fragments giving two possibilities:

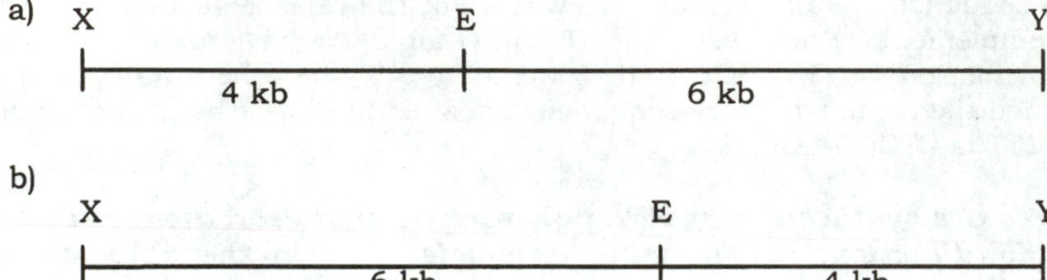

BamHI produces 1.0, 3.5, and 5.5 kb fragments giving six possibilities:

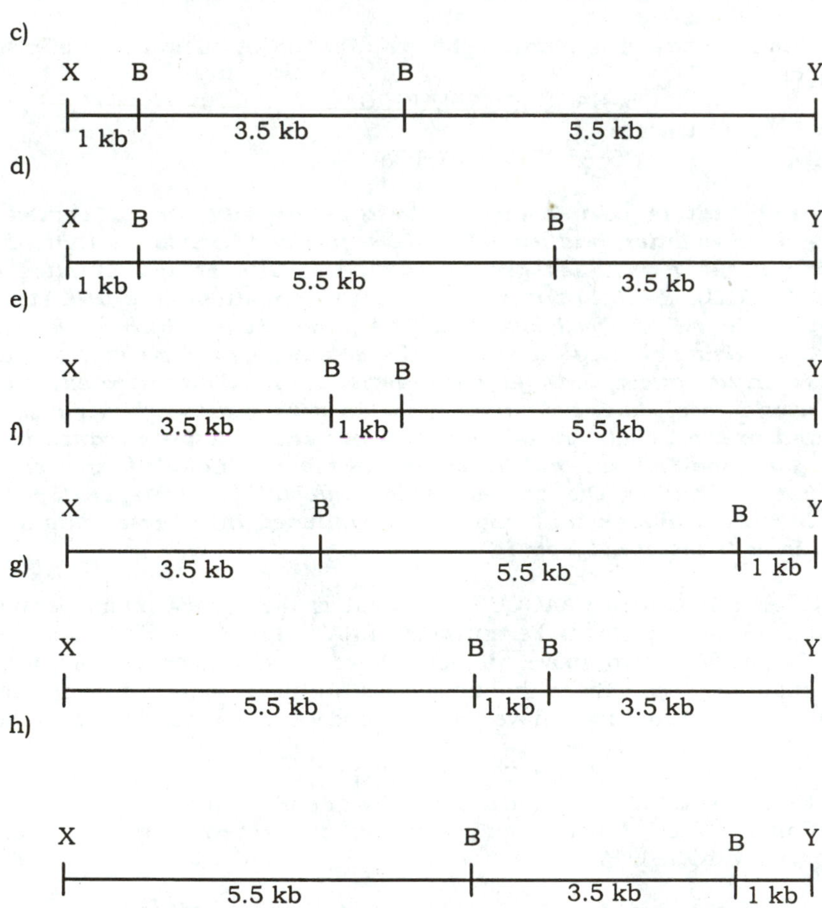

The double digest produces 0.5, 1.0, 3.0, and 5.5 kb fragments. One way to think about this problem is to consider what must happen to the digestion products of one enzyme upon treatment with the other enzyme to generate the final products. For example, consider the EcoRI products. To generate a 5.5 kb fragment from them there must be a BamHI site on the 6 kb EcoRI fragment and it must lie within 0.5 kb of either end. The 4.0 kb fragment must have a BamHI site located 1.0 kb from one end of it. The final map can be constructed by identifying the ways these fragments must be joined to be consistent with the data. The only possibilities are:

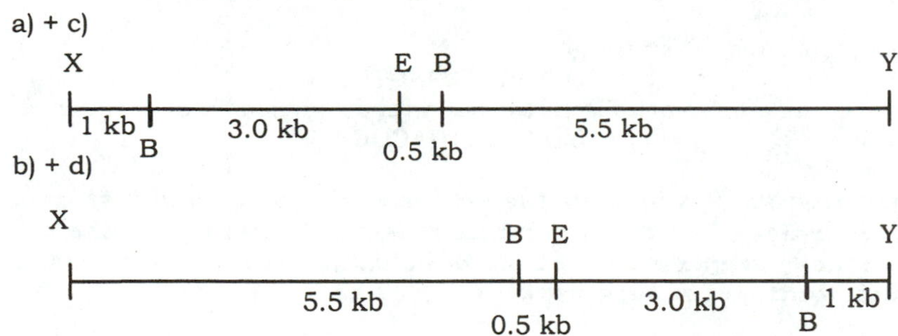

7. The oligonucleotide d-ATGCCTGACT was subjected to sequencing by Sanger's dideoxy method and the products were analyzed by electrophoresis on a polyacrylamide gel. Draw a diagram of the gel banding patterns obtained.

Answer: In order to sequence the oligonucleotide d-ATGCCTGACT by the Sanger dideoxy method, reactions, are set up such that the oligonucleotide serves as a template for primer extension. As the primer is extended (using dNTP's and a DNA polymerase) sequence-specific terminations occur, producing sets of extended primers of different lengths. The sequence actually read in Sanger sequencing is the sequence of the extended primer strand (which is complementary to the template strand).

ddG ddA ddT ddC

Product sequence	Template sequence
3'	5'
T	A
A	T
C	G
G	C
G	C
A	T
C	G
T	A
G	C
A	T
5'	3'

8. The result of sequence determination of an oligonucleotide as performed by the Sanger dideoxy chain termination method is displayed below:

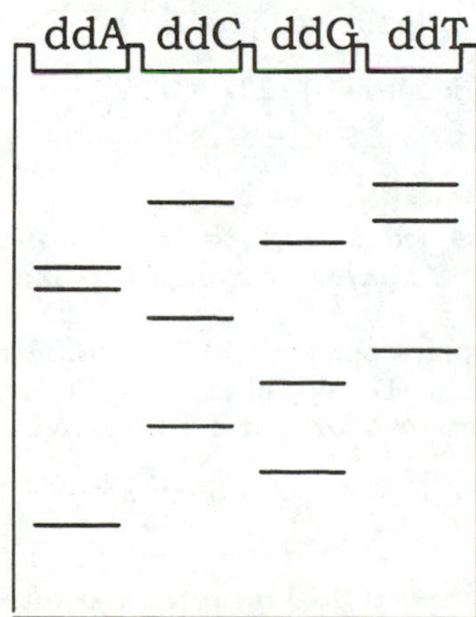

What is the sequence of the original oligonucleotide?

Answer: The sequence read from the gel is:

5'-AGCGTCAAGTCT-3'

It was generated using a template oligonucleotide with a complementary sequence:

5'-AGACTTGACGCT-3'

9. X-ray diffraction studies indicate the existence of a novel double-stranded DNA helical conformation in which ΔZ (the rise per base pair) = 0.32 nm and P (the pitch) = 3.36 nm. What are the other parameters of this novel helix: (a) the number of base pairs per turn, (b) Δ ø (the mean rotation per base pair), and (c) c (the true repeat)?

Answer: For dsDNA with $\Delta Z = 0.32$ nm/base pair, P = 3.36 nm/turn

(a) The number of base pairs per turn is

$$\frac{3.36\frac{\text{nm}}{\text{turn}}}{0.32\frac{\text{nm}}{\text{base pair}}} = 10.5 \text{ per turn}$$

(b) The mean rotation per base is

$$\frac{360\frac{\text{deg}}{\text{turn}}}{10.5\frac{\text{base pairs}}{\text{turn}}} = 34.3^\circ\text{per turn}$$

(c) To determine the true repeat, first find what integer values of X and Y that satisfy the following equation:

$$\text{X turns} \times 10.5\frac{\text{base pairs}}{\text{turn}} = \text{Y base pairs}$$

After two turns and 21 base pairs, the helix repeats itself. The true repeat is

$$2 \text{ turns} \times 3.36\frac{\text{nm}}{\text{turn}} = 6.72 \text{ nm}$$

10. A 41.5 nm-long duplex DNA molecule in the B-conformation adopts the A-conformation upon dehydration. How long is it now? What is its approximate number of base pairs?

Answer:

B-DNA, P = 3.40 nm, $\Delta Z = 0.34$ nm, 10 base pairs/turn

A-DNA, P = 2.46 nm, $\Delta Z = 0.224$ nm, 11 base pairs/turn

For B-DNA, 41.5 nm has

$$\frac{41.5\text{nm}}{0.34\frac{\text{nm}}{\text{base pair}}} = 122 \text{ base pairs that make } \frac{122 \text{ bp}}{10\frac{\text{bp}}{\text{turn}}} = 12.2 \text{ turns}$$

If converted to A-DNA, these 122 base pairs will now make:

$$\frac{122 \text{ bp}}{11\frac{\text{bp}}{\text{turn}}} = 11.1 \text{ turns with an overall length of } 122 \text{ bp} \times 0.224\frac{\text{nm}}{\text{bp}} = 27.3\text{nm } (= 11.1 \text{ turns} \times 2.46\frac{\text{nm}}{\text{turn}})$$

11. If 80% of the base pairs in a duplex DNA molecule (12.5 kbp) are in the B-conformation and 20% are in the Z-conformation, what is the length of the molecule?

Answer: Z-DNA contains 12 base pairs per turn, 0.36 to 0.38 nm rise per base pair and a pitch of 4.86 nm. If 20% of a 12.5 kbp DNA molecule is in Z-conformation and the rest is B-conformation this represents 2.5 kbp in Z-DNA and 10.0 kbp in B-DNA. The overall length is:

$$2.5 \text{ kbp} \times 0.37\frac{\text{nm}}{\text{bp}} + 10.0 \text{ kbp} \times 0.34\frac{\text{nm}}{\text{bp}} = 4{,}325 \text{ nm} = 4.33 \ \mu\text{m}$$

12. There is one nucleosome for every 200 bp of eukaryotic DNA. How many nucleosomes are in a diploid human cell? Nucleosomes can be approximated as disks 11 nm in

diameter and 6 nm long. If all the DNA molecules in a diploid human cell are in the B-conformation, what is the sum of their lengths? If this DNA is now arrayed on nucleosomes in the "beads-on-a-string" motif, what is its approximate total length?

Answer: Human diploid cells consist of 46 chromosomes and 6 x 10^9 bp. The number of nucleosomes is given by:

$$\frac{6\times10^9\text{ bp}}{200\frac{\text{bp}}{\text{nucleosome}}} = 3\times10^7\text{ nucleosomes}$$

The length of B-DNA, 6 x 10^9 bp long, is given by

$$6\times10^9\text{bp}\times0.34\frac{\text{nm}}{\text{bp}} = 2.04\times10^9\text{ nm} = 2.04\text{ meters in length!}$$

This DNA as 3×10^7 bp nucleosomes has a length given by :

$$3\times10^7\text{bp nucleosomes}\times6\frac{\text{nm}}{\text{nucleosome}} = 18\times10^7\text{ nm} = 0.18\text{ m} = 180\text{ mm}$$

13. ***The characteristic secondary structures of tRNA and rRNA molecules are achieved through intrastrand hydrogen bonding. Even for the small tRNA's, remote regions of the primary sequence interact via H bonding when the molecule adopts the cloverleaf pattern. Using Figure 8.51 as a guide, draw the primary structure of tRNA, indicating the positions of various self-complementary regions.***

Answer:

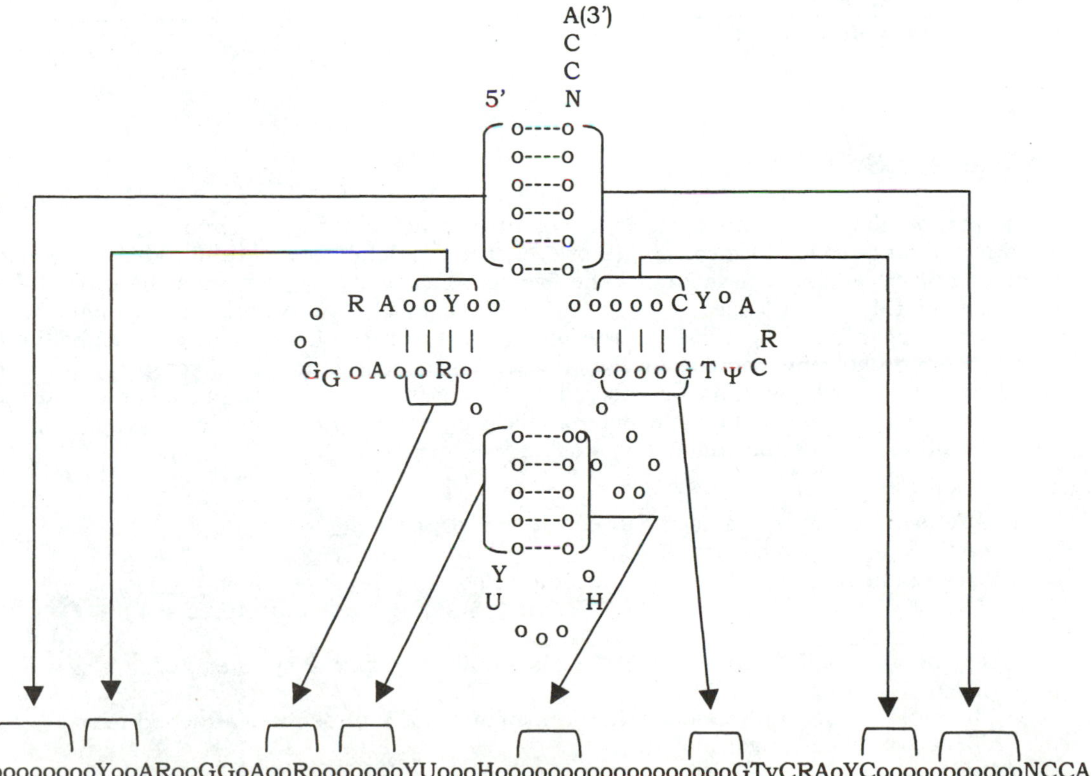

14. ***Using the data in Table 8.2, arrange the DNA's from the following sources in order of increasing T_m: human, salmon, wheat, yeast, E. coli.***

Answer:

The T_m of DNA is proportional to its %(G + C) content which is calculated as:

$$\%(G+C) = \frac{G+C}{G+C+A+T} \times 100\% \text{ but since } G = C \text{ and } A = T$$

$$\%(G+C) = \frac{G}{G+A} \times 100\% \text{ or } = \frac{C}{C+T} \times 100\%$$

From Table 8.2 we are given various molar ratios.

For dsDNA the Adenine to Guanine ratio should equal the Thymine to Cytosine ratio.

By inspecting Table 8.2 we see that this is not always the case.

Therefore, we will calculate %(G + C) using $\frac{A}{G}$ and $\frac{T}{C}$ ratios separately.

Let $\frac{A}{G} = x$ and $\frac{T}{C} = y$

$$\%(G+C) = \frac{1}{1+x} \times 100\% \text{ or } = \frac{1}{1+y} \times 100\%$$

Substituting values of x (first column) or y (second column) we find:

Organism	x	y	%(G+C)
Human	1.56	1.75	39-36
Salmon	1.43	1.43	41
Wheat	1.22	1.18	45-46
Yeast	1.67	1.92	37-34
E. coli	1.05	0.95	49-51

Order of T_m: Yeast < Human < Salmon < Wheat < *E. coli*

Questions for Self Study

1. Fill in the blanks. The two basic kinds of nucleic acids are _____ and ____. They are composed of building blocks termed _____; however, the building blocks are not identical for the two kinds of nucleic acids. One contains the five-carbon sugar _____ whereas the other has a modified form of this sugar called _____. The building blocks all contain nitrogenous bases attached to the sugar by _____ bonds. The nitrogenous bases are either derivatives of the 6-membered heterocyclic ring compound _____ or of purines, a compound composed of a 6-membered heterocyclic ring with a 5-membered _____ ring fused to it. The two common purines are _____ and _____. The 6-membered heterocyclic ring compounds include _____, _____, and _____. A compound with a base attached to a sugar is termed a _____.

2. Answer True or False
   a. ATP is an example of a deoxynucleoside triphosphate _____.
   b. cAMP is a 3'-5' cyclic form of AMP _____.
   c. The α phosphate of GTP is the phosphate closest to the sugar moiety _____.
   d. The only biological function of dCTP is as a building block in synthesis of DNA _____.
   e. The only biological function of CTP is as a building block in synthesis of RNA _____.
   f. The most common ribonucleoside triphosphates have phosphate attached to the 5' carbon of the sugar moiety _____.

3. Chargaff's rules provided an important clue to solve the structure of double-stranded DNA. What are Chargaff's rules?

4. Answer True of False
   a. An A/T base pair and a C/G base pair have about the same physical dimensions _____.

b. If GGGGCCCC represents the sequence of bases in one strand of a double-stranded DNA then the complementary strand must have the sequence CCCCGGGG _____
c. mRNA is single-stranded _____
d. Heterogeneous nuclear RNA or hnRNA are RNA molecules made in the nucleus and processed into mRNA _____
e. rRNA and tRNA are devoid of unmodified nucleosides.

5. DNA and RNA react differently to acid and base conditions. Explain.

6. Of the following statements, which are true for type II restriction enzymes?
a. They are usually exonucleases.
b. Their recognition sequences are usually palindromic.
c. Cleavage is by hydrolysis of both strands.
d. Cleavage produces 3'-phosphates and 5' hydroxyl groups.
e. A single restriction enzyme can produce blunt ends or protruding ends depending on the salt conditions.

7. For chain termination or dideoxy DNA sequencing answer True or False to the following statements
a. When analyzed by electrophoresis, products all have identical 5' ends but different 3' ends _____
b. Sequence of template strand is determined directly by reading the gel from bottom to top _____
c. Requires the use of DNA polymerase _____
d. Chain termination occurs because the 5' end of an extended primer lacks an hydroxyl group _____
e. Products labeled by fluorescent compounds contain these compounds throughout their length _____
f. Products are complementary to the template strand whose sequence is being determined _____

8. Double stranded DNA is composed of canonical base pairs.
a. What are they?
b. How are the two strands of dsDNA oriented relative to each other?
c. In B-form DNA the helix axis is perpendicular to the approximate center of the base pairs and is not in line with the points of attachment of the bases to the sugar. What is the consequence of this on the structure of the helix?

9. Answer the following questions with A (A-form DNA), B (B-form DNA) or Z (Z-DNA).
a. The double helical form adopted by double stranded RNA _____
b. Left-handed helix _____
c. Approximately 10 base pairs per turn _____
d. Plane of base pair perpendicular to helix axis _____
e. Accommodates syn glycosyl bond conformation _____
f. Is stabilized by supercoiling _____
g. Formation is sequence dependent _____
h. Approximately 11 base pairs per turn _____
i. Helix axis runs through the base pairs _____
j. Helix axis runs through the major groove _____

10. Fill in the blanks. The transition from dsDNA to single stand DNA is called _____. This process occurs when DNA is subjected to conditions such as _____ or _____ that disrupt _____ bonds. The transition is accompanied by an _____ in the UV absorbance, or a _____ shift, caused by _____ of bases. The temperature at which one-half of the dsDNA has converted to ssDNA is termed the _____ temperature. This temperature is dependent on the solution conditions. For example, as salt concentration increases, the transition temperature _____. High values of pH and high concentrations of urea will _____ the transition temperature. The reverse process, namely the transition from ssDNA to dsDNA is termed _____.

11. Order the following terms with respect to structural complexity: miniband, chromosome, histone HI, nucleosome, solenoid, base pair, histone octamer.

12. On the diagram of the secondary structure of tRNA shown below indicate the location of the following features:

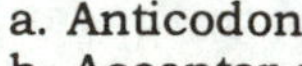

a. Anticodon
b. Acceptor stem
c. CCA
d. D loop
e. Variable loop
f. Location of highly conserved bases involved in tertiary interactions
g. Anticodon stem
h. 5' end
I. 3' end

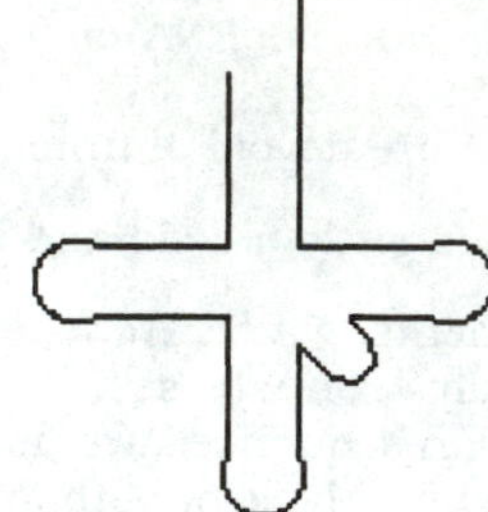

Answers

1. DNA; RNA; nucleotides (or nucleoside monophosphates); ribose; deoxyribose; glycosidic; pyrimidine; imidazole; adenine; guanine; cytosine; uracil; thymine; nucleoside.

2. a.F; b.T; c.T; d.T; e.F; f.T.

3. [A]= [T]; [G] = [C]; [pyrimidines] = [purines]

4. a.T; b.F (sequences are always written 5' to 3'); c.T; d.T; e.F.

5. RNA is relatively resistant to dilute acid whereas DNA undergoes hydrolysis of glycosidic bonds to purines. DNA is not susceptible to alkaline hydrolysis whereas RNA is readily hydrolyzed to nucleotides in alkaline solution.

6. b and c.

7. a.T; b.F; c.T; d.F; e.F; f.T.

8. a. AT and GC; b. The two strands are antiparallel.; c. The helix has a major groove and a minor groove.

9. a. A; b. Z; c. B; d. B; e. Z; f. Z; g. Z; h. A; I. B; j. A

10. denaturation (or melting); pH; temperature (or changes in ionic strength, urea concentration, formamide concentration); hydrogen; increase; hyperchromic; unstacking; melting (or transition or T_m); increases; decrease; renaturation (or reannealing).

11. base pair < histone HI < histone octamer < nucleosome < solenoid < miniband < chromosome.

12.

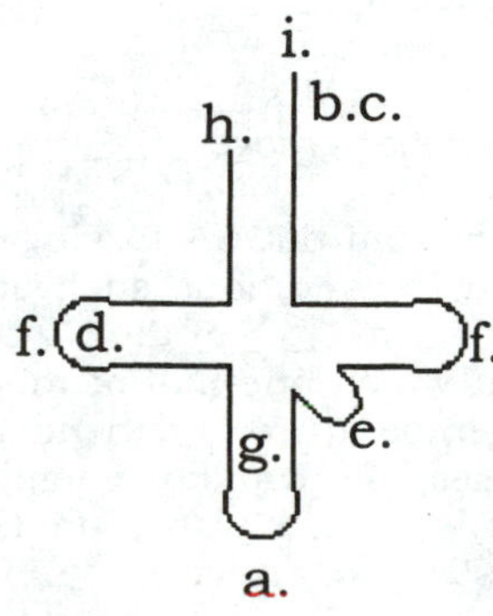

Additional Problems

1. Nucleotides are an important class of biomolecules used as components of the nucleic acids, DNA and RNA. Describe the structure of the nucleoside monophosphates found in DNA and RNA. In your description be sure to describe the three chemical groups that make up a nucleotide and be certain to indicate any difference between deoxyribonucleotides and ribonucleotides.

2. Draw a base pair involving either G and C or A and T.

3. DNA methylation is known to most commonly occur at position 6 in A and position 5 in C. However, the formation of 5-methylcytosine can create so-called mutational hot-spots. Cytosine can undergo deamination at position 4. In the unmethylated form, this deamination can be corrected, because it is easily recognized. Deamination of 5-methylcytosine, however, can create problems. Explain.

4 In the following sequence of DNA, underline a 4-base palindrome, a 6-base palindrome and a purine-rich sequence.

GAGAAATATAGATCAGAGTTAACTC

5. In the following compounds, the order of solubility is adenine < deoxyadenosine < adenosine < adenosine triphosphate. Explain what each is (i.e. base, nucleoside or nucleotide), how they differ and why they show this relative solubility.

6. When a solution of double-stranded DNA is heated, a sharp transition is observed in the ultraviolet absorption properties of the solution. The solution begins to absorb greater amounts of light at high temperatures, corresponding to the process of denaturation. What physical changes in the double-stranded DNA molecules occur during this process of denaturation?

7. The restriction endonuclease *Bam*HI recognizes the sequence GGATCC and cleaves between the Gs. The enzyme *Bgl* II ("Bagel two") recognizes AGATCT and cuts between AG. What kinds of overhangs are generated (i.e., 5' or 3')? Are the overhangs complementary to one another? If the cleavage products are joined together (i.e., a *Bam*HI fragment joined to a *Bgl* II fragment), is the hybrid molecule cleavable by either endonuclease?

Abbreviated Answers

1. The nucleoside monophosphates found in DNA and RNA are composed of a phosphate group, a sugar moiety, and a nitrogenous base. The sugar is a 5-carbon compound, deoxyribose for DNA and ribose for RNA. The phosphate group is attached to the 5'-carbon of the sugar. Attached to the 1'-carbon is a nitrogenous base. Bases are either purines, adenine or guanine, or pyrimidines, cytosine in DNA and RNA, or thymine in DNA only or uracil in RNA only. (Thymine is 5-methyluracil.)

2.

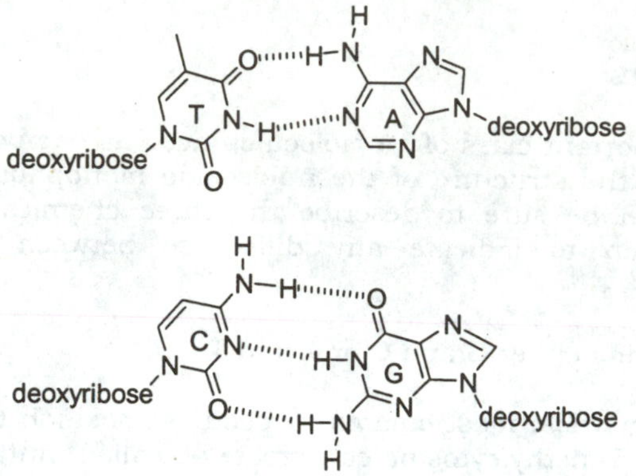

3. Deamination of cytosine produces uracil, a base not commonly found in DNA. Cells have a repair system that scans DNA for uracils and removes them. 5-Methylcytosine causes problems for this repair process. Deamination of 5-methylcytosine produces 5-methyluracil also known as thymine. Thymine is a natural component of DNA.

4. Four-base palindromes are indicated by ~~strikethrough~~ or double ~~strikethrough~~, six-base palindrome is underlined, and purine-rich region is in bold.

GAGA~~**AA**TATAGATCA~~GAGTTAACTC

5. Adenine is a base. In general, the bases are poorly soluble in aqueous solution. Deoxyadenosine is a nucleoside composed of adenine and the 5-carbon sugar deoxyribose. Adenosine is a ribonucleoside containing the 5-carbon ribose sugar. Both deoxyadenosine and adenosine are more soluble than adenine by virtue of the hydroxyl groups on their sugar moieties. Adenosine triphosphate is ATP, a ribonucleoside triphosphate. The presence of the triphosphate group greatly increases solubility.

6. The forces holding double-stranded DNA together include hydrogen bonds and base stacking. Denaturation of double-stranded DNA results in strand-separation and unstacking of bases. Base stacking occurs by an interaction of π-electrons located above and below the base-planes. One of the consequences of base-stacking is a decrease in the molar extinction coefficient. Stacked bases have a lower extinction coefficient than unstacked bases. Thus, denaturation is accompanied by an increase in ultraviolet light absorption, a phenomenon known as the hyperchromic effect. The ultraviolet absorbance of denatured DNA is approximately 40% higher than that of native DNA.

```
   ↓
5'-GGATCC-3'          5'-G            5'-GATCC-3'
3'-CCTAGG-5'   ——→    3'-CCTAG-5'  +       G-5'
        ↑
```

7. *Bam*HI produces 5'-overhangs as shown below:

```
   ↓
5'-AGATCT-3'          5'-A            5'-GATCT-3'
3'-TCTAGA-5'   ——→    3'-TCTAG-5'  +       A-5'
        ↑
```

Bgl II also produces 5'-overhangs as shown below:

The overhangs are compatible but the hybrid site formed by joining a *Bam*HI half site with a *Bgl*II half site is not recognized by either. The junction is no longer a 6-base palindrome.

```
5'-G               5'-GATCT-3'          5'-GGATCT-3'
3'-CCTAG-5'   +         A-5'    ——→     3'-CCTAGA-5'

  BamHI              BglII                 Hybrid
half-site          half-site                site
```

Summary

The nucleic acids, ribonucleic acid (RNA) and deoxyribonucleic acid (DNA), are important biopolymers of nucleotides, compounds containing nitrogenous bases, a five-carbon sugar (ribose or deoxyribose) and phosphate. This chapter describes the basic biochemistry of nucleic acids and nucleotides. The nitrogenous bases come in two types, pyrimidines and purines. Pyrimidines are 6-membered, heterocyclic, aromatic ring structures containing two nitrogens. There are three principal pyrimidines: cytosine, uracil and 5-methyluracil or thymine. Cytosine is found in both DNA and RNA whereas uracil is in RNA and thymine in DNA. The general structure of a purine is a 5-membered imidazole ring fused to a pyrimidine ring. In DNA and RNA, there are two common purines, namely adenine and guanine. Both purines and pyrimidines contain numerous groups that can participate in hydrogen bonds as donors or acceptors or both. In fact, complementary groups exist on adenine and uracil (or thymine) and on guanine and cytosine such that they can form hydrogen-bonded pairs. This base-pairing is the foundation for the structure of double-stranded DNA.

Because the bases have extensive, conjugated double bonds, they absorb light strongly in the UV range. The conjugated bond system allows delocalization of π-electrons, forming electron clouds above and below the base plane. Delocalized π-electrons can interact by base stacking, another force stabilizing double-stranded DNA. Base stacking also affects the efficiency of electronic transitions of π-electrons such that stacked bases absorb less UV light than unstacked bases. The transition from an ordered nucleic acid structure stabilized by hydrogen bonds and base stacking to an unordered structure is accompanied by an increase in UV absorbance. This transition is known as denaturation and it results in a hyperchromic shift in UV absorption.

Purines and pyrimidines have low water solubility which improves when they are attached to either ribose or deoxyribose sugars through N-glycosidic bonds. The resulting compounds are known as nucleosides. The common nucleosides are cytidine, uridine, thymidine, adenosine and guanosine. Phosphorylated derivatives of nucleosides are known as nucleotides, with phosphoric acid esterified normally to the 5' carbon of the sugar. DNA and RNA are polymers of deoxyribonucleoside monophosphates and ribonucleoside monophosphates, respectively. But, nucleoside monophosphates are not used directly to biosynthesize DNA and RNA. Rather, triphosphate derivatives serve this purpose. The nucleotides used to biosynthesize DNA are deoxyadenosine 5'-triphosphate (dATP); deoxyguanosine 5'-triphosphate (dGTP); deoxycytosine 5'-triphosphate (dCTP); and, deoxy thymidine 5'-triphosphate (TTP or dTTP). The corresponding nucleoside triphosphates for RNA are adenosine 5'-triphosphate (ATP); guanosine 5'-triphosphate (GTP); cytosine 5'-triphosphate (CTP); and uridine 5'-triphosphate (UTP).

The deoxyribonucleoside triphosphates are used exclusively as building blocks for DNA. However, the ribonucleoside triphosphates, in addition to serving as the building blocks for RNA, have additional uses. ATP is the major energy currency of the cell, in addition to being a component of coenzymes such as NAD^+ and FAD. GTP is used during protein synthesis and in cell signaling; CTP is involved in phospholipid synthesis; and, UTP is involved in various aspects of carbohydrate metabolism.

Nucleic acids are linear polymers of nucleoside monophosphates linked by phosphodiester bonds between the 3'-hydroxyl of one nucleotide and the 5' phosphate of another. Because of this, nucleic acids are vectorial molecules, with two distinct ends, the 5'-end and the 3'-end. By convention, the sequence of nucleotides in a nucleic acid is represented by the one-letter abbreviations of the bases starting from the 5'-end and continuing toward the 3'-end.

The only biological role of DNA is as genetic material, and the vast majority of organisms use double-stranded DNA for this purpose. In double-stranded DNA (dsDNA), two strands of deoxyribonucleoside monophosphate polymers are joined together in an antiparallel fashion by hydrogen bonds. The hydrogen bonds occur between pairs of complementary bases: A and T; and, G and C. Compositional analysis of dsDNA reveals: [A] = [T]; [C] = [G]; and, [purine] = [pyrimidine]. These relationships are known as Chargaff's rules. Depending on the organism, genomic DNA may be a single DNA molecule (or chromosome) or may be divided into several discrete DNA molecules (chromosomes).

RNA serves a number of biological roles including informational, catalytic and structural. As an informational molecule, messenger RNA functions to bring genetic information, encoded in a sequence of bases in DNA, to the ribosome, the site of cellular protein synthesis. This information is used to direct the sequence of amino acids to be joined to form a protein. mRNA production begins with transcription, in which an RNA copy of a sequence of bases along one strand of DNA is made. In eukaryotic organisms, RNA transcription is localized to the nucleus. The primary transcripts, the initial products of transcription also known as heterogeneous nuclear RNA (hnRNA), are processed into mRNA in a series of reactions. Processing includes (1) Addition of a G residue to the 5'-end of the primary transcript in a process known as capping; (2) Cleavage of the primary transcripts to produce a shortened 3'-end at which adenylic acid residues are added to form polyA tails; and, (3) Removal of various internal sequences (introns or intervening sequences) by cleavage and subsequent ligation of exons).

The process of protein synthesis reveals structural and catalytic aspects of RNA's. Ribosomes are ribonucleoprotein complexes of ribosomal proteins and unique RNA molecules known as ribosomal RNA (rRNA). The amino acids used by the ribosome during protein synthesis are attached to the 3'-end of small RNA molecules known as transfer RNA's (tRNA). Transfer RNA's are used to decode a sequence of three adjacent nucleotides (a codon) in terms of a unique amino acid. In this process, tRNA's form hydrogen bonds between codons on mRNA and a three-base sequence, the anticodon, on tRNA. In addition to tRNA's, cells contain other relatively small, stable RNA molecules. In particular, a class of RNA molecules known as small nuclear RNA's or snRNA's are responsible for mRNA processing in the nucleus of eukaryotic cells.

Apart from the number of strands, there are only minor differences between DNA and RNA: deoxyribose versus ribose sugars, and, thymine versus uracil. The absence of a 2'-OH group in deoxyribose makes DNA stable against base-catalyzed hydrolysis. In fact, DNA is a very stable molecule ideally suited as genetic material. The choice between thymine and uracil arises as a result of the tendency of cytosine to deaminate to uracil. Since thymine is found in DNA, any uracil in DNA must result from deamination of cytosine. Cells have enzymes to remove uracil in DNA, replacing it with thymine and thus preventing mutations.

There is a large number of enzymes, termed nucleases that catalyze cleavage of phosphodiesterase bonds by hydrolysis. These include DNA-specific enzymes (DNase's), RNA-specific enzymes (RNase's) or nonspecific nucleases that attack either internal phosphodiester bonds (endonucleases) or phosphodiester bonds on the end of a polymer (5'-exonucleases or 3'-exonucleases). Further, the phosphodiester bond may be attacked on the *a* side producing a 5'-phosphate or on the *b* side producing a 3'-phosphate. Restriction endonucleases are DNA-specific endonucleases. The type II restriction endonucleases recognizes specific sequences, often palindromic, and cleave within the sequence. Type II restriction endonucleases are important tools in genetic engineering and molecular biology.

The primary structure of a nucleic acid is the sequence of base. By convention the sequence is reported starting at the 5'-end. One important procedure available for determining the primary structure, or nucleotide sequence, of DNA is the Sanger chain termination method. In the chain termination method, a DNA strand of unknown sequence is enzymatically copied in four separate reaction mixtures using DNA polymerase, dATP, dGTP, dCTP and dTTP, plus one of the four possible dideoxynucleoside triphosphate analogs. Incorporation of a dideoxynucleotide into the growing chain terminates subsequent extension of the chain. Represented among the products in the four reaction mixtures is a "ladder" of DNA strands, each one residue longer than the next. The sequence is revealed by considering which dideoxy analog specifically ended the strand. Analysis of the fragments by electrophoretic separation discloses the sequence. Modified chain termination chemistry is used in automated instruments for rapid sequencing of DNA in which dideoxynucleotides derivatized with fluorescent groups are used to determine the DNA sequence.

The double helix is the dominant secondary structure of DNA. Two DNA strands of opposite polarity and complementary base sequence are paired through formation of interchain hydrogen bonds. Exclusive pairing of A only with T and G only with C creates spatially equivalent units, the canonical, or "Watson-Crick", A:T and G:C base pairs. The double helix is stabilized not only by these hydrogen bonds between base pairs, but also by π,π interactions as the flat hydrophobic faces of the aromatic base pairs stack upon one another down the helix center. Hydrogen bonds between the polar sugar-phosphates of the two strands and the surrounding water contribute to helix stability. The negatively charged phosphates along the backbone are localized to the exterior of the helix where they can enter stabilizing ionic interactions with cations in solution. The prevalent double-helical form of DNA in solution is the B conformation, in which the base pairs are arranged virtually at right angles to the helix axis, and each base pair circumscribes an angle of 36° around the helix. Thus there are about 10 base pairs per helix turn. Each base pair contributes 0.34 nm to the length of the helix, whose pitch then is 3.4 nm. Because the N-glycosidic bonds of the paired bases are not directly across the helix diameter from one another, the sugar-PO_4 backbones are not spaced symmetrically about the helix circumference and the helix has a major and a minor groove.

DNA can adopt double-helical forms other than the B conformation. An alternative right-handed form is A-DNA, a "shorter, squatter" helix formed upon slight dehydration of B-DNA. A-DNA has 11 base pairs per turn. DNA:RNA hybrids and double-stranded regions of RNA chains are typically of the A conformation. Z-DNA is a left-handed double helix. Typically, Z-DNA or regions of DNA in the Z-conformation have an alternating purine-pyrimidine nucleotide sequence, as in GpCpGpCpGpC. Along either strand, the GpC dinucleotide is conformationally different than the CpG next in line. GpC dinucleotides circumscribe an angle of -45° about the helix axis, but CpGs cover only -15°. Thus, the GpCs are the 'horizontal' zigs and the CpGs are the 'vertical' zags of this "zigzag" or Z-DNA. Z-DNA has 12 base pairs per turn and a single narrow groove.

The double helix in solution is a dynamic, flexible structure, best described in terms of tertiary structure as a semi-rigid random coil. Even its short-range structure is not truly represented as a smooth, featureless barberpole-like molecule. Bases differ sufficiently in structural character that a short sequence of them lends a structural "signature" to its localized segment of DNA, generating the possibility for sequence-specific recognition by DNA-binding proteins.

The significant tertiary structures of DNA are supercoils and cruciforms. Supercoils can be formed in DNA whose ends are fixed, as in circular DNA molecules. Both positive and negative supercoils render the DNA more compact in structure, as compared to "relaxed" DNA, which lacks supercoils. All naturally occurring circular DNA molecules are negatively supercoiled. Negative supercoiling stabilizes Z-DNA. Cruciforms arise by intrastrand hydrogen bonding in DNA if its nucleotide sequence contains "inverted repeats" or palindromes.

Environmental extremes of temperature, pH or ionic strength can disrupt the hydrogen bonds between the base pairs, leading to separation of the DNA strands, so-called "denaturation" or "melting" of DNA. The strands will "re-anneal" if appropriate conditions for H-bond formation are restored.

Chromosomes are nucleoprotein complexes, consisting principally of DNA and a set of basic proteins known as histones. The histones are organized into nucleosomes, octameric units composed of four pairs of the histones H2A, H2B, H3 and H4, around which the DNA is wound. Nucleosomes are then wound into solenoids to form a 30 nm filament, a fundamental structural motif in chromosome structure. This 30 nm filament is believed to arrange into loops of DNA perhaps a million base pairs long. A cytologically distinct structure in chromosomes, called the mini-band, may consist of 18 loops ordered around the circumference of the chromosome axis. Chromosomes contain hundreds of mini-bands.

DNA can be chemically synthesized in the laboratory using phosphoramidite derivatives of the nucleotides and solid-phase techniques of polymerization. Genes greater than 1,000 base pairs in length have been assembled from oligonucleotide precursors made in this fashion, opening possibilities for the controlled manipulation of the nucleotide sequence of a gene and a facile analysis of the effects brought about by such changes.

RNA molecules are usually single-stranded. Nevertheless, RNA species have ornate, highly conserved secondary structures derived from extensive intrachain sequence complementarity in the strands. The single strands of tRNA, upon alignment of intrastrand complementary regions, form a cloverleaf secondary structure consisting of three H-bonded loops and an H-bonded stem. Hydrogen-bonding interactions between the loops then create a "bent" or L-shaped tertiary structure. Ribosomal RNA molecules also possess a high degree of intrastrand sequence complementarity, yielding a complex secondary structure on alignment. The folding patterns observed in the various rRNA species appear to have been highly conserved across the phylogenetic spectrum from archaea and eubacteria to higher eukaryotes, suggesting a common design and purpose for these molecules that has persisted over evolutionary time. mRNA displays secondary structural properties as well, but little is known about the details of such structures in these rather labile RNA's.

Chapter 9

Recombinant DNA: Cloning and Creation of Chimeric Genes

Chapter Outline

- Genetic engineering: Application of recombinant DNA technology
- Cloning vectors: Requirements
  - Origin of replication
  - Selectable marker: Typically antibiotic resistance
  - Cloning site: Polylinker in region not required for vector function
- Vector Types: dsDNA
  - Plasmids: Circular elements that can accept up to 10kbp of foreign DNA
  - Bacteriophage λ: Linear element used to clone 20kbp inserts
  - Cosmids: λ cohesive ends separated by 40kbp insert DNA
  - Shuttle vector: Origins of replication active in two different organisms
  - Yeast artificial chromosomes (YACs): Linear DNA with autonomously replicating sequence (ori) telomeres, and centromere
- Chimeric or recombinant plasmids
  - Linearized vector and compatible foreign insert DNA
  - DNA ligase joins DNA
  - Transformation incorporates DNA into host cell
- DNA libraries: A collection of recombinant plasmids
  - N = [ln(1-P)/ln(1-f)]
    - P is probability that library of size N has DNA fragment of interest
    - f = (insert size)/(genome size)
- Library types
  - Genomic libraries: Genomic DNA randomly fragmented: Fragments cloned
  - cDNA: mRNA directed synthesis of DNA
    - PolyA mRNA isolated using oligo(dT)
    - Reverse transcriptase produces DNA from mRNA
  - Expression libraries: Vector expresses cloned insert as RNA and perhaps as protein
    - Antibody screens
    - Blue/white screens
- Library screening
  - Southern analysis: DNA
    - Replica plate onto nitrocellulose
    - Lyse and denature using alkaline pH
    - Probe with complementary nucleic acid probe
      - Degenerate oligonucleotides: Made using knowledge of protein sequence
      - Heterologous probe: DNA fragment from gene in related organism
  - Express insert as fusion proteins: Screen for protein using antibodies
    - Transcription of insert requires RNA polymerase promoter
    - Translation of insert requires ribosome binding site
- Reporter gene: Used to clone promoters

- PCR: In vitro amplification of DNA
  - Thermostable DNA polymerase used to replicate DNA
  - Two primers complementary to regions flanking the gene to be amplified
- Site-specific in vitro mutagenesis: Specific alteration in DNA sequence
- Gene Therapy: Gene replacement using expression cassette

Chapter Objectives

Cloning Vectors

Understand the purpose of vectors in recombinant DNA technology. They are the vehicles by which foreign DNA is incorporated into cells in a form that will be amplified by the cell. Some important vectors include the following:

1) Plasmids: small, circular, double-stranded DNA elements containing a selectable marker, typically a gene conferring resistance to an antibiotic.

2) Bacteriophage λ: a linear double-stranded DNA-containing virus used to clone large (up to 16 kbp) DNA fragments.

3) Cosmid: a hybrid cloning vector, containing a plasmid origin of replication and a λ *cos* site, capable of carrying DNA inserts up to 40 kbp.

4) YACs: yeast artificial chromosomes, the "jumbo jets" of genetic engineering, are used to clone extremely large DNA fragments (100's of kbp).

Formation of Recombinants

Generally, vectors contain multiple restriction endonuclease recognition sites in a single region known as the polylinker. To produce a recombinant or chimeric vector, the vector is digested with a restriction enzyme, which recognizes a site in the polylinker, and the DNA to be inserted is digested with an enzyme that will produce compatible ends. The DNA's are joined using DNA ligase and the recombinants are introduced into cells by transformation.

Detecting Recombinants

Producing recombinants is an easy task, but finding the correct recombinant in a library of recombinants is often difficult. Know the techniques available for this purpose including Southern analysis, selecting for expression, screening for expression, and production of hybrid proteins for antibody screening. Reporter gene constructs are used to clone DNA elements involved in regulation of gene expression.

PCR

PCR is a powerful technique for amplifying DNA *in vitro*. Understand the principles behind this technique. You should be able to show that as a function of cycle-number, the full-length product defined by flanking oligonucleotide primers, is amplified exponentially.

Problems and Solutions

1. A DNA fragment isolated from an* EcoRI *digest of genomic DNA was combined with a plasmid vector linearized by* EcoRI *digestion so sticky ends could anneal. Phage T4 DNA ligase was then added to the mixture. List all possible products of the ligation reaction.

Answer: The ligation products will include linear and circular DNA molecules made with either the genomic fragment or the vector DNA or a combination of both. Because both DNA's have the same "sticky ends", they may join in any orientation. (Depending on the ratio of genomic fragment DNA to vector DNA, recombinants between the two DNA's may be favored.) Only ligation products with vector DNA will successfully transform cells. Further, ligated DNA products with more than one copy of vector DNA are unstable in cells. And, linear DNA is rapidly broken down when transformed into bacterial cells.

2. The nucleotide sequence of a polylinker in a particular plasmid vector is

-GAATTCCCGGGGATCCTCTAGAGTCGACCTGCAGGCATGC-

This polylinker contains restriction sites for* BamHI, EcoRI, PstI, SalI, SmaI, SphI, *and* XbaI. *Indicate the location of each restriction site in this sequence. (See Table 8.3 of restriction enzymes for their cleavage sites.)

Answer: Restriction endonuclease recognition sites are often palindromic sequences and so it may be useful to search for palindromes in the sequence. They are readily identified by scanning the sequence for two adjacent nucleotides that are a complementary pair: AT, TA, GC, CG. These sequences are the middle sequences of all uninterrupted palindromes. Once one of these dinucleotides is found the sequences surrounding the dinucleotide are inspected to see if a palindrome exists. By inspecting the nucleotides on either side of the dinucleotide, a palindrome is indicated by the flanking nucleotides also forming complementary pairs. For example, in the sequence given, an AT dinucleotide is encountered at positions 3 and 4. This dinucleotide is flanked by AT (positions 2 & 5) and GC (positions 1 & 6) indicating a 6-base palindrome (GAATTC). EcoRI recognizes this sequence. The next dinucleotide found is CG at position 8 & 9. Clearly these two positions are in the palindrome CCCGGG recognized by *SmaI*. This *SmaI* site overlaps another palindrome GGATCC (centered on the dinucleotide AT) that is recognized by *Bam*HI. The next dinucleotide is TA in TCTAGA recognized by *XbaI*. The *SalI* site is next; its sequence is GTCGAC. This is followed by CTGCAG and GCATGC recognized by *PstI* and *SphI*. The location of sites is as follows:

```
GAATTCCCGGGGATCCTCTAGAGTCGACCTGCAGGCATGC
GAATTC          TCTAGA            GCATGC
 EcoRI           XbaI              SphI
    CCCGGG          GTCGAC
     SmaI            SalI
        GGATCC            CTGCAG
         BamHI             PstI
```

3. A vector has a polylinker containing restriction sites in the following order: HindIII, SacI, XhoI, BglII, XbaI, and ClaI.

a. Give a possible nucleotide sequence for the polylinker.

b. The vector is digested with HindIII and ClaI. A DNA segment contains a HindIII restriction site fragment 650 bases upstream from a ClaI site. This DNA fragment is digested with HindIII and ClaI, and the resulting HindIII-ClaI fragment is directionally cloned into the HindIII-ClaI digested vector. Give the nucleotide sequence at each end of the vector and the insert and show that the insert can be cloned into the vector in only one orientation.

Answer: The sequence of a polylinker with sites for *Hind*III, *SacI*, *XhoI*, *BglII* *XbaI* and *ClaI* will have the following six-base recognition sites:

Site	Enzyme
AAGCTT	*Hind* III
GAGCTC	*Sac* I
CTCGAG	*Xho* I
AGATCT	*Bgl* II
TCTAGA	*Xba* I
ATCGAT	*Cla* I

We can derive the sequence of the polylinker by joining the sequences of the restriction endonuclease recognition sites in order. The sites for Hind III and Sac I are adjacent to each other thus AAGCTT and GAGCTC are continuous. The next six-base sequence is CTCGAG recognized by Xho I. Notice that CTC is already present on the end of the Sac I site thus we need only include GAG to create an Xho I site. The polylinker so far is AAGCTTGAGCTCGAG. The 3' end already has the first two bases of the recognition site for Bgl II so we can complete this site by adding ATCT giving AAGCTTGAGCTCGAGATCT. The 3' end of this sequence already contains half of the Xba I recognition sequence. To complete it we add AGA giving us AAGCTTGAGCTCGAGATCTAGA. Finally, we must include the recognition site for Cla I, which starts with A. Since the 3' end already is A, we can complete the sequence by adding TCGAT giving the following:

AAGCTTGAGCTCGAGATCTAGATCGAT.

Because some restriction sites overlap it will not be possible to use certain combinations of sites for directional cloning because digestion with one enzyme will in certain cases disrupt a sequence recognized by another enzyme. In addition, some restriction enzymes do not cut very well when their recognition site is located at the end of a DNA molecule.

b. Hind III recognizes AAGCTT and cuts after the first A whereas Cla I recognizes ATCGAT and cuts after the first T. Digestion of the polylinker with both enzymes is shown below with both strands shown. Since the polylinker is located in presumably a circular vector, the double digest will produce two fragments, a small fragment (the middle fragment in the digested sequence shown below) and a large fragment (whose ends are shown on the far left and on the far right in the digested sequence shown below).

```
5' AAGCTTGAGCTCGAGATCTAGATCGAT 3'
3' TTCGAACTCGAGCTCTAGATCTAGCTA 5'

   Hind III                  Cla I

5' A        AGCTTGAGCTCGAGATCTAGAT       CGAT 3'
3' TTCGA        ACTCGAGCTCTAGATCTAGC       TA 5'
```

Because the insert DNA was digested with the same two enzymes it will look as follows:

```
AGCTT------650bp------AT
    A-----------------TAGC
```

It is clear that the insert can replace the vector's small fragment only if it is in the same orientation.

4. *Yeast* (Saccharomyces cerevisiae) *has a genome size of 12.1 x 10^7 bp. If a genomic library of yeast DNA was constructed in a bacteriophage λ vector capable of carrying 16 kbp inserts, how many individual clones would have to be screened to have a 99% probability of finding a particular fragment?*

Answer: For a genome 1.21 x 10^7 bp, the number of 16 kbp cloned inserts that would have to be screened to have a 99% probability of finding a particular fragment is given by:

$$N = \frac{\ln(1-P)}{\ln(1-f)} \text{ where } P = 0.99 \text{ and } f = \frac{16 \text{ kbp}}{1.21 \times 10^7} = \frac{16 \times 10^3}{1.21 \times 10^7} = 1.32 \times 10^{-3}$$

$$N = \frac{\ln(1-0.99)}{\ln(1-1.32 \times 10^{-3})} = 3,480$$

5. *The South American lungfish has a genome size of 10.2 x 10^{10} bp. If a genomic library of lungfish DNA were constructed in a cosmid vector capable of carrying inserts averaging 45 kbp in size, how many individual clones would have to be screened to have a 99% probability of finding a particular DNA fragment?*

Answer:

$$N = \frac{\ln(1-P)}{\ln(1-f)} \text{ where } P = 0.99 \text{ and } f = \frac{45\text{kbp}}{1.02 \times 10^{11}} = \frac{45 \times 10^3}{1.02 \times 10^{11}} = 4.41 \times 10^{-7}$$

$$N = \frac{\ln(1-0.99)}{\ln(1-4.41 \times 10^{-7})} = 1.04 \times 10^7 = 10.4 \text{ million}$$

6. *Given the following short DNA duplex of sequence (5'→3')*
ATGCCGTAGTCGATCATTACGATAGCATAGCACAGGGATCACACATGCACACACATGACATAGGACAGATAGCAT
what oligonucleotide primers (17-mers) would be required for PCR amplification of this duplex?

Answer: Amplification by PCR requires two oligonucleotides (17-mers). The oligonucleotide used to amplify from the 5'-side has a sequence identical to the first 17 nucleotides of the fragment:

5' ATGCCGTAGTCGATCAT 3'

The oligonucleotide for the 3'-end is a 17-mer with a sequence complementary to the last 17 bases. Its sequence, written 5' to 3' is:

5' ATGCTATCTGTCCTATG 3'.

(In selecting oligonucleotides for PCR, care must be taken to insure that the oligonucleotides do not bind at alternative sites in the template DNA, they not have large regions of self complementarity, and their 3'-ends are G/C rich.)

7. Figure 9.5 b shows a polylinker that falls within the β-galactosidase coding region of the lacZ gene. This polylinker serves as a cloning site in a fusion protein expression vector where the cloned insert is expressed as a β-galactosidase fusion protein. Assume the vector polylinker was cleaved with BamHI and then ligated with an insert whose sequence reads

GATCCATTTATCCACCGGAGAGCTGGTATCCCCAAAAGACGGCC...

What is the amino acid sequence of the fusion protein? Where is the junction between β-galactosidase and the sequence encoded by the insert? (Consult the genetic code table on the inside front cover to decipher the amino acid sequence.)

Answer: *Bam*HI recognizes GGATCC and cuts after the first G giving

```
G          GATCC
CCTAG          G
```

Once ligated with insert, the sequence becomes

GGATCCATTTATCCACCGGAGAGCTGGTATCCCCAAAAGACGGCC....

What is the amino acid sequence of the fusion protein? To answer this we must know the reading frame at the *Bam*HI site. This is shown in Figure 9.5 b. The reading frame is such that GAT in the *Bam*HI site is in frame. This gives:

```
G|GAT|CCA|TTT|ATC|CAC|CGG|AGA|GCT|GGT|ATC|CCC|AAA|AGA|CGG|CC....
  Asp Pro Phe Ile His Arg Arg Ala Gly Ile Pro Lys Arg Arg Pro
```

8. The amino acid sequence across a region of interest in a protein is

Asn-Ser-Gly-Met-His-Pro-Gly-Lys-Leu-Ala-Ser-Trp-Phe-Val-Gly-Asn-Ser

The nucleotide sequence encoding this region begins and ends with an EcoRI site, making it easy to clone out the sequence and amplify it by polymerase chain reaction (PCR). Give the nucleotide sequence of this region. Suppose you wished to change the middle Ser residue to a Cys to study the effects of this change on the protein's activity. What would be the sequence of the mutant oligonucleotide you would use for PCR amplification?

Answer: We are asked for the nucleotide sequence of a gene that codes for the polypeptide NSGMHPGKLASWFVGNS. To do this we will have to use the genetic code to convert the amino acid sequence into a nucleotide sequence. Because the code is redundant (most of the amino acids are coded for by more than one triplet codon) we will not be able to determine the exact sequence. In reading this answer keep in mind that in addition to A, G, C, and T (or U) additional symbols are used to describe nucleotides. Three additional symbols are used here: N (any base - AGCT-), Y (either pyrimidine, C or T), and R (either purine, A or G). Also, three amino acids, Ser, Leu, and Arg are coded by six codons.

Since the sequence begins and ends with an *Eco*RI site, GAATTC must appear at both ends. The codon for Asn (N), AAY, is part of an *Eco*RI site as AAT. Thus, the 5'-end of the mRNA must begin with (G)AAU. (In the DNA fragment G would have been separated from the rest of the sequence by *Eco*RI digestion because the enzyme cleaves between G and A.) The next amino acid, Ser, has six codons, UCN and AGY. Because we need to complete an EcoRI site UCN must be coding for Ser at this position. The amino acids up to Leu are coded by either two or four codons. Leu is coded by six codons, UUR and CUN. The next serine codon is the one we are asked to convert to Cys by site-directed mutagenesis. The codons of serine are UCN and AGY; Cys codons are UGY. Let us assume that the internal serine is AGY, one base different from the cysteine codon. The 3' end codes for Ser also but here we will assume that the codon set UCN is used with the UC forming the last two bases of the *Eco*RI site. The mRNA and DNA sequences (T's replacing U's) are show below.

```
mRNA                                     CUN
5'(G)AAU-UCN-GGN-AUG-CAY-CCN-GGN-AAR-UUR-GCN-AGY-UGG-UUY-GUN-GGG-AAU-UCN     3'
     Asn-Ser-Gly-Met-His-Pro-Gly-Lys-Leu-Ala-Ser-Trp-Phe-Val-Gly-Asn-Ser

DNA                                      CTN
5'(G)AAT-TCN-GGN-ATG-CAY-CCN-GGN-AAR-TTR-GCN-AGY-TGG-TTY-GTN-GGG-AAT-TCN 3'
```

To mutate the Ser codon to Cys we must design a primer that is complementary to regions flanking the serine codon. Because the Ser codon is closer to the 3' end than the 5' end the primer will be complementary to this end of the sequence. The 5' end of the primer must contain an EcoRI recognition site, GAATTC, a few nucleotides in from the very end: 5' NNNNGAATTC. We included the additional bases to insure that the EcoRI site is cleaved with high efficiency. (Restriction endonucleases often cleave with low efficiency when their recognition site is on the very end of a DNA fragment.) The oligonucleotide is shown directly below the DNA coding sequence.

```
DNA
5′(G)AAT-TCN-GGN-ATG-CAY-CCN-GGN-AAR-TTR-GCN-AGY-TGG-TTY-GTN-GGG-AAT-TCN 3′
Primer Sequence             3′ CCN-TTY-AAR-CGN-ACR-ACC-AAR-CAN-CCC-TTA-AGN-NNN  5′
```

The primer is complementary to the DNA at every position except the first position of the Ser codon to be mutated. Here the primer is complementary to a Cys codon. Thus, annealing the primer to the DNA will produce a one-base mismatch. The PCR amplification products generated using this primer will code for Cys at this position. The primer is designed to anneal to regions flanking the target base. The 3' end of the primer extends 12 bases from the target base and ends in a CC. Although there are many choices for the 3' end of the primer, there are two constraints to keep in mind. The primer must be designed to compensate for the one-base mismatch so it has to extend several bases on either side of the target. It is wise to have the 3' end of the primer include combinations of G's or C's. This will insure that the 3' end of the primer binds tightly to the DNA. The 3' end is the "business end" because it is extended by the DNA polymerase used in PCR.

To produce a mutation by PCR, we can use this primer and a second primer with a sequence identical to a sequence upstream (5' side) of the *Eco*RI site on the 5'-end of the fragment. Using this primer and the mutagenic primer, a PCR fragment can be amplified that has *Eco*RI sites on each end and contains the desired mutation. The sequence of the mutagenic primer is given below.

5' NNN-NGA-ATT-CCC-NAC-RAA-CCA-RCA-NGC-RAA-YTT-NCC 3'

Questions for Self Study

1. What three common features do all cloning vectors share?

2. How is the enzyme T4 DNA ligase used to produce chimeric plasmids?

3. Match the terms in the columns below:

a. Bacteriophage λ	1. A plasmid capable of propagation in two different organisms
b. Cosmid	2. Antibiotic resistance gene.
c. Reporter gene	3. Vector used to clone extremely large DNA molecules.
d. Shuttle Vector	4. Used to clone regulatory regions of genes.
e. Transformation	5. Incorporation of plasmids into cells.
f. YAC	6. A bacterial virus used as a cloning vector.
g. Selectable marker	7. A hybrid vector containing *cos* sites.

4. Which of the following apply to Southern analysis?
    a. Can be used to detect specific recombinants.
    b. Probe could be DNA from a different organism.
    c. Uses antibodies to detect gene expression.
    d. Used labeled RNA or DNA as a probe.
    e. Is used to identify RNA.

5. What is the difference between fusion protein expression and reporter gene expression in genetic engineering?

6. Polymerase chain reaction or PCR is an *in vitro* technique used to amplify specific sequences of DNA. It consists of repeated cycles of a three-step process. What are the three steps that constitute a cycle and what happens during each step?

Answers

1. An origin of replication, a selectable marker, a cloning site.

2. T4 DNA ligase is used to join covalently foreign DNA into a cloning site on a plasmid.

3. a. 6; b. 7; c. 4; d. 1; e. 5; f. 3; g. 2

4. a., b., d.

5. In fusion gene expression, a protein-coding region under study is cloned into an expression vector to produce a recombinant sequence that is expressed as a hybrid or fusion protein. In this case the experimenter is interested in studying the expressed protein. Reporter genes are used to identify transcriptional regulatory elements like promoters and enhancers.

6. Denaturation: dsDNA template is converted to ssDNA by high temperature. Annealing: Temperature is lowered to allow DNA primers to bind to template DNA. Polymerization (extension): Primer extension occurs at the temperature optimum for thermostable DNA polymerase.

Additional Problems

1. An experimenter decided to use PCR to isolate a particular gene from an exotic organism. The gene under study had previously been characterized in several different organisms. The experimenter was able to locate two conserved regions within the gene, to make oligonucleotides complementary to these two regions, and to set up PCR amplification. The experimenter programmed the thermocycler and left for the weekend. After two cycles, a power failure occurred, erasing the memory in the thermocycler. A colleague, realizing the importance of the experiment, decided to restart the thermocycler. However, the colleague knew nothing about PCR, and so the person asked you what to do.
a. Describe a typical PCR cycle and explain the purpose of each step in the cycle.
b. The person takes your advice and is in fact so excited by your description of PCR that the person runs a gel. No products were found. Desperate for results, the person decides to change one of the temperatures in the cycle and performs a second PCR run. Which temperature, which direction, and why?
c. The adjustment results in production of only one faint band, not an impressive yield for such a marvelous procedure. The person decides to make one more attempt and triples the total number of cycles. This time several bands are produced. Can the results be trusted? Explain.

2. One of the constraints of standard PCR is to know the sequence of two separate regions of a gene. Can you suggest a method that eliminates the necessity of having to know two separated sequences within a gene?

3. In question 5 above, you were asked to calculate the genome library size of lungfish DNA necessary to have a 99% probability of finding a particular DNA fragment. The answer came out to 10 million. If the library is plated onto Petri dishes, with approximately 1000 colonies per plate, calculate the number of plates required, the total surface area, and the height if the plates were stacked on top of each other. (A typical Petri dish is 100 x 15 mm.)

4a. A biochemist is attempting to clone a eukaryotic gene and, using Southern analysis of an *Eco*RI digest of total genomic DNA, learns that the gene is located on a 2.5 kbp *Eco*RI fragment. Further, it was discovered that the *Eco*RI fragment lacks *Bam*HI restriction sites. Suggest an appropriate vector to use in this case.
b. The strategy the biochemist decides to take is to digest total genomic DNA with *Eco*RI, separate the fragments by agarose gel electrophoresis, cut out of the gel the 2.5 kbp fragment, elute the DNA from the gel slice, and ligate the DNA into a suitable vector. "Piece of cake!" the biochemist exclaims, "This will be a one-step cloning and we'll have our gene without having to screen our recombinants by Southern analysis." This is an optimistic statement. Can you explain why?

5. In one version of oligonucleotide-directed, site-specific mutagenesis, single-stranded recombinant M13 DNA is used as a template for second-strand synthesis using a mutagenic oligonucleotide. (M13 is a bacterial virus whose life cycle includes a circular dsDNA genomic form, which can be manipulated much like a conventional plasmid, and a circular ssDNA genomic form.) The mutagenic oligonucleotide is complementary to several bases on both sides of the target base but at the target base it produces a single-base mismatch. The oligonucleotide is used *in vitro* to synthesize a complementary strand whose ends are ligated together by DNA ligase. Thus, a single-stranded M13 is converted to a double-stranded, closed, circular form (often referred to as the RF or replicative form). Suggest a simple assay to check for second-strand synthesis and for DNA ligase activity.

Abbreviated Answers

1a. In a typical PCR reaction, double-stranded template DNA and two primers, complementary to two sequences flanking the region to be amplified, are mixed with a thermostable DNA polymerase and dNTP's. The mixture is first heated to 90-95°C to denature the template DNA. Then, the temperature is lowered to between 40°C and 60°C to allow primer binding to the template. Finally, the temperature is adjusted to the temperature optimum for the DNA polymerase to allow primer extension to occur (72-75°C). The process of heating to denature, cooling to allow primer binding, and heating to extend primer represents a cycle that is repeated a number of times, the result of which is an exponential amplification of the DNA between the two primers.

b. Failure to amplify may be a result of inefficient primer-binding to the template because the temperature of the second stage of the cycle, namely primer binding, exceeds the T_m of the primer. One possibility is to lower the primer binding temperature.

c. The production of only a faint band may be parent nature's way of telling you that something is not quite right with the reaction. It may be the case that the primers are inappropriate for the template DNA. Typically, PCR reactions are set up to produce a generous amplification after a specified number of cycles. Increasing substantially the cycle number may allow for amplification of DNA contaminating the template DNA. In addition, increasing the cycle-number provides additional opportunities for random mispriming to occur.

2. In a process known as inverse-PCR, DNA can be amplified from a region of known sequence outward in both directions. Template DNA is digested with a restriction enzyme known to cut outside of the region of interest. (Southern analysis can be used to screen for an appropriate enzyme.) The fragmented DNA is circularized by ligation and then linearized by digestion with an enzyme that cuts at a single location within the DNA. The enzyme used in this case must cut within the region of known sequence producing a linear DNA whose ends contain the previously characterized sequence. Primers whose sequences are derived from the characterized region of DNA are now used to amplify as shown below.

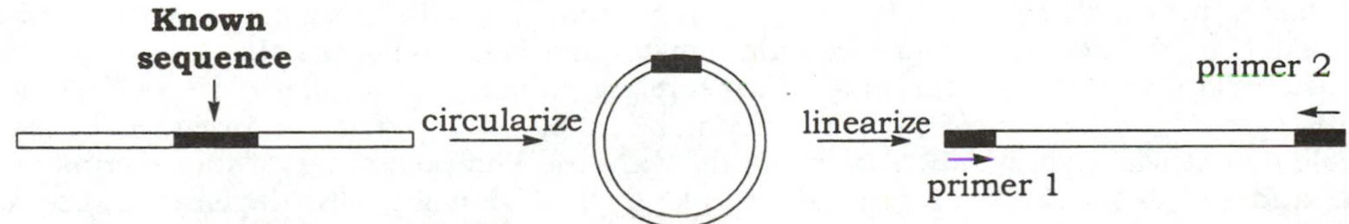

In anchored-PCR, template DNA is fragmented by restriction digestion and the ends ligated to a primer whose sequence is known. Amplification occurs between this primer and an internal, previously characterized sequence within the gene. This is shown below.

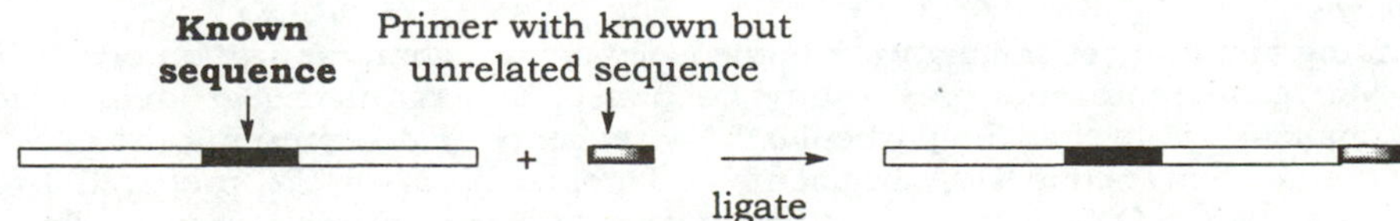

3.

$$\frac{10\times10^{6}\frac{\text{colonies}}{\text{library}}}{1{,}000\frac{\text{colonies}}{\text{plate}}}=10{,}000\text{ plates}$$

$$10{,}000\text{ plates}\times\frac{15\times10^{-3}\text{m}}{\text{plate}}=150\text{ m!}$$

$$10{,}000\text{ plates}\times\pi\times\left(\frac{100\times10^{-3}\text{m}}{2}\right)^{2}=78.5\text{ m}^{2}$$

4a. The 2.5 kbp fragment is a small fragment that could be cloned in a plasmid. Cosmids, λ, and YACs are all more appropriate for much larger fragments.

b. The cloning strategy is sound; however, the biochemist may be a bit optimistic in thinking that the 2.5 kbp fragment identified by Southern analysis is a unique fragment. Eukaryotic genomes are large, greater than 10×10^6 bp for yeast, a simple eukaryote, and a digest with a restriction enzyme with a six-base recognition sequence is expected to generate a very large number of fragments. For example, the probability of finding a particular 6-base sequence is $(1/4)^6 = (1/4{,}096)$. For a genome of 10×10^6 bp one might expect to have around $10 \times 10^6 \times (1/4{,}096) = 2{,}441$ sites. Therefore, an enzyme with a six-base recognition sequence is expected to generate thousands of bands, which will form essentially a continuum upon electrophoresis. The fact that Southern analysis highlights a single band only means that, of all of the bands in the continuum at that particular size, there is one (or possibly more) with the sequence of interest. The biochemist will have to screen the products of the cloning procedure.

5. One easy assay is to examine the products by agarose gel electrophoresis. Single-stranded template M13 has a faster electrophoretic mobility than double-stranded M13. Therefore, second-strand synthesis will be accompanied by a gel shift of the M13 template to a slower migrating form. If second-strand synthesis is successful, a double-stranded, nicked circle is produced. Successful ligation will convert the nicked circle to a covalently-closed double-stranded molecule. Electrophoresis in the presence of ethidium bromide is capable of distinguishing between these two. Ethidium bromide binding to covalently-closed, double-stranded DNA will induce positive supercoils. Thus, the mobility of the ligated product will be slightly faster than relaxed circles in the presence of ethidium.

Summary

Plasmids are circular, extrachromosomal DNA molecules. Plasmids have the ability to self-replicate and often carry genes for novel metabolic capacities (such as antibiotic resistance) that allow host cells to grow under otherwise adverse conditions. These facts make plasmids ideal vehicles for the amplification of foreign DNA segments. The discovery of restriction endonucleases provided an essential tool for generating big DNA fragments with predictable features, such as defined ends. In particular, restriction endonucleases that create 5'- and 3'-overhangs, "sticky ends", ease the task of assembling chimeric plasmids *in vitro*. The circular plasmid is opened, or "linearized", by restriction endonuclease digestion, a foreign DNA segment is annealed to it via hybridization of its ends with the complementary "sticky ends" of the plasmid, and the two DNA sequences are covalently ligated using DNA ligase, a process that recloses the plasmid circle. Using these techniques and related developments, essentially any DNA sequence can be cloned.

Since the size of foreign DNA inserts in plasmids is inherently limited to 10 kbp, other vector systems for the propagation of larger DNA segments have emerged, such as bacteriophage λ-based cosmid cloning systems and even the creation of artificial yeast chromosomes (YACs) for cloning purposes.

The overriding aim of most cloning work is the isolation and characterization of specific genes. Since any given gene represents only a tiny fraction of an organism's genome or total DNA content, the challenge is to find the particular DNA sequence that represents the gene among a large population of extraneous DNA segments. Genomic libraries are prepared by partially digesting the organism's DNA and cloning the fragments into a suitable vector. The library is then screened for the gene of interest, using probes designed to uniquely identify the gene.

Typically, a probe is a single-stranded DNA molecule whose nucleotide sequence is complementary to some region of the target gene. Libraries can also be constructed by copying the mRNA population isolated from a selected cell type into DNA using reverse transcriptase. The product, cDNA, is converted to duplex DNA and cloned into an appropriate vector. cDNA libraries cloned into expression vectors where the DNA inserts are both transcribed and translated in recipient cells can be screened for protein products of the gene using immunological techniques and other methods. The properties of gene regulatory sequences can be investigated by cloning them into sites adjacent to genes encoding easily assayed protein products, so-called reporter genes. Expression of the reporter gene serves as an index of the transcriptional efficiency of the cloned regulatory sequence.

Polymerase chain reaction (PCR) is a technique for greatly amplifying the amount of any DNA segment in a genome, provided sequence information is available about regions flanking the desired segment. This information is used in synthesizing pairs of complementary oligonucleotides that anneal next to the segment (one oligonucleotide at each 3'-end). These oligonucleotides serve to prime synthesis of the region by DNA polymerase. The resultant duplex DNA is melted by heating to 95°C, and the solution is cooled to initiate another cycle of synthesis. Repeating this thermal cycle and synthesis regime for 25 cycles leads to a million-fold amplification in the concentration of the original DNA sequence. PCR has been automated through the invention of thermal cyclers that carry out the entire process in less than 4 hours. PCR is thus an automated, cell-free, amplification procedure for generation of specific DNA sequences.

With cloned genes in hand, alteration of the nucleotide sequences of these genes in order to observe the consequences of these "mutations" on the biological function of the gene or gene product became feasible. *In vitro* mutagenesis has been used to change the nucleotide sequence in a systematic way so that the effects of specific amino acid substitutions within a protein can be tested. Such manipulations have proved very valuable for elucidating the role of individual amino acids in the structure and function of proteins.

The success of recombinant DNA technology now verges on the ability to engineer at will the genetic constitution of organisms for desired ends. The commercial production of therapeutic biomolecules in microbial cultures is already established. For example, human insulin is produced in quantity in *E. coli* cells. Agricultural crops with desired attributes, such as enhanced resistance to frost or longer shelf-life, are in cultivation. The rat growth hormone gene has been cloned and transferred into mouse embryos, creating transgenic mice that at adulthood are twice normal size. Already, transgenic versions of domestic animals such as pigs, sheep and even fish have been developed for human benefit.

Chapter 10

Enzymes: Their Kinetics, Specificity and Regulation

• •

Chapter Outline

- Enzymes
  - Biological catalysts that function in dilute, aqueous solutions under mild conditions (e.g., pH and low temperatures) to increase reaction rate
  - Catalytic power: Ratio of catalyzed rate to uncatalyzed rate
  - Specificity
  - Regulation of catalysis
    - Enzyme levels and types regulated genetically
    - Inhibitors and activators modify enzyme activity
- Non-protein components
  - Cofactors: Inorganic molecules
  - Coenzymes: Organic molecules: Often vitamins
  - Prosthetic group: Firmly bound coenzyme
  - Holoenzyme: Apoenzyme plus prosthetic group
- Chemical kinetics
  - Rate law: $v = -dA/dt = k[A]^n$
    - k = rate constant
    - n = order of reaction
  - Molecularity: Number of molecules that must simultaneously react
    - Unimolecular: n = 1: k = first order rate constant
    - Bimolecular: k = second order rate constant
      - n = 2, or
      - $v = k[A][B]$
- Reaction rates
  - Limited by activation barrier: Free energy needed to reach transition state
  - Arrhenius equation $k = Ae^{-\Delta G/RT}$
  - Influenced by
    - Temperature
    - Catalysts
- Enzyme kinetics
  - Saturation: Zero-order kinetics at high [S]
  - Michaelis-Menten: $v = (V_{max}[S])/(K_m+[S])$
    - $Km = (k_{-1}+k_2)/\ k_1$
    - $Km = [S]$ when $v = V_{max}/2$
    - Conditions applicable
      - $[S]_{initial} > [Enzyme]$
      - pH, temperature, ionic strength, [enzyme] constant
      - Initial rate measured
        - [S] essentially equal to [So]

  - [P] essentially zero

- Enzyme parameters
  - One international unit: Amount to catalyze formation of one micromole of product in one minute
  - Turnover number = $k_{cat} = k_2 = V_{max}/[E_T]$
  - k_1 sets upper limit on catalytic efficiency: Reaction can go no faster than rate at which E and S form ES
- Plots
  - Direct plot: v versus [S]: $v = (V_{max}[S])/(K_m+[S])$
    - Rectangular hyperbola
    - Asymptotically approaches V_{max} when [S] high
    - K_m = [S] when $v = V_{max}/2$
  - Lineweaver-Burk (double-reciprocal): 1/v versus 1/[S]: $1/v = (K_m/V_{max})(1/[S])+1/V_m$
    - Linear
    - Slope = K_m/V_{max}
    - y-intercept = $1/V_{max}$
    - x-intercept = $-1/K_m$
- Temperature dependence of enzyme-catalyzed reactions
  - Below 50°C: Q_{10}: Ratio of activities at two temperatures 10° apart: For typical enzyme $Q_{10}=2$
  - Above 50°C: Typically enzyme denatures
- Inhibition
  - Reversible inhibition: Noncovalent
    - Competitive inhibition: Inhibitor and substrate compete for same binding site
      - V_{max} unchanged
      - $K_m = K_m(1+[I]/K_I)$
    - Noncompetitive inhibition
      - Pure noncompetitive inhibition: S and I bind at different sites
      - Mixed noncompetitive inhibition: I binding influences S binding
  - Irreversible inhibition: Covalent
    - Suicide substrate or Trojan Horse substrate
    - Site-specific affinity label
- Bisubstrate reactions: E + A + B → AEB → PEQ → E + P + Q
  - Sequential or single displacement
    - Random: Either substrate binds to enzyme, either product is released
    - Ordered: Leading substrate binds first followed by second substrate
  - Ping-Pong or double-displacement: Leading substrate binds: Enzyme modified: Product released Second substrate binds: Enzyme unmodified: Second product released
- Catalytic biomolecules
  - Enzymes: Proteins
  - Abzymes: Antibodies
  - Ribozymes: RNA
- Enzyme regulation
  - Approach to equilibrium
  - Km of enzymes in the range of in vivo substrate concentrations
  - Genetic controls
  - Covalent modification
  - Allosteric regulation
  - Others
    - Zymogens: Proenzymes or zymogens: Activated by proteolysis
      - Proinsulin: Insulin
      - Chymotrypsinogen: Chymotrypsin
      - Blood clotting factors: Serine protease cascade leading to fibrinogen to fibrin

- Isozymes: Lactate dehydrogenase: A_4, A_3B_1, A_2B_2, A_1B_3, B_4
- Modular proteins: cAMP-dependent protein kinase: R_2C_2

- General properties of regulatory proteins
  - Kinetic properties
    - Do not follow simple Michaelis-Menten kinetics
    - Activity sigmoidal: Higher order dependence on substrate concentration
    - Cooperativity
  - Allosteric inhibition
  - Often regulated by activation
  - Oligomeric organization
  - Effectors alter distribution of conformational isomers
- Cooperativity models
  - Monod, Wyman, Changeux (1965): Symmetry model
    - Two conformations: T (tense or taut) and R (relaxed)
    - R state high affinity, T state lower affinity
    - Positive homotropic effectors: Substrate binding shifts equilibrium to R
    - Heterotropic effectors
      - Positive effector: Binds to R state and shifts equilibrium toward R
      - Negative effector: Binds to T state and shifts equilibrium toward T
  - Koshland, Nemethy, Filmer (1966) Sequential model: Induced fit: S-binding induces conformational change
- Glycogen Phosphorylase
  - Structure
    - Homodimer: 846-amino acids per monomer
    - Glycogen binding site
    - Phosphate binding site: Phosphate is positive homotropic effector
    - Serine 14: Site of covalent phosphorylation of enzyme: Phosphoprotein active and insensitive to allosteric regulation
    - ATP/AMP binding site: Heterotropic effectors: ATP inhibits: AMP activates
    - Glucose-6-phosphate binding site: Negative heterotropic regulator
    - Caffeine: Allosteric inhibitor
- Enzyme cascade
  - Hormone binds to cell surface receptor releasing GDP from heterotrimeric G-protein
  - GTP binds to G_α subunit causing it to dissociate as G_α:GTP complex
  - G_α:GTP complex stimulates adenylyl cyclase
    - GTPase activity produces inactive G_α:GDP complex
    - G_α:GDP complex must return to membrane for reactivation
  - Adenylyl cyclase produces cAMP
  - cAMP binds to regulatory subunits of R_2C_2 cAMP-dependent protein kinase
    - R subunist dissociate
    - C subunits become active
  - Protein kinase phosphorylates glycogen phosphorylase kinase
  - Glycogen phosphorylase kinase phosphorylates serine 14 of glycogen phosphorylase
  - Protein phosphates removed by phosphoprotein phosphatase I

Chapter Objectives

Michaelis-Menten enzyme kinetics

One of the keys to understanding Michaelis-Menten enzyme kinetics is to remember the model and its assumptions. The enzyme-substrate complex is in rapid equilibrium with free substrate. Product formation involves a fast catalytic step followed by a slow release of product. The velocity, υ, is the initial velocity of the reaction measured immediately upon addition of substrate when product concentration is very small. Typically the total enzyme concentration is small relative to substrate concentration and is kept constant. The Michaelis-Menten equation is not a complicated equation and is easy to understand.

$$\upsilon = \frac{V_{max}[S]}{K_m + [S]} = \frac{k_2[E_T][S]}{\frac{k_{-1}+k_2}{k_1} + [S]}$$

It is easy to remember that K_m has units of concentration because the denominator of the Michaelis-Menten equation is the term K_m + [S]. V_{max} is equal to the product of k_2 and $[E_T]$. V_{max} is reached when substrate concentration is large. How large? Large relative to K_m. When the substrate concentration is large relative to K_m the enzyme is saturated and essentially all of it is tied up as ES complex. The concentration of ES complex is equal to the total enzyme concentration and the rate therefore is equal to $k_2 \times [E_T]$. The equation is an example of a rectangular hyperbola. In a plot of υ vs. [S], V_{max} is approached asymptotically as [S] increases and K_m is the substrate concentration that supports $\upsilon = V_{max}/2$.

Graphical representation

The Lineweaver-Burk double-reciprocal plot uses $1/\upsilon$ and $1/[S]$ as variables because they are linearly related giving rise to straight lines whose slopes and intercepts are easily determined. A minor disadvantage to double-reciprocal plots is that they use inverse space: as [S] increases, $1/[S]$ decreases. With this in mind it is easy to remember that the $1/\upsilon$ intercept is equal to $1/V_{max}$, because the intercept occurs at the smallest real value of $1/\upsilon$. Smaller values of $1/\upsilon$ occur only to the left of the $1/\upsilon$-axis, where $1/[S]$ is negative. (Negative concentrations do not exist in the real world.) The $1/[S]$ intercept is equal to $-1/K_m$.

Enzymes

Enzymes are proteins that act as biological catalysts but unlike inorganic catalysts, enzymes exhibit incredible specificity and can be regulated in numerous ways. Know the factors controlling enzymatic activity including: product build-up and approach to equilibrium, availability of substrates and cofactors, regulation of enzyme production and degradation, covalent modification, allosteric regulation, biosynthesis of enzymes as zymogens, isozymes, and regulation by modulator proteins. Keep in mind that enzymes as proteins may be capable of assuming different conformations and that the substrate-binding site may undergo local conformational change upon substrate binding.

Allosteric Regulation

In order to understand allosteric regulation, you must first understand standard Michaelis-Menten kinetics. Plots of υ versus [S] are hyperbolic, approaching V_{max} as [S] becomes large and having K_m defined as the substrate concentration where $\upsilon = 0.5 \times V_{max}$. Allosteric proteins exhibit sigmoidal plots of υ versus [S]. Inhibitors or activators that bind at sites distinct from the substrate-binding site yet still influence catalysis may influence the shape of the sigmoidal curve.

Allosteric Models

Be familiar with the two models for allosteric behavior: the symmetry model of Monod, Wyman, and Changeux; and, the sequential model of Koshland, Nemethy, and Filmer. The symmetry model postulates the existence of two conformational states, **R** (relaxed) and **T** (taut), which differ in their affinity for substrate with **R** having a higher affinity than **T**. Cooperativity arises because substrate binding shifts the **R/T** equilibrium towards the **R** state. Allosteric activators and inhibitors bind at specific sites distinct from the substrate-binding site but influence substrate binding by influencing the **R/T** equilibrium. Inhibitor binding favors the **T** state whereas activator binding favors the **R** state. In the sequential model, ligand-binding sites are in communication with each other such that the state of occupancy of a ligand-binding site influences the state of occupancy of other sites.

Inhibitors

Competitive inhibition occurs whenever an inhibitor competes with a substrate for the substrate binding site in a reversible manner. How will the presence of a competitive inhibitor affect enzyme kinetics? Clearly, at high [S], S will out-compete the inhibitor. Thus, V_{max} is unaffected. The apparent K_m is increased to $K_m(1 + [I]/K_I)$. (The term $[I]/K_I$ is the inhibitor concentration normalized to the inhibitor dissociation constant, K_I.)

In noncompetitive inhibition, the inhibitor binds reversibly to a site different than the substrate binding site but inhibitor-bound enzyme is inactive. The easiest case to remember is pure noncompetitive inhibition. Here, I binds to its enzyme binding site without regard to the state of occupancy of the substrate binding site. In other words, E and ES bind I with equal affinity. The consequence is a decrease in the apparent V_{max} to $V_{max}/(1 + [I]/K_I)$; however, K_m is unaffected. Mixed noncompetitive inhibition occurs when I binding to E and ES differ. Both K_m and V_{max} are affected.

Bisubstrate reactions

Single-displacement or sequential reactions occur as follows:

$$A + B + E \rightarrow AEB \rightarrow PEQ \rightarrow E + P + Q$$

The two substrates may bind in a specific order or in random order.
Ping-Pong or double-displacement reactions proceed as follows:

$$A + E \rightarrow EA \rightarrow E'P \rightarrow E' + P;\ E' + B \rightarrow E'B \rightarrow EQ \rightarrow E + Q$$

Problems and Solutions

1. According to the Michaelis-Menten equation, what is the υ/V_{max} ratio when $[S] = 4K_m$?

Answer:

$$\upsilon = \frac{V_{max}[S]}{K_m + [S]}$$

When $[S] = 4K_m$

$$\upsilon = \frac{V_{max} \times 4K_m}{K_m + 4K_m} = \frac{4V_{max}}{5}, \text{ or}$$

$$\frac{\upsilon}{V_{max}} = 0.8$$

2. If V_{max} = 100 μmol/mL·sec and K_m = 2 mM, what is the velocity of the reaction when [S] = 20 mM?

Answer: For V_{max} = 100 μmol/sec and K_m = 2 mM, [S] = 20 mM

$$\upsilon = \frac{V_{max}[S]}{K_m + [S]}$$

$$= \frac{100\frac{\mu mol}{mL \cdot sec} \times 20\ mM}{2\ mM + 20\ mM}$$

$$= \frac{100\frac{\mu mol}{mL \cdot sec} \times 20\ mM}{22\ mM}$$

$$= 91\frac{\mu mol}{mL \cdot sec}$$

3. For a Michaelis-Menten reaction, $k_1 = 7 \times 10^7$/ M·sec, $k_{-1} = 1 \times 10^3$/sec and $k_2 = 2 \times 10^4$/sec. What are the values of K_m?

Answer: When $k_1 = 7 \times 10^7$/M·sec, $k_{-1} = 1 \times 10^3$/sec, $k_2 = 2 \times 10^4$/sec

$$K_m = \text{Michaelis - Menten constant} = \frac{k_{-1} + k_2}{k_1}$$

$$= \frac{\frac{1 \times 10^3}{\text{sec}} + \frac{2 \times 10^4}{\text{sec}}}{\frac{7 \times 10^7}{\text{M} \cdot \text{sec}}}$$

$$K_m = 3.0 \times 10^{-4}\ \text{M}$$

4. The following kinetic data were obtained for an enzyme in the absence of inhibitor (1), and in the presence of two different inhibitors (2) and (3) at 5 mM concentration. Assume [E_T] is the same in each experiment.

[S] (mM)	*(1) υ (μmol/mL·sec)*	*(2) υ (μmol/mL·sec)*	*(3) υ (μmol/mL·sec)*
1	*12*	*4.3*	*5.5*
2	*20*	*8*	*9*
4	*29*	*14*	*13*
5	*35*	*21*	*16*
12	*40*	*26*	*18*

a. Determine V_{max} and K_m for the enzyme.

b. Determine the type of inhibition and the K_I for each inhibitor.

Answer: The data may be analyzed using reciprocal variables. For each [S] and corresponding υ we will calculate 1/[S] and 1/υ.

[S] (mM)	1/[S] M-1	v(1) umol/mL sec	1/v(1) mL sec/mol	v(2) umol/mL sec	1/v(2) mL sec/mol	v(3) umol/mL sec	1/v(3) mL sec/mol
1	1000	12	8.33E+04	4.3	2.33E+05	5.5	1.82E+05
2	500	20	5.00E+04	8	1.25E+05	9	1.11E+05
4	250	29	3.45E+04	14	7.14E+04	13	7.69E+04
5	200	35	2.86E+04	21	4.76E+04	16	6.25E+04
12	83	40	2.50E+04	26	3.85E+04	18	5.56E+04

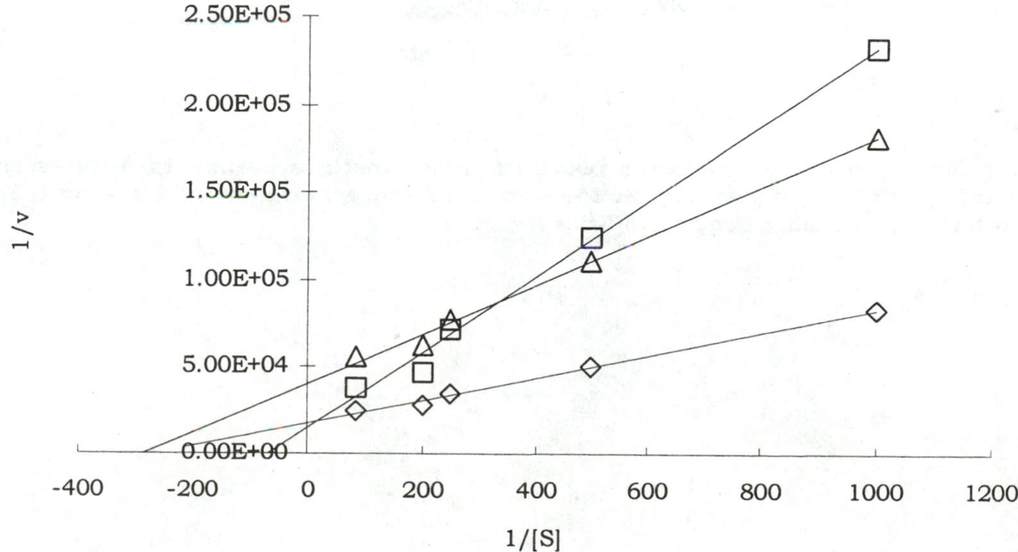

Plots of $1/\upsilon$ vs. $1/[S]$ indicate straight lines given by

(1) $1/\upsilon = 63.1(1/[S]) + 1.96 \times 10^4$
(2) $1/\upsilon = 212.3(1/[S]) + 1.99 \times 10^4$
(3) $1/\upsilon = 137.4(1/[S]) + 4.38 \times 10^4$

In general,

$$\frac{1}{\upsilon} = \frac{K_m}{V_{max}} \times \frac{1}{[S]} + \frac{1}{V_{max}}$$

Thus, the y-intercept is equal to $1/V_{max}$ and the x-intercept is equal to $-1/K_m$.

(a)

Condition	Vmax (μmol/mL·sec)	Km (mM)
No inhibitor	51	3.2
5 mM inhibitor (2)	50.3	10.7
5 mM inhibitor (3)	22.8	3.1

(b) Inhibitor (2) increases the apparent K_m of the enzyme without affecting V_{max}. This is characteristic of a competitive inhibitor. In this case K_I is calculated as follows

$$K_{m,app} = K_m(1 + \frac{[I]}{K_I}) \text{ or}$$

$$10.7 \text{ mM} = 3.2 \text{ mM}(1 + \frac{5 \text{ mM}}{K_I})$$

Solving for K_I we find

$$K_I = \frac{5 \text{ mM}}{(\frac{10 \text{ mM}}{3.2 \text{ mM}} - 1)} = 2.35 \text{ mM}$$

Inhibitor (3) decreases V_{max} but leaves K_m unchanged, an example of noncompetitive inhibition. In this case K_I is calculated as follows

$$\frac{1}{V_{max,app}} = \frac{1}{V_{max}}(1 + \frac{[I]}{K_I}) \text{ or}$$

$$\frac{1}{22.8} = \frac{1}{51}(1 + \frac{[I]}{K_I})$$

Solving for K_I we find

$$K_I = \frac{5 \text{ mM}}{(\frac{51}{22.8} - 1)} = 4.04 \text{ mM}$$

5. The following graphical patterns obtained from kinetic experiments have several possible interpretations depending on the nature of the experiment and the variables being plotted. Give at least two possibilities for each.

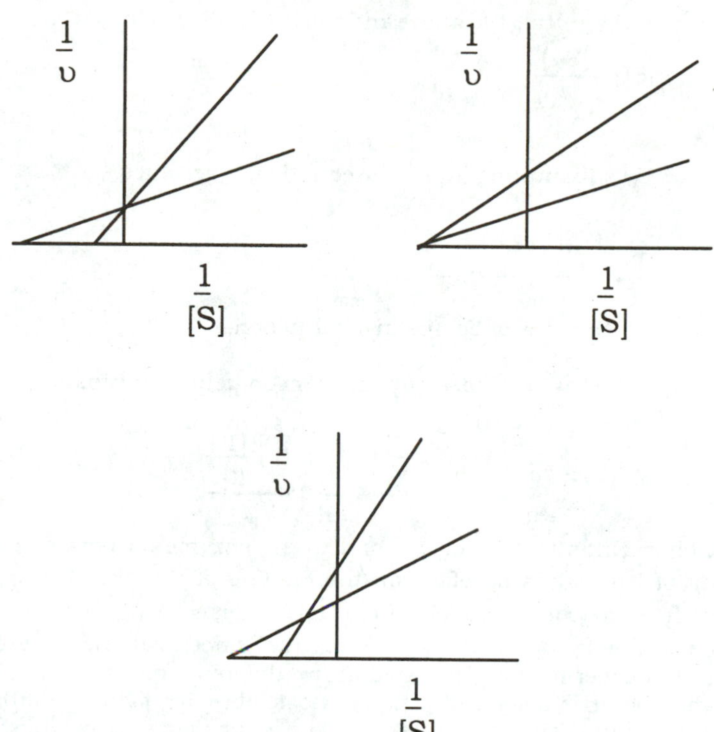

1
υ
1
[S]

Answer: In the top left graph, the results may be due to a competitive inhibitor. The line with the steeper slope is from enzyme kinetic measurements in the presence of an inhibitor whereas the other line is enzyme kinetic data without inhibitor. In competitive inhibition, V_{max}, the reciprocal of the $1/\upsilon$ intercept, is independent of the presence of inhibitor. However, $K_{m,app}$ increases, as reflected in the decrease (in absolute magnitude) of the $1/[S]$ intercept which equals $1/K_{m,app}$. In competitive inhibition, I and S compete for the same enzyme-binding site. The apparent K_m increases because, in the presence of inhibitor, higher concentrations of substrate are required to half-saturate the enzyme. The lines are described by the following equation:

$$\frac{1}{\upsilon}=\frac{K_m(1+\frac{[I]}{K_I})}{V_{max}}\times\frac{1}{[S]}+\frac{1}{V_{max}}$$

An alternative interpretation of the data is that the inhibitor binds to and forms a complex with the substrate. In this case the equation

$$\frac{1}{\upsilon}=\frac{K_m}{V_{max}}\times\frac{1}{[S]}+\frac{1}{V_{max}}$$

must be modified to take this inhibitor-substrate interaction into account. In this equation, [S] is the initial concentration of substrate available to the enzyme. The concentration of substrate used to construct the plots is the total substrate concentration, $[S_T]$. Under normal conditions, $[S_T]$ = [S], but in this case $[S_T]$ = [S] + [SI] because some of the substrate is complexed to I. We can derive an expression for [S] as follows:

From $[S_T]=[S]+[SI]$, we have

$[S]=[S_T]-[SI]$, and using

$K_I=\frac{[I][S]}{[SI]}$, we see that

$$[SI]=\frac{[I][S]}{K_I}$$

Substituting [SI] above and solving for [S] we find,

$$[S] = \frac{[S_T]}{1 + \frac{[I]}{K_I}}$$

By substituting into $\frac{1}{\upsilon}$ the equation, it becomes :

$$\frac{1}{\upsilon} = \frac{K_m(1 + \frac{[I]}{K_I})}{V_{max}} \times \frac{1}{[S_T]} + \frac{1}{V_{max}}$$

This equation is identical to the case of competitive inhibition.

The top right hand graph may be a case of pure, noncompetitive inhibition, which is described by the following equation

$$\frac{1}{\upsilon} = \frac{K_m(1 + \frac{[I]}{K_I})}{V_{max}} \times \frac{1}{[S]} + \frac{(1 + \frac{[I]}{K_I})}{V_{max}}$$

In this case the inhibitor binds to the enzyme and to the enzyme-substrate complex with equal affinity. The binding of I to E has no effect on the binding of S to E. Thus, K_m is unaffected because the inhibitor-free enzyme is capable of normal catalysis. V_{max} is decreased to $V_{max}/(1 + [I]/K_I)$. In effect, the inhibitor lowers the active enzyme concentration. The data can also be explained as the result of experiments conducted at two different enzyme concentrations. (In this case no inhibition whatsoever is involved.) Another possibility, involving an inhibitor, is that the inhibitor binds irreversibly to E and inactivates it. This is the same as lowering the total enzyme concentration. Finally, we may be looking at a case of random, single-displacement bisubstrate reaction. Binding of substrate A is not affected by binding of B and vice versa.

The bottom, center graph may be an example of mixed noncompetitive inhibition in which the inhibitor I binds to both E and ES but unequally. In this case, I binds to ES with higher affinity (i.e., lower K'_I) than to S. V_{max} is decreased and $K_{m,app}$ is increased. Alternatively, this may be an example of a single-displacement bisubstrate mechanism characterized by the following equation

$$\frac{1}{\upsilon} = \frac{1}{V_{max}}(K_m^A + \frac{K_S^A K_m^B}{[B]}) \times \frac{1}{[A]} + \frac{1}{V_{max}}(1 + \frac{K_m^B}{[B]})$$

6. Liver alcohol dehydrogenase (ADH) is relatively nonspecific and will oxidize ethanol or other alcohols, including methanol. Methanol oxidation yields formaldehyde, which is quite toxic, causing, among other things, blindness. Mistaking it for the cheap wine he usually prefers, my dog Clancy ingested about 50 mL of windshield washer fluid (a solution 50% in methanol). Knowing that methanol would be excreted eventually by Clancy's kidneys if its oxidation could be blocked, and realizing that, in terms of methanol oxidation by ADH, ethanol would act as a competitive inhibitor, I decided to offer Clancy some wine. How much of Clancy's favorite vintage (12% ethanol) must he consume in order to lower the activity of his ADH on methanol to 5% of its normal value if the K_m values of canine ADH for ethanol and methanol are 1 millimolar and 10 millimolar, respectively? (The K_I for ethanol in its role as competitive inhibitor of methanol oxidation by ADH is the same as its K_m). Both the methanol and ethanol will quickly distribute throughout Clancy's body fluids, which amount to about 15 L. Assume the densities of 50% methanol and the wine are both 0.9.

Answer: The K_m values of alcohol dehydrogenase for ethanol and methanol are 1 mM and 10 mM. How many moles of methanol are in 50 mL of a 50% solution (v/v)? The solution was made by adding 25 mL methanol and adjusting the volume to 50 mL with water. This amount of methanol (i.e., 25 mL) weighs 25 mL x 0.9 g/mL = 22.5 g. The molecular composition of methanol is CH_4O with a weight of 32 g/mol. Thus, Clancy consumed 22.5 g ÷ 32 g/mol) = 0.7 moles of methanol. Fortunately for him, he diluted it into 15 L of body fluid giving a final concentration of 0.7 moles

÷15 L = 46.9 mM. This concentration is well above the Km of alcohol dehydrogenase for methanol at 10 mM. We expect the enzyme to function at

$$\upsilon = \frac{V_{max} \times 46.9 \text{ mM}}{10 \text{ mM} + 46.9 \text{ mM}}$$
$$= 0.82V_{max}$$

Now, how much ethanol must Clancy consume to lower υ to 5% of 0.82 V_{max} or to about .041 V_{max}?

$$\upsilon = \frac{V_{max}[S]}{K_m(1 + \frac{[I]}{K_I}) + [S]}, \text{ or}$$

$$0.041V_{max} = \frac{V_{max} \times 46.9 \text{ mM}}{10 \text{ mM}(1 + \frac{[\text{Ethanol}]}{1.0 \text{ mM}}) + 46.9 \text{ mM}}$$

Solving for [Ethanol], we find [Ethanol] = 109 mM

To raise his alcohol concentration to 109 mM, he must drink:

$$109 \text{ mM} \times 15 \text{ L} = 1.63 \text{ moles of ethanol.}$$

If Clancy gets his alcohol (ethanol) from wine, how much wine must he drink? The M_r of ethanol (C_2H_6O) is 46, so 1.63 moles represents:

$$1.63 \text{ mol} \times 46\frac{\text{g}}{\text{mol}} = 75\text{g ethanol.}$$

At a density of 0.9 g/mL, this represents

$$75 \text{ g pure ethanol} = 12\% \text{ X grams of wine}$$

$$\text{Grams of wine} = \frac{75\text{g}}{.12} = 625 \text{ g}$$

$$\frac{625\text{g}}{0.9\frac{\text{g}}{\text{mL}}} = 694 \text{ mL of wine}$$

Solving for V_T,

V_T= 694 mL of 12% ethanol. Clancy needs about one 750-mL bottle of wine.

7. List six general ways in which enzyme activity is controlled.

Answer: Enzyme activity may be controlled by:

1. Accumulation of product as the reaction approaches equilibrium. As substrate is converted to product, [P] increases.
2. Availability of substrates and cofactors.
3. Regulation of the amounts of enzyme synthesized or degraded by cells.
4. Covalent modifications catalyzed by modifying enzymes or converter enzymes may activate or inhibit an enzyme or alter its kinetic properties.
5. Allosteric regulation leading to either inhibition or activation.
6. Regulation by interaction with modulator proteins.
7. Activation or inactivation by proteolysis.

8. Why do you suppose proteolytic enzymes are often synthesized as inactive zymogens?

Answer: Proteolytic enzymes hydrolyze peptide bonds of proteins, the most diverse and abundant class of biopolymers. Clearly cells must take care in the biosynthesis of proteases to guard against inadvertent digestion of cellular proteins. Thus, proteolytic enzymes are typically produced as inactive zymogens that are activated only at the site at which their action is required.

9. Draw a Lineweaver-Burk plots for the following: A Monod-Wyman-Changeux allosteric K enzyme system, showing separate curves for the kinetic response in (1) the absence of any effectors; (2) the presence of allosteric activator A; and (3) the presence of allosteric

inhibitor I. Also draw a similar set of curves for a Monod-Wyman-Changeux allosteric V enzyme system.

Answer: The Monod-Wyman-Changeux model considers a multimeric enzyme with n substrate binding sites. The enzyme can exist in two states, R for relaxed and T for taut or tense, that differ in their affinity for substrate. The R state has high affinity whereas, in the extreme case, the T state has no affinity for substrate. Under given conditions (such as temperature, pH, ionic strength, etc.) R and T are in equilibrium and the ratio of their concentrations is governed by an equilibrium constant. Cooperativity occurs because substrate preferentially binds to R thus shifting the R to T equilibrium towards R.

In the absence of cooperativity a multisubunit enzyme would obey simple Michaelis-Menten kinetics with each substrate binding site acting independently. In the cooperative case, substrate binding leads to an increase in the population of high affinity sites, which manifests itself as a higher order dependence of velocity on substrate concentration. In a purely cooperative system, the dependence order will be equal to the number of substrate binding sites in the multimeric protein.

Michaelis-Menten kinetics is governed by

$$v = \frac{Vmax[S]}{Km + [S]}$$

Cooperativity is described by

$$v = \frac{Vmax[S]^n}{Km^n + [S]^n}$$

The cooperativity equation may be further generalized to take into account heterotropic interactions. Heterotropic interactions refer to binding of molecules at sites other than substrate-binding sites that influence the R to T equilibrium. In an allosteric K enzyme system, heterotropic interactions will influence Km, the substrate concentration at which the system is functioning at half Vmax. A heterotropic inhibitor shifts the equilibrium towards the T state and thus increases the apparent Km whereas a heterotropic activator shifts the equilibrium towards the R state and thus decreases the apparent Km. In order to include heterotropic interactions we replace Km with the following:

$$Km, app = Km\frac{(1+\beta)}{(1+\gamma)}$$

Where $\beta = [I]/K_I$ and $\gamma = [A]/K_A$, and [I] and [A] refer to inhibitor and activator concentrations, and K_I and K_A are inhibitor and activator dissociation binding constants. The expression for v becomes:

$$v = \frac{Vmax[S]^n}{(Km\frac{(1+\beta)}{(1+\gamma)})^n + [S]^n}$$

In the Lineweaver-Burk plot shown below the solid line represents cooperative kinetics in the absence of both inhibitor and activator for n = 2.0. The dashed line (below) is cooperative kinetics in the presence of activator. The dotted line (above) is cooperative kinetics in the absence of inhibitor.

The concentration at which the system reaches half-maximum velocity (i.e., 1/v = 2) is lower in the presence of activator and higher in the presence of inhibitor.

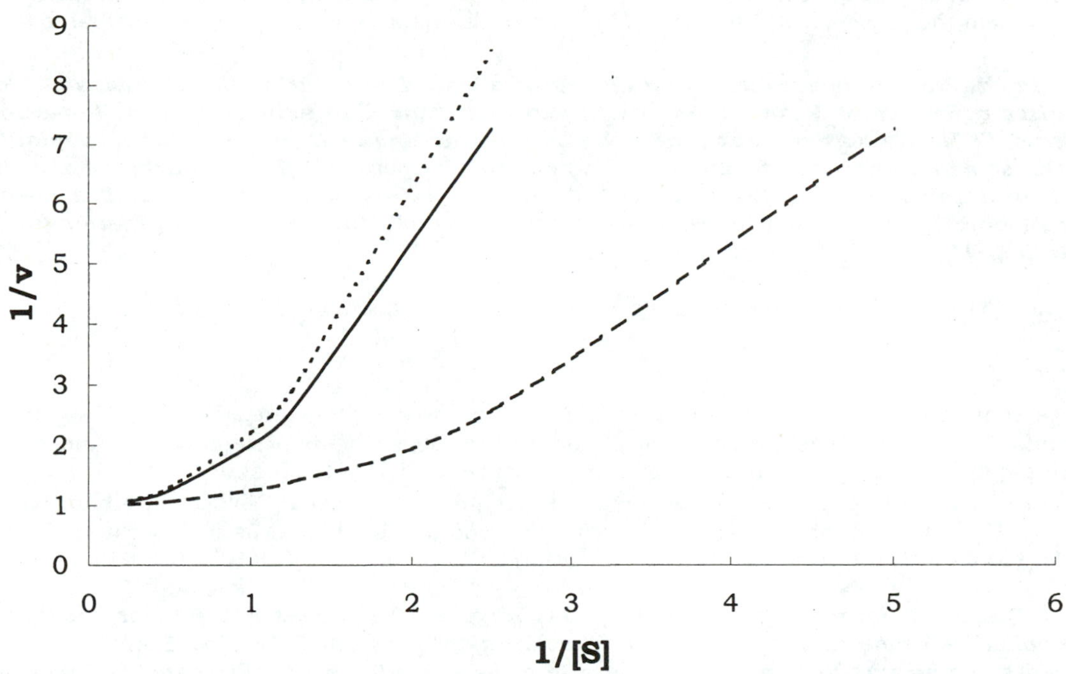

In an allosteric V enzyme system, heterotropic interactions will influence Vmax; Km apparent does not change. The Michaelis-Menten equation may be used with Vmax being modified by the term (1 + β)/1 + γ) or

$$v = \frac{Vmax\frac{(1 + \beta)}{(1+\gamma)}[S]}{Km + [S]}$$

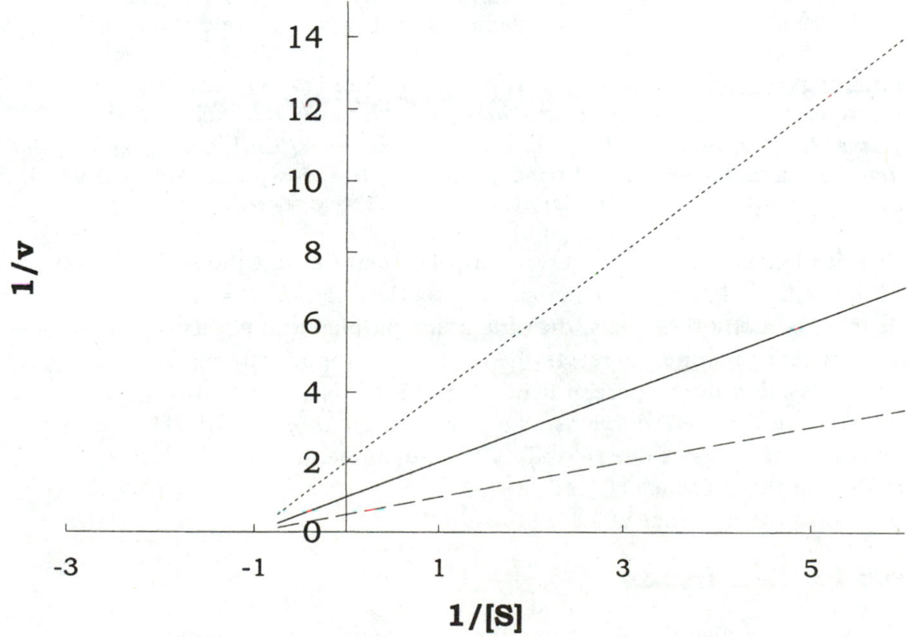

In the plot shown above the solid line is in the absence of inhibitor and activator, the dashed line (below) is in the presence of activator, and the dotted line (above) is in the presence of inhibitor.

10. In the Monod-Wyman-Changeux model for allosteric regulation, what values of L and relative affinities of R and T for A will lead activator A to exhibit positive homotropic effects? (That is, under what conditions will the binding of A enhance further A-binding, in the same manner that S-binding shows positive cooperativity?) Similarly, what values of L and relative affinities of R and T for I will lead inhibitor I to exhibit positive homotropic effects? (That is, under what conditions will the binding of I promote further I-binding?)

Answer: The term L is the equilibrium constant for **R** to **T** transition:

$$R_o \leftrightarrows T_o, \text{ and } L = \frac{[T_o]}{[R_o]}$$

Large values of L indicate that the **T** state is favored over the **R** state, thus $[T_o]>[R_o]$. In this situation, A will show positive homotropic binding. Because A binds preferentially to the **R** state, A-binding will favor a substantial increase in the concentration of the **R** state.

For small values of L, the **R** state is favored over the **T** state, thus $[T_o]<[R_o]$. Inhibitor binding to the **T** state will show positive homotropic binding in this situation because I binds preferentially to the **T** state.

11. The cAMP formed by adenylyl cyclase (Figure 10.42) does not persist because 5'-phosphodiesterase activity prevalent in cells hydrolyzes cAMP to give 5'-AMP. Caffeine inhibits 5'-phosphodiesterase activity. Describe the effects on glycogen phosphorylase activity that arise as a consequence of drinking lots of caffeinated coffee.

Answer: By inhibiting phosphodiesterase activity cAMP levels will slowly rise or will remain elevated for longer periods of time. This will prolong stimulation of cAMP-dependent protein kinase leading to persistent phosphorylation of glycogen phosphorylase kinase and in turn glycogen phosphorylase. Caffeine is an allosteric inhibitor of glycogen phosphorylase b, the unphosphorylated form of the enzyme that is relatively inactive and sensitive to allosteric regulation.

(Caffeine can influence three physiological conditions: Functioning of adenosine receptors, intracellular calcium levels, and cyclic nucleotide levels. Caffeine is a competitive inhibitor of adenosine receptors and is effective at low levels. In contrast, intracellular calcium levels and cyclic nucleotide levels are increased by caffeine but only at high levels of caffeine.)

12. Enzymes have evolved such that their K_m values (or $K_{0.5}$ values) for substrate(s) are roughly equal to the in vivo concentration(s) of the substrate(s). Assume that glycogen phosphorylase is assayed at $[P_i] \approx K_{0.5}$ in the absence and presence of AMP or ATP. Estimate from Figure 10.38 the relative glycogen phosphorylase activity when (a) neither AMP or ATP is present, (b) AMP is present, and (c) ATP is present.

Answer: $K_{0.5}$ is defined as the concentration of substrate at which the enzyme functions at $0.5\times V_{max}$. From Figure 10.38b we can estimate the location of $K_{0.5}$ on the x-axis by measuring the distance to V_{max} along the υ-axis, drawing a line parallel to the x-axis at a value of υ equal to $0.5\times V_{max}$, and, where this line intersects the -ATP curve (upper curve), drawing a perpendicular. The point of intersection of the perpendicular on the x-axis is equal to $K_{0.5}$. The intersection of this perpendicular on the +ATP curve, is υ for $= K_{0.5}$ in the presence of ATP.

a. $\upsilon = 0.5\times V_{max}$ in the absence of ATP and AMP (by definition).
b. $\upsilon \approx 0.85\times V_{max}$ in the presence of AMP.
c. $\upsilon \approx 0.12\times V_{max}$ in the presence of ATP.

Questions for Self Study

1. Enzymes have three distinctive features that distinguish them from chemical catalysts. What are they?

2. Match the terms in the first column with the descriptions in the second column.

a. k in $\upsilon = k[A]$	1. K_m
b. $d[P]/dt$ or $-d[S]/dt$	2. V_{max}
c. $d[P]/dt = -d[S]/dt$	3. First-order rate constant
d. k in $\upsilon = k[A][B]$	4. Michaelis constant
e. k_{-1}/k_1	5. V_{max}/E_T
f. $(k_{-1} + k_2)/k_1$	6. The velocity or rate of a reaction.
g. υ at $[S] = \infty$	7. Enzyme:substrate dissociation constant
h. [S] when $v = V_{max}/2$	8. $1/V_{max}$
i. k_{cat}	9. Equilibrium or steady state
j. $1/\upsilon$ when $1/[S] = 0$	10. Second-order rate constant

3. Fill in the blanks. The energy barrier preventing thermodynamically favorable reactions from occurring is known as the _____. In general, catalysts _____ this energy barrier by providing an alternate reaction pathway leading from substrate to product. The intermediate along this reaction pathway that has the highest free energy is known as the _____.

4. Answer True or False

   a. For competitive inhibition the apparent K_m is larger than K_m by the factor $(1 + [I]/K_I)$. _____

   b. For noncompetitive inhibition the apparent V_{max} is larger than V_{max} by the factor $(1 + [I]/K_I)$. _____

   c. A plot of V_{max} versus temperature is linear for a typical enzyme. _____

   d. Ordered, single-displacement reactions refer to bisubstrate reactions in which two substrates bind to the enzyme before catalysis occurs. _____

   e. Enzyme modification occurs during double-displacement or Ping-Pong reaction. _____

5. What is a ribozyme? What is an abzyme?

6. What is the essential difference between the "lock and key" hypothesis of enzyme substrate interaction and the "induced fit" hypothesis?

7. How are the proteins, trypsinogen, chymotrypsinogen, pepsinogen, procarboxypeptidase, proelastase, related to active enzymes?

8. Lactate dehydrogenase is a tetramer that catalyzes the interconversion of lactate and pyruvate. There are two different subunits, A and B that can interact to form either homotetramers or heterotetramers with different kinetic properties. Write the five possible tetrameric combinations. What kind of enzyme regulation does lactate dehydrogenase typify?

9. The enzyme aspartate transcarbamoylase catalyzes the condensation of carbamoyl phosphate and aspartic acid to form N-carbamoylasparate. A plot of initial velocity versus aspartate concentration is not a rectangular hyperbola as found for enzymes that obey Michaelis-Menten kinetics. Rather, the plot is sigmoidal. Further, CTP shifts the curve to the right along the substrate axis and makes the sigmoidal shape more pronounced. ATP shifts the curve to the left and makes the sigmoidal shape less pronounced. What kind of enzyme is aspartate transcarbamoylase and what roles do CTP and ATP play in enzyme activity?

10. Match the items in the left hand column with items in the right hand column.

a. Symmetry Model	1. Second noncatalytic substrate binding site.
b. Monod, Wyman, and Changeux	2. Substrate binding concentration dependence.
c. K system	3. Negative effector.
d. V system	4. Conformation change due to substrate binding
e. Koshland, Nemethy, and Filmer	5. $K_{0.5}$ changes in response to effectors.
f. Induced fit	6. V_{max} regulated.
g. Hill coefficient	7. Sequential model.
h. Increases [R conformation]	8. $R_o \rightarrow T_o$

i. Decreases [R conformation]	9. Two conformational states.
j. Homotropic effector	10. Positive effector.

Answers

1. Enormous catalytic power, specificity, and the ability to be regulated.

2. a. 3; b. 6; c. 9; d. 10; e. 7; f. 1 or 4; g. 2; h. 1 or 4; i. 5; j. 8.

3. Energy of activation; lower; transition state.

4. a. T; b. F; c. F; d. T; e. T

5. A ribozyme is a catalytic RNA. An abzyme is a catalytic antibody.

6. The "lock and key" hypothesis depicts the enzyme as a rigid template into which the substrate fits. In the "induced fit" hypothesis enzymes are highly flexible, dynamic molecules and substrate binding modifies the shape of the enzyme's active site.

7. These proteins are precursors of the enzymes, trypsin, chymotrypsin, pepsin, carboxypeptidase, and elastase. They are converted to active enzymes by hydrolysis of peptide bonds.

8. A_4, A_3B, A_2B_2, AB_3, B_4; isozymes.

9. Aspartate transcarbamoylase is a regulatory enzyme. CTP is a negative heterotropic effector and ATP is a positive heterotropic effector.

10. a. 8 or 9; b. 8 or 9; c. 5; d. 6; e. 7; f. 4; g. 2; h. 10; i. 3; j. 1.

Additional Problems

1. A researcher was studying the kinetic properties of β-galactosidase using an assay in which o-nitrophenol-β-galactoside (ONPG), a colorless substrate, is converted to galactose and o-nitrophenolate, a brightly-colored, yellow compound. Upon addition of substrate to 0.25 mM substrate to a fixed amount of enzyme, o-nitrophenolate (ONP) production was monitored as a function of time by spectrophotometry at λ = 410 nm. The following data were collected:

Time (sec)	$A_{410\ nm}$
0	0.000
15	0.158
30	0.273
45	0.360
60	0.429
75	0.484
90	0.529
150	0.652
210	0.724
270	0.771
330	0.805
390	0.830
450	0.849
510	0.864

a. Convert $A_{410\ nm}$ to concentration of o-nitrophenolate, [ONP], using $\varepsilon = 3.76\ mM^{-1}cm^{-1}$ as the extinction coefficient.
b. Plot [ONP] versus time.
c. Determine the initial velocity of the reaction.
d. Explain why the curve is nearly linear initially and later approaches a plateau.

e. Describe how K_m and V_{max} for β-galactosidase can be determined with additional experimentation.

2. Under what conditions can an enzyme assay be used to determine the relative amounts of an enzyme present?

3. For many reactions, it is often difficult to measure product formation, so coupled assays are often used. In a coupled enzyme assay, the activity of an enzyme is determined by measuring the activity of a second enzyme that uses as substrate the product of a reaction catalyzed by the first enzyme. The utility of coupled assays is that the product of the second enzyme is easy or convenient to measure. You are asked to design a coupled assay to measure the relative amounts of a particular enzyme in several samples. In qualitative terms with respect to K_m and V_{max} for both enzymes, explain how an appropriate coupled assay is set up.

4. In the case of a competitive inhibitor, explain how the inhibitor dissociation constant, K_I, can be determined from enzyme kinetic data.

5. Is K_m a good measure of the dissociation constant for substrate binding to enzyme? Explain.

6. Derive an expression for υ as a function of [S] for mixed, noncompetitive inhibition.

7. For certain inhibitor/enzyme combinations the presence of inhibitor decreases V_{max}; however, it also decreases $K_{m,app}$. This seems to suggest that in the presence of an inhibitor the remaining active enzyme's performance is improved since it saturates at a lower substrate concentration. Can you explain this paradox?

8. Enzyme activity may be controlled in several ways (see answer to question 1 above). It is clear that some of these regulatory mechanisms are effective nearly instantly whereas others are slowly activated. With respect to duration, certain controls are short-lived whereas others are long-lasting. Compare and contrast enzyme control mechanisms with respect to these parameters.

Abbreviated Answers

1a. The concentration of ONP, [ONP], is calculated using Beer's Law, $A = \varepsilon \times C \times l$, where ε is the extinction coefficient, C is the concentration, and l is the path length (1 cm in this case). $[ONP] = A_{410\ nm}/(\varepsilon \times 1\ cm)$

Time (sec)	A_{410nm}	[ONP] (mM)
0	0	0
15	0.158	0.042
30	0.273	0.073
45	0.36	0.096
60	0.429	0.114
75	0.484	0.129
90	0.529	0.141
150	0.652	0.173
210	0.724	0.193
270	0.771	0.205
330	0.805	0.214
390	0.83	0.221
450	0.849	0.226
510	0.864	0.23

b. The plot of [ONP] versus time is shown below

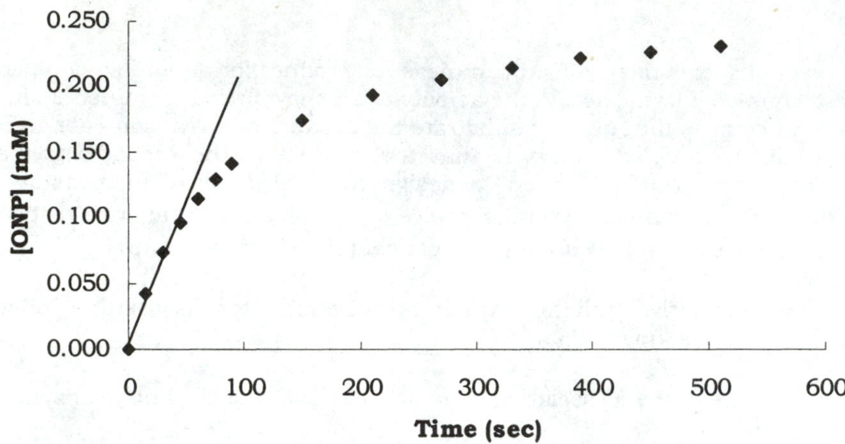

c. The initial velocity is equal to the slope of a line drawn through the initial points as shown above and has a value of 0.00188 mM/sec.
d. The enzyme is converting substrate to product and the curve is the time course of this process. The initial points are linear because the enzyme is obeying Michaelis-Menten kinetics. At later times, two things happen to cause the rate of product formation, given by the slope of a tangent drawn along the graph, to decrease: [S] decreases; and, [P] increases. The curve is approaching 2.5 mM, the initial value of substrate, indicating that nearly all of the initial substrate has been converted to product.
e. By repeating the experiment at different initial substrate concentrations, the initial velocities at varying substrate concentrations can be measured. These data are used to determine K_m and V_{max}.

2. At substrate concentrations high relative to K_m, an enzyme functions at or near V_{max}. V_{max}, given by $V_{max} = k_2 E_T$, is proportional to the total enzyme concentration.

3. The following reaction sequence represents a coupled enzyme assay.

$$S \xrightarrow{E1} P1 \xrightarrow{E2} P2$$

We are interested in determining the relative amount of E_1 by measuring the rate of production of P_2. The concentration of the substrate for E_1, namely S, must be large relative to K_{m1} in order for E_1 to function at V_{max}. Also, the rate at which E_2 produces P_2 must be constant. This requires that $[P_1]$ be at a steady-state concentration. The rate of change of $[P_1]$ is given by

$$\frac{d[P_1]}{dt} = V_{max,E_1} - \frac{V_{max,E_2}[P_1]}{K_{m2} + [P_1]}$$

For $[P_1]$ to be at steady state, $\frac{d[P_1]}{dt} = 0$, and solving for $[P_1]$ we find :

$[P_1]_{steady\ state} = \dfrac{K_{m2} \times V_{max,E_1}}{V_{max,E_2} - V_{max,E_1}}$ which is positive only when

$V_{max,E_2} > V_{max,E_1}$

Thus, [S] must be high enough to saturate E_1 and $k_2[E_{2,T}]$ must be greater than $V_{max,E1}$.

4. For competitive inhibition, $K_{m,app} = K_m(1 + [I]/K_I)$. By measuring $K_{m,app}$ at several different inhibitor concentrations and plotting $K_{m,app}$ against [I], the slope of the resulting straight line divided by the y-intercept (equal to K_m) is $1/K_I$.

5. K_m and K_d are defined as follows:

$$K_m = \frac{k_{-1} + k_2}{k_1}, \text{ and } K_D = \frac{k_{-1}}{k_1}$$

When k_{-1} is large relative to k_2, $K_m \approx K_D$

6. The following reactions occur for mixed, noncompetitive inhibition:

$$E + S \rightarrow ES \rightarrow P;\ E + I \rightarrow EI;\ ES + I \rightarrow IES$$

Binding of I to E and to ES are governed by:

$$K_I = \frac{[E][I]}{[EI]};\ K_I' = \frac{[ES][I]}{[IES]}$$

These equations can be rearranged to give:

$$[EI] = \frac{[E][I]}{K_I};\ [IES] = \frac{[ES][I]}{K_I'}$$

The rate of change of [ES] is given by:

$$\frac{d[ES]}{dt} = -k_2[ES] - k_{-1}[ES] + k_1[E][S]$$

Using this equation and making the steady-state assumption we find that

$$[ES] = \frac{k_1[E][S]}{k_2 + k_{-1}} = \frac{[E][S]}{K_m}$$

The total concentration of enzyme, E_T, is given by:

$$[E_T] = [E] + [ES] + [EI] + [IES]$$

Substituting the expressions for [ES], [EI], and [IES] above we find that:

$$[E_T] = [E] + \frac{[E][S]}{K_m} + \frac{[E][I]}{K_I} + \frac{[E][S][I]}{K_m K_I'}, \text{ and solving for [E] we find :}$$

$$[E] = \frac{\dfrac{[E_T] \times K_m}{(1 + \frac{[I]}{K_I'})}}{K_m \dfrac{(1 + \frac{[I]}{K_I})}{(1 + \frac{[I]}{K_I'})} + [S]}$$

The rate of product formation is given by:

$$\upsilon = k_2[ES] = k_2 \frac{[E][S]}{K_m}$$

Finally, substituting the expression for [E] above we find:

$$\upsilon = \frac{\dfrac{k_2[E_T]}{(1 + \frac{[I]}{K_I'})} \times [S]}{K_m \times \dfrac{(1 + \frac{[I]}{K_I})}{(1 + \frac{[I]}{K_I'})} + [S]} = \frac{V_{max,app}[S]}{K_{m,app} + [S]}$$

Clearly, $V_{max,app}$ is always less than V_{max}. $K_{m,app}$ may be less than, equal to, or greater than K_m depending on the magnitude of K_I relative to K'_I.

7. I binds to ES with greater affinity (lower dissociation constant) than to E. Therefore, the addition of I is expected to affect the E + S ⇆ ES equilibrium by shifting it to the right allowing more ES to form at lower [S].

8. Variations in [product], [S], and [cofactors] have immediate consequences for enzyme activity. The same is true for allosteric regulation. Further, these controls last as long as the signals are present. Genetic controls, such as induction and repression leading to an increase or decrease in enzyme synthesis, are fast, although not immediate, and can be long lasting. Clearly, degradation is long lasting. However, a delicate balance between synthesis and degradation can lead to a very responsive system of control. Covalent modification may lead to rapid and long lasting changes in enzyme activity. Balancing the activity of modifying enzymes against enzymes that can reverse these modifications creates a sensitive system of enzyme control. Finally, zymogens and isozymes are examples of enzyme control mechanisms that can be extremely long lasting.

Summary

Enzymes are the catalysts of metabolism, allowing living systems to achieve kinetic control over the thermodynamic potential within organic reactions. Enzymes have three distinctive attributes that are essential to their biological purpose: enormous catalytic power, great selectivity in catalyzing only very specific reactions, and the ability to be regulated so that their activity is compatible with the momentary needs of the cell. Formally, enzymes are classified by a system of nomenclature based on the particular reaction they catalyze, although certain trivial names for enzymes often enjoy common usage. Enzymes may be simple proteins or they may be proteins complexed with nonprotein components called cofactors. Cofactors include metal ions and organic molecules known as coenzymes.

An enzyme accelerates the rate of a process by lowering the thermodynamic barrier to reaction, known as the free energy of activation. The reactant in an enzyme-catalyzed reaction is called the substrate. The fact that graphs of enzyme activity as a function of substrate concentration show a limiting plateau or saturation level was the important clue that an enzyme (E) actually binds its substrate (S) to form an enzyme:substrate (ES) complex that can react to yield product (P). Enzyme kinetic analysis is based on the notion put forth by Michaelis and Menten that E and S reversibly interact, and ES can react to form P. Briggs and Haldane extended this model by assuming that the system $E + S \leftrightarrows ES \rightarrow P$ quickly reaches a steady state condition where $d[ES]/dt = 0$. The standard Michaelis-Menten equation for an enzymatic reaction is then derived:

$$\upsilon = \frac{V_{max}[S]}{K_m + [S]}$$

V_{max} is the maximal velocity attained when [S] is so great as to not be rate-limiting; K_m is a constant reflecting the relative affinity and reactivity of the enzyme-substrate interaction. Because graphs of υ versus [S] are not linear, the Michaelis-Menten expression is often rearranged to give equations that yield linear relationships between rate and substrate concentration.

Because their activity is highly dependent on maintenance of their native protein structure, enzymes are very sensitive to pH and temperature. Enzyme inhibition has been an important means for investigating enzymes and their mode of action. Such studies have also contributed to human health through the science of pharmacology, since many drugs exert their action as inhibitors of enzymes. Some inhibitors act by competing with the substrate for binding to the active site of the enzyme, altering the apparent K_m. Other inhibitors uniquely affect the apparent V_{max} while still others affect both $K_{m,app}$ and $V_{max,app}$. Inhibitors, which react covalently with the enzyme, can lead to irreversible inhibition of its activity.

Most enzymatic reactions involve more than one substrate. These multi-substrate reactions can be analyzed kinetically to determine whether two (or more) substrates are bound before reaction occurs (so-called sequential or single-displacement reactions), or whether one substrate binds and reacts with the enzyme before the other substrate is bound (Ping-Pong or double-displacement reactions).

Recent discoveries have broadened our view of biocatalysis. Certain RNA molecules have the essential features of enzymes: rate enhancement, catalytic turnover and specificity of reaction. This finding has important implications to our view of molecular or pre-biotic evolution. Further, biocatalysts in the form of antibodies have been "tailor-made" by using transition state analogs of a reaction as antigens. The antibodies elicited in response to these analogs act as biocatalysts by facilitating entry of the reactants into a reactive transition-state intermediate. The creation of "designer enzymes" has become a realistic possibility.

Enzymes display an extraordinary specificity with regard to the substrates they act upon and the reaction they catalyze. Molecular recognition through structural complementarity is the basis of this specificity. The active sites of enzymes are pockets or clefts in the protein structure whose shape and charge distribution are complementary to the substrate's own properties. In turn, substrate binding induces changes in protein conformation so that the substrate is embraced by the enzyme and catalytic groups are brought to bear. In effect, the enzyme:substrate complex mimics the transition state intermediate of the reaction.

Regulation of enzyme activity can be achieved at several levels: enzyme synthesis or degradation, reversible covalent modifications catalyzed by converter enzymes, noncovalent interactions with cellular metabolites (allosteric regulation), zymogen activation, isozymes or modular proteins.

Allosteric regulation occurs when regulatory metabolites modulate the activity of allosteric enzymes by binding at sites distinct from the active site. Such effectors may activate (positively regulate) or inhibit (negatively regulate) enzyme activity. Often, an allosteric enzyme catalyzes the first step in an essential metabolic pathway. The end product of the pathway may be a feedback inhibitor (negative regulator) of this allosteric enzyme. Allosteric enzymes are characterized by sigmoid υ versus [S] kinetic patterns. The sigmoid shape of such kinetic curves suggests that the enzymatic rate is proportional to $[S]^n$, where $n > 1$. Allosteric enzymes are oligomeric and possess multiple ligand-binding sites.

The Monod, Wyman and Changeux model for the behavior of allosteric proteins postulates that allosteric proteins are oligomers composed of identical subunits. The protein can exist in either of two conformational states, the **T** or "taut" state, and the **R** or "relaxed" state. The distribution of the protein between the two conformational states is characterized by the equilibrium constant, L: $L = \mathbf{T_0}/\mathbf{R_0}$. The different states have different affinities for the various ligands. Usually, it is assumed that the substrate S binds "only" to the **R** state; the positive allosteric effector A binds "only" to the **R** state; and the negative allosteric effector I binds "only" to the **T** state. Given these parameters, the sigmoid rate response of allosteric enzymes to substrate concentration and the effects exerted by allosteric effectors are understood. The system described above is called a "*K*" system, because the substrate concentration giving half-maximal velocity, defined as $K_{0.5}$, varies, while V remains constant. In the "*V*" system of allosteric control, **R** and **T** states have the same affinity for S, but the **R** state is catalytically active, while the **T** state is not. A and I still differ in their relative affinities for **R** and **T**. Both the *K* and the *V* systems assume that the oligomeric allosteric protein undergoes a concerted conformational transition so that all subunits have identical conformations at the same time. The Koshland, Nemethy and Filmer model rests on the "induced fit" hypothesis. That is, via tightly linked subunit interactions, the binding of ligand to one subunit can influence the ligand-binding affinity of neighboring subunits. Therefore, an oligomeric protein could pass through a sequential series of conformations with differing ligand affinities, and different subunits may not necessarily have the same conformation at the same time.

Glycogen phosphorylase is a homodimeric protein that catalyzes the catabolism of glycogen into glucose-1-phosphate. The enzyme is inhibited by two allosteric inhibitors, ATP and glucose-6-phosphate, which both function as negative heterotropic effectors by decreasing the affinity of P_i. AMP acts as a positive heterotropic effector by binding to the enzyme at the ATP-binding site. AMP binding not only blocks ATP binding but it increases the enzyme's affinity for P_i. These allosteric regulators all function on phosphorylase b. Phosphorylase a, a more active form of the enzyme, is relatively insensitive to allosteric regulation. The physical basis for the difference in these two forms is the presence of a phosphate group on serine 14. The phosphate group may be removed by phosphoprotein phosphatase I and reattached by glycogen phosphorylase kinase, an enzyme that is itself activated by phosphorylation. These covalent modifications are the result of an enzymatic cascade that is initiated with hormone binding to cell surface receptors, transduced into release of a GTP-binding protein that stimulates adenylyl cyclase, which in turn produces cAMP that stimulates cAMP-dependent protein kinase leading to phosphorylation of glycogen phosphorylase kinase.

Chapter 11

Mechanisms of Enzyme Action

Chapter Outline

- Transition state intermediate
  - Energy difference between ES and $EX^{\ddagger}$ is smaller than that between S and $X^{\ddagger}$ therefore catalyzed rate of product formation will be much faster
  - Enzyme binds substrate and transition state but stabilizes transition state more than substrate
- Catalytic mechanisms or factors
  - Destabilization of ES complex
    - Entropy loss upon formation of ES complex
      - ES complex unstable due to loss of degrees of freedom
      - Entropy loss balanced by enzyme:substrate and enzyme:transition state interactions
      - Transition state binding favored over substrate
    - Destabilization of S because E designed to bind transition state
      - Structural strain
      - Desolvation: Desolvated ionic groups show increased reactivity
      - Electrostatic effects: Charge proximity
  - Stabilization of $EX^{\ddagger}$ complex: Transition state analogs are potent inhibitors
  - Covalent catalysis
    - Nucleophilic catalysis
      - Nucleophilic centers: Amines, carboxylates, hydroxyls, imidazole, thiol
      - Electrophilic targets: Phosphoryl, acyl, glycosyl groups
    - Electrophilic catalysis: Coenzymes
  - General acid-base catalysis
    - Specific acid-base catalysis: H^+ or OH^- involved: No dependence on buffer concentration
    - General acid-base catalysis
      - AH (protonated acid) donates proton
      - B^- (unprotonated base) accepts proton
  - Metal ion catalysis
    - Metalloenzymes: Metal ion tightly bound
    - Metal activated enzymes: Metal ion loosely bound
    - Metal functions
      - Stabilize formation of negative charge
      - Powerful nucleophile at neutral pH
  - Proximity and orientation: Raises effective concentration of reacting substrates
- Serine proteases
  - Enzyme diversity
    - Trypsin, chymotrypsin, elastase: Digestive enzymes
    - Thrombin: Blood clotting
    - Subtilisin: Bacterial enzyme
    - Tissue plasminogen activator: Blood clot breakdown: Converts plasminogen to plasmin

- Acetylcholinesterase (not a protease): Degrades neurotransmitter acetylcholine
- Digestive enzyme specificity: Cleavage on carbonyl side
  - Chymotrypsin: Phe, Trp, Tyr: Hydrophobic pocket near catalytic triad
  - Trypsin: Lys, Arg: Aspartic acid at bottom of hydrophobic pocket
  - Elastase: Small neutral amino acids: Mouth of pocket closed by threonine and valine
- Burst kinetics: Rapid release of amino end: Carboxyl end attached to enzyme and slowly hydrolyzed
- Catalytic triad: His57, Asp102, Ser195
  - Ser195: Acyl-enzyme intermediate formed by hydroxylate attack on carbonyl carbon of peptide bond
  - His57
    - General base: Deprotonates Ser195: Deprotonates water
    - General acid: Protonates carbonyl oxygen: Protonates peptide nitrogen
  - Asp102: Orients His57

- Aspartic proteases
  - Enzyme diversity
    - Pepsin, chymosin: Digestive enzymes
    - Cathepsin D: Protein degradation in lysosomes
    - Renin: Blood pressure regulation
    - AIDS HIV-1 protease: Viral life cycle
  - Catalytic dyad: Two aspartic acids
    - Unprotonated Asp: General base: Deprotonates water
    - Protonated Asp: General acid: Protonates carbonyl oxygen

Chapter Objectives

General Considerations

To appreciate why uncatalyzed reactions can be slow, you must understand that even the most favorable path from reactant to product passes through some transition state intermediate. Formation of the transition state intermediate is an energy-requiring process, and the energy necessary for its formation, the activation energy, is an energy barrier that must be overcome for the reaction to occur. Enzyme-catalyzed reactions also have an activation energy but it is not the energy difference between the substrate and the transition state that matters; rather it is the energy difference between the enzyme-substrate complex and the enzyme-transition state complex. Enzymes increase reaction rates because the energy barrier between ES and $EX^{\ddagger}$ is less than the barrier between S and $X^{\ddagger}$ (See Figure 11.1).

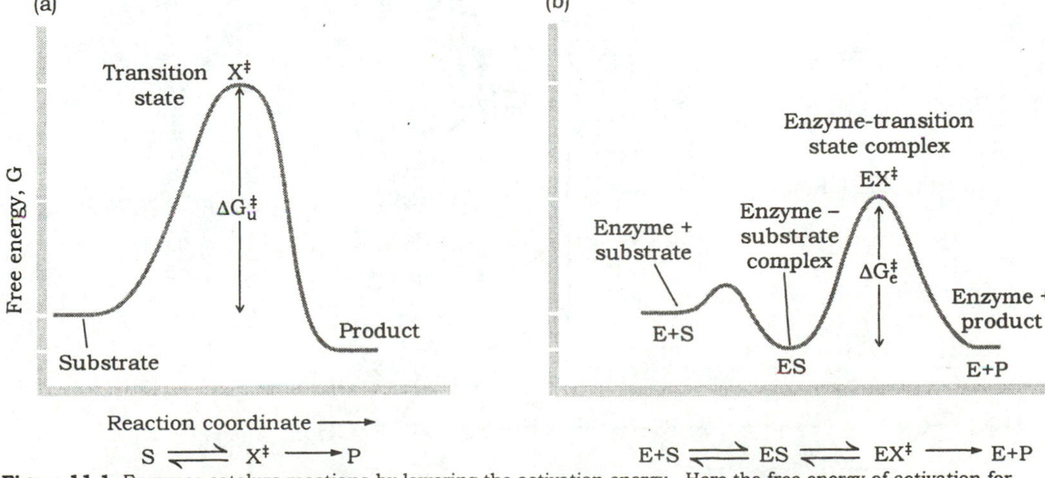

Figure 11.1 Enzymes catalyze reactions by lowering the activation energy. Here the free energy of activation for (a) the enzyme-catalyzed reaction, $\Delta G_e^{\ddagger}$, is smaller than that for (b) the noncatalyzed reaction, $\Delta G_u^{\ddagger}$.

Understand the catalytic mechanisms used by enzymes to achieve enormous rate accelerations, which include: 1. Entropy loss in ES formation; 2. Destabilization of ES due to strain, desolvation or electrostatic effects; 3. Covalent catalysis; 4. General acid or base catalysis; 5. Metal ion catalysis; and 6. Proximity and orientation.

Serine Proteases - Covalent Catalysis

The serine proteases and related enzymes are arguably the best-characterized group of enzymes. The key to catalysis is a catalytic triad of His-57, Asp-102, and Ser-195 (in chymotrypsin). Understand the role of serine in catalysis, the fact that it is transiently, covalently modified during catalysis accounting for the burst kinetics, a rapid release of one product followed by a slow release of the second product. The mechanism based on this catalytic triad is found in a number of enzymes, including trypsin, chymotrypsin, elastase, thrombin, subtilisin, tissue plasminogen activator, and acetylcholinesterase, to name a few. We have already encountered a few of the digestive serine proteases in Chapter 4 as tools used in primary sequence determination of proteins. Substrate specificity is determined by a substrate-binding pocket, which is a separate feature from the catalytic triad.

Aspartic Proteases - General Acid/Base Catalysis

Pepsin, chymosin, cathepsin D, and HIV-1 protease are examples of aspartic proteases. The mechanism of action is acid/base catalysis with two aspartic acid residues functioning as a general acid and a general base. The pH profile of enzymes with this type of catalysis is bell-shaped, indicating the involvement of a proton donor and acceptor.

Problems and Solutions

1. Tosyl-L-phenylalanine chloromethylketone (TPCK) specifically inhibits chymotrypsin by covalently labeling His^{57}.

a. Propose a mechanism for the inactivation reaction, indicating the structure of the product(s).

b. State why this inhibitor is specific for chymotrypsin.

c. Propose a reagent based on the structure of TPCK that might be an effective inhibitor of trypsin.

Answer: The structure of tosyl-L-phenylalanine chloromethylketone (TPCK) and the reaction mechanism leading to covalent labeling of His^{57} are shown below:

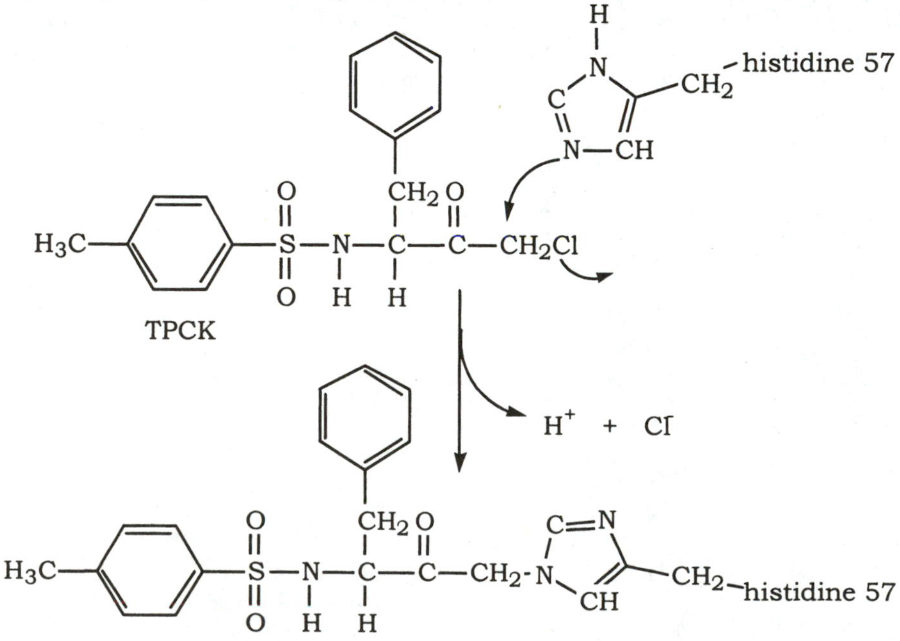

H
N
C
CH_2 histidine 57
N—CH
O
CH_2 O
H_3C—⟨ring⟩—S—N—C—C—CH_2Cl
O H H
TPCK

H^+ + Cl^-

O
CH_2 O
C=N
H_3C—⟨ring⟩—S—N—C—C—CH_2-N
CH
CH_2 histidine 57
O H H

b. The phenyl group of TPCK binds to the specificity pocket of chymotrypsin by hydrophobic interactions. This brings the chloromethylketo group into the active site.

c. Trypsin shows specificity for peptide bonds in which the carbonyl carbon is from a positively charged amino acid such as arginine or lysine. Replacing the phenylalanine residue of TPCK with arginine or lysine may produce reagents specific for trypsin.

2. In this chapter, the experiment in which Craik and Rutter replaced Asp^{102} with Asn in trypsin (reducing activity 10,000-fold) was discussed.
a. On the basis of your knowledge of the catalytic triad structure in trypsin, suggest a structure for the "uncatalytic triad" of Asn-His-Ser in this mutant enzyme.
b. Explain why the structure you have proposed explains the reduced activity of the mutant trypsin.
c. See the original journal articles (Sprang, et al., 1987. Science 237:905-909 and Craik, et al., 1987. Science 237:909-13) to see what Craik and Rutter's answer to this question was.

Answer: a. Catalysis involves ionization of Ser-195, which is accomplished in the case of wild-type trypsin by loss of a proton to His-57 whose pK_a is influenced by Asp-102 as shown below. In the case of Asn-102 substitution, the amino group of Asn-102 functions as a hydrogen-bond donor thus preventing His-57 from accepting a proton from Ser-195.

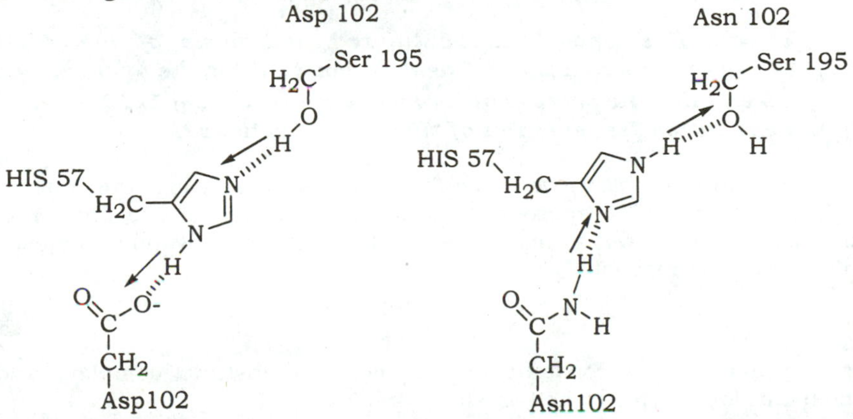

Asp 102　　　　　　　　Asn 102

Ser 195　HIS 57　Asp102　　　　Ser 195　HIS 57　Asn102

b. In the wild-type enzyme, Asp-102 functions as a hydrogen-bond donor and acceptor. In the mutant enzyme, Asn-102 predominantly functions as a hydrogen-bond donor as shown above.

3. Pepstatin (below) is an extremely potent inhibitor of the monomeric aspartic proteases, with K_i values of less than 1 nM.
a. Based on the structure of pepstatin, suggest an explanation for the strongly inhibitory properties of this peptide.
b. Would pepstatin be expected to also inhibit the HIV-1 protease? Explain your answer.

Answer: The structure of pepstatin is shown below.

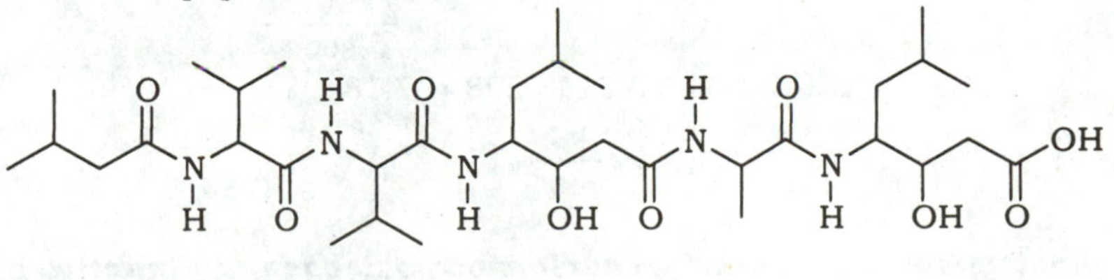

It contains 5 peptide bonds joining together the following groups: methylbutanoic acid; valine; valine; 4-amino-3-hydroxy methylheptanoic acid; alanine; and, 4-amino-3-hydroxymethyl heptanoic acid. Thus, pepstatin has a structure very similar to a 6-residue polypeptide chain of hydrophobic amino acids. The active site of pepsin can accommodate a chain of up to 7 residues in length. Further, pepsin cleaves peptide bonds between hydrophobic amino acids. Pepstatin binds tightly to the active site because of the nature of its hydrophobic side chains. Assuming that the catalytic site is in the middle of the hydrophobic binding pocket, pepstatin will be positioned with a carbon-carbon single bond at the active site.

```
 H  H    H
 |  |    |
—C—C—C—N—
 |  |  ||
OH  H  O
```

Thus, a hydroxyl group replaces a carbonyl. The high pK_a of the hydroxyl will resist catalysis. Furthermore, the hydroxylated carbon mimics the tetrahedral amide hydrate transition state.

The HIV-1 protease cleaves between Tyr and Pro in the sequence Ser-Gln-Asn-Tyr-Pro-Ile-Val. The structure of HIV-1 protease is similar to that of mammalian aspartic proteases. The active site is a cleft between two identical subunits with two aspartic acids contributed from both subunits at the heart of the catalytic mechanism. The mammalian aspartic protease preferentially cleaves between hydrophobic amino acids residues. The sequence shown above has only two hydrophobic residues on the C terminus. We might expect that the catalytic pocket of HIV-1 protease to be lined with hydrophilic groups that will accommodate the relatively hydrophilic N-terminal residues. Thus, we do not expect the highly hydrophobic pepstatin to inhibit HIV-I protease. The inhibitor dissociation constants (K_I) of pepsin and HIV-1 for pepstatin are 1 nM and 1 μM, respectively, reflecting the fact that pepstatin binds tightly to pepsin but relatively poorly to HIV-1

4. The k_{cat} for alkaline phosphatase-catalyzed hydrolysis of methylphosphate is approximately 14/sec at pH 8 and 25°C. The rate constant for the uncatalyzed hydrolysis of methylphosphate under the same conditions is approximately 1×10^{-15}/sec. What is the difference in the activation free energies of these two reactions?

Answer: To solve this problem we need an expression relating the rate constants to thermodynamic properties. The appropriate equation is derived in the answer to problem 6 (below). Using this equation, we can calculate the difference in activation energies between the uncatalyzed and catalyzed reaction:

$$\frac{k_e}{k_u} = \frac{K_S}{K_T}$$

Where K_S is the dissociation constant for the enzyme-substrate complex, and K_T is the dissociation constant for enzyme-transition state complex.

In general,

$$K = e^{\frac{-\Delta G}{RT}}, \text{and}$$

$$\frac{k_e}{k_u} = \frac{K_S}{K_T} = \frac{e^{\frac{-\Delta G_S}{RT}}}{e^{\frac{-\Delta G_T}{RT}}} = e^{\frac{(\Delta\Delta_T - \Delta G_S)}{RT}} \quad \text{or}$$

$$\Delta G_T - \Delta G_S = RT\ln\frac{k_e}{k_u} = RT\ln\frac{\frac{14}{\text{sec}}}{\frac{1\times10^{-15}}{\text{sec}}}$$

$$\Delta G_T - \Delta G_S = 8.314\times298\times37.18$$

$$\Delta G_T - \Delta G_S = 92\frac{\text{kJ}}{\text{mol}}$$

5. Active α-chymotrypsin is produced from chymotrypsinogen, an inactive precursor. The first intermediate –π-chymotrypsin– displays chymotrypsin activity. Suggest proteolytic enzymes that might carry out these cleavage reactions effectively.

Answer: Conversion of chymotrypsinogen to π-chymotrypsin involves a single peptide bond cleavage at arginine 15 of chymotrypsinogen. Trypsin specificity for arginine and lysine can produce this cleavage. Chymotrypsin itself catalyzes the conversion of π-chymotrypsin to α-chymotrypsin.

6. Based on the reaction scheme shown below, derive an expression for k_e/k_u, the ratio of the rate constants for the catalyzed and uncatalyzed reactions, respectively, in terms of the free energies of activation for the catalyzed ($\Delta G^{\ddagger}_e$) and the uncatalyzed ($\Delta G^{\ddagger}_u$) reactions.

$$S \overset{K_u^{\ddagger}}{\leftrightarrows} X^{\ddagger} \overset{k_u'}{\rightarrow} P$$

$$E + S \underset{K_S}{\leftrightarrows} ES \underset{K_e^{\ddagger}}{\leftrightarrows} EX^{\ddagger} \underset{k_e'}{\rightarrow} EP \rightarrow E + P$$

Answer:

The enzyme-catalyzed rate is given by $\upsilon_e = k_e[ES] = k_e'[EX^{\ddagger}]$, but

$K_e^{\ddagger} = \dfrac{[EX^{\ddagger}]}{[ES]}$ and $[EX^{\ddagger}] = K_e^{\ddagger} \times [ES]$ thus

$k_e[ES] = k_e'[EX^{\ddagger}] = k_e' \times K_e^{\ddagger} \times [ES]$ or

$k_e = k_e' \times K_e^{\ddagger}$

The catalyzed rate is given by $\upsilon_u = k_u[S] = k_u'[X^{\ddagger}]$, but

$K_u^{\ddagger} = \dfrac{[X^{\ddagger}]}{[S]}$ and $[X^{\ddagger}] = K_u^{\ddagger} \times [S]$ so

$k_u[S] = k_u'[X^{\ddagger}] = k_u' \times K_u^{\ddagger} \times [S]$ or

$k_u = k_u' \times K_u^{\ddagger}$

The ratio $\dfrac{k_e}{k_u} = \dfrac{k_e' \times K_e^{\ddagger}}{k_u' \times K_u^{\ddagger}}$

If we assume that $k_e' \approx k_u'$
(i.e., the rate of conversion of transition state
to product is independent of path.)

$$\frac{k_e}{k_u} = \frac{K_e^{\ddagger}}{K_u^{\ddagger}}$$

In general, $\Delta G = -RT\ln K$, or $K = e^{\frac{-\Delta G}{RT}}$, thus

$$\frac{k_e}{k_u} = \frac{e^{\frac{-\Delta G_e^{\ddagger}}{RT}}}{e^{\frac{-\Delta G_u^{\ddagger}}{RT}}} = e^{\frac{(G_u^{\ddagger} - G_e^{\ddagger})}{RT}}$$

Questions for Self Study

1. List the six catalytic mechanisms or factors that contribute to the performance of enzymes.

2. What is a transition-state analog?

3. Match an item in the first column to one in the second column.

a. Covalent catalysis	1. Sensitive to changes in [metal ion].
b. Specific acid-base catalysis	2. Formation of modified amino acid residue.
c. General acid-base catalysis	3. Tightly bound Zn^{2+}.
d. Metalloenzyme	4. Catalysis involves specific H^+ donor.
e. Metal activated enzyme	5. Rate is pH dependent but [Buffer] independent.

4. What three amino acids form the catalytic triad of the serine proteases?

5. The aspartic proteases show a bell-shaped curve of enzyme activity versus pH centered around pH = 3. How is this information consistent with the involvement of two aspartic acid residues in catalysis?

Answers

1. Entropy loss in ES formation; destabilization of ES due to strain, desolvation or electrostatic effects; covalent catalysis; general acid or base catalysis; metal ion catalysis; proximity and orientation.

2. A transition-state analog is a molecule that is chemically and structurally similar to the transition state intermediate.

3. a. 2; b. 5; c. 4; d. 3; e. 1.

4. Aspartic acid, histidine, and serine.

5. The fact that the center of the curve is around pH = 3.0 indicates that an acidic amino acid such as aspartic acid or glutamic acid is involved. The bell-shape is consistent with the involvement of both a general acid and a general base i.e., two distinct groups.

Additional Problems

1. The pH profile of pepsin is bell-shaped, indicating the involvement of a general acid and a general base in catalysis with pK_a's around 4.3 and 1.4, respectively. The two candidates for these roles are two aspartic acid residues, Asp-32 and Asp-215. However, the pK_a's of aspartic acid residues in polypeptides range from 4.2 to 4.6. How might the pK_a of an aspartic acid residue be influenced by protein structure?

2. State the specificity exhibited by the serine proteases trypsin, chymotrypsin, and elastase, and explain how the protein determines this specificity.

3. a. *p*-Nitrophenyl acetate is slowly hydrolyzed by chymotrypsin to *p*-nitrophenol and acetic acid. One nice property of this compound is that the product, *p*-nitrophenol, when ionized, is brightly yellow-colored, making it convenient to follow enzyme activity spectrophotometrically. Draw [*p*-nitrophenylate] produced as a function of time.
b. On the same graph, draw [acetate] production as a function of time.
c. Why are the two plots different?

4. Discuss the importance of histidine in enzyme catalysis.

5. a. Papain is a sulfhydryl protease from papaya that shows broad specificity. (It is used as a component of meat tenderizers.) During catalysis, an acyl-enzyme intermediate is formed as an acylthioester. The active residue is Cys^{25}, which is readily modified by iodoacetate, a potent inhibitor of papain. How might the role of Cys^{25} in papain be similar to Ser^{195} in chymotrypsin?
b. Suggest a mechanism by which iodoacetate inhibits papain.

Abbreviated Answers

1. The pK_a of the side-chain carboxyl group of aspartic acid is around 4.3. Thus, the general acid in pepsin has a pK_a in the normal range of pK_a's exhibited by aspartic acids. For an aspartic acid to function as a general base with pK_a = 1.4, its pK_a must be lowered considerably. How might this be achieved? In a protein, the polypeptide chain may fold in a manner to place an aspartic acid in an environment that influences its pK_a. One way to lower the pK_a is to have positive-charge residues nearby stabilizing the negatively charged carboxylate form of the aspartic acid side chain.

2. Chymotrypsin cleaves peptide bonds in which the carbonyl group is donated by amino acids with bulky, hydrophobic groups such as Phe, Trp, and Tyr. Trypsin's specificity is for Arg and Lys, large, positively charged amino acids. Elastase hydrolyzes after Ala and Gly. The substrate-binding pocket in chymotrypsin is a deep, hydrophobic pocket adjacent to the catalytic site. A similar pocket is found in trypsin but with an aspartic acid residue located deep in the pocket (in a position occupied by a serine residue in chymotrypsin). In elastase, a valine residue and a threonine residue fill the substrate-binding pocket. In chymotrypsin and trypsin these positions are occupied by glycine residues.

3. a and b

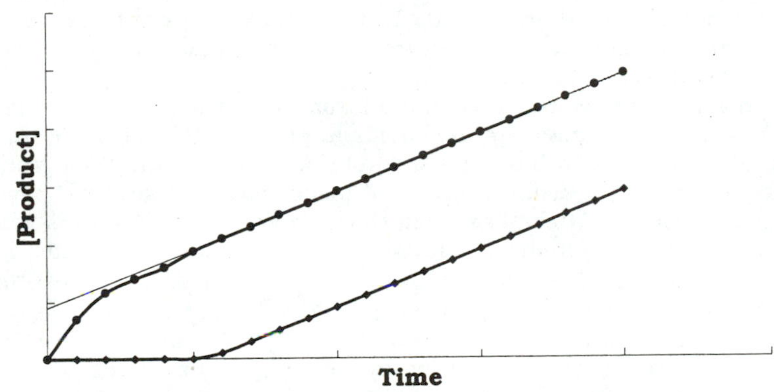

c. Initially, rapid production of *p*-nitrophenolate occurs without accompanying acetate production. Only at later times are the rates of production of *p*-nitrophenolate and acetate identical. This "burst" of activity produces a molar equivalent of *p*-nitrophenolate. Hydrolysis of *p*-nitrophenylacetate is a two-step process. First, the acetate group is transferred from the substrate to Ser^{195} on the enzyme and *p*-nitrophenolate is released. Subsequently, acetate is hydrolyzed from Ser^{195}. The second step is the rate-limiting step. The acetate production curve exhibits an initial lag corresponding to formation of acyl-Ser^{195}.

4. Histidine is the only amino acid with a pK_a around neutral pH. It can function as a general acid or general base, as seen for His^{57} in the serine proteases.

5a. In chymotrypsin catalysis involves a nucleophilic attack of the carbonyl carbon of the peptide bond by Ser^{195} forming an acylester intermediate. This is facilitated by loss of a proton by Ser^{195} to nearby His^{57}. In papain, the active-site residue is a cysteine. Cysteine has a much lower pK_a than serine and will more readily function as a nucleophile. An acylthioester is formed as an intermediate in catalysis in papain.

b.

I–CH$_2$–C(=O)OH + S^-–CH$_2$–Cys 25 ⟶ HO–C(=O)–CH$_2$–S–CH$_2$–Cys 25 + I^-

Summary

Enzymes accelerate reaction rates by reducing the activation energies for biochemical reactions. To do so, enzymes must stabilize the transition state complex $EX^{‡}$, more than they stabilize the enzyme-substrate complex, ES. Put another way, enzymes are designed by nature to bind the transition state more tightly than the substrate or product. Enzyme-catalyzed reactions are typically 10^7 to 10^{14} times faster than their uncatalyzed counterparts. The catalytic mechanisms, which contribute to the prowess of enzymes, include a) entropy loss in ES formation, b) destabilization of ES due to strain, desolvation or electrostatic effects, c) covalent catalysis, general acid-base catalysis, metal ion catalysis, and proximity and orientation effects. When a substrate binds to an enzyme, the (usually large) intrinsic binding energy is compensated by entropy loss and destabilization in the ES complex, but not in the $EX^{\ddagger}$ complex. The transition state for any reaction is a 'moving target' - a transient species, which exists for only 10^{-14} - 10^{-13}s. The short lifetime of the transition state prevents quantitative studies of this complex, but the dissociation constant for the $EX^{\ddagger}$ complex can be estimated to be on the order of 10^{-15} M. Transition state analogs are stable molecules that resemble the transition state for a particular reaction, and that therefore bind very tightly to their respective enzymes. Since they are only approximations of the transition state structure, transition state analogs cannot, however, bind as tightly as the transition state itself.

Some enzyme reactions derive rate acceleration from the formation of a covalent bond with the substrate. Many of the side chains of amino acids in proteins offer suitable nucleophilic centers for catalysis, including amines, carboxylates, hydroxyls, imidazole and thiol groups. Electrophilic catalysis may also occur, but usually involves coenzyme adducts that possess or generate electrophilic centers. General acid/base catalysis, in which a proton is transferred from a suitable donor/acceptor to or from the substrate in the transition state, may also provide significant rate accelerations. General acid/base catalysis is most effective when the pK_a of the donating or accepting group is near the ambient pH. Metal ions may also exert catalytic power in enzyme reactions. Metals may act as electrophilic catalysts, stabilizing the increased electron density or negative charge, which may develop in a reaction, or may act to coordinate, and thus increase the acidity of nucleophilic groups at the active site. Enzymes also provide significant rate acceleration via proximity and orientation effects, i.e. by providing an active site environment in which all the reacting groups are proximal and oriented optimally for the desired reaction.

The serine proteases, including trypsin, chymotrypsin, elastase, thrombin, subtilisin, tissue plasminogen activator and many similar enzymes, cleave peptide chains in a mechanism involving a covalent intermediate and general acid/base catalysis. The active sites of serine proteases contain a highly conserved catalytic triad consisting of Asp102, His57 and Ser195. Asp102 functions to immobilize and orient His57, which accepts a proton from Ser195, facilitating attack by Ser195 on the acyl carbon of the scissile bond. The mechanisms of the serine proteases were elucidated in studies of hydrolysis of model organic esters. The observation of burst kinetics, the rapid release of one of the reaction products in an amount equal to the enzyme concentration, followed by a slower, steady state product release, provided evidence for a multi-step reaction sequence and the eventual identification of covalent acylenzyme intermediates.

The aspartic proteases are so-named for a pair of symmetrically-placed aspartate residues at the active site, which act as a catalytic dyad, alternately acting as general acid/base groups to catalyze hydrolysis of peptides via an amide dihydrate intermediate. Aspartic protease polypeptides consist of two homologous domains, which fold to produce a tertiary structure with approximate two-fold symmetry. The HIV-1 protease, which cleaves the polyproteins produced by the AIDS virus, is an aspartic protease and is a dimer of identical subunits, which mimics the two-lobed monomeric structure of pepsin and other aspartic proteases.

Chapter 12

Proteins of the Blood

• •

Chapter Outline

- Albumin: Human serum albumin
  - Functions
    - Stabilize physical environment
      - Osmotic pressure
      - pH Buffer
    - Transportation of biomolecules
      - Fatty acids from adipose tissue to muscle and liver
      - Bilirubin and heme from spleen to liver
      - Steroids from adrenal glands to target tissues
      - Thyroxine from thyroid to target tissues
    - Protective function: Binds toxic metabolic by-products, xenobiotics, etc.
  - Fatty acid binding characteristics
    - Five binding sites
    - Carboxylate group forms salt bridge with Lys or Arg
    - Long chained fatty acids and steroids bind with highest affinity
- Lipoproteins
  - Function: Transportation of phospholipids, triacylglycerols (fats), cholesterol
  - Classification: Based on density
    - High density lipoproteins: Synthesized in endoplasmic reticulum of liver
    - Low density lipoproteins: Derived from IDL (see below)
    - Intermediate density lipoproteins (intermediate between low and very low): Derived from VLDL (see below)
    - Very low density lipoproteins: Synthesized in endoplasmic reticulum of liver
    - Chylomicrons: Synthesized in intestines
  - Structure
    - Core: Triacylglycerols and cholesterol esters (fatty acid esters) form globule
    - Phospholipid monolayer surrounds core with polar heads out
    - Proteins wrapped around outside
  - Systemic lipid movement
    - VLDL synthesized by liver released into circulatory system
    - Chylomicrons assembled by intestines and released into lymphatic and circulatory system
    - Lipases in tissue remove lipids from lipoprotein complexes: VLDV to IDL to LDL
    - LDL removed from circulation by LDL receptors on liver plasma membrane
    - HDL: Over time accumulates cholesterol from many sources: Returns cholesterol to liver
- Immunoglobulins
  - Immune response: Recognition of foreign substance (antigen) by γ globulin or other immune protein
  - Immunoglobulin G: IgG or γ globulin: Antibodies secreted by B cells: $\alpha_2\beta_2$ tetramer
    - Heavy chain (α): 50 kD, 446 amino acid residues
      - V_H: First 108 residues: Variable and hypervariable positions: One disulfide

        - C_{H1}, C_{H2}, C_{H3}: 109 to 214: One disulfide in each
    - Light chain (β): 25 kD, 214 amino acid residues
        - V_L: First 108 residues: Variable and hypervariable positions: One disulfide
        - C_L: 109 to 214: One disulfide and C-terminal cysteine disulfide bonded to H chain
- Fibrinogen: Glycoprotein precursor of fibrin: Protein meshwork of blood clots
  - Three chains: Aα, Bγ, and γ
  - Each half of symmetric protein contains one copy of each chain crosslinked by disulfide bonds
    - E region: N-termini of subunit chains: Halves joined via disulfides linking E regions
    - D region: C-termini of subunit chains
    - Between E and D coiled-coil alpha helices
  - Clot formation
    - N terminal 16 amino acid residue removed by thrombin
    - Removal exposes GPR "knob"
    - "Knob" fits into D region "holes"
    - Factor XIII: Transglutaminase that crosslinks structure
  - Clot dissolution
    - TPA activates conversion of plasminogen to plasmin
    - Plasmin cleaves C termini of α-chains and N termini of β-chains
- Oxygen binding proteins hemoglobin and myoglobin
  - Hemoglobin: Transports oxygen from lung to tissue: Sigmoidal oxygen binding: $\alpha_2\beta_2$ tetramer
  - Myoglobin: Oxygen storage protein: Simple oxygen binding curve: Monomer
- Oxygen binding group
  - Ferrous iron
  - Protoporphyrin ring: Four ring nitrogens coordinate iron
  - Histidine F8 5th iron coordinate
  - 6th Iron coordinate
    - In oxy-protein: Oxygen
    - In deoxy-protein: Nothing
    - In met-protein: Water: Ferric iron in met: Does not bind oxygen
- Myoglobin: Monomeric protein: Function
  - Stores oxygen
  - Cradles heme group: Increases effective solubility
  - Protects iron from oxidation
  - Provides specificity to oxygen binding pocket: HisE7 sterically hinders CO binding
- Hemoglobin: Dimer of dimers: $(\alpha\beta)_2$
  - Sigmoidal oxygen binding: Cooperativity: Oxygen is positive homotropic regulator
    - Oxygen binding moves iron closer to protoporphyrin plane
    - Iron moves histidine F8
    - F8 movement transmitted to other subunits within tetramer
  - Protons, carbon dioxide, chloride, 2,3-BPG negative heterotropic effectors
    - Bohr effect: Oxygen release accompanied by uptake of about 2 protons: Low pH favors deoxyHb
    - Carbon dioxide
      - Contributes to lowering of pH via carbonic anhydrase reaction producing carbonic acid
      - Carbamate anion formed at free amino group
    - 2,3-BPG: Binds to β chains: Favors T state
  - Physiology
    - Fetal hemoglobin: Binds 2,3-BPG poorly: High oxygen affinity
    - Sickle cell anemia: Amino acid substitution (Glu to Val) responsible for hemoglobin polymerization

Chapter Objectives

Blood is a complex aqueous mixture of cells, proteins, metabolites, and small organic and inorganic compounds responsible for a number of physiological properties of living organisms. This chapter highlights a few important properties of blood by focusing on key proteins responsible for these properties. These include transport of lipids by serum albumin and lipoproteins, recognition of foreign substances by antibodies, protection from loss of blood by blood clotting proteins, and transport of carbon dioxide and oxygen by hemoglobin.

Human Serum Albumin

This protein binds and transports fatty acids and other lipid-like molecules. Its five fatty acid binding sites and high serum concentration combine to provide a high capacity transport system for a host of molecules including long-chain fatty acids.

Lipoproteins

The key point to appreciate here is the interplay of various lipoprotein complexes. These macromolecular complexes are responsible for movement of fats (triacylglycerols) and cholesterol often as fatty acid esters. These neutral lipids form globules at the core of lipoprotein particles. The core is surrounded by a phospholipid monolayer, which in turn interacts with specific proteins. The proteins have two functions: they catalyze various reactions resulting in movement of lipid out of or into a lipoprotein particle, and they serve as molecular labels recognized by receptor proteins.

VLDL, synthesized in the liver, is converted during circulation to IDL then to LDL, which is returned to the liver. HDL acquires cholesterol during its lifetime in circulation. Chylomicrons transport lipids from the digestive system.

Immunoglubulin G

These proteins are responsible for recognition of foreign molecules and this recognition is often the first stage of an immune response. These Y-shaped molecules contain two antigen recognition sites. In a pure population of immunoglobulins (so called monoclonal antibodies) the antigen binding sites are directed toward a single determinant. Reactions between antibody and antigen may lead to crosslinking of two antigen molecules by a single antibody. A large foreign particle (e.g., protein) may have several antigenic determinants and an immune response to this agent may result in production of several antibodies (polyclonal antibodies) directed towards different determinants on the same molecule. Recognition of antigen by polyclonal antibodies may lead to formation of antigen:antibody aggregates.

Fibrin

These elongated proteins are synthesized as highly soluble precursors termed fibrinogen. In response to various signals, a cascade of events is initiated leading to proteolysis of fibrinogen to fibrin. In addition to its low solubility and ability to form specific aggregates, fibrin is crosslinked covalently to form clots. Clot removal requires further proteolytic cleavage of fibrin.

Myoglobin and Hemoglobin

These were the first proteins whose structures were solved by x-ray crystallography and they serve as a paradigm for understanding allosteric regulation. Be sure to understand the physiological function of these two proteins: myoglobin is an oxygen-storage protein; whereas, hemoglobin is an oxygen-transporting protein. Comparison of the structures of myoglobin and hemoglobin is a classic study in the four levels of protein structure: amino acid sequence, secondary structure, three-dimensional structure, and, for hemoglobin, quaternary structure. To a first approximation, hemoglobin appears to be simply a tetramer of myoglobin-like subunits but the subunits are in fact not identical and hemoglobin binds O_2 cooperatively. Be sure to understand cooperativity and the affects of H^+, CO_2 and 2,3-bisphosphoglycerate on O_2-binding curves for hemoglobin.

Problems and Solutions

1. Assume an adult human with 5 L of blood (containing 0.75 mM serum albumin) has every fatty acid-binding site on albumin occupied by palmitates ($H_3C(CH_2)_{14}$-COO^-). How

many grams of fatty acid (in the form of palmitate) is being transported in this person's blood?

Answer: Serum albumin has five principal fatty acid-binding sites and so 5 L of 0.75 mM serum albumin has the potential for carrying:

$$5\ \text{L} \times 0.75 \times 10^{-3}\ \frac{\text{mole HSA}}{\text{L}} \times \frac{5\ \text{FA binding sites}}{\text{mole HSA}} = 0.01875\ \text{mole palmitate}$$

The molecular weight of palmitate ($C_{16}H_{31}O_2$) is calculated as follows:

$$16\,\text{C} \times 12\,\frac{\text{grams}}{\text{mole C}} + 31\,\text{H} \times 1.0\,\frac{\text{grams}}{\text{mole H}} + 2\,\text{O} \times 16\,\frac{\text{grams}}{\text{mole O}} = 255\,\frac{\text{grams}}{\text{mole}}$$

The 0.01875 mole palmitate calculated above corresponds to the following weight in grams:

$$0.01875\ \text{mole palmitate} \times 255\,\frac{\text{gram}}{\text{mole}} = 4.78\text{g}$$

The solubility of palmitic acid in water is 0.00087 grams per 100 grams water or approximately 0.0087 grams per liter. If the 5 L of blood had to transport fatty acids as unbound molecules the solubility would limit the total amount carried to

$$5\ \text{L} \times 0.0087\,\frac{\text{gram}}{\text{L}} = 0.0435\ \text{gram}$$

2. Assume the density of the proteins in a LDL is 1.4 g/mL and the density of the lipids is 0.9 g/mL. If the density of the LDL is 1.02 g/mL, what fraction of its weight is protein and what fraction is lipid?

Answer: Let f_p = the fractional weight of protein and f_l = the fractional weight of lipid.

$$f_p + f_l = 1.0, \text{ and}$$
$$1.4 \times f_p + 0.9 \times f_l = 1.02$$

We can solve for f_p using these two equations as follows:

$f_l = 1.0 - f_p$, which may be substituted as shown below:

$$1.4 \times f_p + 0.9 \times (1.0 - f_p) = 1.02$$

Solving for f_p we find :

$$f_p = \frac{1.02 - 0.9}{1.4 - 0.9} = 0.24 \text{ and}$$
$$f_l = 1.0 - f_p = 1.0 - 0.24 = 0.76$$

3. Drug X binds to serum albumin with a dissociation constant (Kd) of 10^{-6}M. If the total concentration of drug X in the bloodstream is 10 uM, what is the concentration of free (unbound) drug X?

Answer: The total amount of drug in the bloodstream is partitioned into free (unbound) drug, X, and serum albumin-complexed drug, SA-X. Or,

$$[\text{X}] + [\text{SA - X}] = \text{Total drug} = 10\mu\text{M} \quad (1)$$

In a similar fashion serum albumin is partitioned into drug-free serum albumin [SA] and drug-complexed serum albumin [SA-X]. From question 1 above we know that the total serum albumin concentration is 0.75 mM. Or,

$$[\text{SA}] + [\text{SA - X}] = 0.75\ \text{mM} \quad (2)$$

The interaction between drug and serum albumin, assuming only a single drug-binding site on serum albumin, is governed by the following equilibrium (written as a dissociation of bound drug):

$$\text{SA-X} \leftrightarrows \text{X} + \text{SA}$$

The equilibrium dissociation constant for this reaction is written as:

$$K_D = \frac{[\text{X}][\text{SA}]}{[\text{SA - X}]} \quad (3)$$

By subtracting equation (1) from equation (2) and solving for [SA] we find:

$$[SA] = 0.75\text{mM} - 10\ \mu\text{M} + [X]$$

And substituting this expression into equation (3) and solving equation (3) for [SA-X] we have:

$$[SA - X] = \frac{[X][SA]}{K_D} = \frac{[X]\left(0.75\ \text{mM} - 10\ \mu\text{M} + [X]\right)}{K_D}$$

Substituting the expression for [SA-X] into equation (1) and solving for [X] we find:

$$[X] + \frac{[X](0.75\ \text{mM} - 10\ \mu\text{M} + [X])}{K_D} = 10\ \mu\text{M, or}$$

$$[X]^2 + (0.75\ \text{mM} - 10\ \mu\text{M} + K_D) \times [X] - 10\ \mu\text{M} \times K_D = 0 \text{ or}$$

$$[X]^2 + (0.00074)[X] - 1 \times 10^{-11} = 0, \text{ a quadratic whose roots are}$$

$$[X] = \frac{-0.00074 \pm \sqrt{(0.00074)^2 - 4(-1 \times 10^{-11})}}{2}$$

$$[X] = 1.35 \times 10^{-8}\text{M or } 0.135\ \text{nM}$$

4. An individual has a blood cholesterol level of 200 mg/DL and 5 L of blood. Assume that half of the lipid (by weight) in LDLs is cholesterol and also that half of this cholesterol is associated with LDL. How many grams of LDL are in this person's blood? (Use the answer to question 2 in solving this problem.)

Answer: Because 1 dL = 0.1 L, blood cholesterol level at 200 mg/dL corresponds to 2000 mg/L or 2 g/L. Therefore, 5 L of blood contains 10 g of cholesterol. If half of this cholesterol is associated with LDL, the total LDL-associated cholesterol is 0.5×10 g = 5 g. Since half of the lipid (by weight) in LDLs is cholesterol, let X = the weight of lipid in LDL. Therefore, 0.5X = 5 g. Solving for X we find that X = 10 g. So, 10 g of lipid are associated with LDL. From problem 2 we determined that LDLs are 76% by weight in lipids, Thus,

$$76\%\ \text{LDL} = 10\ \text{g or}$$

$$\text{LDL} = \frac{10\ \text{g}}{.76} = 13.2\ \text{gram}$$

5. The equation $\frac{Y}{1-Y} = (\frac{pO_2}{P_{50}})^n$ allows the calculation of Y (the fractional saturation of hemoglobin with O_2, given P_{50} and n. Let P_{50} = 26 torr and n = 2.8. Calculate Y in the lungs where pO_2 = 100 torr and Y in the capillaries where pO_2 = 40 torr. What is the efficiency of O_2 delivery under these conditions (expressed as $Y_{lungs} - Y_{capillaries}$)? Repeat the calculations, but for n = 1. Compare the values for $Y_{lungs} - Y_{capillaries}$ for n = 2.8 versus $Y_{lungs} - Y_{capillaries}$ for n = 1 to determine the effect of cooperative O_2-binding of oxygen delivery by hemoglobin.

Answer: Y = fractional saturation of hemoglobin. For P_{50} = 26 torr, n = 2.8, calculate Y at pO_2 = 100 torr and at pO_2 = 40 torr.

$$\frac{Y}{1-Y} = (\frac{pO_2}{P_{50}})^n$$

$$Y = (1-Y)(\frac{pO_2}{P_{50}})^n = (\frac{pO_2}{P_{50}})^n - (\frac{pO_2}{P_{50}})^n \times Y, \text{ and solving for Y we find:}$$

$$Y = \frac{(\frac{pO_2}{P_{50}})^n}{1 + (\frac{pO_2}{P_{50}})^n} = \frac{(pO_2)^n}{(pO_2)^n + (P_{50})^n}$$

For n = 2.8, P_{50} = 26 torr: Y_{lungs} = 0.98 at pO_2 = 100.

$Y_{capillaries}$ = 0.77 at pO_2 = 40.

Efficiency of O_2 delivery = $Y_{lungs} - Y_{capillaries}$ = 0.21

For n = 1.0, P50 = 26 torr: Y_{lungs} = 0.79 at pO2 = 100.
$Y_{capillaries}$ = 0.61 at pO2 = 40.
Efficiency of O2 delivery = Ylungs - Ycapillaries = 0.18

At an oxygen tension of 100 torr, hemoglobin with n = 2.8 is nearly completely saturated. Under the same condition hemoglobin with n = 1.0 is only 79% saturated. Thus, for the case of n = 2.8, hemoglobin can carry a maximum load of oxygen away from the lung. In the tissues at a pO2 = 40 torr, hemoglobin with n = 2.8 versus n = 1.0 can change its fractional saturation by 0.21 and 0.18, respectively. When expressed as a percentage of the initial load ([efficiency ÷ Ylung] x 100%), O2 delivery by hemoglobin with n = 2.8 and n = 1.0 is 21% and 23%, a modest difference.

6. If no precautions are taken, blood that has been stored for some time becomes depleted in 2,3-BPG. What will happen if such blood is used in a transfusion?

Answer: The compound 2,3-bisphosphoglycerate (2,3-BPG) binds to a single site on the hemoglobin tetramer. The binding site is located on the two-fold symmetry axis and is made up of 8 basic groups, 4 each from the two β chains. 2,3-BPG binds with ionic interactions and stabilizes hemoglobin in the T_o state. Because 2,3-BPG is a side product of glycolysis, it can be metabolized along this pathway when glycolytic intermediates are low. Blood stored for long periods may be low in 2,3-BPG for this reason. In red cells depleted of 2,3-BPG, hemoglobin will bind O_2 avidly and not release it to the tissues. Transfusions with such blood could actually lead to suffocation.

Questions for Self Study

1. Match an item in the first column to one in the second column.

a. HDL	1. Assembled in intestines from dietary fats
b. Chylomicron	2. Synthesized in liver at low density
c. LDL	3. Lipoprotein lipase product
d. IDL	4. Density decreases with time
e. VLDL	5. Uptake is through coated pits

2. Describe the structure of Immunoglobulin G. In your description be sure to indicate the location of regions responsible for features found in all IgG proteins and regions that make IgG molecules unique.

3. During conversion of fibrinogen to fibrin proteolysis releases the N termini of Aα and Bβ polypeptide chains. The released fragments, fibrinopeptide A and fibrinopeptide B, result in drastic solubility changes in the remaining protein, fibrin, which results in protein aggregation and clot formation. In general terms what properties might fibrinopeptide A and fibrinopeptide B have to provide high solubility and to prevent inadvertent aggregation of fibrinogen?

4. Fill in the blanks. Oxygen binding to myoglobin is _____ by pH whereas decreasing pH _____ the affinity of hemoglobin for oxygen. This is known as the _____ effect. One physiological source of protons that will decrease pH in tissue is produced by the enzyme carbonic anhydrase, which catalyzes the following reaction: _____. One of the substrates of this reaction, namely _____, by itself lowers the oxygen affinity of hemoglobin by reacting with free α-amino groups to produce a _____, a negatively charged species that can form salt bridges and stabilize the _____ form of hemoglobin. Another compound that will stabilize this form of hemoglobin is _____, which binds at a single site on hemoglobin.

5. Fetal hemoglobin has (higher or lower) affinity for oxygen than does adult hemoglobin because it has (higher or lower) affinity for 2,3-bisphosphoglycerate.

Answers

1. a. 4; b. 1; c. 5; d. 3; e. 2.

2. IgG's are $\alpha_2\beta_2$ tetramers composed of two 25 kD light chains and two 50 kD heavy chains. Each light chain consists of an N terminal variable region and a C terminal constant region. The variable region is responsible in part for antigen recognition and contains hypervariable positions forming complementarity-determining regions. The heavy chains are composed of a variable N terminal region and three constant regions. The variable and constant regions of both chains are folded into collapsed β-barrel domains stabilized by disulfide bonds. Disulfide bonds also join light chains to heavy chains and heavy chains to each other to form Y-shaped molecules with antigen binding sites on the arms.

3. Fibrinopeptide A and fibrinopeptide B might be expected to be highly polar peptides and in fact they are. Each contains a number of aspartic and glutamic acid residues making the peptide fragments negatively charged, highly soluble, and unlikely to interact.

4. unaffected; lowers; Bohr; $CO_2 + H_2O \rightarrow H_2CO_3 \rightarrow H^+ + HCO_3^-$; carbon dioxide; carbamate; T (or tense or taut); 2,3-bisphosphoglycerate.

5. Higher; lower

Additional Problems

1. The heme-binding site of both myoglobin and hemoglobin is a pocket in the hydrophobic core of these proteins. This pocket is lined with hydrophobic amino acid side-chains with the exception of histidine F8 (the 8th residue of helix F). Generally, hydrophilic residues such as histidine are not located in hydrophobic interiors. What are the important functions of histidine F8 that require it to be buried in a hydrophobic environment?

2. The histidine discussed above is the so-called distal histidine. A second histidine, the proximal histidine, is located on the opposite side of the heme ring. How does the proximal histidine influence ligand specificity?

3. Free heme is an effective oxygen-binding molecule but it has two shortcomings: its iron is readily oxidized, and it forms a heme:O_2:heme complex that has low solubility. How has incorporation of heme into a protein corrected these heme flaws?

4. Discuss the role of amino acid substitution in adapting fetal hemoglobin and adult hemoglobin to different physiological functions.

5. It is rather easy to determine if an accidental cut involves a vein or an artery. Arterial cuts produce crimson-colored blood whereas venous cuts produce darker, bluish-tinted blood. Explain.

6a. A lecturer wants to demonstrate the spectral changes upon oxygenation and deoxygenation of hemoglobin. So, pure hemoglobin, obtained as a lyophilized (freeze-dried) powder from a biochemical supply company, is resuspended in a buffered salt solution. The resulting solution is brownish-red in color indicating a mixture of oxyhemoglobin and met-hemoglobin. What is met-hemoglobin and how can it be converted to oxyhemoglobin?
b. The conversion is made to oxyhemoglobin and the lecturer now decides to produce deoxyhemoglobin by lowering the O_2-tension by placing the solution under partial vacuum. This is only moderately successful. Can you suggest modifications to the procedure that will favor deoxyhemoglobin formation?

Abbreviated Answers

1. This histidine, known as the distal histidine because it is on the opposite side of the oxygen-binding site where the so-called proximal histidine is located (see following question), is one of the iron coordinates in both myoglobin and hemoglobin. In hemoglobin, movement of the distal

histidine in a single subunit of Hb can influence the accessibility of the oxygen-binding site in the other subunits.

2. The proximal histidine is located close enough to the heme iron so as to prevent carbon monoxide from binding to hemoglobin with optimum geometry.

3. By forming a protein:heme complex, the apparent solubility of heme is greatly increased. Thus, higher concentrations can be maintained without the problem of limited solubility, especially in the presence of oxygen. Because the heme is partially buried in a hydrophobic pocket, it is protected from oxidation. Oxidation produces met-Hb, which does not transport oxygen.

4. Oxygen is supplied to the fetal blood system via the placenta where gas exchange occurs between the fetus and the mother's circulatory system. Thus, the fetus must successfully compete with adult, maternal hemoglobin for oxygen. This is possible because fetal hemoglobin has a higher affinity for oxygen than does adult hemoglobin. The β-chains of adult-Hb are replaced by γ-chains in fetal-HB. A critical function of β-chains is formation of a 2,3-BPG binding-site at the β/β interface of the (αβ) dimer junction. The binding site is composed of eight positive charges, four from each β chain of which three are from side chains and the fourth is from the N-terminus. In fetal-HB, the 2,3-BPG binding-site is considerably weaker due to an amino acid substitution, Ser-143 for His-143, at the 2,3-BPG binding site. As a consequence fetal-HB binds 2,3-BPG less avidly and therefore has a high oxygen affinity.

5. In arterial blood, hemoglobin is oxygenated whereas in venous blood, deoxy-hemoglobin predominates. The change in color upon change in oxygenation state is due to an alteration in the iron coordination. In oxyHb, iron is six-fold coordinated with the sixth position occupied by oxygen. In deoxyHb, iron is five-fold coordinated.

6a. Met-hemoglobin is hemoglobin with its iron atom in the +3 oxidation state as ferric (Fe^{3+}) iron. It can be converted to oxyhemoglobin by treatment with a reducing agent such as sodium hydrosulfite (sodium dithionite).
b. A couple of things come to mind. Since it is pure hemoglobin, it may lack 2,3-bisphosphoglycerate (2,3-BPG), accounting for the tenacity of oxygen binding. Alternatively, the pH of the solution could be lowered and CO_2added to decrease oxygen affinity.

Summary

Blood is a complex mixture of a variety of cells including erythrocytes, leukocytes, lymphocytes, and macrophages. The fluid portion of blood, the blood plasma, is a rich mixture of proteins, metabolites, nutrients, ions, etc., which are responsible for important physiological functions. The must abundant plasma protein is albumin. Human serum albumin is an extremely soluble protein capable of binding up to 5 long-chain fatty acids. High concentrations of serum albumin are responsible for the osmotic pressure of plasma. (Children suffering from marasmus, a condition of dietary starvation, exhibit a "pot belly" due to abdominal swelling as a consequence of low serum protein levels.) Albumin is also responsible for transportation of fatty acids, eicosanoids, bile acids, steroid hormones and can sequester lipophilic toxins, drugs, and xenobiotics.

Neutral lipids such as triacylglcerol and cholesterol and phospholipids are transported in the circulatory system as lipoproteins. The core structure is a triacylglycerol globule with cholesterol esters surrounded by a monolayer of phospholipids and proteins. The ratio of protein to lipid serves as a crude classification system. Very low density lipoprotein is synthesized in the endoplasmic reticulum of the liver and released into the blood. This particle moves lipids to muscle and adipose cells and in the process is converted to IDL and LDL as lipids are removed. LDL returns to the liver by binding to LDL receptors on the plasma membrane of liver cells. LDL complexed to LDL receptors are removed from the plasma membrane via formation of coated pits and coated vesicles, which ultimately fuse with lysosomes containing lysosomal acid lipases responsible for lipid degradation. High density lipoproteins enter the circulatory system with no cholesterol but over time accumulate cholesterol esters from various sources. Thus, their density decreases over time because of accumulation of cholesterol, which can be returned to the liver.

The second most abundant class of serum proteins is the globulins. Immunoglobulin G, secreted by B-cells, recognizes and binds antigens, foreign molecules that often are signals indicating a compromise in the organism's protective barriers. IgG's are $\alpha_2\beta_2$ tetramers composed of two light chains and two heavy chains. Both chains contain collapsed β-barrel domains stabilized by disulfide bonds. The chains are further joined by disulfide bonds into Y-shaped molecules with two identical antigen binding sites. The antigen binding sites are composed of amino acid residues from variable regions of heavy and light chains. Within the variable regions are hypervariable positions occupied by amino acids designed to provide a binding site with high affinity for a particular antigen.

Since blood is a collection of cells with specific functions it may be thought of as an organ, a collection of cells serving a physiological purpose. Unlike an organ, however, blood is largely a fluid lacking tissues and so it functions within the confines of the circulatory system. Whenever the integrity of the circulatory system is compromised blood has the ability for form clots. Blood clotting involves a cascade of events culminating in the proteolytic processing of fibrinogen, a soluble glycoprotein, into fibrin, the basic meshwork protein of blood clots. Fibrin has low solubility and is capable of associating into aggregates, which can be crosslinked covalently into a durable structure.

Cells are supplied with oxygen and relieved of carbon dioxide by erythrocytes, which serve as packages of highly concentrated hemoglobin. In muscle tissue, a related protein myoglobin provides a local source of oxygen. Comparison of these two oxygen-binding hemoproteins illustrates the differences between normal and allosteric proteins. Myoglobin (Mb) is a monomer that has one heme and binds one O_2. Hemoglobin (Hb) is an $\alpha_2\beta_2$ tetramer, has 4 hemes and binds 4 O_2. Mb functions as an oxygen storage protein in muscle, while Hb serves as the vertebrate oxygen-transport protein. The oxygen-binding curve for Mb resembles the substrate saturation curves seen for normal, "Michaelis-Menten" enzymes, but the Hb O_2-binding curve is sigmoid, indicating that the binding of O_2 to Hb is cooperative. Cooperative binding means that the binding of one O_2 to an Hb molecule enhances the binding of additional O_2 molecules by the same Hb. O_2-binding by these hemoproteins draws the heme Fe^{2+} atom closer into the porphyrin ring system. This small shift in the position of the iron atom is not particularly important to the biological function of Mb, but in Hb, it triggers the molecular events that underlie its allosteric properties and it has profound consequences for the physiological function of Hb.

Oxygenation markedly affects the quaternary structure of the Hb molecule. In a functional sense, the Hb molecule is composed of two equivalent αβ-dimers. Oxygenation causes the two αβ-dimers to rotate relative to one another. Specific H-bonds and salt bridges which stabilize the deoxy form of Hb are ruptured when O_2 binds. Hb adopts either of two conformational states, the deoxy-, or **T**, state, and the oxy-, or **R**, state. The transition from the deoxy Hb conformational state to the oxyHb state opens the molecule so that O_2 now gains access to previously inaccessible heme groups of the β-chains. This mechanism explains the cooperative O_2-binding exhibited by Hb.

H^+, CO_2, and 2,3-bisphosphoglyceric acid (BPG) all promote dissociation of O_2 from Hb by binding preferentially to the deoxy state. BPG binds one-to-one with Hb. The BPG-binding site lies within the central cavity of Hb. BPG electrostatically crosslinks the β-chains, stabilizing the deoxyHb conformation. Fetal Hb (Hb *F*) differs from normal Hb (Hb *A*) in having two γ-chains in place of the β-chains. An important difference is that each γ-chain has one less positively charged amino acid R-group than the β-chain in the BPG-binding site, so Hb *F* binds BPG less effectively than Hb *A*. Since BPG is bound less effectively, O_2 is bound more avidly by Hb *F*, thereby allowing the fetus to abstract O_2 from the maternal blood circulation. In sickle-cell Hb (Hb *S*), Glu at position β6 is replaced by Val, introducing a new hydrophobic site on the surface of Hb molecules that makes them stick together to form filaments.

Chapter 13

Molecular Motors

• •

Chapter Outline

- Molecular motors: Energy transducing molecules that convert chemical energy into mechanical energy
- Cytoskeleton
  - Microtubules: Tubulin polymers: Hollow tubules
  - Microfilaments: Actin polymers: Thin filaments
  - Intermediate filaments: Protein polymers larger in diameter than thin filaments
- Microtubule-based systems
  - Microtubules
    - Hollow tubes composed of the heterodimeric protein tubulin (110 kD)
      - α tubulin (55 kD)
      - β tubulin (55kD)
    - Tubulin heterodimers form polar structure
      - 13 Monomers per turn: 30 nm Diameter
      - (+) end and (-) end
        - During microtubule formation dimer adds onto (+) end
        - (+) End near cell periphery: (-) End toward centriole
        - In vitro: Treadmilling: Subunit adds onto (+) end and one falls off (-) end at steady state: GTP hydrolysis accompanies (+) end addition
      - Two general ways microtubules involved in motility
        - As cytoskeletal elements along which vesicles move: (+) to (-) or vice versa
        - As cellular appendage: Cilia or eukaryotic flagella
- Cellular appendages
  - Cilia: Short, numerous arm-like cellular projections: Two functions
    - Movement of individual cells
    - Movement of fluid across surface of tissues composed of ciliated cells
  - Flagella: Long arm-like cellular projections: Function in cell motility
  - Basic structure of cilia or flagella is axoneme
    - Nine outer doublet microtubules
    - Two central pair microtubules
    - Axoneme surrounded by extension of plasma membrane
    - Doublet microtubules bend as a consequence of motor molecules attempting to slide adjacent doublet microtubules
- Microtubule motor molecules
  - Dynein: Two types: Axonemal dynein and cytosolic dynein
    - Axonemal dynein: Arms transiently connecting doublet microtubules in axonemes of cilia and flagella
    - Cytosolic dynein: Movement of organelles and vesicles from (+) to (-) end along cytoskeleton
  - Kinesin
    - Movement of organelles and vesicles from (-) to (+) end along cytoskeleton

        - Movement opposite to that of dynein
- DNA based motor: Helicase
  - DNA unwound by helicase action
  - Helicase highly processive: Remains bound to DNA for long distances
  - Two types
    - Hexameric helicases form ringlike structures that slides along DNA
    - Rep helicase
      - Homodimeric
      - Movement is "hand-over-hand"
        - One subunit of dimer binds ssDNA in absence of ATP
        - ATP induces second subunit binding to dsDNA
        - ATP hydrolysis induces DNA unwinding leaving dimer now bound to ssDNA
        - ADP and Pi dissociation releases DNA from one subunit
        - Subunit now free to bind dsDNA (in presence of ATP) to restart cycle
- Actin/myosin-based systems
  - Skeletal muscle: Multinucleated cells: Striated
  - Cardiac muscle: Myocytes: Striated
- Skeletal muscle cell anatomy
  - Fibrils: Skeletal muscle cells: Elongated
  - Myofibrils: Linear arrays of muscle proteins
  - Sarcomere: Repeat unit along myofibril
    - Z line to Z line
    - A band: Electron dense: Centered in sarcomere: Thick filaments and thin filaments
    - H zone: Centered in A band: Thick filaments
    - M disc: Centered in H zone
    - I band: Centered on Z-line: Thin filaments: Polarity opposite on either side of Z line
  - Sarcoplasmic reticulum: Membrane system surrounding myofibrils
    - Longitudinal tubules: Run along the length of myofibril
    - Terminal cisternae: Cap longitudinal tubules and contact t-tubules
    - Triad: Junction between two terminal cisternae and t-tubule
  - Transverse (or t) tubules: Extensions of plasma membrane (sarcolemma) that contact myofibrils
- Thin filaments
  - Filamentous (F) actin: Polymers of globular (G) actin
    - Polar filaments
    - Stimulates ATPase activity of myosin
  - Tropomyosin: Heterodimeric protein: Coiled coils of alpha helix
    - Dimers form head to tail polymers
    - Polymers bind to F actin
    - Stoichiometry: 7 Actin per tropomyosin heterodimer
  - Troponin
    - Troponin T: Binds to tropomyosin
    - Troponin I: Binds to tropomyosin and actin
    - Troponin C: Binds to Troponin I
- Thick filaments: Bipolar myosin filaments
  - Myosin composition
    - Two heavy chains
    - Two LC1 (light chains): Essential light chains
    - Two LC2: Regulatory light chains
  - Myosin structure
    - Heavy chain tails associate to form filaments
    - Heads: Heavy and light chains: ATPase activity
- Muscle contraction: Sliding filament model
  - Muscle tension (when accompanied by muscle shortening i.e., nonisometric contractions) is a result of thin filaments sliding relative to thick filaments

- Sliding depends on ATP and calcium
    - ATP: Myosin heads are actin-activated ATPase
        - ATP dissociates myosin from actin
        - Myosin hydrolyzes ATP but requires actin to release products
        - Release of products is accompanied by conformational change
        - ATP binding reverses conformational change
    - Calcium
        - Released from intracellular stores
        - Binds to Tn-C
        - Tn-C/calcium complex binds to Tn-I
        - Tn-I vacates myosin binding site on thin filament
- Rotary motors in bacteria
    - Extracellular protein filaments: Polymers of flagellin: Function as propellers
    - Membrane localized motors
        - Attached to flagellin
        - Use proton gradient for energy source
        - Reversal of motor causes change in direction of swimming

Chapter Objectives

Microtubules

Microtubules are hollow cylindrical structures composed of the heterodimeric protein tubulin. Know the basic structure of the microtubule, the fact that the wall is composed of 13 protofilaments made up of a head-to-tail arrangement of tubulin dimers and that the structure has two distinct ends, the plus end and the minus end. These designations come from the fact that subunit addition to a microtubule occurs predominantly on the plus end. Microtubules function as cytoskeletal elements and as the principle polymeric structure in cilia and eukaryotic flagella. In cilia and flagella, microtubules are arranged in a characteristic 9 + 2 array known as the axoneme. A key, non-tubulin component of the axoneme is the microtubule-dependent ATPase dynein. The axoneme is a mechanochemical energy transducer converting the chemical energy released upon hydrolysis of ATP by dynein into the energy of axonemal movement. Cytoplasmic microtubules are involved in a number of cell functions including movement of particles within cells. This is best typified by axonal transport in neurons. Axonal microtubules like cytoplasmic microtubules are arranged with their plus ends towards the periphery of the cell or, in this case, towards the axon ends, and their minus ends towards the cell body. Movement of organelles and vesicles from the plus end of a microtubule to the minus end is mediated by cytosolic dyneins. Movement in the opposite direction is dependent on kinesins, ATPases similar to cytosolic dyneins.

Muscle Cell Anatomy

Skeletal muscle cells contain fiber bundles termed myofibrils composed predominantly of actin and myosin filaments. The myofibrils are divided into function units called sarcomeres. Know how the basic anatomy of a sarcomere is composed of actin filaments and myosin filaments and how the two interdigitate. A sarcomere is a banded structure bordered by two Z lines. The A band marks the location of myosin filaments and is centered in the sarcomere. The bipolar myosin filaments contain a central portion devoid of myosin heads. The actin filaments are located in the I band and in part of the A band as they interdigitate with the myosin fibers. The region of the A band not containing interdigitated actin filaments is termed the H zone. The central bare region of the myosin bipolar filaments is centered in the H zone. During contraction, the I band shortens because actin and myosin filaments increase their regions of overlap. As a consequence, the H band shortens. The t-tubules are extensions of the sarcolemma that reach into the muscle cell interior and terminate close to the sarcoplasmic reticulum. Calcium ion release from the sarcoplasmic reticulum is the key event in initiation of muscle contraction and calcium release is set off by electrical signals propagated from the cell surface to the sarcoplasmic reticulum via the t-tubules.

Thin Filaments

The thin filaments are filaments of globular actin studded at regular intervals with troponin complex of TnT, TnI, and TnC and lined with tropomyosin. Myosin binding sites on F-actin can be covered and exposed by movement of tropomyosin in a calcium-dependent manner mediated by TnC. The thin filaments are attached to the Z lines and have opposite polarities on each side of the Z line.

Thick Filaments

Myosin is the principal component of the thick filaments. A myosin molecule is composed of two heavy chains and two pairs of light chains, LC1 and LC2. Myosin is an asymmetric molecule with two heads (on one end) and a tail. The thick filaments are bundles of myosins, interacting in a tail-to-tail fashion to give rise to a "dumbbell" like filament with a bare zone in the middle (tails only) and the heads sticking out around each end. The heads are the business end of the molecule; they have ATPase activity and are responsible for muscle contraction

Sliding Filaments and Myosin ATPase Cycle

Myosin is an actin-activated ATPase that cycles between two conformational states. You should be familiar with the ATPase cycle of myosin and how it leads to force generation. In the resting state, the nucleotide-binding site of myosin is filled with ADP and P_i. Calcium, released upon stimulation, binds to TnC and induces a conformational change that causes movement of tropomyosin resulting in exposure of myosin binding sites on actin. Myosin binds to actin, releases P_i and subsequently undergoes a conformational change and release of ADP. The conformational change results in movement of the actin filament relative to myosin in a power stroke. ATP will now bind to the nucleotide-binding site causing myosin to dissociate from actin. Myosin now hydrolyzes ATP as it returns to the resting conformation.

Calcium Ion Regulation

Calcium release is mediated through a series of calcium channels. Uptake of calcium leading to muscle relaxation is by calcium pumps.

Problems and Solutions

1. The cheetah is generally regarded as nature's fastest mammal, but another amazing animal almost as fast as the cheetah) is the pronghorn antelope, which roams the plains of Wyoming. Whereas the cheetah can maintain its top speed of 70 mph for only a few seconds, the pronghorn antelope can run at 60 mph for about an hour! (It is thought to have evolved to do so in order to elude now-extinct ancestral cheetahs that lived in North America.) What differences would you expect in the muscle structure and anatomy of pronghorn antelopes that could account for their remarkable speed and endurance?

Answer: Clearly the pronghorn antelope must be a master of oxidative phosphorylation and because of this must be capable of supplying muscle with oxygen-rich blood. The organs involved in oxygen homeostasis, namely the circulatory system, including the heart, blood, and blood vessels, are optimized for this task. The animal has a large, powerful heart, blood with a high hematocrit and red cells with high amounts of hemoglobin, and well vascularized muscle and lung tissue. The animal is light and compact and of efficient design. In addition, muscle cells contain large numbers of mitochondria and numerous short muscle fibers.

2. An ATP analog, β,γ-methylene-ATP, in which a -CH_2- group replaces the oxygen atom between the β- and γ- phosphorus atoms, is a potent inhibitor of muscle contraction. At which step in the contraction cycle would you expect β,γ-methylene-ATP to block contraction?

Answer: β,γ–methylene-ATP is a nonhydrolyzable analog of ATP that binds to ATP-binding sites and is capable of inducing conformational changes associated with ATP binding but is not destroyed by hydrolysis. During muscle contraction, myosin heads interact with actin in a cyclic manner that results in muscle contraction. Myosin in the resting state, not in contact with actin, has bound ADP and P_i. Binding of actin stimulates release of P_i, which is followed by a conformational change of the myosin heads and ADP release. The conformational change, called the power stroke, is responsible for force generation. Binding of ATP or the ATP analog will subsequently cause myosin to dissociate from actin. To reestablish the myosin conformation prior to the power stroke, ATP hydrolysis is required. The nonhydrolyzable ATP analog will cause myosin dissociation but without subsequent hydrolysis no conformational change occurs to reestablish the myosin head in its original conformation.

3. ATP stores in muscle are augmented or supplemented by stores of phosphocreatine. During periods of contraction, phosphocreatine is hydrolyzed to drive the synthesis of needed ATP in the creatine kinase reaction:

$$\textbf{Phosphocreatine + ADP} \leftrightarrows \textbf{creatine + ATP}$$

Muscle cells contain two different isozymes of creatine kinase, one in the mitochondria and one in the sarcoplasm. Explain.

Answer: Creatine is phosphorylated at the expense of ATP in the mitochondria and then exported to the sarcoplasm. In the sarcoplasm, it functions as a store of high energy phosphate bonds to be used during intense muscle activity to replenish ATP supplies. Creatine phosphorylation in the mitochondria is accomplished by a mitochondrial creatine kinase. Phosphorylation of ADP by phosphocreatine in the sarcoplasm is accomplished by a sarcoplasmic enzyme. Free creatine and phosphocreatine thus form a circuit with phosphocreatine leaving the mitochondria and creatine entering.

4. Rigor is a muscle condition in which muscle fibers, depleted of ATP and phosphocreatine, develop a state of extreme rigidity and cannot be easily extended. (In death, this state is called rigor mortis, the rigor of death.) From what you have learned about muscle contraction, explain the state of rigor in molecular terms.

Answer: Rigor occurs when the myosin head is in contact with actin in the absence of ATP. When ATP levels are low, sarcoplasmic levels of calcium ion increase thereby causing a conformational change in troponin that leads to tropomyosin movement and uncovering of myosin-binding sites on actin. Myosin binds, releases P_i, and undergoes a conformational change leading to force generation and release of ADP. However, in the absence of ATP, myosin remains attached to actin in the contracted state leading to stiff, contracted muscles typical of rigor mortis.

5. Skeletal muscle can generate approximately 3 to 4 kg of tension or force per square centimeter of cross-sectional area. This number is roughly the same for all mammals. Because many human muscles have large cross-sectional areas, the force that these muscles can (and must) generate is prodigious. The gluteus maximus (on which you are probably sitting as you read this) can generate a tension of 1200 kg! Estimate the cross-sectional areas of all of the muscles in your body and the total force that your skeletal muscles could generate if they all contracted at once.

Answer: The average adult male has approximately 5,500 cm^2 of skeletal muscle in cross section. If muscle generates 3 to 4 kg of tension per cm^2 then 55,000 cm^2 can generate between 1.65×10^3 and 2.20×10^3 kg of force.

Questions for Self Study

1. The following diagram is a cross section of a flagellum showing the structure of an axoneme but with the labels left out. Label the following: A-tubule, B-tubule, inner dynein arm, outer dynein arm, radial spoke, spoke head, plasma membrane, nexin, central singlets.

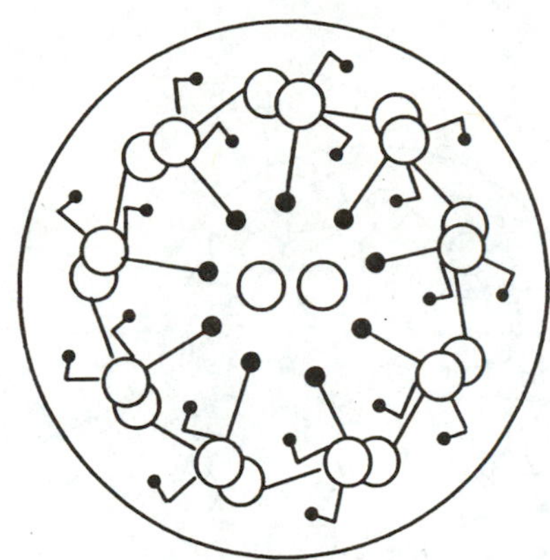

2. Fill in the blanks. Cytoskeletal elements composed of hollow, cylindrical structures 24 nM in diameter are known as _____. These polymers are made up of the protein _____, a dimeric protein composed of _____ and _____. These cytoskeletal elements are the major structural elements of cilia and flagella. An ATPase, _____, is responsible for production of a sliding motion between polymers that is converted to a bending motion typical of ciliary or flagellar motion. A cytosolic variety of this ATPase is found in axons where it mediates the movement of organelles and vesicles from the plus end of a microtubule to the minus end. (The plus end is the end at which dimer subunits are ___ to the microtubule.) Another ATPase, _____, mediates movement of organelles and vesicles in the opposite direction.

3. On the microscopic level, skeletal and cardiac muscle display light and dark bands. What is the general name given to muscle with banding patterns?

4. Fill in the blanks. The basic structural units of skeletal muscle contraction are _____. They are arranged as linear arrays into small fibers called _____. The fibers are covered by a specialized endoplasmic reticulum called _____. The ends of the fibers are bounded by extension of the muscle cell plasma membrane or _____. The extensions form _____ that connect the specialized sarcoplasmic reticulum with the plasma membrane.

5. In the diagram presented below label the following: I band, Z line, A band, H zone, location of thin filaments only, location of thick filaments only, overlap of thin and thick filaments.

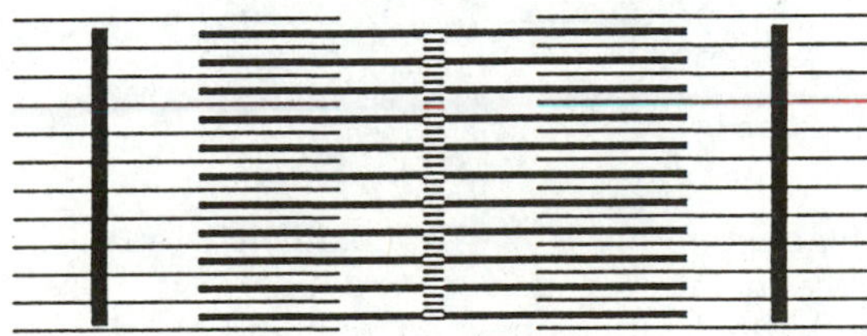

6. Sort the terms below into thick filament or thin filament components.

a. G-actin	b. Heavy chains	c. Troponin I	d. Regulatory light chain
e. Tropomyosin	f. S1 fragments	g. HMM	h. Creatine kinase
i. β-Actinin	j. Troponin C	k. Light meromyosin	l. Troponin T
m. Essential light chain	n. Myosin		

7. In skeletal muscle, cycles of contraction and relaxation are a result of a calcium cycle. Describe this calcium cycle.

Answers

1.

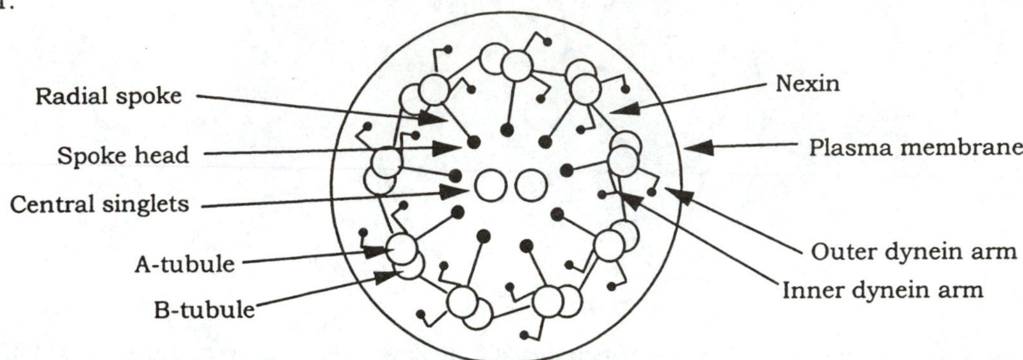

2. Microtubules; tubulin; α-tubulin; β-tubulin; dynein; added; kinesin.

3. Striated.

4. Sarcomeres; myofibrils; sarcoplasmic reticulum; sarcolemma; transverse (or T) tubules.

5.

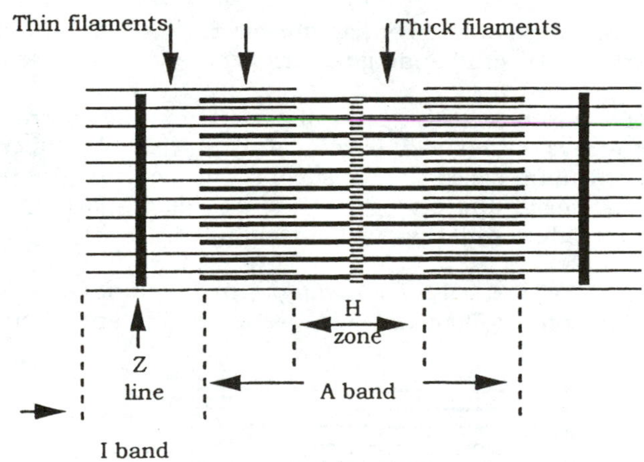

6.

Thick filament terms	Thin filament terms
b. Heavy chains	a. G-actin
d. Regulatory light chain	c. Troponin I
f. S1 fragments	e. Tropomyosin
g. HMM	i. β-Actinin
h. Creatine kinase	j. Troponin C
k. Light meromyosin	l. Troponin T
m. Essential light chain	
n. Myosin	

7. Calcium is released from the sarcoplasmic reticulum in response to electrical stimulation. This results in an increase in the concentration of sarcoplasmic calcium, which leads to calcium binding to troponin C, and movement of troponin I and tropomyosin. The result of tropomyosin movement is to expose myosin-binding sites on actin. Relaxation occurs when the sarcoplasmic calcium ion concentration is lowered as a result of active transport of calcium back into the sarcoplasmic reticulum by calcium pumps.

Additional Problems

1. Design a simple experiment to determine the association constant for assembly of tubulin dimers to the ends of a microtubule.

2. The concentration of tubulin dimers determined above is the critical dimer concentration for assembly of microtubules. At equilibrium, dimers are being added to the polymer but if dimer assembly causes the free tubulin dimer concentration to fall below the critical concentration, the microtubule will disassemble, thereby releasing free dimers to increase the concentration of dimers. The plus end and the minus end of microtubules have different critical dimer concentrations with the minus end being higher than the plus end. How does this lead to treadmilling?

3. The tenderloin is a muscle that runs along the back of an animal and is the source of filet mignon, a tender cut of meat, so tender in fact that when cooked, it can be cut with a fork. Can you suggest a reason why this muscle is so tender?

4. The graph below shows tension developed by a sarcomere held at various fixed lengths where 100 refers to the natural resting length of the sarcomere. Based on your knowledge of sarcomere microanatomy, explain the line segments "ab", "bc", "cd", and "de".

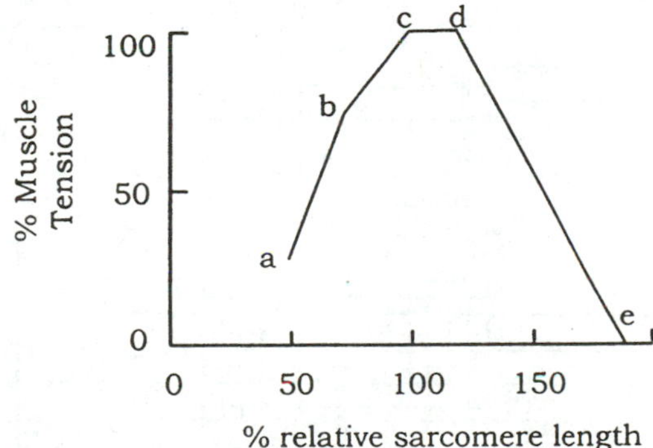

5. Myosin heads can be used to stain actin filaments in nonmuscle cells for the purpose of identification of a filament as an actin filament and to establish the direction of polarity. Explain.

6. It is customary to sever the spinal cord of a recently sacrificed fish in order to prevent the fish from becoming stiff. Why?

Abbreviated Answers

1. If we assume that the thermodynamics of elongation is more favorable than initiation, then, in conditions that favor microtubule formation, microtubules will be in equilibrium with free tubulin dimers or:

$$MT_{n-1} + T \leftrightarrows MT_n \text{ where}$$

$$K = \frac{[MT_n]}{[MT_{n-1}][T]}$$

Recognizing that $[MT_{n-1}] = [MT_n]$ we see that $K_A = [T]^{-1}$. If we could measure the concentration of free tubulin dimer in equilibrium with microtubules we could determine K. This can be accomplished by simply centrifuging the sample at a speed that will sediment microtubules but not tubulin dimers. By measuring the concentration of tubulin in the supernatant, we can determine K.

2. Consider a solution of tubulin polymerizing into microtubules. As the concentration of free dimers decreases to the critical concentration of minus end assembly, addition of dimers to the minus end stops. However, since the critical concentration of plus end assembly is lower than

minus end assembly, dimers will continue to add to the plus end. This lowers the concentration of free subunits below the critical concentration of minus end assembly, causing subunits to dissociate from the minus end. The system reaches a steady state at which minus end disassembly is balanced by plus end assembly.

3. There are a couple of reasons. First, the muscle is composed of a relatively small number of very large muscle fibers with little connective tissue. Second, the muscle is not an extremely active muscle like the muscles of the shoulder. Finally, the filet is a cross sectional cut out of the muscle across the long axis of the muscle fibers. The meat can be easily cut into smaller pieces by simply separating muscle fibers with a fork.

4. At position "e", the sarcomere has been stretched so that thick and thin filaments no longer overlap and no tension is possible. Moving from "e" to "d", the length corresponds to an increase in overlap between thick and thin filaments and therefore a decrease in sarcomere length. At "d", actin filaments maximally overlap the region of thick filaments containing heads. At point "c", the thin filaments from opposite sides of the sarcomere come in contact. Moving from "d" to "c" no change in the number of cross bridges overlapping with thin filaments occurs and the sarcomere can generate maximum tension. Sarcomere length between "c" and "b" results in overlap of thin filaments from opposite sides of the sarcomere. The drop in tension in moving from "b" to "a" results when actin filaments from opposite sides of the sarcomere begin to overlap with myosin heads on the opposite side of the thick filament. The following diagrams illustrate these points.

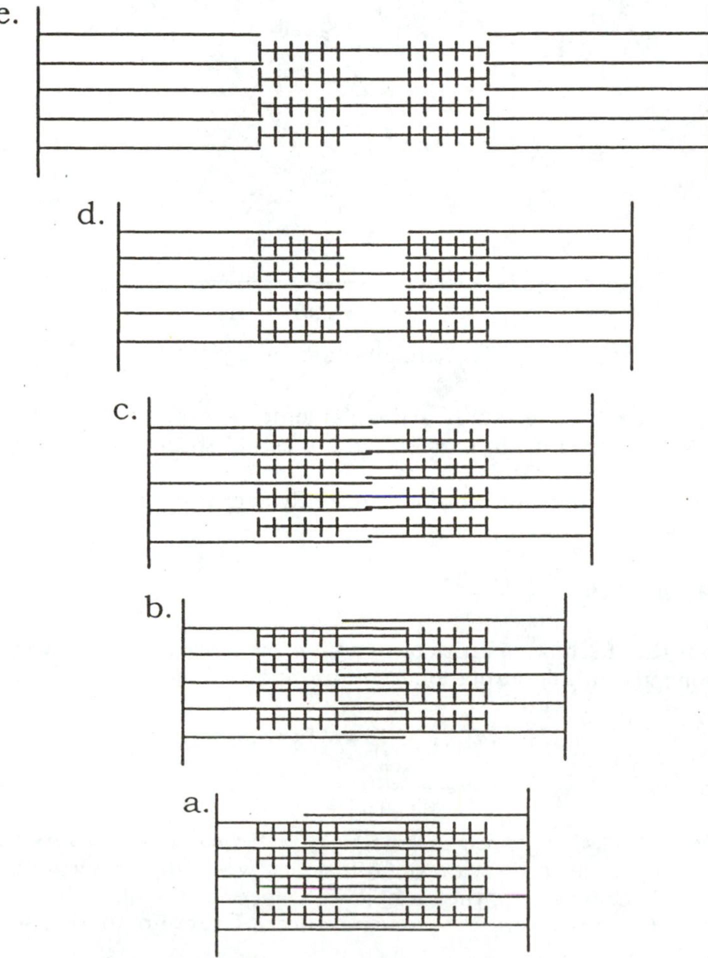

5. The S1 fragment of myosin will bind to actin fibers and decorate them in a characteristic arrowhead pattern. The tips of the arrowheads point away from the Z line in a sarcomere.

6. Fish become stiff because of rigor mortis as a result of a release of calcium and depletion of ATP stores. After death, the nerves quickly deteriorate and begin to randomly fire leading to

muscle stimulation and rapid depletion of ATP stores. Severing the spinal cord and icing the fish can delay the depletion of ATP stores.

Summary

All organisms (at least at some stage in life) are capable of movement or motion. The energy for these processes is chemical energy that is transduced into mechanical energy by motor proteins. Motor proteins found in microtubule-based motile systems and those found in actin-dependent systems utilize hydrolysis of ATP whereas the rotary motor found in bacterial flagella uses the energy of a proton gradient.

Microtubules are self-assembling structures that are fundamental components of cilia and eukaryotic flagella. Microtubules are formed by polymerization of tubulin, a dimeric protein comprised of similar 55 kD subunits. By the GTP-dependent process of treadmilling, tubulin units are added at the plus end of the microtubule and removed from the minus end. With their plus ends oriented toward the cell periphery and the minus ends toward the center of the cell, microtubules form a significant part of the cytoskeleton. The axoneme structure of cilia and flagella consist of a complex bundle of microtubule fibers, including two central microtubules surrounded by nine pairs of joined doublet microtubules, all surrounded by plasma membrane. The motion effected by these structures results from the ATP-driven sliding or walking of dyneins along one microtubule while remaining firmly attached to an adjacent microtubule. Within eukaryotic cells, cellular dyneins and kinesin proteins are responsible for microtubule-based transport of organelles and vesicles.

DNA helicases move along DNA molecules converting double stranded DNA into single stranded DNA. This process is continuous for long distances along a DNA molecule and movement of this type is termed processive movement. Some helicases form ringlike structures that slide along a DNA molecule. The Rep helicase of *E. coli* is a homodimeric protein that uses a "hand-over-hand" movement based on negative cooperativity of DNA binding sites within a dimer.

Actin and myosin are two proteins responsible for various kinds of motion in a number of cell types. The best characterized system is muscle. Muscle contraction is essential to higher organisms for limb movement, locomotion, beating of the heart and other functions. Skeletal muscle cells -the muscle fibers- are multinucleate whereas cardiac cells contain single nuclei. Skeletal and cardiac muscle are striated. The sarcolemmal membrane contains extensions called transverse tubules which surround the sarcomeres and which enable the sarcolemmal membrane to contact the ends of each myofibril in the muscle fiber. Between t-tubules, the sarcomere is covered by the sarcoplasmic reticulum membrane. The terminal cisternae of the SR are joined to the t-tubules by foot structures. The trigger signal for muscle contraction is an increase in $[Ca^{2+}]$ cytoplasmic. Relaxation occurs when $[Ca^{2+}]$ is reduced.

Electron micrographs of banding patterns in skeletal muscle are consistent with a model for contraction in which thick and thin filaments slide along each other. Actin is the principal component of muscle thin filaments. It consists of a monomeric protein, G actin, which polymerizes to form long filaments of right-handed helical F-actin. Other components of thin filaments include tropomyosin and the troponin complex. Myosin is the principal component of muscle thick filaments and consists of a globular head and a long, helical tail. The tails intertwine to form a left-handed coiled coil. The myosin heads exhibit ATPase activity, and it is the hydrolysis of ATP by the myosin heads that drives muscle contraction. Repeating structures in the myosin tails, including 7-residue, 28-residue, and 196-residue repeating units facilitate the formation of myosin-solvent and myosin-myosin interactions, which stabilize the packing of myosin molecules in the thick filaments.

The molecular events of skeletal muscle contraction are powered by the ATPase activity of myosin. Actin increases the ATPase activity of myosin by 200-fold. Further, myosin and actin spontaneously associate to form an actomyosin complex, but ATP decreases the affinity of myosin for actin. In the absence of actin, ATP is rapidly cleaved at the myosin active site, but the hydrolysis products - ADP and P_i - are not released. Actin activates myosin ATPase by stimulating the release of P_i and then ADP. Thus, ATP hydrolysis and the association and dissociation of actin and myosin are coupled. The coupling involves a conformational change in the myosin head, driven by the free energy of hydrolysis of ATP, so that dissociation of myosin and actin, hydrolysis of ATP and rebinding of myosin and actin occur with a net movement of the myosin head along the actin filament. Each such conformation change moves the thick filament approximately 100 Å along the thin filament.

Regulation of contraction by Ca^{2+} is achieved by interaction with the troponin complex.

Binding of Ca^{2+} to troponin C increases the binding of troponin C to troponin I, simultaneously decreasing the interaction of troponin I with actin. As a result, tropomyosin slides deeper into the actin thin filament groove, exposing myosin binding sites on actin and initiating the muscle contraction cycle.

Flagellated bacterial cells produce motility by rotating extracellular protein filaments, termed flagella and composed of the protein flagellin, using a rotary motor, which harnesses the energy of proton gradients to produce movement.

Chapter 14

The Organization of Metabolism

Chapter Outline

- Metabolic diversity
  - Autotrophs: Utilize carbon dioxide as carbon source
  - Heterotrophs: Require carbon in organic molecules
  - Phototrophs: Use light as source of energy
  - Chemotrophs: Use chemical reactions as source of energy
- Role of oxygen
  - Obligate aerobes: Require oxygen
  - Obligate anaerobes: No need for oxygen: Oxygen poisonous
  - Facultative anaerobes: Adapt to anaerobic conditions
- Metabolic processes: Two types
  - Catabolism
    - Exergonic: Energy transduced into ATP
    - Oxidative degradation of complex molecules into simpler compounds
    - Electrons captured by NAD^+
  - Anabolism
    - Complex biomolecules produced from simpler precursors
    - Synthesis driven by energy typically ATP hydrolysis
    - NADPH serves as source of electrons
  - Regulation
    - Anabolism and catabolism separately regulated
    - Competing pathways often localized in different cellular compartments
    - Anabolic and catabolic pathways (involving same end points) differ in at least one or more steps
      - Allows pathways to be regulated separately
      - Avoids simultaneous functioning of both pathways
  - Connections
    - ATP from catabolism used in anabolism
    - NADPH electrons: Two sources
      - From NADH
      - From fuel molecules
    - Key catabolic intermediates used to synthesize building blocks for anabolism
- Enzymes organization
  - Separate, soluble entities: Intermediates supplied by diffusion
  - Multienzyme complexes: Intermediates passed between components of complex
  - Membrane-bound enzymes or complexes
- Experimental methods
  - Metabolic inhibitors
  - Genetic mutations
  - Isotopic traces
    - Radioactive isotopes: Detected when isotope decays by release of energy
    - Stable, heavy isotopes: Alter density of product

- NMR: Signal sensitive to electronic environment of atomic nucleus
- Nutrition: Use of foods by organisms: Five components
  - Proteins: Source of amino acids
    - Essential amino acids
    - Degradation of amino acids
      - Glucogenic: Some amino acids can be used to produce glucose
      - Ketogenic: Some amino acids can be used to produce fatty acids
  - Carbohydrates: Metabolic energy derived from catabolism of carbohydrates
  - Lipids
    - Metabolic energy sources
    - Components of membranes
    - Sources of essential fatty acids: Linoleic and linolenic acids
  - Fiber: Cellulose, hemicellulose, lignins
  - Vitamins: Two classes
    - Water soluble: Used as coenzymes or coenzyme precursors (except vitamin C)
      - Vitamin B_1: Thiamine
        - Thiazole ring-substituted pyrimidine used as thiamine pyrophosphate
        - Coenzyme in carbohydrate metabolism
      - Nicotinamide: Niacin: NAD and NADP
        - Involved in electron transfers
        - Dehydrogenases
      - Riboflavin: Vitamin B_2: FMN and FAD
        - Ribitol plus flavin (isoalloxazine ring)
        - One electron and two electron reactions
      - Pantothenic acid: Coenzyme A and acyl carrier protein
      - Pyridoxine: Vitamin B_6: Pyridoxal phosphate
        - Coenzyme that functions as electron sink
        - Variety of reactions: Transaminations, decarboxylations, eliminations, racemizations, aldol reactions
      - Vitamin B_{12}: Cyanocobalamine
        - Corrin ring plus cobalt ion
        - Coenzyme forms: 5'-Deoxyadenosylcobalamine and methylcobalamine
        - Reactions: Intramolecular rearrangements, methyl group transfer, ribonucleotide to deoxyribonucleotide (in certain bacteria)
      - Vitamin C: Ascorbic acid: Reducing agent
      - Biotin: Mobile carrier of carboxyl groups
      - Lipoic acid: Acyl group carrier
      - Folic acid
        - Coenzyme form tetrahydrofolate
        - One carbon pool: Reduced forms of carbon
    - Fat soluble
      - Vitamin A: Retinol: Involved in vision and intercellular communication
      - Vitamin D: Ergocalciferol and cholecalciferol: Calcium and phosphate homeostasis
      - Vitamin E: Tocopherol: Antioxidant
      - Vitamin K: Naphthoquinone: γ–Carboxyglutamyl formation in blood clotting proteins

Chapter Objectives

Understand the terms metabolism, catabolism, and anabolism. Know the classification of organisms with respect to carbon and energy requirements into photoautotrophs, photoheterotrophs, chemoautotrophs, and chemoheterotrophs. (What are you? Your house plants?) Understand the terms obligate aerobes, facultative anaerobes, and obligate anaerobes.

The remainder of the text will concentrate on details of metabolism and take a step-by-step tour of individual pathways. This chapter gives the big picture. For example, oxygen moves in a cycle from photoautotrophic cells as O_2 to heterotrophic cells, from heterotrophic cells to heterotrophic cells as oxy-compounds, and from heterotrophic cells back to photoautotrophic cells as CO_2 and H_2O. Carbon cycles between photoautotrophs and heterotrophs as CO_2 and organic compounds.

Understand the role of ATP in metabolism. ATP is an energy-rich compound used to coupled exergonic to endergonic processes. In general, catabolic pathways are exergonic and produce ATP whereas anabolic pathways are endergonic and consume ATP.

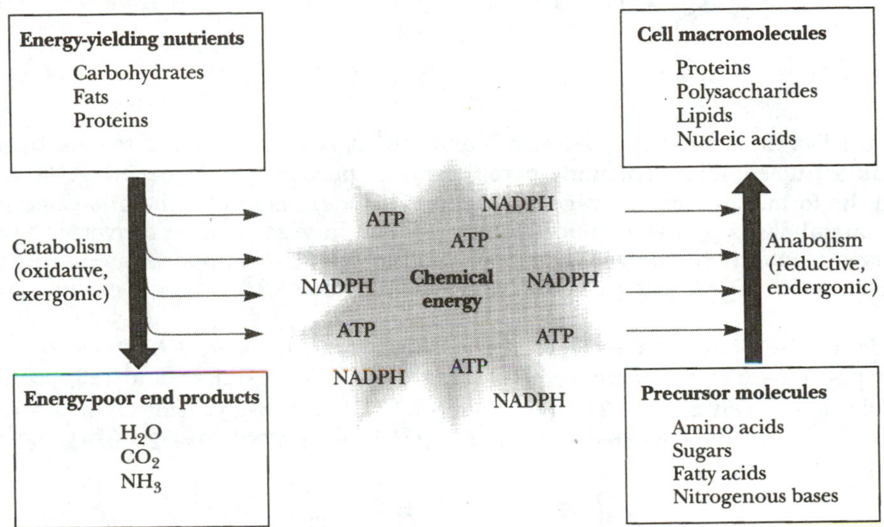

Figure 14.3 Energy relationships between the pathways of catabolism and anabolism. Oxidative, exergonic pathways of catabolism release free energy and reducing power that are captured in the form of ATP and HADPH, respectively. Anabolic processes are endergonic, consuming chemical energy in the form of ATP and using NADPH as a source of high-energy electrons for reductive purposes.

The cofactors NAD^+ and NADPH play key roles in oxidation-reduction reactions. Catabolic pathways are often oxidative in nature with NAD^+ serving as the oxidant. In contrast, anabolic pathways are reductive with NADPH serving as a reductant (see Figure 18.11).

Intermediary metabolism can be overwhelming at first glance but appreciate the fact that the pathways converge to or diverge from a small number of key intermediates. In catabolism, large macromolecules such as proteins, polysaccharides, and lipids, are broken down into their component building blocks, which are catabolized into a small set of common metabolic intermediates, which may be degraded to simple end products (e.g., CO_2, NH_3, H_2O). Alternatively, these intermediates can be used in anabolism to resynthesize macromolecules. Resynthesis pathways may share a number of intermediates with degradative pathways, but at least one step is different. This allows for independent regulation of the competing actions. How is regulation achieved? Clearly reactions must be sensitive to concentrations of substrates and products but more sophisticated controls are often imposed at key reactions. These controls include allosteric regulation, covalent modification, and alteration in the level of enzyme or the type of enzyme.

Vitamins

The vitamins can be divided into two groups based on solubility: water-soluble and fat-soluble. The water-soluble vitamins include: thiamine (vitamin B_1), nicotinic acid (niacin), riboflavin, pantothenic acid, pyridoxine (vitamin B_6), cyanocobalamin (vitamin B_{12}), ascorbic acid (vitamin C), biotin, lipoic acid, and folic acid. We have already encountered some of these (i.e., vitamin C in hydroxylation of collagen) and, in intermediary metabolism, we will study many vitamin-dependent reactions. The fat-soluble vitamins are vitamin A, vitamin D, tocopherol (vitamin E) and vitamin K (naphthoquinone).

Water Soluble Vitamins

Thiamine and Thiamine Pyrophosphate
The structure of thiamine pyrophosphate is shown below.

Water Soluble Vitamins
Thiamine and Thiamine Pyrophosphate
The structure of thiamine pyrophosphate is shown below.

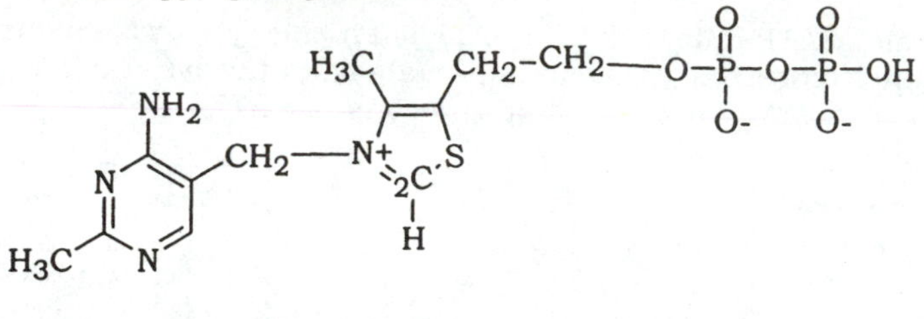

The carbon in the five-membered ring between N and S (C-2) is readily ionized to a carbanion that can function as a nucleophile. Thiamine pyrophosphate participates in decarboxylations of α-keto acids and the formation and cleavage of α-hydroxyketones; examples include yeast pyruvate decarboxylase (metabolizes pyruvate to acetaldehyde which in yeast can be converted to ethanol) and transketolase reactions (which will be encountered in the pentose phosphate shunt pathway and in the Calvin cycle). The key to catalysis is nucleophilic attack by C-2 on a carbonyl carbon.

Nicotinamide is the "business end" of the coenzymes NAD^+ and $NADP^+$, two important molecules that participate in oxidation/reduction reactions. The oxidized and reduced forms of the nicotinamide are shown below. The N-R bond of NAD^+ is a N-glycosidic bond to ribose-5'-monophosphate in phosphate diester linkage to AMP. In $NADP^+$ the 2'-carbon of AMP is phosphorylated.

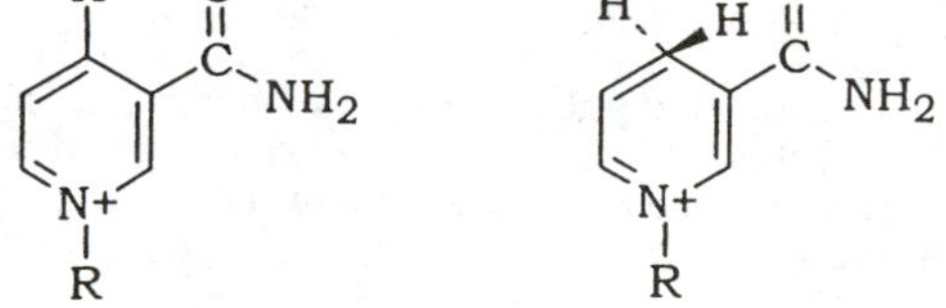

In metabolism we will see that NAD^+ is important in oxidation of substrates, whereas NADPH is an important reducing agent. These reactions involve hydride (two electrons and a proton) transfer. Dehydrogenases and reductases are nicotinamide adenine nucleotide-requiring enzymes.

Riboflavin is found in FMN and FAD and participates in oxidation/reduction reactions. Riboflavin comprises two components, ribitol and flavin. The flavin portion can participate in oxidation/reduction reactions. The fully oxidized and reduced structures of riboflavin are shown below.

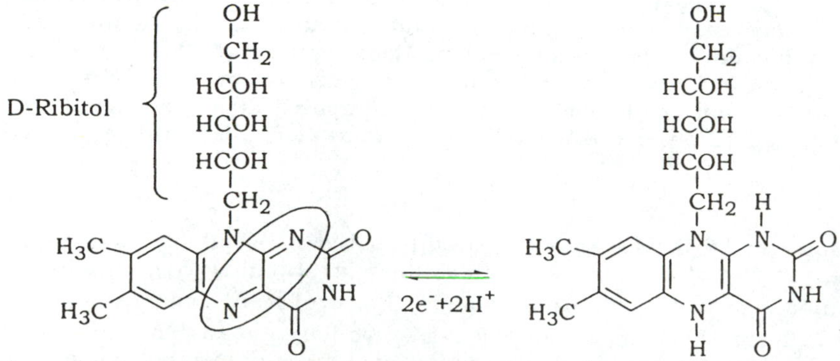

Pantothenic acid is a constituent of acyl carrier proteins and coenzyme A. Coenzyme A functions as an acyl group carrier. Its structure is shown below.

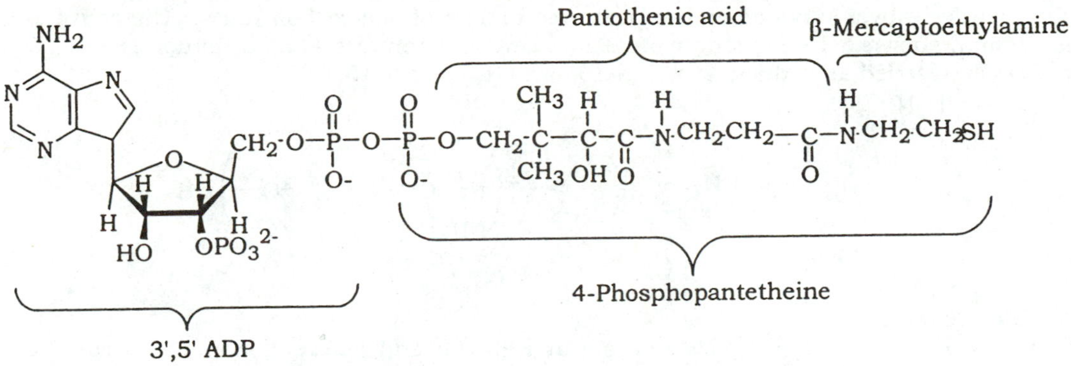

Pantothenic acid

β-Mercaptoethylamine

4-Phosphopantetheine

3',5' ADP

The acyl group is attached in thiolester linkage to the sulfhydryl of coenzyme-A.

Pyridoxal-5-phosphate (PLP) is the biologically active form of vitamin B_6. It participates in a number of reactions involving amino acids including: transamination, α-decarboxylation, β-decarboxylation, β-elimination, γ-elimination, racemization and aldol reactions. The key to catalysis is the ability of pyridoxal-5-phosphate to form a stable Schiff base adduct with amino acids. The structure of PLP attached to a lysine residue is shown below.

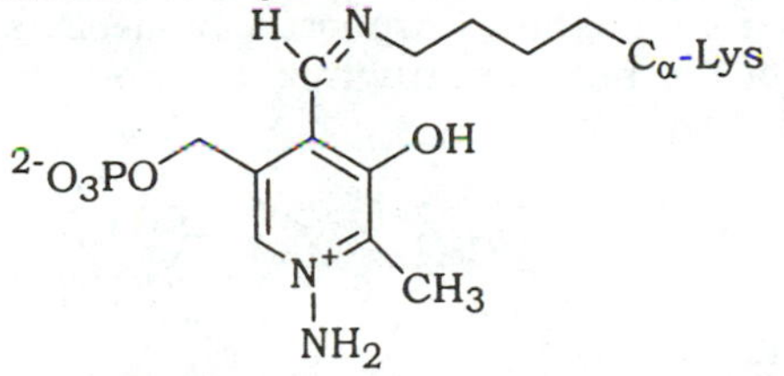

Vitamin B_{12} or cyanocobalamin participates in three types of reactions: intramolecular rearrangements, reductions of ribonucleotides to deoxyribonucleotides, and methyl group transfers.

Ascorbic acid is a reducing agent that participates in a number of important biochemical reactions. The structures of L-ascorbic acid, and dehydroxy-L-ascorbic acid are below.

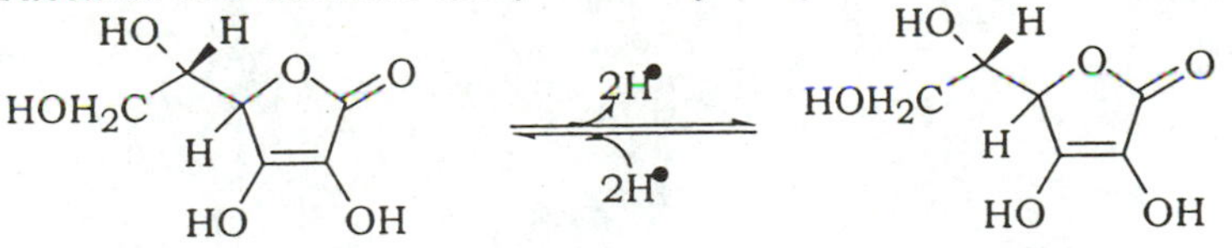

Biotin functions as a carrier of carboxyl groups usually derived from bicarbonate. The structure of N-carboxybiotin covalently linked to a protein via a lysine is shown below.

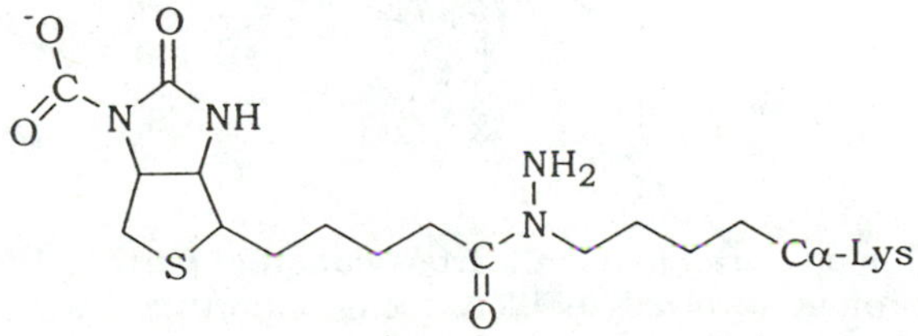

Lipoic acid, shown below, is an acyl group carrier. The acyl group is attached to lipoic acid in thiolester linkage. Upon deacylation, the sulfhydryls must be oxidized to a disulfide in order to function.

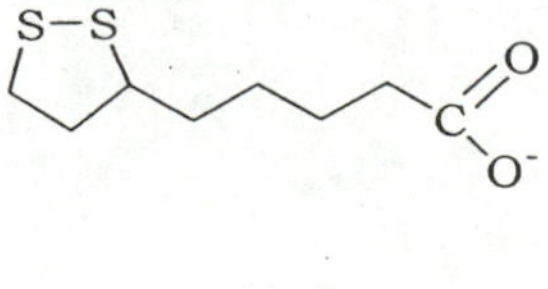

Folic acid derivatives function as acceptors and donors of one-carbon units. The active form of this vitamin, shown below, is tetrahydrofolate. One-carbon units at all oxidation levels except that of CO_2 are carried as adducts to N-5, N-10, or to N-5 and N-10.

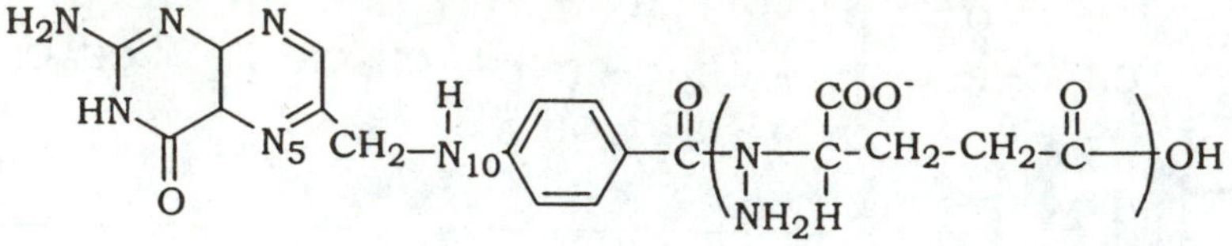

Fat Soluble Vitamins

Vitamin A or retinol plays a role in vision, and growth and differentiation. Its structure is shown below.

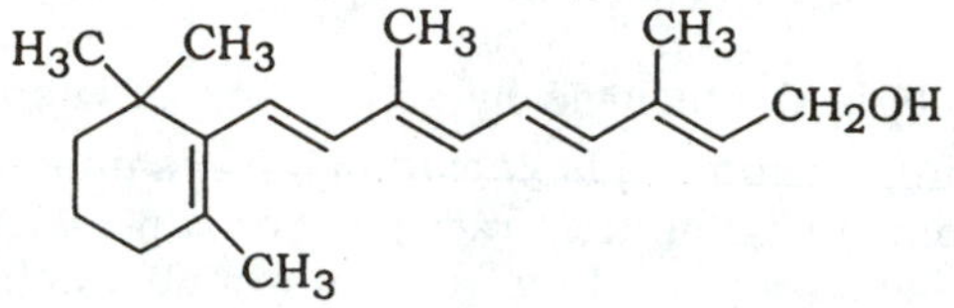

Vitamin D is involved in calcium and phosphorus homeostasis. It is produced from 7-dehydrocholesterol by ultraviolet light. The structure of one active form of vitamin D, 1,25-dihydroxyvitamin D_3, is shown below.

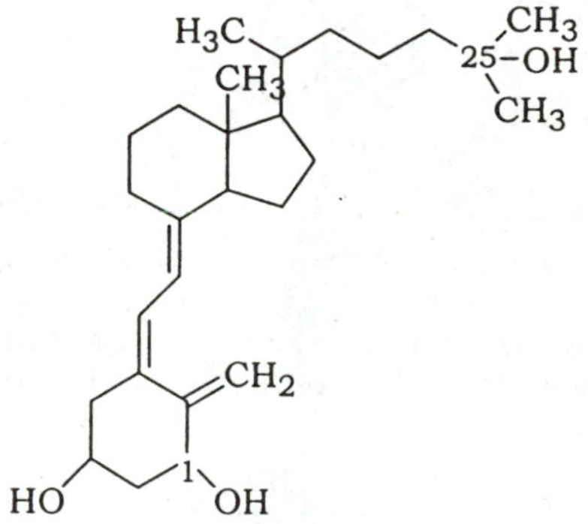

Vitamin E or α-tocopherol, shown below, is a potent antioxidant.

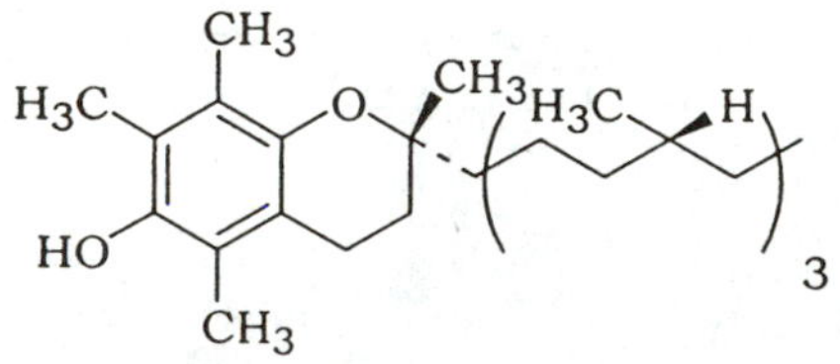

Vitamin K is required for post-translational carboxylation of glutamyl residues on the terminal end of prothrombin, a protein involved in blood coagulation cascade. Biological function is dependent on this modification. The structure of vitamin K_1 or phylloquinone is shown below.

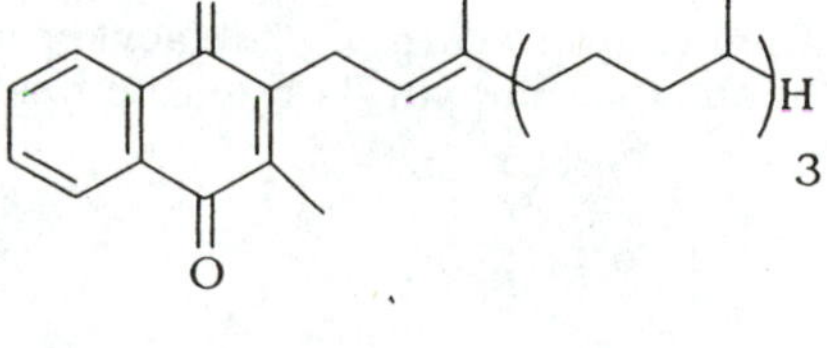

Problems and Solutions

1. Estimates indicate that 3×10^{14} kg of CO_2 are cycled through the biosphere annually. How many human equivalents (70-kg persons composed of 18% carbon by weight) could be produced each year from this amount of CO_2?

Answer: The amount of carbon in a 70 kg person is:

$$0.18 \times 70\ \text{kg} = 12.6\ \text{kg}$$

$$\text{Carbon dioxide is } \frac{12}{12+16+16} \times 100\% = 27.3\% \text{ by weight in carbon}$$

So, 3×10^{14} kg of CO_2 represents:

$$3 \times 10^{14}\ \text{kg} \times 0.273 = 8.18 \times 10^{13}\ \text{kg}$$

This represents

$$\frac{8.18 \times 10^{13}\ \text{kg}}{12.6 \frac{\text{kg}}{\text{person}}} = 6.5 \times 10^{12}\ \text{people}$$

That is, 6,500,000,000,000 or 6.5 trillion people.

2. Define the difference in carbon and energy metabolism between photoautotrophs and photoheterotrophs, and between chemoautotrophs and chemoheterotrophs.

Answer: Autotrophs can utilize CO_2 to synthesize biomolecules, whereas heterotrophs require more complex organic compounds. For example, an autotroph utilizes CO_2 as a carbon source to produce reduced carbon compounds like monosaccharides. Heterotrophs require compounds already containing reduced carbons to synthesize biomolecules.

In terms of energy, phototrophs can transduce light energy into chemical energy. Chemotrophs rely on chemical reactions, principally oxidation/reduction reactions during which a compound is oxidized in a series of steps. During this process energy is transduced into other chemical forms.

Photoautotrophs are at the base of the food pyramid (Figure 1.2, Garrett and Grisham) and capture the energy of sunlight to reduce CO_2 into carbohydrates. Photoheterotrophs utilize sunlight but are incapable of fixing CO_2. Thus, they require a source of reduced carbon. Chemoautotrophs can utilize CO_2 and do so using a chemical reaction, such as oxidation of an inorganic compound, as an energy source. Chemoheterotrophs require carbon in reduced form for both synthesis of biomolecules and for energy transduction.

3. What are the features that generally distinguish pathways of catabolism from pathways of anabolism?

Answer: Catabolic pathways are oxidative and result in the conversion of complex molecules into simple compounds. The reactions are overall thermodynamically spontaneous (negative ΔG). Catabolic pathways can result in production of a high-energy compound, such as ATP from ADP and P_i, and/or support reduction of NAD^+ to NADH.

Anabolic pathways are generally reductive in nature and must be driven thermodynamically, often by coupling key steps in a pathway to hydrolysis of ATP. The electrons for reduction often are supplied by NADPH, a reduced, phosphorylated form of NAD^+. Anabolic pathways convert simple compounds into more complex biomolecules.

4. What are the three principal modes of enzyme organization in metabolic pathways.

Answer: A metabolic pathway consists of all of the individual reactions involved in transforming a substrate into an end product. The reactions or steps are almost always catalyzed by enzymes. The enzymes may be separate proteins localized within a cell or cellular compartment but otherwise soluble. The product of one reaction is made available to the next reaction by diffusion. Alternatively, two or more enzymes may be organized into a multienzyme complex. Here, the enzymes catalyze sequential steps in the pathway. In this organization, the product of one enzyme is released directly to the next enzyme in the pathway. Finally, enzymes and enzyme complexes may be confined to a particular membrane within the cell. The purpose may be to provide a bridge between compartments, or to localize an enzymatic activity in a two-dimensional surface. Substrates confined to move in or along the membrane are only required to do a two-dimensional search to encounter enzyme.

5. Why do metabolic pathways have so many different steps?

Answer: Metabolic pathways consist of a number of chemical reactions and each reaction represents a distinct stage in the process of converting substrate into end-product. Each stage is catalyzed by an enzyme and produces an intermediate. This arrangement provides a number of control points along the path. By regulating the activity of enzymes, the flow of molecules through the pathway can be controlled. In addition, the intermediates formed may serve as branch points connecting different pathways. Finally, having individual steps may allow for the coupling of different pathways and for efficient utilization of energy and reductive power between pathways. For example, two-electron oxidations and reductions in different pathways may be linked by steps utilizing NAD^+ or NADPH. Energy transduction and utilization may be more efficient when metabolic steps are tailored to energy release sufficient to drive ATP synthesis or to use ATP hydrolysis. For example, under cellular conditions, ADP phosphorylation requires about 40 to 50 kJ/mol. In many cases, conversions of metabolic intermediates, sufficiently exergonic to release this amount of energy, are coupled to ADP phosphorylation. The overall reaction, namely ATP synthesis and conversion of intermediates, is still exergonic but only moderately so because a portion of the energy is stored as ATP. Metabolic pathways have evolved to include steps sufficiently exergonic to synthesize ATP in an efficient manner.

6. Why is the pathway for the biosynthesis of a biomolecule at least partially different from the pathway for its catabolism? Why is the pathway for the biosynthesis of a biomolecule inherently more complex than the pathway for its degradation?

Answer: Catabolic pathways are exergonic with one or more steps having large negative ΔG values. These steps represent thermodynamic barriers to an anabolic sequence functioning by reversal of the catabolic pathway. Thus, separate reactions are required to bypass them. For example, in one step in glycolysis, phosphofructokinase catalyzes the transfer of a phosphate group from ATP to fructose-6-phosphate to produce ADP and fructose-1,6-bisphosphate. In gluconeogenesis, the antithesis of glycolysis, the conversion of fructose-1,6-bisphosphate to fructose-6-phosphate is accomplished by a simple hydrolysis reaction catalyzed by the enzyme fructose-1,6-bisphosphatase. Under physiological concentrations of ATP, ADP and P_i, the kinase reaction in glycolysis favors production of fructose-1,6-bisphosphate whereas the phosphatase reaction favors production of fructose-6-phosphate. Parallel catabolic and anabolic pathways must include at least one different step to allow for independent regulation of each pathway.

Biosynthesis of complex biomolecules from simple compounds is an endergonic process. Thus, for anabolic pathways to function they must be coupled to highly exergonic reactions such as hydrolysis of ATP. These coupling reactions increase the complexity of anabolic pathways. For example, conversion of phosphoenolpyruvate (PEP) to pyruvate with phosphorylation of ADP to ATP is a single reaction in glycolysis. The reversal of this step in gluconeogenesis proceeds in two stages: Pyruvate is first carboxylated to oxaloacetate, with hydrolysis of ATP driving the reaction; second, oxaloacetate is converted to PEP with GTP serving as a phosphoryl group donor.

7. What are the metabolic roles of ATP, NAD^+, and NADPH?

Answer: The living state requires a constant source of energy for biosynthesis, for movement, for replication, in short, for all of the activities characterizing the living state. The immediate source of energy depends on the cell type (see problem 2) and includes light energy, chemical energy, or both. However, independent of the source of energy, all cells transduce energy into the chemical energy of the phosphoric anhydride bonds of ATP. ATP is the energy currency of cells. Hydrolysis of the high-energy bonds of ATP may be used to drive biosynthetic pathways and to support all of the other activities of the living state. Thus, ATP serves to couple exergonic processes to endergonic processes. For example, ATP synthesized during catabolic, exergonic pathways is used in hydrolysis reactions to drive anabolic, endergonic pathways.

The major catabolic pathways common to virtually all cells are based on oxidation of carbon compounds. For example, the conversion of glucose to C_2O and H_2O includes three oxidative sequences. In glycolysis, glucose is converted to pyruvate. Next, pyruvate is oxidized to acetyl-coenzyme A. Finally, acetyl-coenzyme A is metabolized to CO_2 and H_2O in the citric acid cycle. These three oxidative sequences are accompanied by electron transfers and NAD^+ serves as the primary electron acceptor.

In contrast, anabolic pathways are often reductive; the immediate reductant or electron donor is NADPH. Thus, this reduced, phosphorylated form of NAD^+ plays a role in metabolism exactly opposite to that of NAD^+.

8. *What are the advantages of compartmentalizing particular metabolic pathways (see Chapter 1) within specific organelles?*

Answer: Compartmentalization provides an easy solution to the problem of regulating two opposing pathways. For example, the catabolic pathway for fatty acid metabolism, known as β–oxidation, occurs in the mitochondria. In contrast, fatty acid biosynthesis largely occurs in the cytosol. Compartments may also provide a particular set of conditions favorable to only certain reactions. There are enzymes known as acid hydrolases that are localized in lysosomes. These enzymes include proteases, nucleases, lipases and glycosidases and function at acidic pH. The enzymes are confined by the membrane barrier of lysosomes and their activity is enhanced by the acidic pH maintained within lysosomes. If inadvertently released into the cytosol, these enzymes fail to cause damage to cellular components because the cytosol is maintained at a higher pH value.

9. *Maple syrup urine disease (MSUD) is an autosomal recessive genetic disease characterized by progressive neurological dysfunction and a sweet, burnt-sugar or maple-syrup smell in the urine. Affected individuals carry high levels of branched-chain amino acids (leucine, isoleucine, and valine) and their respective branched-chain, α–keto acids in cells and body fluids. The genetic defect has been traced to the mitochondrial branched-chain, α–ketoacid dehydrogenase (BCKD). Affected individuals exhibit mutations in their BCKD, but these mutant enzyme molecules have normal activity. Nonetheless, treatment of MSUD patients with substantial doses of thiamine can alleviate the symptoms of the disease. Suggest an explanation for the symptoms described and for the role of thiamine in the amelioration of the symptoms of MSUD.*

Answer: The branched chain amino acids, valine, leucine, and isoleucine, are metabolized by transamination reactions followed by oxidative decarboxylation of the resulting keto acids to produce acyl derivatives of CoA.

$$^{+}H_3N{-}\underset{COO^-}{\overset{R}{C}}{-}H \longrightarrow \underset{COO^-}{\overset{R}{C}}{=}O \xrightarrow[\text{CoA-SH}\quad CO_2]{NAD^+\quad NADH + H^+} \underset{\underset{CoA}{S}}{\overset{R}{C}}{=}O$$

In maple syrup urine disease, high levels of α–keto acids of the branched chain amino acids accumulate in body fluids indicating that deamination occurs but it is not followed by decarboxylation. This is due to a defect in branched-chain α-keto acid dehydrogenase. Several specific mutations that cause this disease have been identified. Some patients suffering from MSUD respond to high levels of thiamine (100 - 200 mg/day) apparently because thiamine increases the biological half-life of branched-chain α-keto acid dehydrogenase. However not all MSUD patients respond to thiamine.

Questions for Self Study

1. The entries in the last four columns of Table 14.2 have been rearranged in an incorrect order. Correct them.

Classification	Carbon Source	Energy Source	Electron Donors	Examples
Phototrophs	Organic compounds	Oxidation-reduction reactions	Organic compounds	Nonsulfur purple bacteria
Photoheterotrophs	CO_2	Oxidation-reduction reactions	H_2O, H_2S, S, other inorganic compounds	All animals, most microorganism, nonphotosynthetic plant tissue such as roots, photosynthetic cells in the dark
Chemotrophs	Organic compounds	Light	Organic compounds e.g., glucose	Green plants, algae, cyanobacteria, photosynthetic bacteria
Chemoheterotrophs	CO_2	Light	Inorganic compounds: H_2, H_2S, NH_4^+, NO_2^-, Fe^{2+}, Mn^{2+}	Nitrifying bacteria; hydrogen, sulfur, and iron bacteria

2. Match terms in the two columns

   a. Obligate aerobes
   b. Facultative anaerobes
   c. Obligate anaerobes
   d. Aerobes
   e. Anaerobes

   1. Use O_2 as electron acceptor.
   2. Cannot use O_2 as electron acceptor.
   3. O_2 required for life.
   4. Use O_2 but can adapt to lack of O_2.
   5. Cannot use or tolerate O_2.

3. For each of the items listed below are they indicative of an anabolic (A) pathway or a catabolic (C) pathway?
   a. Consumes ATP.
   b. Involves the cofactor NAD^+.
   c. Results in reduction of compounds using NADPH.
   d. Energy-yielding.
   e. Production of glucose by photoautotroph.
   f. Results in decrease in molecular complexity.

4. Metabolic regulation of anabolic and catabolic pathways is accomplished by regulating the activities of enzymes. What three levels of enzymatic control are most common?

5. Anabolic pathways are endergonic whereas catabolic pathways are exergonic. What role does ATP play in coupling the two kinds of pathways?

6. How are radioactive isotopes and stable, heavy isotopes used in biochemical research?

7. Fill in the blanks. Vitamin B_1 or _____ is a coenzyme used in reactions in which bonds to carbonyl carbons, such as _____ or _____ are synthesized or cleaved. This vitamin is involved in decarboxylation of α-keto acids and the formation and cleavage of α-hydroxyketones. The enzyme pyruvate decarboxylase is a vitamin B_1-dependent enzyme that catalyzes _____ of the α-keto acid _____. Transketolase moves a two-carbon unit from a _____ to an _____. The mechanism of action of this vitamin depends on formation of a _____ in the substituted thiazole ring. Dietary deficiencies of this vitamin lead to the nervous system disease _____.

8. Identify the following in the compound shown below.

a. AMP
b. Ribose
c. Position of phosphate in NADP $^+$
d. Nicotinamide
e. Pro-chiral carbon
f. Hydride ion binding site
g. Phosphodiester bond
h. High energy bond
i. Positively charged nitrogen
j. Nicotinic acid
k. Pyridine ring

H O
C–NH2
N
O–CH2
-O
P
O
O
O
H H
H H
OH OH
NH2
O
P
-O
N
N
O–CH2
N
N
O
H H
H H
OH OH

9. The nicotinamide coenzymes play important roles as electron carriers in oxidation-reduction reactions. What other vitamin is used in a similar manner? With respect to interactions with proteins, how are cofactors of this vitamin different than the nicotinamide coenzymes?

10. Pantothenic acid is a constituent of both a complex coenzyme and a protein. What are they? Both are involved in metabolism of what two-carbon unit?

11. Pyridoxal phosphate (vitamin B_6) catalyzes a number of reactions involving amino acids including decarboxylations, eliminations, racemization and aldol reactions.
   a. Pyridoxal phosphate can interact with the amino group of amino acids through what type of adduct?
   b. How might the agent $NaBH_4$ modify this adduct?
   c. What are the consequences (usually) of this modification?
   d. Why can pyridoxal phosphate exhibit such a wide range of activities?

12. Of the vitamins, Vitamin B_{12} or cyanocobalamin (C), biotin (B), and folic acid (F), which best fits the following statements?
   a. Derivative of the highly colored compounds pteridine and pterin.
   b. Intramolecular rearrangements.
   c. Linked to a protein via amide linkage with the ε-amino group of a lysine residue.
   d. Contains cobalt.
   e. Modifying enzyme inhibited by the anticancer agent amethopterin (methotrexate).
   f. Prevents pernicious anemia.
   g. Binds tightly to the egg white protein avidin.
   h. Methyl group transfers.
   i. Mobile carboxyl group carrier.
   j. Carrier of a variety of one-carbon units.

13. Lipoic acid and vitamin C can both participate in what kind of simple chemical reaction?

14. For the fat-soluble vitamins A, D, E and K, which best fits the following statements?
   a. Cholesterol derivative.
   b. Isoprenoid compound.
   c. α-Tocopherol.
   d. Requires action of ultraviolet light for synthesis.
   e. Deficiencies cause problems with blood clotting.
   f. Deficiency causes rickets.
   g. Potent antioxidant.
   h. Involved in carboxylation of glutamic acid residues on certain proteins.

Transported to peripheral tissue from liver complexed to retinol-binding protein.
j. Is a component of rhodopsin.

Answers

1.

Classification	Carbon Source	Energy Source	Electron Donors	Examples
Phototrophs	CO_2	Light	H_2O, H_2S, S, other inorganic compounds	Green plants, algae, cyanobacteria, photosynthetic bacteria
Photoheterotrophs	Organic compounds	Light	Organic compounds	Nonsulfur purple bacteria
Chemotrophs	CO_2	Oxidation-reduction reactions	Inorganic compounds: H_2, H_2S, NH_4^+, NO_2^-, Fe^{2+}, Mn^{2+}	Nitrifying bacteria; hydrogen, sulfur, and iron bacteria
Chemoheterotrophs	Organic compounds	Oxidation-reduction reactions	Organic compounds e.g., glucose	All animals, most microorganism, nonphotosynthetic plant tissue such as roots, photosynthetic cells in the dark

2. a. 3; b. 4; c. 5; d. 1; e. 2.

3. a. A; b. C; c. A; d. C; e. A; f. C.

4. Allosteric regulation, covalent modification, and enzyme synthesis and degradation.

5. Steps in anabolic pathways are often driven by hydrolysis of ATP. The result of catabolism is the phosphorylation of ADP to ATP.

6. Both types of isotopes are used to label compounds. Radioactive isotopes are unstable and release energy upon decay. The presence of a radioactively labeled compound can be detected by the energy emitted upon radioactive decay. Stable, heavy isotopes increase the density or mass of a compound into which they are incorporated.

7. Thiamine; aldehydes; ketones; decarboxylation; pyruvate; ketose; aldose; carbanion; beriberi.

8.

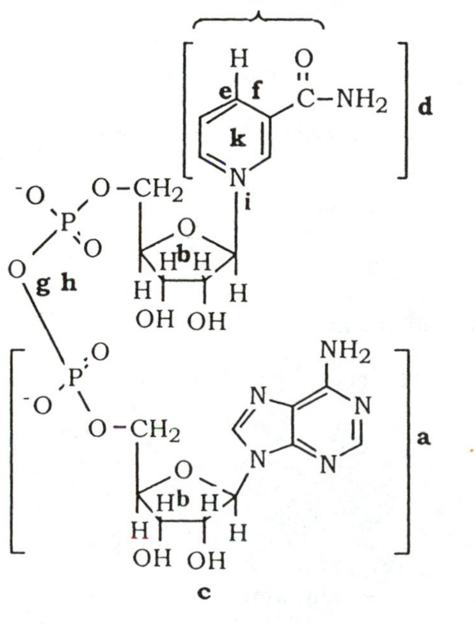

9. Riboflavin or vitamin B_2 is a precursor of flavin mononucleotide and flavin adenine dinucleotide. The flavoproteins, enzymes that bind flavins, bind the coenzymes tightly. Thus, cellular free flavin concentrations are low and most of the flavin is protein-bound.

10. Coenzyme A (CoA) and acyl carrier protein; acetyl group.

11. a. Schiff base; b. reduce it; c. inactivation; d. serves as an effective electron sink to stabilize reaction intermediates.

12. a. F; b. C; c. B; d. C; e. F; f. C; g. B; h. C or F; i. B; j. F.

13. Oxidation-reduction.

14. a. D; b. A, D, E, or K; c. E; d. D; e. K; f. D; g. E; h. K; i. A; j. A.

Additional Problems

1. Table 14.3 lists the properties of various radioisotopes. For ^{14}C the following information is given: Type -radioactive, Radiation Type -β^-, Half-life- 5700 yr. Explain this information.

2. Calculate the percent density difference between double-stranded DNA containing ^{14}N (the normal isotope) and ^{15}N (the heavy isotope). Assume that the average molecular weight of a base-pair in ds-DNA is 625 and that the DNA is 50% A-T rich.

3. A gasoline-powered, internal-combustion engine and a polar bear metabolizing blubber run very similar reactions overall to produce energy; however, the details are quite different. Explain.

4. NMR spectroscopy has been used to measure cellular pH by determining the chemical shift of cellular P_i. Phosphorus NMR is particularly convenient because ^{31}P has a large magnetic moment and is the naturally occurring isotope of phosphorus. But, why is inorganic phosphate studied? Why not one of the phosphates of ATP? Typically [ATP] is greater than [P_i].

5. Draw the structure of γ-carboxyglutamic acid. Which fat-soluble vitamin is necessary for production of this modified amino acid? Suggest a biological role for this modified amino acid.

6. Deficiencies in vitamin D cause rickets. Rickets is a rare disease in countries with sunny climates. Explain.

7. The enzyme alcohol dehydrogenase catalyzes the following reaction:

$$CH_3{-}CH_2OH \xrightarrow{NAD^+ \;\rightarrow\; NADH + H^+} CH_3{-}C(H){=}O$$

The carbon oxidation producing an aldehyde is accompanied by reduction of the coenzyme NAD^+. Account for the change in charge for NAD^+ in this reaction.

8. Most of the carbons in pantothenic acid derive from three amino acids or amino acid precursors. What are they?

9. Sodium borohydride, $NaBH_4$, is a potent inhibitor of pyridoxal phosphate-utilizing enzymes that form a Schiff base during catalysis. Why?

10a. Over a leisurely breakfast with a friend, read the ingredients in your favorite, vitamin-enriched breakfast cereal and explain as many of the additives as possible.

b. There are many who forsake breakfast, but understanding the need for a balanced diet, will accompany their morning coffee (or tea) with a megavitamin (usually an earthy brown-colored pill of enormous size). What is one problem with this approach to nutrition?

Abbreviated Answers

1. The designation "radioactive" indicates that ^{14}C is unstable and will be converted into another isotope or element. Radiation type β^- informs us that the conversion occurs by beta emission; an electron is emitted from the nucleus. The atomic mass remains constant (the atomic mass is the sum of neutrons and protons); however, the atomic number increases by 1 to 7 and ^{14}C is converted to ^{14}N, a stable isotope. The emitted electron may be detected by liquid scintillation counting or by exposure to film. In liquid scintillation counting, the electron excites fluorescent molecules that emit visible light upon returning to their ground state. The amount of light is measured using a photomultiplier. In film exposure, the electron reduces silver grains in the film. Under certain conditions, the number of grains reduced is proportional to the number of decay events. The half-life refers to the time it takes for one-half of the radioisotope to decay. So, after 5700 years only half of the starting material remains.

2. Our first task is to determine the number of nitrogens in an "average base-pair". A, G, C, and T have 5, 5, 3, and 2 nitrogens, respectively. Therefore A:T and G:C base-pairs have 7 and 8 nitrogens and an average base-pair in a 50% A-T rich ds-DNA has 7.5 nitrogens. The molecular weight of a ^{14}N-containing base-pair is 625. For a ^{15}N-containing base-pair, the molecular weight is:

$$625 - (14 \times 7.5) + (15 \times 7.5) = 632.5$$

The percent density difference id given by:

$$\%\ \text{Difference} = \frac{632.5 - 625}{625} \times 100\% = 1.2\%$$

3. The engine and the polar bear are both relying on carbon oxidation reactions for energy and the general equation that describes both reactions is:

$$(\text{-}CH_2\text{-})_n + O_2 \leftrightarrows nCO_2 + nH_2O)$$

In an engine, gasoline vapor mixed with oxygen is ignited by a spark to initiate carbon oxidation. (The reaction is often not complete and CO or carbon monoxide and other partially oxidized carbon compounds are produced.) The polar bear takes a different approach. Fat, with an average carbon not unlike a typical hydrocarbon found in gasoline, is metabolized via β-oxidation, the citric acid cycle, and electron transport to produce CO_2; O_2 is consumed during electron transport to produce H_2O.

4. The chemical shift of phosphorus depends on, among other things, the local magnetic environment in which the phosphorus is located. Protonation of phosphate groups will result in a chemical shift. The advantage to using inorganic phosphate to monitor intracellular pH is that phosphorus has a pK_a around neutral pH. The pK_as of phosphate in nucleotides are approximately one pH unit lower.

5. γ-Carboxyglutamate (Gla) has an additional carboxylate group attached to the γ-carbon of the side chain of glutamic acid as shown below

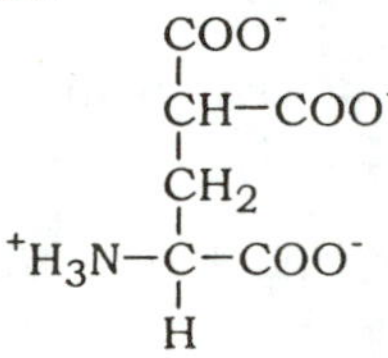

```
         COO⁻
         |
         CH—COO⁻
         |
         CH₂
         |
   ⁺H₃N—C—COO⁻
         |
         H
```

This amino acid is found in a small number of proteins where it functions as a calcium-binding site. It is not incorporated into a protein during protein synthesis but is a result of post-translational modification in a vitamin K-dependent reaction.

6. Rickets is caused by a deficiency of vitamin D, a fat soluble vitamin produced by UV radiation of 7-dehydrocholesterol, an intermediate in cholesterol synthesis. Before the days of vitamin-enriched milk, vitamin D-deficiencies were a common problem in winter months. Countries with sunny climates rarely have problems with rickets because sufficient sunlight is available to convert 7-dehydrocholesterol to vitamin D to meet metabolic requirements.

7. NAD^+ is reduced to NADH by accepting 2 electrons and one proton as a hydride ion. The addition of this negative charge accounts for the neutral charge on NADH.

8.

Pantetheine

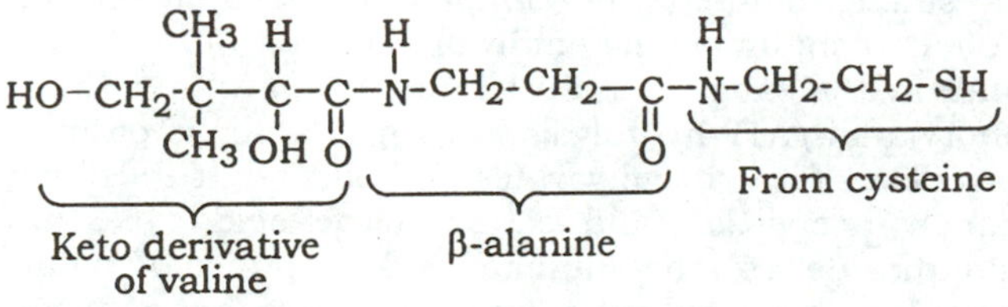

$$HO{-}CH_2{-}C(CH_3)_2{-}CH(OH){-}C(=O){-}NH{-}CH_2{-}CH_2{-}C(=O){-}NH{-}CH_2{-}CH_2{-}SH$$

Keto derivative of valine (HO–CH₂–C(CH₃)₂–CH(OH)–C(=O)–) · β-alanine (–NH–CH₂–CH₂–C(=O)–) · From cysteine (–NH–CH₂–CH₂–SH)

9. Sodium borohydride will reduce the Schiff base as shown below.

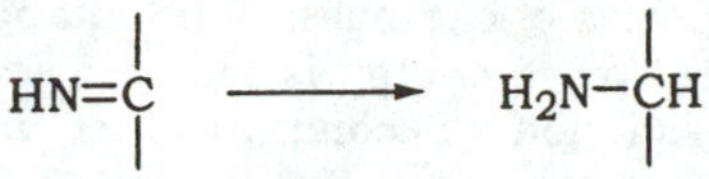

$$HN{=}C< \longrightarrow H_2N{-}CH<$$

10a. Here is what my cereal box says.

Ingredients: Corn, sugar, malt flavoring, corn syrup, **Vitamins and Iron**: ascorbic acid (vitamin C), iron, niacinamide, pyridoxine hydrochloride (vitamin B_6), riboflavin (vitamin B_2), vitamin A palmitate, thiamin hydrochloride (vitamin B_1), folic acid, and vitamin D. To maintain quality, BHT has been added to the packaging

The only ingredient you should need help on is BHT or butylated hydroxytoluene, an antioxidant used as a preservative.

b. Many of the water soluble vitamins are simply lost upon urination. (In fact it is possible that for one or two urinations after taking a megavitamin, the urine will be brightly yellow-colored due to loss of riboflavin.) It makes more sense to have several well balanced meals daily.

Summary

From a chemical perspective, living organisms are open thermodynamic systems far displaced from equilibrium. Further, cellular constituents are characteristically in a chemically more reduced state than the inanimate matter that surrounds them. Nevertheless, cells must somehow extract energy and reducing power from the environment and use this energy and reducing power to maintain the complex chemical activities that collectively constitute the living state. The sum of all these processes is metabolism.

Maps of intermediary metabolism portray hundreds of enzymatic reactions, but these represent only the principal metabolic pathways. It is estimated that the cells of higher organisms carry out literally thousands of metabolic reactions, each catalyzed by a specific enzyme. Despite the great variations and adaptations found in organisms, virtually all have the same basic set of major metabolic pathways, a fact providing strong evidence that they evolved from a common ancestor. Still, considerable metabolic diversity exists between organisms in terms of the sources of carbon and energy that they exploit. Photoautotrophs require only light as a source of energy and CO_2 as a source of carbon; photoheterotrophs use light as a sole source of energy but must have an organic supply of carbon in order to synthesize their complement of biomolecules. Chemoautotrophs can live on CO_2 as their sole source of carbon provided an oxidizable inorganic substrate, such as Fe^{2+}, NO_2^- or NH_4^+, is available to supply energy. Chemoheterotrophs are organisms needing an organic carbon source for both energy and biomolecular synthesis.

The flows of energy, carbon and oxygen within the biosphere are intimately related. Phototrophs use solar energy to fix CO_2 into more reduced organic compounds; the electrons for this reduction come from water and O_2 is evolved. Heterotrophs feed on the organic products of photosynthesis, harvesting the energy in these compounds in oxidative reactions that release carbon dioxide, consume O_2, and form H_2O.

Metabolism consists of two major realms: the degradative pathways of catabolism and the biosynthetic pathways of anabolism. A metabolic pathway is composed of a sequence of enzymatic reactions, transforming some precursor to a specific end product via a consecutive series of intermediates. The enzymes of a particular pathway may exist either as physically separate entities, as a stable multi-enzyme complex, or as a membrane-associated system. Metabolic pathways often contain many individual steps for energetic reasons: energy transactions typically fall in the range of 0-50 kJ/mol, to fit within the prevailing free energy change for ADP phosphorylation/ATP hydrolysis in cells. Catabolic pathways exist to degrade a wide variety of fuel molecules - fats, carbohydrates, proteins - but these pathways converge to a few end products, namely acetyl-CoA and certain citric acid cycle intermediates. Water, ammonium and carbon dioxide are the ultimate end products of catabolism. In contrast, anabolic pathways diverge from a limited set of building blocks (amino acids, sugars, nucleotides and acetyl-CoA) to the end results of their synthetic activity, the astounding variety of complex biomolecules. Corresponding pathways of catabolism and anabolism differ for essential reasons. First, thermodynamics dictates that more energy is required for the synthesis of a biomolecule than can be realized from its degradation. Second, in order to have independent regulation of opposing metabolic sequences, the sequences must consist, at least in part, of unique reactions mediated by separate enzymes that serve as distinct sites for regulation.

ATP is the energy currency of the cell. ATP is a major product of catabolism; in turn, ATP is a principal substrate for anabolism, providing the chemical energy that serves as the driving force for biosynthesis. Electrons released in oxidative reactions of catabolism are collected in the pyridine nucleotide coenzymes, NADH and NADPH. The electrons carried in these two coenzymes have very different fates. Those in NADH are passed to O_2 to form H_2O in an energetic process known as oxidative phosphorylation, which couples NADH oxidation to ATP synthesis. In contrast, the electrons in NADPH provide the reducing power necessary to drive reductive anabolic reactions.

Metabolism is a tightly regulated, integrated process. Metabolic regulation is achieved by controlling the activity of key enzymes on one of a number of levels: allosteric regulation, covalent modification and enzyme synthesis/degradation.

The reactions comprising various metabolic pathways have been revealed through a growing arsenal of experimental methods. Early approaches relied on cell-free assays of metabolic intermediates and the enzymatic reactions transforming them. Enzyme inhibitors were useful tools in these studies. Genetic mutations result in specific metabolic blocks that are very informative about metabolic reactions. Isotopic tracers are particularly good probes of metabolism since isotopically labeled compounds have essentially the same chemical behavior as their unlabeled counterparts. Isotopes can be traced through their radioactivity or their difference in mass (or density). NMR spectroscopy has the power to reveal detailed information about the chemical fate of substances in a noninvasive, nondestructive way. Evidence regarding the compartmentation of metabolic processes within specific organelles has established that the flow of intermediates in metabolism is channeled spatially as well as chemically.

Nutrition is the use of foods by organisms. Nutrients include proteins, carbohydrates, lipids, fibers, and vitamins. Proteins serve as a source of essential amino acids. Excess amino acids may be converted to glucose, fatty acids, or completely degraded. Catabolism of carbohydrates represent an important source of metabolic energy. Lipid degradation is also an important source of metabolic energy; however, lipids also provide essential fatty acids. Fibers, including cellulose, hemicellulose, and lignin, play important roles in digestion and absorption. Vitamins are nutrients, which are required in the diet, usually in trace amounts, if they cannot be synthesized by the organism itself. Except for vitamin C, the water soluble vitamins are all components or precursors of coenzymes, low molecular weight species that bring unique functionalities to certain enzymes. The fat soluble vitamins are not related to coenzymes, but still have essential roles in various biological processes.

Thiamine, known as vitamin B_1, was the first vitamin to be discovered, and in its active form of thiamine pyrophosphate, is involved in reactions of carbohydrate metabolism in which bonds to carbonyl carbons are synthesized or cleaved. The cationic imine nitrogen of TPP provides electrostatic stabilization of the carbanion formed upon removal of the C-2 proton, and, once TPP attack on the substrate has occurred, the cationic imine can act as an effective electron sink to stabilize the negative charge, which develops on the carbon under attack.

The flavin dinucleotide coenzymes (derived from riboflavin), the pyridine dinucleotides (derived from niacin) and coenzyme A (derived from pantothenic acid) all contain adenine nucleotides which facilitate recognition and binding by appropriate enzymes. Nicotinamide adenine dinucleotide and nicotinamide adenine dinucleotide phosphate act as electron carriers in redox reactions. NAD^+ is an electron acceptor in oxidative (catabolic) pathways, and NADPH is an electron donor in reductive (biosynthetic) pathways. These reactions involve direct hydride transfer to or from the coenzyme and the enzymes carrying out such reactions are known as dehydrogenases. The flavin coenzymes, FMN and FAD, fill unique roles in organisms because they are stronger oxidizing agents than NAD^+ and $NADP^+$, because they can be reduced by both one-electron and two-electron pathways, and because they can be reoxidized easily by molecular oxygen. Coenzyme A is a constituent of acyl carrier proteins and it functions to 1) activate acyl groups for transfer by nucleophilic attack, and 2) activate the α-hydrogen of the acyl group for abstraction as a proton.

Pyridoxal-5-phosphate, a form of vitamin B_6, participates in a wide variety of reactions involving amino acids, including transaminations, decarboxylations, eliminations, racemizations and aldol reactions. The versatility of PLP is due to its ability to 1) form stable Schiff base (aldimine) adducts with α-amino groups of amino acids and 2) act as an effective electron sink to stabilize reaction intermediates.

Vitamin B_{12} is converted in the body to two active forms: 5'-deoxyadenosylcobalamin and methylcobalamin. The B_{12} coenzymes participate in three types of reactions, including 1) intramolecular rearrangements, 2) reduction of ribonucleotides to deoxyribonucleotides (in certain bacteria) and 3) methyl group transfers. Activation of B_{12} to 5'-deoxyadenosylcobalamin involves sequential reduction of Co^{3+} to Co^{2+}, an extremely powerful nucleophile that attacks the C-5 carbon of ATP. Intramolecular rearrangements catalyzed by B_{12} involve homolytic cleavage of the Co-carbon bond of the coenzyme, which produces a reactive $-CH_2\cdot$ radical capable of hydrogen atom abstraction from the substrate.

Ascorbic acid or vitamin C functions as an effective reducing agent and electron carrier in numerous biological processes. It is required for the metabolism of tyrosine in the brain, for the mobilization of iron from the spleen, for stimulation of the immune system and for the mediation of histamine metabolism and the allergic response.

Biotin functions as a mobile carboxyl group carrier and participates in a variety of enzymatic carboxylation reactions. Bound to enzymes via a covalent bond to the ε-amino group of a lysine residue, it acquires carboxyl groups at one subsite of an enzyme's active site and delivers them to a substrate acceptor at another subsite. Biotin carboxylations are driven by ATP hydrolysis and involve the formation of a carbonyl-phosphate intermediate.

Lipoic acid also exists in nature covalently linked to lysine residues on enzymes. It functions as an acyl group carrier and is found primarily in pyruvate dehydrogenase and α-ketoglutarate dehydrogenase, two multienzyme complexes involved in carbohydrate metabolism. Lipoic acid couples acyl group transfer and electron transfer during oxidation and decarboxylation of α-keto acids.

Folate coenzymes are important acceptors and donors of one carbon units for all oxidation levels of carbon except that of CO_2 (for which biotin is the relevant carrier). Tetrahydrofolate is formed in two successive reductions of folate by dihydrofolate reductase. This enzyme is the apparent site of action of several important anti-cancer agents, including amethopterin or methotrexate and aminopterin, which are potent inhibitors of dihydrofolate reductase.

Vitamin A or retinol is essential for vision and plays a role in stimulating growth and differentiation in tissues. Absorbed by mucosal cells in the intestines, it is stored and processed in the liver, and is delivered to other tissues complexed to retinol-binding proteins. In the eyes, retinol is oxidized by a specific dehydrogenase to all-*trans* retinal and then to 11-*cis* retinal by a specific isomerase. The aldehyde group of retinal forms a Schiff base with a lysine on opsin to form light-sensitive rhodopsin.

The term vitamin D refers to a family of closely related molecules including ergocalciferol and cholecalciferol. These vitamins are produced in the skin of animals by the action of sunlight on sterol precursors. The D vitamins regulate calcium and phosphate metabolism and homeostasis. Calcium homeostasis involves coordination of calcium absorption in the intestine, deposition in the bones and excretion by the kidneys.

Vitamin E or α-tocopherol is a potent anti-oxidant and its function in animals and humans is presumed to be based on this property. Vitamin K or naphthoquinone is required for the activity of glutamyl carboxylase, an enzyme which carboxylates glutamate residues on the amino terminal end of prothrombin, a postribosomal modification essential for the proper function of the blood clotting process.

Chapter 15

Glycolysis

• •

Chapter Outline

- Glycolysis: Embden-Meyerhof pathway: Glucose to pyruvate
- Energy consuming phase: Five reactions: Two ATP consumed
  - Production of glucose-6-P: ATP consumed: 1st Priming step
    - Hexokinase
      - Low Km for glucose
      - Reacts with many different 6-carbon sugars
      - Glucose-6-P allosteric inhibitor
    - Glucokinase: Liver enzyme
      - High Km for glucose: Functions only when glucose abundant
      - No allosteric inhibition
      - Enzyme levels regulated by insulin
    - Advantages to phosphorylated sugar
      - Impermeable: Remains inside cell
      - Glucose concentration gradient maintained
      - Glucose carbon skeleton committed to metabolism
  - Fructose-6-P formation: Aldose-ketose isomerization
    - Prepares C-1 for second phosphorylation
    - Prepares C-3 for eventual cleavage
  - Phosphofructokinase reaction: Fructose-1,6-bisphosphate: ATP consumed: 2nd Priming step
    - Commits carbon skeleton to glycolysis
    - Allosteric regulation
      - Inhibitors
        - ATP: Increases Km for Fructose-6-P
        - Citrate
      - Activators
        - AMP: Reverses inhibition of ATP: Formed by adenylate kinase from ADP
        - Fructose-2,6-bisphosphate: Increase affinity for F-6-P: Decrease ATP inhibition
  - Fructose bisphosphate aldolase: Formation of DHAP and glyceraldehyde-3-P
    - Reverse of Aldol condensation
    - Two classes of aldolases
      - Class I: Lysine forms Schiff base with carbonyl carbon: $NaBH_4$ inhibits
      - Class II: Zinc-containing enzyme: EDTA inhibits
  - Trios phosphate isomerase
    - DHAP/glyceraldehyde interconversion
    - Ketose to aldose isomerization
- Energy yielding phase: Five reactions: Four ATP produced per glucose
  - Glyceraldehyde-3-phosphate dehydrogenase: 1,3-Bisphosphoglycerate
    - Oxidation/reduction
      - NAD^+ reduced to NADH

        - Glyceraldehyde oxidized to glycerate
    - Carboxylic acid phosphoric acid anhydride formed
    - Site of arsenate poisoning
  - Phosphoglycerate kinase: ATP produced
    - Substrate level phosphorylation
    - 1,3-Bisphosphoglycerate also used to form 2,3-BPG by mutase
    - Site of action of arsenate: Phosphate analog but labile to hydrolysis
  - 3-Phosphoglycerate to 2-phosphoglycerate conversion: Catalyzed by mutase
  - Enolase reaction
    - 2-Phosphoglycerate to PEP
    - Dehydration reaction
  - Pyruvate kinase
    - ATP produced in substrate-level phosphorylation
    - Regulation of enzyme
      - Allosteric regulation
        - AMP activates
        - Fructose-1,6-bisphosphate activates
        - ATP, acetyl CoA, alanine inhibit
      - Covalent inhibition by phosphorylation
        - Glucagon raises levels of cAMP
        - cAMP-dependent protein kinase phosphorylates pyruvate kinase
        - Phosphorylated kinase has high Km for PEP and inhibited by ATP and alanine
- Metabolism of other sugars
  - Fructose
    - Fructokinase at C-1: Aldolase forms DHAP and glyceraldehyde: Triose kinase to produce glyceraldehyde-3-P
    - Hexokinase at C:6 to produce fructose-6-P
  - Mannose: Hexokinase: Mannose-6-P isomerase to fructose-6-P
  - Galactose: Leloir pathway
    - Galactokinase at C-1 to produce galactose-1-P
    - Galactose-1-P + UDP-glucose to glucose-1-P and UDP-galactose
    - UDP-galactose epimerase: UDP-glucose
  - Glycerol
    - Glycerol kinase
    - Glycerol phosphate dehydrogenase to DHAP
- Fate of NADH and pyruvate
  - Aerobic respiration
    - Pyruvate to acetyl-Coenzyme A to citric acid cycle to carbon dioxide
    - NADH to electron transport
  - Anaerobic metabolism: Fermentation
    - Lactic acid fermentation: Pyruvate to lactic acid regenerates NAD^+
    - Ethanol fermentation
      - Pyruvate to acetaldehyde
      - Acetaldehyde to ethanol regenerates NAD^+

Chapter Objectives

Glycolysis

Glycolysis is the metabolic pathway leading from glucose to two molecules of pyruvate. Before jumping into the details it may be instructive to look at some basic chemistry. Glucose and pyruvic acid are shown below.

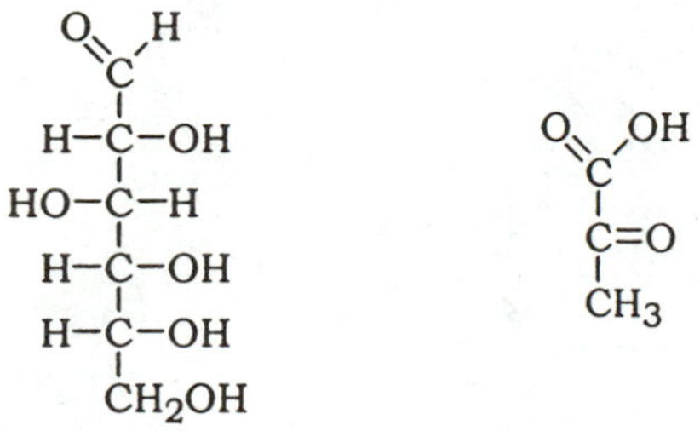

```
    O   H
     \\ /
      C
      |
   H-C-OH              O   OH
      |                 \\ /
  HO-C-H                 C
      |                  |
   H-C-OH                C=O
      |                  |
   H-C-OH                CH3
      |
     CH2OH
```

The carboxylate carbon and the keto carbon of pyruvate are at a more oxidized state whereas the methyl carbon is at a more reduced state relative to an "average" carbon of glucose. Therefore, this conversion, from glucose to pyruvate, must be accompanied by an additional reduction reaction. (In effect, the oxidation of one of the carbons of pyruvate is balanced by reduction of another carbon.) What gets reduced is NAD^+. When pyruvate is further metabolized to lactic acid or ethanol and CO_2, NADH is used to reduce the carbonyl carbon to an alcoholic carbon. The net result is regeneration of NAD^+, which allows glycolysis to continue. Glycolysis produces ATP and the energy for the reaction is the result of an internal oxidation/reduction.

Understand the basic organization of glycolysis into an energy-consuming phase and an energy-producing phase (see Figure 15.1). The two kinases, hexokinase (and glucokinase) and phosphofructokinase are key enzymes in phase I. Know how each is regulated but focus on phosphofructokinase. Phase II is the pay-back portion of glycolysis. Key enzymes in this phase include: glyceraldehyde-3-phosphate dehydrogenase, responsible for oxidation of the carbon skeleton; phosphoglycerate kinase, which catalyzes a substrate-level phosphorylation of ADP; and pyruvate kinase, responsible for the second substrate-level phosphorylation.

The Fate of Pyruvate

The ability to produce pyruvate from glucose depends on availability of the NAD^+. But, this coenzyme is in limiting amounts in cells and, thus, must be recycled for glycolysis to continue. In fermentation, the end product of glycolysis, namely pyruvate, is used as a substrate to recycle NAD^+. Understand alcoholic fermentation and lactic acid fermentation. An advantage to relying on these processes is the ability to produce ATP rapidly. The disadvantages include the production of problematic waste products containing reduced carbons.

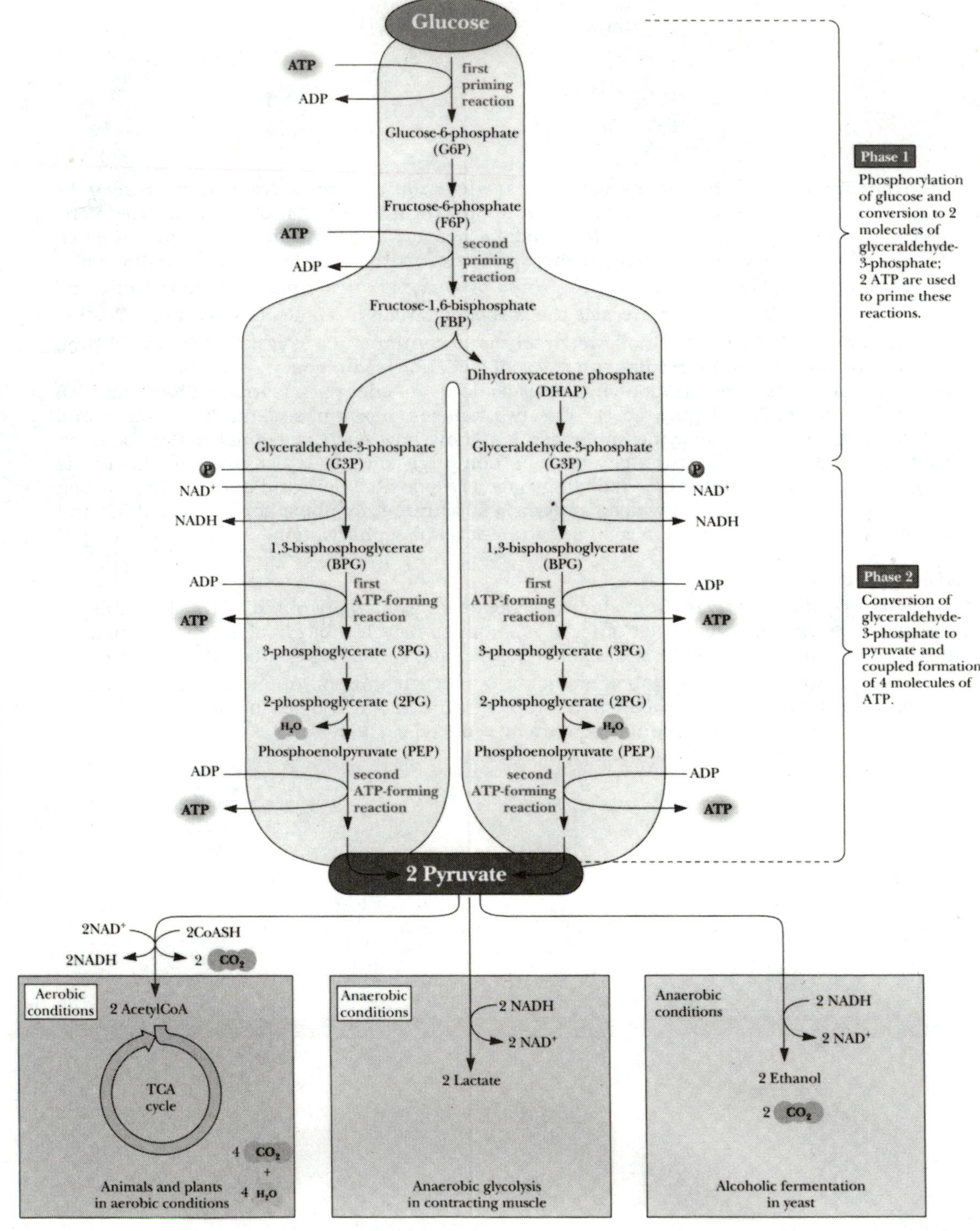

Figure 15.1 The glycolytic pathway

Problems and Solutions

1. List the reactions of glycolysis that:
   a. are energy-consuming (under standard state conditions).
   b. are energy-yielding (under standard state conditions).
   c. consume ATP.
   d. yield ATP.
   e. are strongly influenced by changes in concentration of substrate and product because of their molecularity.
   f. are at or near equilibrium in the erythrocyte (see Table 15.2).

Answer: Energy-consuming reactions have positive values of ΔG whereas energy-yielding reactions have negative values of ΔG. Under standard conditions the same rules apply but to values of ΔG°'. Using Table 15.1, we can classify reactions into these two types as follows:

Energy-consuming reaction	Enzyme	ΔG°' kJ/mol
Glucose-6-phosphate → fructose-6-phosphate	phosphoglucoisomerase	+1.67
Fructose-1,6-bisphosphate → dihydroxyacetone-P + glyceraldehyde-P	fructose biphosphate aldolase	+23.9
Dihydroxyacetone-P → glyceraldehyde-P	triose phosphate isomerase	+7.56
Glyceraldehyde-P + P_i + NAD^+ → 1,3-bisphosphoglycerate+ NADH + H^+	glyceraldehyde-3-P dehydrogenase	+6.30
3-Phosphoglycerate → 2-phosphoglycerate	phosphoglycerate mutase	+4.4
2-Phosphoglycerate → phosphoenolpyruvate + H_2O	enolase	+1.8

b.

Energy-yielding reaction	Enzyme	ΔG°' kJ/mol
α-D-glucose + ATP^{4-} → glucose-6-phosphate + ADP^{3-}	hexokinase, glucokinase	-16.7
Fructose-6-phosphate + ATP^{4-} → fructose-1,6-bisphosphate + ADP^{3-} + H^+	phosphofructokinase	-14.2
1,3-Bisphosphoglycerate + ADP^{3-} → 3-P-glycerate + ATP^{4-}	phosphoglycerate kinase	-18.9
Phosphoenolpyruvate + ADP^{3-} + H^+ → 1,3-bisphosphoglycerate+ NADH + H^+	pyruvate kinase	-31.7
Pyruvate + NADH + H^+ → lactate + NAD^+	lactate dehydrogenase	-25.2

c.

ATP-utilizing reaction	Enzyme	ΔG°' kJ/mol
α-D-glucose + ATP^{4-} → glucose-6-phosphate + ADP^{3-}	hexokinase, glucokinase	-16.7
Fructose-6-phosphate + ATP^{4-} → fructose-1,6-bisphosphate + ADP^{3-} + H^+	phosphofructokinase	-14.2

d.

ATP-producing reaction	Enzyme	ΔG°' kJ/mol
1,3-Bisphosphoglycerate + ADP^{3-} → 3-P-glycerate + ATP^{4-}	phosphoglycerate kinase	-18.9
Phosphoenolpyruvate + ADP^{3-} + H^+ → 1,3-bisphosphoglycerate+ NADH + H^+	pyruvate kinase	-31.7

e. What reactions are strongly influenced by concentration? In general,

$$\Delta G = \Delta G^{\circ\prime} + RT \times \ln \frac{\prod_i [P_i]}{\prod_i [R_i]}$$

Reactions that generate an increase or decrease in the number of products over reactants are expected to be sensitive to concentration changes. This is the case for reactions catalyzed by two enzymes: fructose bisphosphate aldolase and glyceraldehyde-3-phosphate dehydrogenase.

f. Reactions that are at or near equilibrium have values of ΔG close to zero. Using this criterion, the reactions catalyzed by the following enzymes must be close to equilibrium: phosphoglucoisomerase, phosphofructokinase, fructose bisphosphate aldolase, triose phosphate isomerase, phosphoglycerate kinase, phosphoglycerate mutase, and enolase.

2. Determine the anticipated location in pyruvate of labeled carbons if glucose molecules labeled (in separate experiments) with ^{14}C at each position of the carbon skeleton proceed through the glycolytic pathway.

Answer: Starting with glucose labeled at a particular carbon, the label will remain at that carbon until the fructose bisphosphate aldolase reaction. This reaction produces glyceraldehyde-3-phosphate and dihydroxyacetone phosphate. Carbons 1 (the aldehyde), 2 and 3 of glyceraldehyde derive from carbons 4, 5 and 6 of glucose. The phosphorylated carbon of dihydroxyacetone phosphate derives from carbon 1 of glucose and the ketone carbon from carbon 2. However, the next two reactions are an isomerization of the two triose phosphates into glyceraldehyde-3-phosphate and conversion to 1,3-bisphosphoglycerate. The isomerization interconverts the triose phosphates. Thus, carbon 3 of glyceraldehyde-3-phosphate derives from both carbons 1 and 6 of glucose; carbon 2 is from carbons 2 and 5 of glucose; and, carbon 1 is from carbons 3 and 4 of glucose. The steps from glyceraldehyde to pyruvate convert carbon 1 of glyceraldehyde-3-phosphate into the carboxyl carbon of pyruvate. So, the carboxyl carbon of pyruvate will acquire label from carbons 3 and 4 of glucose. The keto carbon of pyruvate is from carbon 2 of glyceraldehyde-3-phosphate, which derives from carbons 2 and 5 of glucose. Finally, the methyl carbon of pyruvate derives from carbons 1 and 6 of glucose.

$_1CH_2OPO_3^-$ — $_2C{=}O$ — HO-$_3C$-H — H-$_4C$-OH — H-$_5C$-OH — $_6CH_2OPO_3^-$ ⟶

DHAP: $_1CH_2OPO_3^-$ — $_2C{=}O$ — $_3CH_2OH$

⇅

Glyceraldehyde-3-phosphate: H–$_4C{=}O$ — H-$_5C$-OH — $_6CH_2OPO_3^-$

⟶ H–$_{3,4}C{=}O$ — H-$_{2,5}C$-OH — $_{1,6}CH_2OPO_3^-$

3. In an erythrocyte undergoing glycolysis, what would be the effect of a sudden increase in the concentration of (a) ATP? (b) AMP? (c) fructose-1,6-bisphosphate? (d) fructose-2,6-bisphosphate? (e) citrate? (f) glucose-6-phosphate?

Answer: Erythrocytes lack mitochondria and derive energy largely from glycolysis. Glycolysis is regulated at the two steps catalyzed by phosphofructokinase and pyruvate kinase. Phosphofructokinase is a key enzyme that serves as a "valve" regulating the movement of glucose units into glycolysis. Prior to formation of fructose-1,6-bisphosphate, there are other options (e.g., pentose phosphate pathway, glycogen synthesis) to glycolysis. The phosphofructokinase reaction commits glucose to glycolysis. This enzyme is subject to a number of regulatory mechanisms. At low ATP levels, the enzyme has a low K_m for fructose-6-phosphate; however, at high ATP levels, the enzyme behaves cooperatively and has an increased K_m for fructose-6-phosphate and a sigmoidal dependence of activity on fructose-6-phosphate concentration. This behavior is a result of allosteric regulation by ATP. Phosphofructokinase has an additional ATP binding site distinct from the substrate site. ATP binding to the allosteric site causes activity to be less sensitive to fructose-6-phosphate. This allosteric inhibition by ATP is reversed by AMP binding competitively. The inhibitory effect of ATP is enhanced by citrate. This citric acid cycle intermediate is produced in the mitochondria, and, when mitochondrial concentrations are high, citrate may be exported to the cytosol to be used in fatty acid biosynthesis. Cytosolic citrate also functions as an inhibitor of glycolysis by binding to phosphofructokinase and enhancing the allosteric effects of ATP. Since erythrocytes lack mitochondria, citrate is not expected to play an important role in regulation of glycolysis. Phosphofructokinase is also allosterically regulated by fructose-2,6-bisphosphate whose concentration is regulated by the enzymatic activities: phosphofructokinase 2 and fructose-2,6-bisphosphatase. High concentrations of fructose-2,6-bisphosphate is a signal to increase glycolysis by increasing the affinity of phosphofructokinase for its substrate, fructose-6-phosphate, and by decreasing the inhibitory effects of ATP.

The second key enzyme in regulation of glycolysis is pyruvate kinase. This enzyme has allosteric binding sites for AMP and fructose-2,6-bisphosphate leading to activation of the enzyme. Further, the enzyme is allosterically inhibited by ATP (and by alanine, and acetyl-coenzyme A).

With these points in mind we predict that increased concentrations of ATP will lead to inhibition of glycolysis, and increased concentrations of AMP will stimulate glycolysis. An increase in fructose-1,6-bisphosphate will stimulate glycolysis. Fructose-2,6-bisphosphate is a signal to increase the rate of glycolysis. Citrate will inhibit glycolysis. Finally, increased concentrations of glucose-6-phosphate may stimulate glycolysis. Glucose-6-phosphate is isomerized to fructose-6-phosphate, a substrate of phosphofructokinase. Depending on the ATP, AMP, citrate and fructose-2,6-bisphosphate levels, phosphofructokinase may be quite responsive to increased levels of fructose-6-phosphate. (The first reaction of glycolysis, phosphorylation of glucose to glucose-6-phosphate may be sensitive to glucose-6-phosphate levels as well. In muscle, the enzyme catalyzing this reaction is inhibited by glucose-6-phosphate, whereas in liver the enzyme is not.)

4. Discuss the cycling of NADH and NAD$^+$ in glycolysis and the related fermentation reactions.

Answer: Nicotinamide adenine dinucleotide plays a role in metabolism as an enzyme cofactor in oxidation-reduction reactions. It is found in limited concentrations and cycles between reduced (NADH) and oxidized (NAD$^+$) forms. Conversion of glucose to two molecules of pyruvate results in the net production of two molecules of ATP and the reduction of two molecules of NAD$^+$ to NADH. To continue the process of glycolysis, NAD$^+$ must be regenerated. Under anaerobic conditions, this is accomplished by using pyruvate or a derivative of pyruvate (acetaldehyde) to oxidize NADH back to NAD$^+$. What is the net effect? Glucose is converted to two molecules of lactic acid (or two molecules of CO_2 and ethanol). The reaction driving the synthesis of ATP, namely glucose to lactate, is an internal oxidation-reduction reaction. This is easily seen by comparing the carbons of glucose to the carbons of pyruvate and lactate:

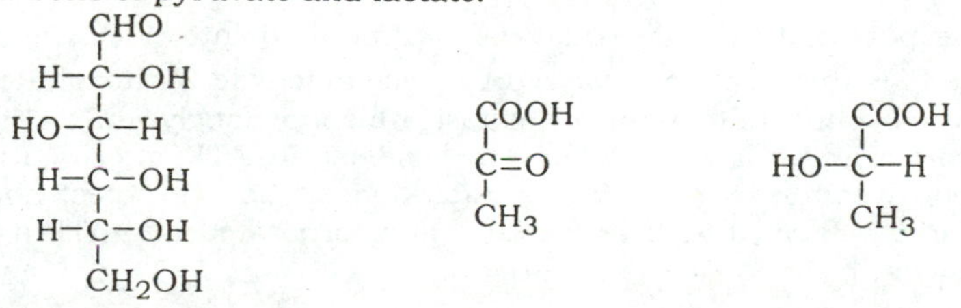

```
     CHO
     |
   H-C-OH
     |
  HO-C-H          COOH           COOH
     |             |              |
   H-C-OH         C=O         HO-C-H
     |             |              |
   H-C-OH         CH3            CH3
     |
     CH2OH

  D-glucose    pyruvic acid   lactic acid
```

The carboxyl carbons of pyruvic and lactic acids are oxidized relative to the carbons in glucose, whereas the methyl carbons are reduced. For lactic acid, carbon 2 is at the same oxidation state of an average carbon atom in glucose. Thus, conversion of glucose to lactic acid is in effect an internal oxidation-reduction. The same is not true for pyruvic acid because the keto carbon is oxidized relative to the average carbon in glucose. During pyruvic acid production NAD$^+$ was reduced. Conversion of pyruvic acid to lactic acid, a reduction, is accompanied by oxidation of NADH to NAD$^+$. Cycling of nicotinamide adenine dinucleotide between its oxidized and reduced forms, in effect, allows an internal oxidation-reduction reaction to occur within glucose half molecules.

5. Write the reactions that permit galactose to be utilized in glycolysis. Write a suitable mechanism for one of these reactions.

Answer: Galactose is first phosphorylated at carbon 1 by galactokinase in the following reaction:

$$\text{D-Galactose} + ATP^{4-} \rightarrow \text{D-galactose-1-phosphate}^{2-} + ADP^{3-} + H^{+}$$

The enzyme galactose-1-phosphate uridylyl transferase reacts galactose-1-phosphate with UDP-glucose to produce UDP-galactose and glucose-1-phosphate:

$$\text{Galactose-1-phosphate} + \text{UDP-glucose} \rightarrow \text{UDP-galactose} + \text{glucose-1-phosphate}$$

Next, UDP-galactose is converted to UDP-glucose by UDP-glucose-4-epimerase:

$$\text{UDP-galactose} \rightarrow \text{UDP-glucose}$$

The glucose moiety of UDP-glucose may be released by the second reaction as glucose-1-phosphate. Alternatively, it may be incorporated into glycogen and subsequently released as glucose-1-phosphate. Finally, glucose-1-phosphate is converted to glucose-6-phosphate by phosphoglucomutase:

$$\text{Glucose-1-phosphate} \rightarrow \text{glucose-6-phosphate.}$$

The UDP-glucose used in the second reaction is produced by UDP-glucose pyrophosphorylase in the following reaction:

$$\text{Glucose-1-phosphate} + UTP \rightarrow \text{UDP-glucose} + PP_i$$

(There is another reaction that produces glucose-1-phosphate, namely, the reaction catalyzed by glycogen phosphorylase: glycogen + P_i → glucose-1-phosphate)

The conversion of UDP-galactose to UDP-glucose by UDP-glucose-4-epimerase might proceed by formation of a 4-keto intermediate.

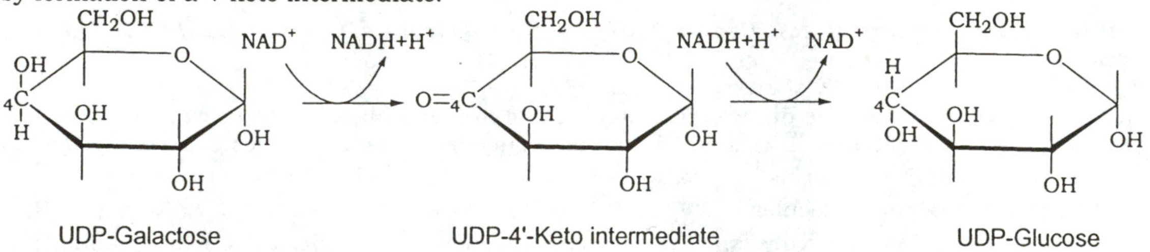

CH_2OH … NAD$^+$ NADH+H$^+$ … $O{=}_4C$ … NADH+H$^+$ NAD$^+$ … CH_2OH

UDP-Galactose → UDP-4'-Keto intermediate → UDP-Glucose

6. If ^{32}P-labeled inorganic phosphate were introduced to erythrocytes undergoing glycolysis, would you expect to detect ^{32}P in glycolytic intermediates? If so, describe the relevant reactions and the ^{32}P incorporation you would observe.

Answer: Ignoring the possibility that ^{32}P may be incorporated into ATP, the only reaction in glycolysis that utilizes P_i is the reaction catalyzed by glyceraldehyde-3-phosphate dehydrogenase in which glyceraldehyde-3-phosphate is converted to 1,3-bisphosphoglycerate. Thus, carbon 1 of 1,3-bisphosphoglycerate will be labeled. The label is lost to ATP after 1,3-bisphosphate is converted to 3-phosphoglycerate by phosphoglycerate kinase. So, we expect no other glycolytic intermediates to be directly labeled. (Once the label is incorporated into ATP it will show up at carbons 1 and 6 of glucose.)

7. Sucrose can enter glycolysis by either of two routes:

Sucrose phosphorylase

$$\textit{Sucrose} + P_i \leftrightharpoons \textit{fructose} + \textit{glucose-1-phosphate}$$

Invertase

$$\textit{Sucrose} + H_2O \leftrightharpoons \textit{fructose} + \textit{glucose}$$

Would either of these reactions offer an advantage over the other in the preparation of sucrose for entry into glycolysis?

Answer: The reaction catalyzed by sucrose phosphorylase produces glucose-1-phosphate, which enters glycolysis as glucose-6-phosphate, by the action of phosphoglucomutase. Therefore, glucose-6-phosphate produced by this route, bypasses the hexokinase reaction, normally employed, in which ATP hydrolysis occurs.

8. What would be the consequences of a Mg^{2+} deficiency for the reactions of glycolysis?

Answer: Kinases actually employ Mg^{2+}-ATP^{4-}, and not ATP^{4-}, as substrate. Therefore, a magnesium ion deficiency will affect the following enzymes: hexokinase (glucokinase), phosphofructokinase, phosphoglycerate kinase, and pyruvate kinase.

9. Triose phosphate isomerase catalyzes the conversion of dihydroxyacetone-P to glyceraldehyde-3-P. The standard free energy change, ΔG°', for this reaction is + 7.6 kJ/mol. However, the observed free energy change (ΔG) for this reaction in erythrocytes is + 2.4 kJ/mol.
a. Calculate the ratio of [dihydroxyacetone-P]/[glyceraldehyde-3-P] in erythrocytes from ΔG.
b. If [dihydroxyacetone-P] = 0.2 mM, what is [glyceraldehyde-3-P]?

Answer: The reaction being examined is:

Dihydroxyacetone-P → glyceraldehyde-3-P

The observed free energy change (ΔG) is calculated from the standard free energy change (ΔG°') using the following equation:

$$\Delta G = \Delta G^{\circ\prime} + RT \times \ln\frac{[\text{Products}]}{[\text{Reactants}]}$$

In this case, we have

$$\Delta G = \Delta G^{\circ\prime} + RT \times \ln\frac{[\text{glyceraldehyde}-3-\text{phosphate}]}{[\text{dihydroxyacetone phosphate}]}$$

Since the question asks us for the ratio of [dihydroxyacetone-P]/glyceraldehyde-3-P], we will rearrange the equation using the property log (x) = – log (1/x) to give

$$\Delta G = \Delta G^{\circ\prime} - RT \times \ln\frac{[\text{dihydroxyacetone phosphate}]}{[\text{glyceraldehyde}-3-\text{phosphate}]}$$

Solving for the desired ratio, we have

$$\frac{[\text{dihydroxyacetone phosphate}]}{[\text{glyceraldehyde}-3-\text{phosphate}]} = e^{\frac{(\Delta G^{\circ\prime}-\Delta G)}{RT}}$$

Substituting the numbers, and using 37°C as the temperature at which erythrocytes find themselves, we have

$$\frac{[\text{dihydroxyacetone phosphate}]}{[\text{glyceraldehyde}-3-\text{phosphate}]} = e^{\frac{(7.6-2.4)}{8.314\times10^{-3}\times310}} = e^{2.018}$$

$$\frac{[\text{dihydroxyacetone phosphate}]}{[\text{glyceraldehyde}-3-\text{phosphate}]} = 7.52$$

This answer tells us that the chemical reaction as written above lies to the left, which we already knew from the positive value for its ΔG.

b. Given that the dihydroxyacetone-P concentration in erythrocytes is 0.2 m*M*, we can calculate the concentration of glyceraldehyde-3-P from the answer for part a above.

$$[\text{glyceraldehyde}-3-\text{phosphate}] = \frac{[\text{dihydroxyacetone phosphate}]}{7.52} = \frac{0.2\text{ mM}}{7.52}$$

$$[\text{glyceraldehyde}-3-\text{phosphate}] = 0.027\text{ mM} = 27\ \mu\text{M}$$

10. Enolase catalyzes the conversion of 2-phosphoglycerate to phosphoenolpyruvate + H_2O. The standard free energy change, ΔG°', for this reaction is + 1.8 kJ/mol. If the concentration of 2-phosphoglycerate is 0.045 mM and the concentration of

phosphoenolpyruvate is 0.034 mM, what is ΔG, the free energy change for the enolase reaction, under these conditions?

Answer: The reaction being examined is:

$$\text{2-Phosphoglycerate} \rightarrow \text{phosphoenolpyruvate} + H_2O$$

The observed free energy change (ΔG) is calculated from the standard free energy change ($\Delta G°'$) using the following equation:

$$\Delta G = \Delta G°' + RT \times \ln\frac{[\text{Products}]}{[\text{Reactants}]}, \text{ or in this case}$$

$$\Delta G = \Delta G°' + RT \times \ln\frac{[\text{phosphoenoypyruvate}]}{[\text{2-phosphoglycerate}]}$$

Substituting the given values for the concentrations and $\Delta G°'$, and using 25°C for the temperature, we have:

$$\Delta G = 1.8\frac{\text{kJ}}{\text{mol}} + 8.314\times10^{-3}\frac{\text{kJ}}{\text{mol}\cdot\text{K}} \times 298\text{K} \times \ln\frac{0.034\text{ mM}}{0.045\text{ mM}}$$

$$\Delta G = 1.8\frac{\text{kJ}}{\text{mol}} - 0.69\frac{\text{kJ}}{\text{mol}} = +1.11\frac{\text{kJ}}{\text{mol}}$$

11. The standard free energy change (ΔG°') for hydrolysis of phosphoenolpyruvate (PEP) is – 61.9 kJ/mol. The standard free energy change (ΔG°') for ATP hydrolysis is – 30.5 kJ/mol.

a. What is the standard free energy change for the pyruvate kinase reaction:

ADP + phosphoenolpyruvate → ATP + pyruvate

b. What is the equilibrium constant for this reaction?

c. Assuming the intracellular concentrations of [ATP] and [ADP] remain fixed at 8 mM and 1 mM, respectively, what will be the ratio of [pyruvate] / [phosphoenolpyruvate] when the pyruvate kinase reaction reaches equilibrium?

Answer: a. The reaction

ADP + phosphoenolpyruvate → ATP + pyruvate

is derived as the sum of two coupled reactions

Reaction 1: Phosphoenolpyruvate + H_2O → pyruvate + P_i;

$\Delta G_1^{°'} = -61.9$ kJ/mol

and

Reaction 2: ADP + P_i → ATP + H_2O;

$\Delta G_2^{°'} = +30.5$ kJ/mol

The value of $\Delta G°'$ for the latter reaction is positive (+) because we have written it in the direction of ATP synthesis, rather than in the direction of hydrolysis, for which the value of $\Delta G°'$ is given in the problem.

The total change in the standard free energy of coupled reactions is given by the sum of the free energy changes of the individual reactions. Accordingly, we have for the reaction in question,

$$\Delta G°'_{\text{Total}} = \Delta G°'_1 + \Delta G°'_2 = -61.9\frac{\text{kJ}}{\text{mol}} + 30.5\frac{\text{kJ}}{\text{mol}} = -31.4\frac{\text{kJ}}{\text{mol}}$$

b. The equilibrium constant, K'_{eq}, is derived from the standard free energy change using the following equation:

$$\Delta G°' = -RT \times \ln K'_{eq}$$

which upon rearranging yields

$$K'_{eq} = e^{-\frac{\Delta G°'}{RT}}$$

Substituting the appropriate numerical values, with the product RT = 2.48 at 25°C, we have

$$K'_{eq} = e^{-\frac{(-31.4)}{2.48}} = 3.15 \times 10^5$$

The equilibrium position of this reaction obviously lies far to the right.

c. The equilibrium constant is the ratio of products to reactants at equilibrium. For this reaction,

$$K'_{eq} = \frac{[ATP][pruvate]}{[ADP][PEP]} = \frac{[ATP]}{[ADP]} \times \frac{[pruvate]}{[PEP]} = \frac{8\ mM}{1\ mM} \times \frac{[pruvate]}{[PEP]}, \text{ or}$$

$$\frac{[pruvate]}{[PEP]} = \frac{K'_{eq}}{8} = \frac{3.15 \times 10^5}{8} = 39{,}412$$

12. The standard free energy change ($\Delta G°'$) for hydrolysis of fructose-1,6-bisphosphate (FBP) to fructose-6-phosphate (F-6-P) and P_i is -16.7 kJ/mol:

FBP + H_2O → fructose-6-P + P_i

The standard free energy change ($\Delta G°'$) for ATP hydrolysis is -30.5 kJ/mol:

ATP + H, → ADP + P_i

a. What is the standard free energy change for the phosphofructokinase reaction:

ATP + fructose-6-phosphate → ADP + FBP

b. What is the equilibrium constant for this reaction?

c. Assuming the intracellular concentrations of [ATP] and [ADP] are maintained constant at 4 mM and 1.6 mM, respectively, in a rat liver cell, what will be the ratio of [FBP]/[fructose-6-P] when the phosphofructokinase reaction reaches equilibrium?

Answer: a. The reaction

ATP + fructose-6-P → ADP + FBP

is derived as the sum of two coupled reactions

Reaction 1: fructose-6-P + P_i → FBP + H_2O;

$\Delta G_1^{°'} = + 16.7$

and

Reaction 2: ATP + H_2O → ADP + P_i;

$\Delta G_2^{°'} = - 30.5$ kJ/mol

The value of $\Delta G°'$ for the former reaction is positive (+) because we have written it in the direction of P_i removal, rather than in the direction of P_i addition, for which the value of $\Delta G°'$ is given in the problem.

The total change in the standard free energy of coupled reactions is given by the sum of the free energy changes of the individual reactions. Accordingly, we have for the reaction in question,

$$\Delta G°'_{Total} = \Delta G°'_1 + \Delta G°'_2 = +16.7\frac{kJ}{mol} - 30.5\frac{kJ}{mol} = -13.8\frac{kJ}{mol}$$

b. The equilibrium constant, K'_{eq}, is derived from the standard free energy change using the following equation:

$$\Delta G°' = -RT \times \ln K'_{eq}$$

which upon rearranging yields

$$K'_{eq} = e^{-\frac{\Delta G°'}{RT}}$$

Substituting the appropriate numerical values, with the product $RT = 2.48$ at 25°C, we have

$$K'_{eq} = e^{-\frac{(-13.8)}{2.48}} = 261$$

The equilibrium position of this reaction lies far to the right.

The equilibrium constant is the ratio of products to reactants at equilibrium. For this reaction,

$$K'_{eq} = \frac{[ADP][FBP]}{[ATP][fructose-6-P]} = \frac{[ADP]}{[ATP]} \times \frac{[FBP]}{[fructose-6-P]}$$

$$K'_{eq} = \frac{1.6mM}{4mM} \times \frac{[FBP]}{[fructose-6-P]} = 0.4 \times \frac{[FBP]}{[fructose-6-P]}, \text{ or}$$

$$\frac{[FBP]}{[fructose-6-P]} = \frac{K'_{eq}}{0.4} = \frac{261}{0.4} = 652$$

$$K'_{eq} = \frac{[ATP][1,3-BPG]}{[ADP][3-PG]} = \frac{[ATP]}{[ADP]} \times \frac{[1,3-BPG]}{[3-PG]}$$

$$K'_{eq} = \frac{[ATP]}{[ADP]} \times \frac{1\ \mu M}{120\ \mu M}, \text{ or}$$

$$\frac{[ATP]}{[ADP]} = 120 \times K'_{eq} = 120 \times 2.2 \times 10^3 = 2.7 \times 10^5$$

Questions for Self Study

1. Answer True or False to the following statements.
   a. Another name for glycolysis is the Embden-Meyerhof pathway. _____
   b. Glycolysis is confined to the mitochondrial matrix. _____
   c. In the first phase of glycolysis, glucose is converted to two molecules of glyceraldehyde-3-phosphate and NAD^+ is reduced. _____
   d. The only difference between hexokinase and glucokinase is sugar specificity. _____
   e. Phosphofructokinase hydrolyzes ATP. _____
   f. Conversion of glucose-6-P to fructose-6-P and conversion of glyceraldehyde-3-P to dihydroxyacetone-P are aldose to ketose conversions. _____

2. Phosphofructokinase is the key enzyme regulating glycolysis. For the following compounds indicate their effects on phosphofructokinase activity.
   a. ATP
   b. AMP
   c. Citrate
   d. Fructose-2,6-bisphosphate
   e. Fructose-6-phosphate

3. The ability of a cell to continue glycolysis in the presence of adequate supplies of glucose and a need for ATP is critically dependent on the reaction catalyzed by glyceraldehyde-3-phosphate dehydrogenase. Why? What is the reaction catalyzed by this enzyme?

4. Which two reactions in the second phase of glycolysis are examples of substrate level phosphorylations?

5. a. The production of lactate or ethanol and carbon dioxide from glucose are examples of what process?
b. What is the purpose of this process?

6. Fill in the blanks. For galactose to enter glycolysis it must first follow the _____. Galactose is first converted to galactose-1-phosphate by _____ in an ATP-dependent reaction. Galactose-1-phosphate is exchanged for the glucose-1-phosphate moiety of a uridine nucleotide derivative producing _____. This compound is acted on by a glucose-4-epimerase converting it to _____.

7. Conversion of 1,3-bisphosphoglycerate to 2,3-bisphosphoglycerate (a allosteric effector of hemoglobin) and the conversion of 3-phosphoglycerate to 2-phosphoglycerate are examples of reactions catalyzed by what kind of enzymes?

Answers

1. a. T; b. F; c. F; d. F; e. F; f. T.

2. a. inhibits; b. blocks ATP inhibition; c. inhibits; d. stimulates; e. stimulates.

3. The enzyme converts glyceraldehyde-3-phosphate and phosphate to 1,3-bisphosphoglycerate with accompanying reduction of NAD^+ to NADH. The reaction is critical because the levels of NAD^+ are limiting.

4. The reactions catalyzed by phosphoglycerate kinase and pyruvate kinase.

5. a. Fermentation; b. Fermentation provides a pathway for regeneration of NAD^+ from NADH that does not involve oxygen or another external electron acceptor.

6. Leloir; galactokinase; UDP-galactose; UDP-glucose.

7. Mutases

Additional Problems

1. Why is it advantageous to phosphorylate glucose immediately after being transported into cells?

2. Describe the basic chemistry of interconversion of glucose-6-phosphate to fructose-6-phosphate and of dihydroxyacetone phosphate to glyceraldehyde-3-phosphate.

3. What reaction is catalyzed by the enzyme fructose bisphosphate aldolase? Class I aldolases are irreversibly inactivated by $NaBH_4$ but inactivation requires the presence of either fructose 1,6-bisphosphate or dihydroxyacetone phosphate. The enzyme can be treated with $NaBH_4$ in the presence of glyceraldehyde-3-phosphate and upon removal of $NaBH_4$ the enzyme is active. Explain.

4. Describe the reaction catalyzed by phosphoglycerate mutase. What other molecules are required for activity?

5. What key differences exist between hexokinase and glucokinase to insure that glucose is properly apportioned between muscle and liver?

6. Draw the structures of glycerol and of glyceraldehyde. Anaerobic organisms, grown in the absence of oxygen, can use glyceraldehyde as a source of metabolic energy but not glycerol. However, metabolism of glycerol starts with glycerol kinase converting glycerol to 3-phosphoglycerol, which is subsequently metabolized to glyceraldehyde-3-phosphate. Why is glycerol catabolism a problem for anaerobic organisms yet conversion of glyceraldehyde to pyruvate possible?

7. Mannose is a 2' epimer of glucose. However, only one additional enzyme is required to get mannose into glycolysis. Outline a reasonable pathway by which mannose enters glycolysis.

Abbreviated Answers

1. Phosphorylation of glucose makes it essentially impermeable to the cell membrane and traps glucose inside the cell. Phosphorylation also keeps the [glucose] low inside cells allowing uptake to occur down a glucose concentration gradient. Finally, phosphorylation commits glucose to some form of metabolism.

2. Both reactions are isomerizations of aldoses to ketoses.

3. Fructose bisphosphate aldolase interconverts fructose-1,6-bisphosphate with dihydroxyacetone phosphate and glyceraldehyde-3-phosphate. Class I aldolases form a Schiff base during catalysis. (See answer to problem 5, above.) The Schiff base forms at the keto carbon of either fructose-1,6-bisphosphate or dihydroxyacetone phosphate and $NaBH_4$ will reduce the Schiff base to a stable adduct. Once modified in this fashion, the enzyme is irreversibly inactivated. Treating the enzyme with either $NaBH_4$ alone or with $NaBH_4$ and glyceraldehyde-3-phosphate has no effect because a Schiff base is not formed.

4. Phosphoglycerate mutase moves the phosphate group from C-3 of 3-phosphoglycerate to C-2 to produce 2-phosphoglycerate. The yeast and rabbit muscle enzymes use 2,3-bisphosphoglycerate as a cofactor. The cofactor transfers a phosphate group to an active-site

histidine to form a phosphohistidine intermediate. The phosphate group is then transferred to C-2 of 3-phosphoglycerate at the active site to form enzyme-bound 2,3-bisphosphoglycerate. The product, 2-phosphoglycerate, is formed when phosphate is subsequently transferred from C-3 to an enzyme histidine. The cofactor 2,3-bis-phosphoglycerate is also used as an allosteric inhibitor of hemoglobin. It is formed from 1,3-bisphosphoglycerate by bisphosphoglycerate mutase.

5. Hexokinase has a low K_m for glucose, approximately 0.1 mM, and is inhibited by glucose-6-phosphate. Thus, cells containing hexokinase, such as muscle, will continue to phosphorylate glucose even when blood-glucose levels are low. However, if glucose is not metabolized immediately and glucose-6-phosphate levels build up, hexokinase is inhibited. Glucokinase, found in the liver, has a much higher K_m for glucose, approximately 10 mM, insuring that liver cells phosphorylate glucose only when blood-glucose levels are high. In addition, glucokinase is not inhibited by glucose-6-phosphate. This allows liver cells to continue to accumulate glucose. In a later chapter we will see that excess glucose is stored as glycogen.

6. The structures of glycerol and glyceraldehyde are shown below.

```
        CH2OH              H   O
         |                  \ //
    HO—C—H                   C
         |                   |
        CH2OH             H—C—OH
                             |
                            CH2OH

      Glycerol          Glyceraldehyde
```

Glycerol is metabolized by first being phosphorylated by glycerol kinase to *sn*-glycerol-3-phosphate, which is converted to dihydroxyacetone phosphate (DHAP) by glycerol phosphate dehydrogenase. DHAP is subsequently isomerized to glyceraldehyde. Since glycerol is converted into glyceraldehyde it is curious that anaerobic organisms can utilize one of these compounds but not the other. The problem lies with the enzyme glycerol phosphate dehydrogenase. This enzyme uses NAD^+ as coenzyme. Thus, glyceraldehyde production by this pathway results in NAD^+ reduction. A second NAD^+ is reduced when glyceraldehyde is metabolized via glycolysis. One NAD^+ is regenerated when pyruvate is converted to lactate or ethanol and CO_2 but since two were reduced, one NADH remains. In the absence of oxygen or some other suitable electron acceptor, NAD^+ levels drop as all of the coenzyme is trapped in the reduced form NADH.

7. Mannose is phosphorylated by hexokinase to produce mannose-6-phosphate, which is isomerized by phosphomannose isomerase to fructose-6-phosphate, a glycolytic intermediate.

Summary

Glycolysis - the degradation of glucose and other simple sugars - is a catabolic process carried out by nearly all cells. An anaerobic process, it provides precursor molecules for aerobic catabolic pathways, such as the critic acid cycle, and serves as an emergency energy source when oxygen is limiting. Glycolysis consists of two phases. The first series of five reactions breaks glucose down to two molecules of glyceraldehyde-3-phosphate. The second phase converts these two molecules of glyceraldehyde-3-P into two molecules of pyruvate. The overall pathway produces two molecules of ATP per glucose consumed. In aerobic organisms, including humans, pyruvate is oxidatively decarboxylated to produce a molecule of acetyl coenzyme A, which fuels the citric acid cycle (and subsequent electron transport). Under anaerobic conditions, the pyruvate formed in glycolysis can be reduced to lactate (in oxygen-starved muscles or in microorganisms) or ethanol (in yeast) in fermentation processes. During glycolysis, a portion of the metabolic energy of the glucose molecule is converted into ATP. The subsequent entry of pyruvate into the TCA cycle provides more energy in the form of ATP and reduced coenzymes.

The initial reaction of glycolysis involves phosphorylation of glucose at C-6, a reaction that consumes an ATP molecule in order to make more ATP later. Hexokinase is the primary catalyst for this reaction, but, when glucose levels rise in the liver, glucokinase becomes active and shares the responsibility for phosphorylation of glucose. The second reaction of glycolysis is the isomerization of glucose-6-phosphate to fructose-6-phosphate, carried out by phosphoglucoisomerase in a reaction involving the transient formation of an ene-diol intermediate. The new primary alcohol group created at C-1 on the carbon skeleton is the site of another phosphorylation in the next reaction, catalyzed by phosphofructokinase. Like the hexokinase/glucokinase reaction, this reaction also consumes a molecule of ATP. The phosphofructokinase reaction is the most elaborately regulated reaction in the glycolytic pathway. ATP is an allosteric inhibitor of phosphofructokinase, whereas AMP can reverse the inhibition due to ATP. Although ATP levels in cells rarely change by more than 10%, the action of adenylate kinase facilitates relatively large changes in AMP levels in response to hydrolysis of ATP. The phosphofructokinase reaction is also inhibited by citrate, providing a critical link between regulation of glycolysis and the citric acid cycle. Fructose-2,6-bisphosphate is a potent allosteric activator that increases the affinity of the enzyme for the substrate fructose-6-phosphate.

The 6-carbon substrate skeleton is cleaved to two 3-carbon units in the fructose bisphosphate aldolase reaction. The products are dihydroxyacetone phosphate and glyceraldehyde-3-phosphate. Two types of fructose bisphosphate aldolase enzymes exist in nature. Animal tissues produce a Class I aldolase that forms a covalent Schiff base adduct with substrate and is thus inhibited by $NaBH_4$. Bacteria and fungi produce a Class II aldolase that does not form Schiff base adducts with substrate, but which contains an active site Zn^{2+} ion which acts as an electrophile to polarize the carbonyl group of the substrate and stabilize the enolate intermediate in the aldolase reaction. The dihydroxyacetone phosphate produced by the aldolase reaction is converted to another molecule of glyceraldehyde-3-phosphate by the fifth reaction of the pathway, triose phosphate isomerase. Although these last two reactions are energetically unfavorable, the earlier priming reactions bring the equilibrium constant for the first five reactions of glycolysis close to 1.

The second phase of glycolysis begins with the oxidation of glyceraldehyde-3-phosphate to produce a high energy phosphate metabolite, 1,3-bisphosphoglycerate and a reduced coenzyme, NADH, in the glyceraldehyde-3-phosphate dehydrogenase reaction. The reaction mechanism involves nucleophilic attack by a cysteine-SH group at the active site on the carbonyl carbon of glyceraldehyde-3-phosphate to form a hemithioacetal intermediate, followed by hydride transfer to NAD^+ and nucleophilic attack by phosphate to displace the product from the enzyme. In the seventh step of the pathway, phosphoglycerate kinase, transfers a phosphoryl group from 1,3-bisphosphoglycerate to ADP to form an ATP, an example of substrate level phosphorylation. Since each glucose molecule sends two molecules of 1,3-bisphosphoglycerate into the phosphoglycerate kinase reaction, this step pays off the ATP debt created by the two priming reactions of glycolysis. The phosphoglycerate mutase reaction converts 3-phosphoglycerate to 2-phosphoglycerate and the enolase reaction then converts 2-phosphoglycerate to phosphoenolpyruvate, a high energy phosphate which can drive the synthesis of another ATP molecule in the pyruvate kinase reaction, the final reaction of the glycolytic pathway. Pyruvate kinase is subject to regulation and possesses allosteric sites for activation by AMP and fructose-2,6-bisphosphate and inhibition by ATP, acetyl-CoA and alanine. Phosphorylation of pyruvate kinase by cAMP-dependent protein kinase makes pyruvate kinase more sensitive to inhibition by ATP and alanine and raises the K_m for PEP. Other sugars, including fructose, mannose and galactose, can enter the glycolytic pathway if they can be converted by appropriate enzymes into glycolysis intermediates.

Chapter 16

The Tricarboxylic Acid Cycle

• •

Chapter Outline

- TCA cycle (tricarboxylic acid): Citric acid cycle: Krebs cycle
  - Input: Acetyl units
  - Output: Carbon dioxide, ATP (or GTP), reduced cofactors
  - Related activities
    - Electron transport: Recycles cofactors
    - Oxidative phosphorylation: Transduces energy into ATP
- Pyruvate dehydrogenase complex: Major source of acetyl units from glucose
  - Reaction
    - Pyruvate to acetyl CoA and carbon dioxide
    - NAD^+ reduced
  - Enzyme complex: Three enzyme activities
    - Pyruvate dehydrogenase
    - Dihydrolipoyl transacetylase
    - Dihydrolipoyl dehydrogenase
  - Coenzymes
    - Thiamine pyrophosphate
    - Coenzyme A
    - Lipoic acid
    - NAD^+
    - FAD: Protein bound
  - Regulation
    - Acetyl-CoA, NADH inhibitors
    - AMP activates: GTP inhibits
    - Covalent modification (mammalian enzyme): Phosphoenzyme inhibited
      - Pyruvate dehydrogenase kinase: Activate by acetyl-CoA and NADH
      - Pyruvate dehydrogenase phosphatase: Calcium activated
- Citrate synthase
  - Perkin condensation: Acetyl CoA + oxaloacetate to citrate
  - Regulation
    - NADH: Allosteric inhibitor
    - Succinyl CoA: Allosteric inhibitor: Intracycle inhibitor
- Aconitase
  - Citrate to aconitate to isocitrate: Dehydration-rehydration
  - Iron-sulfur cluster
  - Site of action of fluoroacetate: Trojan horse inhibitor: Requires conversion to fluorocitrate
- Isocitrate dehydrogenase
  - Oxidative decarboxylation of isocitrate: Forms α-ketoglutarate
  - NAD^+ reduced
  - Regulation
    - NADH, ATP allosteric inhibitors

- ADP, NAD^+ allosteric activator
- Covalent modification (*E. coli* enzyme): Phosphoenzyme inhibited
  - Kinase inhibited by TCA cycle intermediates
  - Phosphatase activated by TCA cycle intermediates

❖ α-Ketoglutarate dehydrogenase: Multienzyme complex: Similar to pyruvate dehydrogenase complex
- ➢ Oxidative decarboxylation of α-ketoglutarate: Produces succinyl-CoA
- ➢ NAD^+ reduced
- ➢ Regulation
  - NADH inhibits

❖ Succinyl-CoA Synthetase: Succinate thiokinase
- ➢ Substrate-level phosphorylation
  - GTP in mammals: Nucleoside diphosphate kinase moves P to ATP
  - ATP in plants and bacteria

❖ Succinate dehydrogenase
- ➢ FAD reduced
- ➢ Membrane-bound enzyme: Covalently bound FAD
- ➢ Electron transport: Electrons moved to coenzyme Q

❖ Fumarase: Hydration of fumarate: Produces malate

❖ Malate dehydrogenase
- ➢ NAD^+ reduced
- ➢ Oxaloacetate produced

❖ TCA intermediates connected to biosynthetic pathways
- ➢ α-Ketoglutarate: Precursor of Glu: Precursor of
  - Pro, Arg, Asn
- ➢ Succinyl CoA: Porphyrin ring synthesis
- ➢ Fumarate: Precursor of Glu
- ➢ Oxaloacetate: Precursor of
  - Asp: Precursor of
    - Pyrimidine nucleotides, Asn, Met, Lys, Thr, Iso
  - PEP: Precursor of
    - Phe, Tyr, Trp (plants and microorganism)
    - 3-Phosphoglycerate: Precursor of
      - ♦ Ser, Gly, Cys
    - Glucose

❖ Anaplerotic reactions: Produce TCA cycle intermediates
- ➢ Pyruvate carboxylase
- ➢ PEP carboxykinase
- ➢ Malic enzyme

❖ Glyoxylate cycle: Plants and bacteria: Uses acetyl units to produce oxaloacetate: Oxaloacetate to pyruvate: Pyruvate to glucose
- ➢ Citric acid cycle to isocitrate
- ➢ Isocitrate lyase produces glyoxylate and succinate
- ➢ Succinate to oxaloacetate via citric acid cycle
- ➢ Malate synthase: Glyoxylate + acetyl CoA to malate

Chapter Objectives

Citric Acid Cycle

The citric acid cycle is a central metabolic cycle to which catabolism of monosaccharides, amino acids, and fatty acids converge. We see that pyruvate, the end product of glycolysis, is introduced as an acetyl unit attached to coenzyme A catalyzed by pyruvate dehydrogenase. (A carbon is lost as CO_2.) The pyruvate dehydrogenase reaction is worth understanding, because we encounter almost identical biochemistry in α-ketoglutarate dehydrogenase catalysis of α-ketoglutarate to succinyl-CoA. Pyruvate is decomposed to CO_2 and a two-carbon acetyl group, which is attached to thiamine pyrophosphate. The acetyl group is transferred to coenzyme A via lipoic acid and in the process, lipoic acid is reduced. To reoxidize lipoic acid, electrons are transferred to protein-bound FAD and then to NAD^+.

The citric acid cycle begins with condensation of acetyl-CoA and oxaloacetate to produce citric acid, a tricarboxylic acid (see Figure 16.4). The cycle then converts citrate to oxaloacetate and two molecules of CO_2. You should understand that the basic chemistry is oxidation. (Input is an acetyl group, outputs are carbon dioxide.) Oxidations are always accompanied by reductions and the reductions that take place include three steps at which NAD^+ is reduced to NADH: isocitrate to α-ketoglutarate conversion catalyzed by isocitrate dehydrogenase, α-ketoglutarate to succinyl-CoA conversion catalyzed by α-ketoglutarate dehydrogenase (similar in reaction mechanism to pyruvate dehydrogenase), and malate to oxaloacetate conversion catalyzed by malate dehydrogenase. The first two reactions are responsible for the two CO_2 molecules produced by the cycle. An additional reduction occurs during the conversion of succinate to fumarate catalyzed by succinate dehydrogenase. This enzyme plays two metabolic roles, one in the TCA cycle and the other as a component of the electron transport chain.

There are a number of important points to remember about the citric acid cycle that will help you with other aspects of biochemistry.

1. When amino acid metabolism is covered we will see that a few of the citric acid cycle intermediates derive from and can be converted into amino acids. Examples include pyruvic acid/alanine, α-ketoglutarate/glutamic acid, and oxaloacetate/aspartic acid.

2. The citric acid cycle has a substrate-level phosphorylation catalyzed by succinyl-CoA synthetase. Synthetases are enzymes that drive product formation with NTP hydrolysis. So, succinyl-CoA synthetase can produce succinyl-CoA using hydrolysis of either GTP or ATP depending on the source of the enzyme. (Citric acid synthase synthesizes citric acid but it does not utilize hydrolysis of ATP to drive the reaction.)

3. The steps from succinate to fumarate to malate to oxaloacetate are almost identical in chemistry to β-oxidation, the principal pathway of fatty acid catabolism we will study in Chapter 24.

Regulation of the Cycle

The citric acid cycle is regulated at pyruvate dehydrogenase by acetyl-CoA, CoA, NAD^+, NADH, and ATP; at citrate synthase by NADH, ATP, and succinyl CoA; at isocitrate dehydrogenase by NAD^+, ADP, and ATP; and, at α–ketoglutarate dehydrogenase by NADH, AMP, and succinyl-CoA. The cycle is sensitive to NAD^+ levels, a measure of electron transport activity, to adenine nucleotides, a measure of energy charge, and to immediate substrates like acetyl-CoA or succinyl-CoA.

Citric Acid Cycle Outputs and Inputs

The whole point of the cycle is to consume acetyl units for energy with CO_2 as a byproduct. However, other pathways can produce TCA cycle intermediates and TCA intermediates can feed into other pathways (Figure 16.14). For example, acetyl units can be derived from catabolism of fatty acids and certain amino acids, whereas citrate is used to move carbons out of the citric acid cycle to synthesize fatty acids. Amino acids and citric acid cycle intermediates are interconvertible. Reactions leading to citric cycle intermediates are known as anaplerotic reactions.

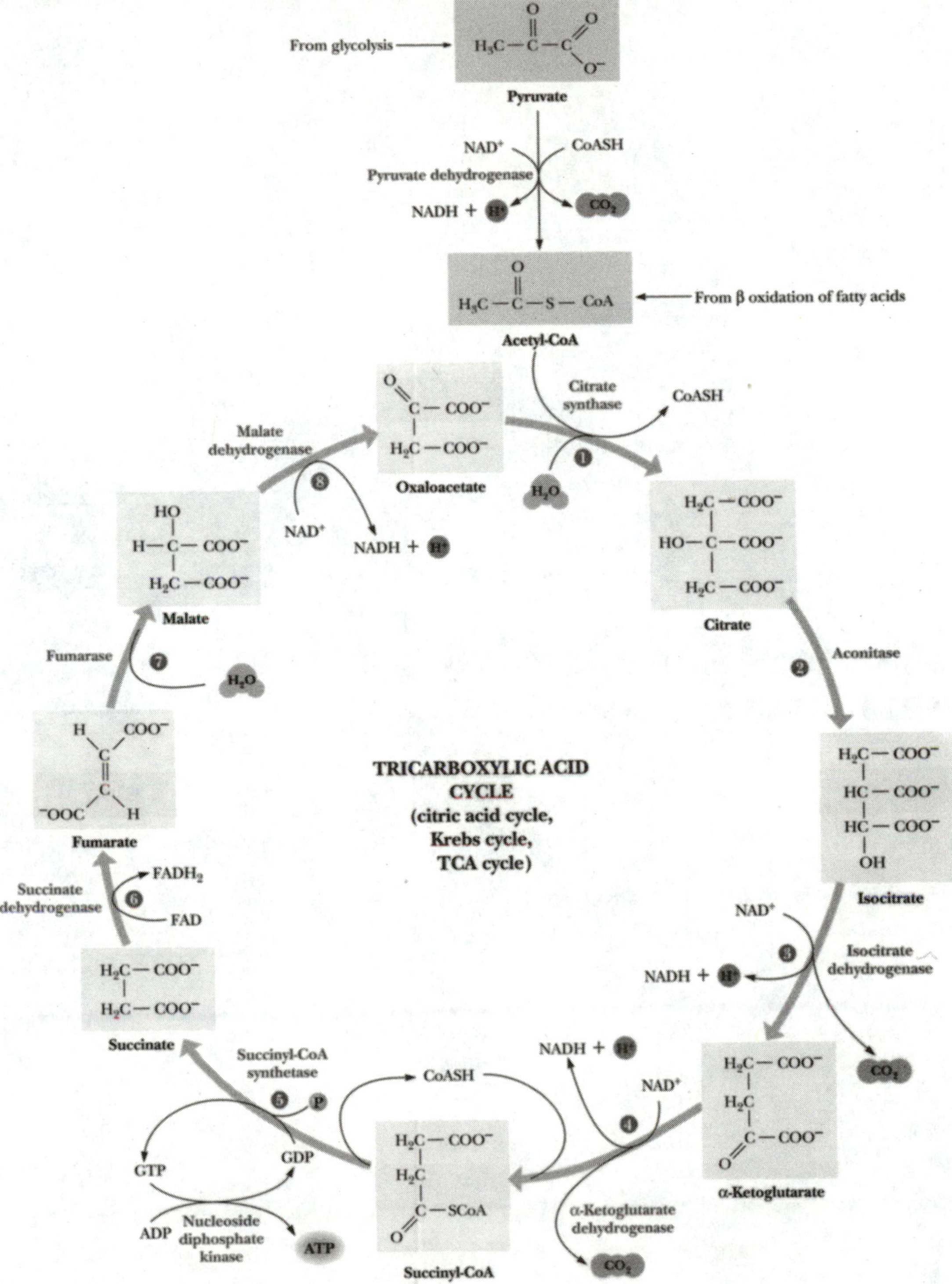

Figure 16.2 The tricarboxylic acid (TCA) cycle.

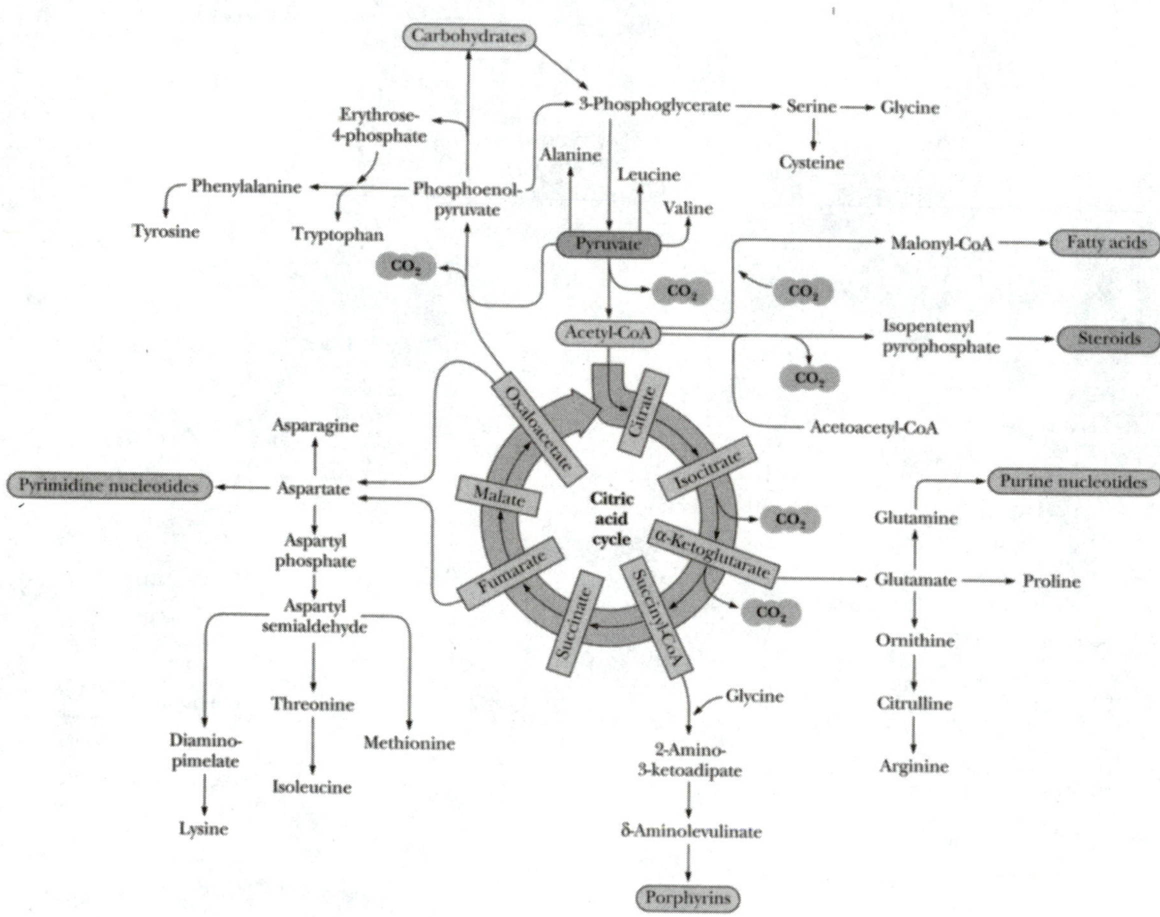

Figure 16.14 TCA cycle provides intermediates for numerous biosynthetic processes in the cell.

Problems and Solutions

1. Describe the labeling pattern that would result from the introduction into the TCA cycle of glutamate labeled at the C_γ with ^{14}C.

Answer: Glutamate may be converted to α-ketoglutarate either by transamination or by the action of glutamate dehydrogenase. In either case, glutamate with label at the C_γ will produce oxaloacetate labeled at the C_γ of oxaloacetate as shown below.

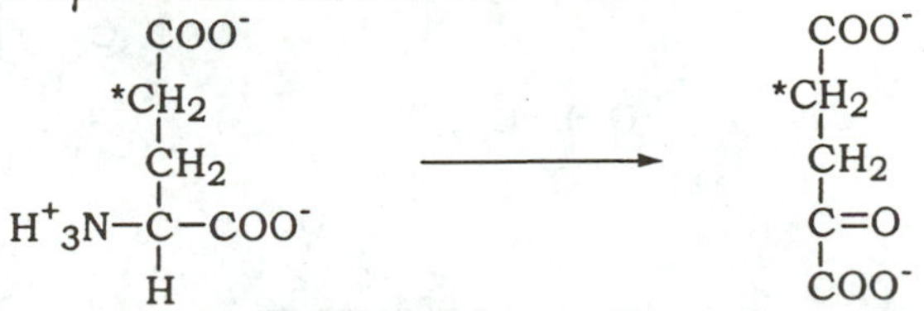

```
          COO-                           COO-
           |                              |
         *CH2                           *CH2
           |                              |
          CH2        ----------->        CH2
           |                              |
   H+3N—C—COO-                           C=O
           |                              |
           H                             COO-
```

The label is not lost in the first cycle and shows up in oxaloacetate at either of two positions.

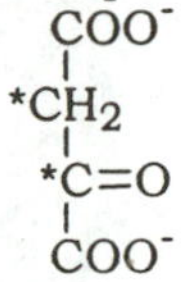

```
  COO-
   |
 *CH2
   |
 *C=O
   |
  COO-
```

No label is lost in the second round as well; however, in each subsequent cycle, 50% of the label is lost. This occurs because at the end of the second round, label is equally distributed among all the carbons of oxaloacetate. Since each cycle releases carbons from oxaloacetate, one-half of the label is lost. The remaining label is redistributed equally among all the carbons and again one-half is lost with subsequent cycles.

2. Describe the effect on the TCA cycle of (a) increasing the concentration of NAD^+, (b) reducing the concentration of ATP, and (c) increasing the concentration of isocitrate.

Answer: The citric acid cycle is tightly coupled to electron transport by the reactions catalyzed by the following enzymes: isocitrate dehydrogenase, α-ketoglutarate dehydrogenase, succinate dehydrogenase and malate dehydrogenase. The *in vivo* reactions catalyzed by isocitrate dehydrogenase and α–ketoglutarate are far from equilibrium, with large negative ΔG' values (See Table 16.1), and are important control points for the citric acid cycle.

Isocitrate dehydrogenase is allosterically activated by ADP and NAD^+ and inhibited by ATP and NADH. Therefore, this enzyme is sensitive to the NAD^+:NADH and ADP:ATP ratios. An increase in NAD^+ or a reduction in the concentration of ATP will stimulate this enzyme and the citric acid cycle. Further, increasing the concentration of isocitrate will increase the citric acid cycle. This occurs, not because isocitrate is a regulator of the cycle, but rather because it is a substrate of isocitrate dehydrogenase, an enzyme catalyzing a reaction far from equilibrium and therefore sensitive to changes in concentration of substrate.

α–Ketoglutarate dehydrogenase is inhibited by NADH and an increase concentration of NAD^+ resulting from a decrease in NADH will relieve this inhibition and stimulate the cycle. Finally, the activity of pyruvate dehydrogenase is sensitive to both NAD^+ and ATP; high NAD^+ and low ATP lead to stimulation of pyruvate dehydrogenase leading to an increased rate of production of acetyl-CoA if pyruvate is available.

3. The serine residue of isocitrate dehydrogenase that is phosphorylated by protein kinase lies within the active site of the enzyme. This situation contrasts with most other examples of covalent modification by protein phosphorylation, where the phosphorylation occurs at a site remote from the active site. What direct effect do you think such active-site phosphorylation might have on the catalytic activity of isocitrate dehydrogenase? (See Barford, D., 1991. Molecular mechanisms for the control of enzymic activity by protein phosphorylation. Biochimica et Biophysica Acta 1133:55-62.).

Answer: Isocitrate dehydrogenase catalyzes the oxidative decarboxylation of isocitrate to α-ketoglutarate with reduction of NAD^+ to NADH. Phosphorylation of Ser^{113} leads to enzyme inactivation because isocitrate no longer binds to the enzyme. Ser^{113} normally forms a hydrogen bond with the γ-carboxyl group of isocitrate. Phosphorylated Ser^{113} interferes with isocitrate binding by electrostatic repulsion of the carboxyl group and by steric hindrance.

4. The first step of the α-ketoglutarate dehydrogenase reaction involves decarboxylation of the substrate and leaves a covalent TPP intermediate. Write a reasonable mechanism for this reaction.

Answer:

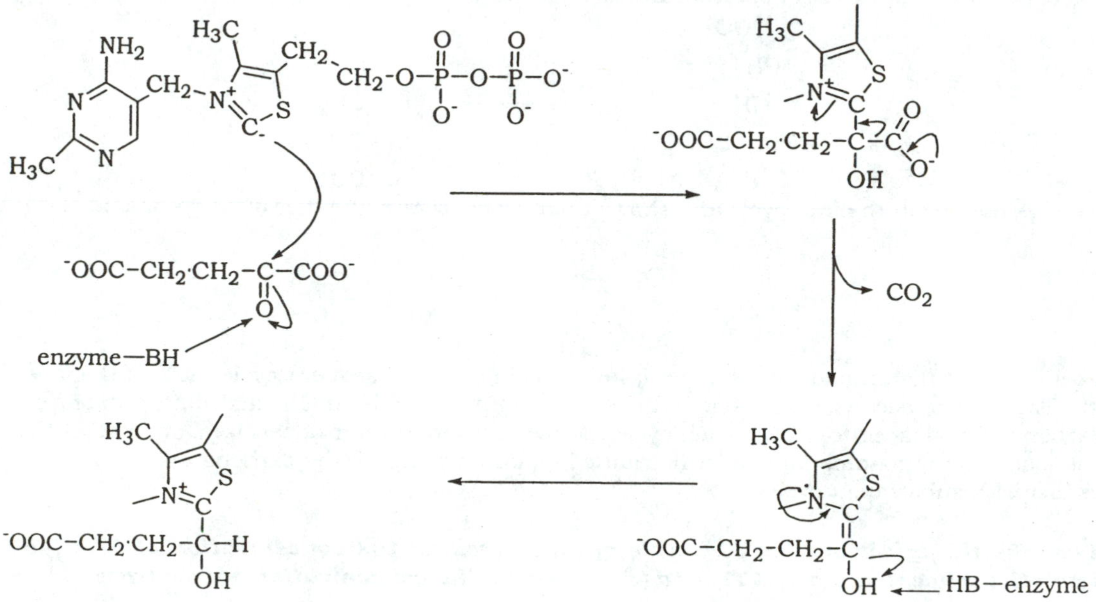

CO_2

enzyme—BH

HB—enzyme

5. In a tissue where the TCA cycle has been inhibited by fluoroacetate, what difference in the concentration of each TCA cycle metabolite would you expect, compared with a normal, uninhibited tissue?

Answer: Fluoroacetate is an example of a suicide inhibitor or a Trojan horse inhibitor. When tested, *in vitro,* against isolated enzymes of the TCA cycle, fluoroacetate has no effect. However, when used on isolated mitochondria or *in vivo,* fluoroacetate is a potent inhibitor of aconitase. Fluoroacetate is converted to fluoroacetyl-coenzyme A by acetyl-CoA synthetase, which is condensed with oxaloacetate by citrate synthase to produce 2R,3S-fluorocitrate. This isomer specifically inhibits aconitase. With aconitase inhibited, we might expect TCA cycle intermediates above aconitase to increase in concentration while intermediates below aconitase to decrease in concentration. In particular, citrate concentrations will increase in concentration and isocitrate and all subsequent metabolites will decrease in concentration.

6. Based on the description in Chapter 14 of the physical properties of FAD and $FADH_2$, suggest a method for the measurement of the enzyme activity of succinate dehydrogenase.

Answer: Succinate dehydrogenase catalyzes the oxidation of succinate to fumarate. This enzyme serves a dual purpose: It is a TCA cycle enzyme and it is a component of the electron transport system. Succinate dehydrogenase is a inner mitochondrial membrane-bound enzyme containing a covalently bound FAD that cycles between oxidized and reduced ($FADH_2$) forms as it passes electrons from succinate to coenzyme Q. A convenient way to measure the activity of this enzyme is to take advantage of the fact that the reduced and oxidized forms of FAD exhibit different absorption spectra. In particular, FAD is yellow in color and has a maximal absorbance at 450 nm whereas $FADH_2$ is colorless. By monitoring the absorbance at 450 nm, the activity of succinate dehydrogenase can be measured.

7. Starting with citrate, isocitrate, α-ketoglutarate, and succinate, state which of the individual carbons of the molecule undergo oxidation in the next step of the TCA cycle. Which of the molecules undergo a net oxidation?

Answer:

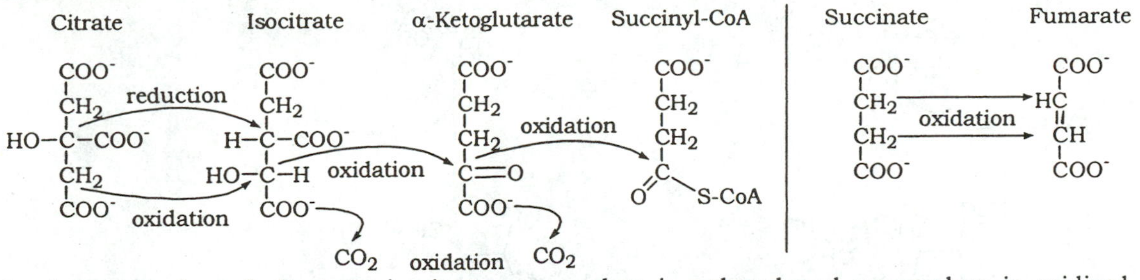

Citrate Isocitrate α-Ketoglutarate Succinyl-CoA | Succinate Fumarate

Citrate: COO^-, CH_2, HO–C–COO^-, CH_2, COO^- → (reduction; oxidation) → Isocitrate: COO^-, CH_2, H–C–COO^-, HO–C–H, COO^- → (oxidation; CO_2) → α-Ketoglutarate: COO^-, CH_2, CH_2, C=O, COO^- → (oxidation; oxidation; CO_2) → Succinyl-CoA: COO^-, CH_2, CH_2, C(=O)S-CoA

Succinate: COO^-, CH_2, CH_2, COO^- → (oxidation) → Fumarate: COO^-, HC, ‖, CH, COO^-

In the conversion of citrate to isocitrate, one carbon is reduced and one carbon is oxidized. Hence, there is no net change in the oxidation state in going from citrate to isocitrate. Production of α-ketoglutarate from isocitrate is the result of an oxidation. The enzyme that catalyzes this reaction, isocitrate dehydrogenase, uses NAD^+ as a cofactor, which is reduced to NADH during the reaction. Likewise, conversion of α–ketoglutarate to succinate is an oxidation. In this case, the conversion is a two-step process. In the first step, α–ketoglutarate is converted to succinyl-CoA by α–ketoglutarate dehydrogenase. This is an oxidation (the carbon in a thiol ester bond is at the same oxidation state as a carbon in an ester bond) and so it must be accompanied by a reduction, in this case NAD^+ is reduced to NADH. Succinyl-CoA to succinate is a hydrolysis reaction that is coupled to production of a high-energy phosphate bond (as GTP or ATP). Finally, succinate to fumarate is catalyzed by succinate dehydrogenase. This enzyme has a enzyme-bound FAD coenzyme that is reduced when succinate is oxidized to fumarate.

8. In addition to fluoroacetate, consider whether other analogs of TCA cycle metabolites or intermediates might be introduced to inhibit other, specific reactions of the cycle. Explain your reasoning.

Answer: Fluoroacetate blocks the citric acid cycle by reacting with oxaloacetate to form fluorocitrate, which is a powerful inhibitor of aconitase. In the aconitase reaction, the elements of water are abstracted from citrate and added back to form isocitrate. Addition of water also occurs in the conversion of fumarate to malate catalyzed by fumarase and we might expect fluorinated derivatives of fumarate or malate to inhibit this enzyme. Malonate, a succinate analog, is a competitive inhibitor of fumarase. 3-Nitro-2-S-hydroxylpropionate, when deprotonated, is a transition state analog of fumarase. These compounds are shown below.

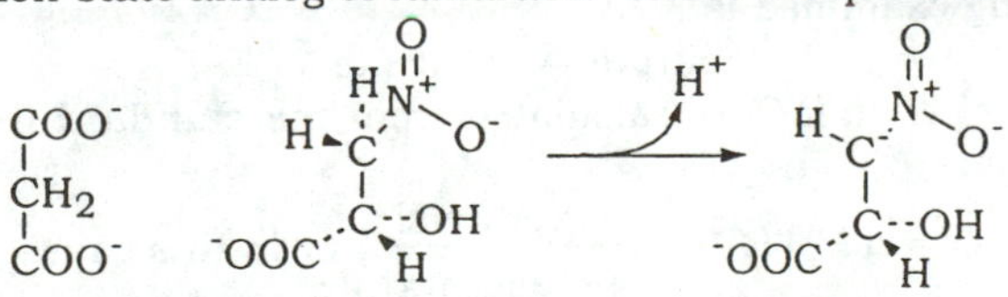

Malonate: COO^-, CH_2, COO^-

S-Nitro-2-S-hydroxypropionate: H–C(H)–N^+(=O)O^-, $^-$OOC–C(H)–OH → (H^+) → H–C–N^+(=O)O^-, $^-$OOC–C(H)–OH

Malonate S-Nitro-2-S-hydroxypropionate

9. Based on the action of thiamine pyrophosphate in catalysis of the pyruvate dehydrogenase reaction, suggest a suitable chemical mechanism for the pyruvate decarboxylase reaction in yeast:

pyruvate → acetaldehyde + CO_2

Answer

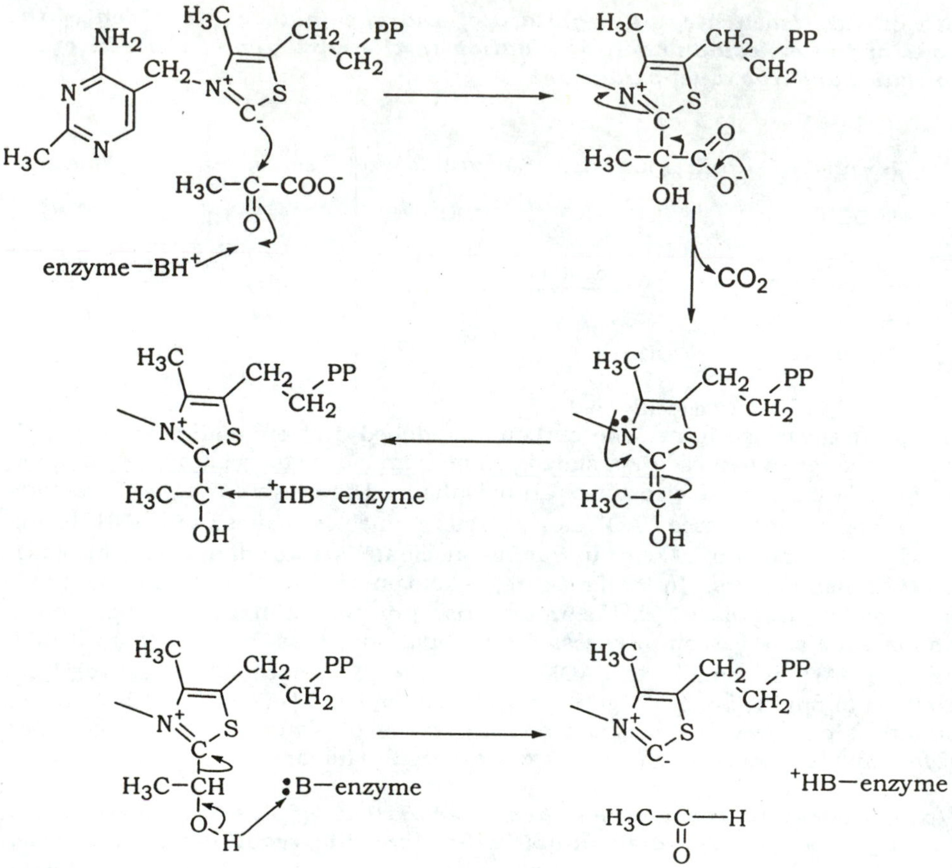

10. Aconitase catalyzes the citric acid cycle reaction:

citrate ⇆ isocitrate

The standard free energy change, $\Delta G°'$, for this reaction is +6.7 kJ/mol. However, the observed free energy change (ΔG) for this reaction in pig heart mitochondria is +0.8 kJ/mol. What is the ratio of [isocitrate]/[citrate] in these mitochondria? If [isocitrate] = 0.03 mM, what is [citrate]?

Answer: The reaction being examined is

citrate → isocitrate

The observed free energy change (ΔG) is calculated from the standard free energy change ($\Delta G°'$) using the following equation:

$$\Delta G = \Delta G°' + RT \times \ln\frac{[\text{Products}]}{[\text{Reactants}]}, \text{ or in this case}$$

$$\Delta G = \Delta G°' + RT \times \ln\frac{[\text{isocitrate}]}{[\text{citrate}]}$$

This equation may be rearranged to give:

$$\frac{[\text{isocitrate}]}{[\text{citrate}]} = e^{\frac{\Delta G - \Delta G°'}{RT}}$$

Substituting the appropriate numerical values, with the product $RT = 2.48$ at 25°C, we have:

$$\frac{[\text{isocitrate}]}{[\text{citrate}]} = e^{\frac{+0.8\frac{\text{kJ}}{\text{mol}} - 6.7\frac{\text{kJ}}{\text{mol}}}{2.48\frac{\text{kJ}}{\text{mol}}}} = e^{\frac{-5.9}{2.48}} = 0.093$$

The concentration of citrate is calculated as follows:

$$\frac{[\text{isocitrate}]}{[\text{citrate}]} = 0.093, \text{ or}$$

$$[\text{citrate}] = \frac{[\text{isocitrate}]}{0.093} = \frac{0.03 \text{ mM}}{0.093} = 0.324 \text{ mM}$$

Questions for Self Study

1. a. Complete the following reaction and name the enzyme complex that catalyzes it.
   Pyruvate + CoA + ___ → acetyl-CoA + ___ + ___ + CO_2.
   b. This enzyme complex uses five coenzymes. Name the coenzymes.

2. The citric acid cycle is also known as the Krebs cycle and the TCA (tricarboxylic acid) cycle. Identify the tricarboxylic acids in the cycle.

3. Fill in the blanks. The citric acid cycle converts two-carbon _____ units into _____. In the process the carbons are _____ and coenzymes are _____. The coenzymes serve as mobile electron carriers connecting the citric acid cycle to _____.

4. α-Ketoglutarate can be produced by oxidative decarboxylation of a citric acid cycle intermediate or by deamination of an amino acid. Identify these two sources of α-ketoglutarate.

5. In an earlier chapter we learned that thiol ester bonds are high-energy bonds. In the citric acid cycle there is one substrate-level phosphorylation (of GDP) driven by cleavage of a thiol ester bond. Identify the reaction and the enzyme. (Hint: The enzyme is named for the reverse reaction.)

6. Succinate dehydrogenase converts succinate to fumarate. This enzyme is a member of which two important metabolic sequences?

7. The citric acid cycle results in net oxidation of carbons. The production of succinate leaves a four-carbon unit in the cycle that is converted to oxaloacetate. Order the compounds shown below from least oxidized to most oxidized and identify them.

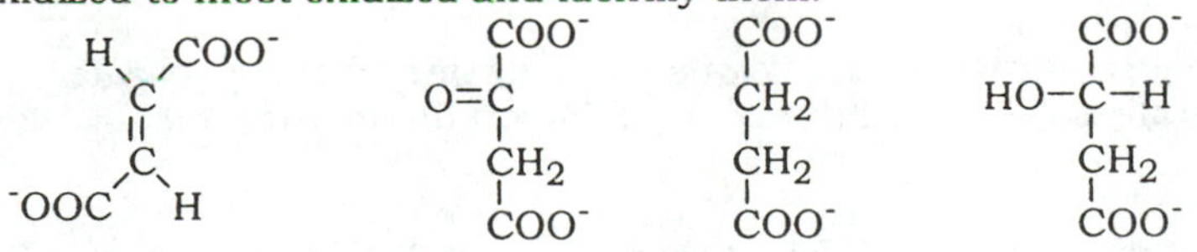

```
 H   COO⁻         COO⁻         COO⁻           COO⁻
  \ /              |            |              |
   C             O=C           CH2          HO-C-H
   ||              |            |              |
   C              CH2          CH2            CH2
  / \              |            |              |
⁻OOC  H           COO⁻         COO⁻           COO⁻
```

8. What is an anaplerotic reaction?

9. The citric acid cycle is regulated by metabolic signals at pyruvate dehydrogenase leading into the cycle and at three places in the cycle, citrate synthase, isocitrate dehydrogenase, and α-ketoglutarate dehydrogenase. List the metabolic signals and state how they affect the cycle.

Answers

1. a.Pyruvate + CoA + NAD^+ → acetyl-CoA + NADH + H^++ CO_2; pyruvate dehydrogenase;
   b. Thiamine pyrophosphate, coenzyme A, lipoic acid, NAD^+, and FAD.

2. Citrate and isocitrate.

3. Acetyl; carbon dioxide; oxidized; reduced; electron transport chain.

4. Isocitrate and glutamic acid.

5. The reaction is conversion of succinyl CoA to succinate and CoA with phosphorylation of GDP to GTP. The enzyme is succinyl-CoA synthetase. (Why is it a synthetase and not a synthase?)

6. Citric acid cycle and electron transport.

7. The compounds are succinate, fumarate, malate, and oxaloacetate. Fumarate and malate are at the same oxidation state.

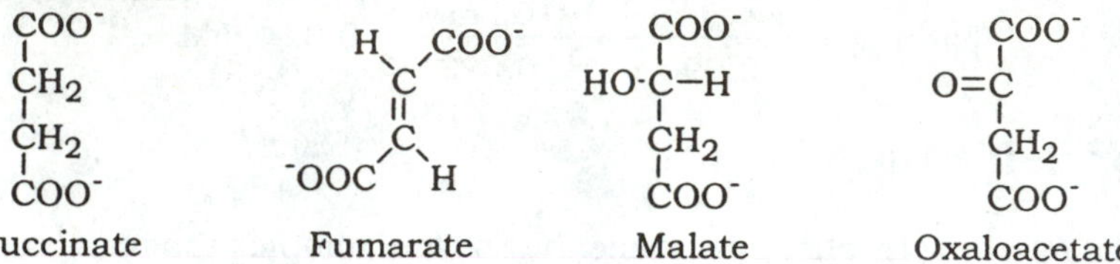

Succinate Fumarate Malate Oxaloacetate

8. It is a reaction that replenishes citric acid cycle intermediates.

9. NADH inhibits; NAD^+ stimulates; ATP inhibits; ADP and AMP stimulate; Succinyl-CoA inhibits; acetyl-CoA inhibits, CoA stimulates.

Additional Problems

1a. Identify the following compounds.

CH_3—C(=O)—COO^-

COO^-—CH_2—C(=O)—COO^-

COO^-—CH_2—CH_2—C(=O)—COO^-

b. If a Schiff base was produced between a side chain lysine and each of these compounds and then subsequently displaced by ammonia, what would each of these compounds be converted to?

2. The four steps in the citric acid cycle from succinate to oxaloacetate are similar in chemistry to the fatty acid catabolic pathway known as β-oxidation. In this pathway, a carbon in a fatty acid chain two positions away from the carboxylic acid group (the first position is α, next β, the last ω or omega) is oxidized to a keto group. Show that this also occurs when succinate is converted to oxaloacetate.

3. In glycolysis, phosphofructokinase is allosterically inhibited by citrate. What is a high concentration of citrate signaling and why is it important to turn off glycolysis under these conditions?

4. How is succinate dehydrogenase different than the other TCA cycle enzymes?

5. Succinyl-CoA synthetase is responsible for the only substrate-level phosphorylation found in the citric acid cycle. The mammalian enzyme uses GDP as a nucleotide and produces GTP. How is the high-energy phosphate bond of GTP utilized by cells?

6. If a labeled carbon atom of citric acid survives one turn of the cycle it shows up in two positions in oxaloacetate. Which steps are responsible for this?

7. Pyruvate carboxylase is a biotin-containing enzyme that reacts pyruvate with CO_2 to produce oxaloacetate. In general terms, describe this reaction. What is the role of biotin in this reaction? What general type of reaction is this an example of?

Abbreviated Answers

1a. Pyruvate, oxaloacetate, and α-ketoglutarate.
b. alanine, aspartic acid, glutamic acid.

2.

Succinate ($^-OOC-CH_2-CH_2-COO^-$) —oxidation→ Fumarate ($^-OOC-CH=CH-COO^-$) —hydration→ Malate ($^-OOC-CH_2-CH(OH)-COO^-$) —oxidation→ Oxaloacetate ($^-OOC-CH_2-C(=O)-COO^-$)

This carbon is β to the lower carboxyl group

3. High concentrations of citrate are an indication that the citric acid cycle is backing up. This can occur because ATP levels are satisfactory, NAD^+ is unavailable, or an alternate source of acetyl-units, perhaps from fatty acid catabolism, is available. There is no need to continue glycolysis and citrate inhibition at phosphofructokinase insures that glucose units are conserved.

4. Succinate dehydrogenase is responsible for the conversion of succinate to fumarate in the citric acid cycle. This enzyme is also part of the electron transport chain and moves electrons to coenzyme Q and is localized in the inner mitochondrial membrane. The other TCA cycle enzymes are located in the matrix.

5. Signal transduction and protein synthesis both use hydrolysis of GTP. In addition, GTP can be used to phosphorylate ADP by nucleoside diphosphate kinase, which catalyzes the following reaction:

$$GTP + ADP \rightarrow ATP + GDP.$$

6. Both succinate and fumarate are symmetrical molecules with two sets of equivalent carbons. Label at succinate (or fumarate) thus shows up in two positions in later stages.

7. Pyruvate carboxylase catalyzes a two-step reaction leading to oxaloacetate formation. In the first step, bicarbonate is attached to biotin in an ATP-dependent reaction. In the second step, the activated carboxyl group is transferred to pyruvate to produce oxaloacetate. Biotin functions as a carrier of activated carboxyl groups. The reaction is an example of an anaplerotic reaction because it leads to synthesis of a citric acid cycle intermediate.

Summary

Pyruvate produced in glycolysis can be oxidatively decarboxylated to acetyl-CoA and then oxidized in the tricarboxylic acid cycle (also known as the citric acid cycle or Krebs cycle). Electrons produced in this oxidative process are collected in reduced coenzymes (NADH and $FADH_2$) then passed through the electron transport pathway to oxygen (O_2), the final electron acceptor.

The oxidative decarboxylation of pyruvate is carried out by pyruvate dehydrogenase, a multienzyme complex composed of three types of protein subunits and five different coenzymes. Acetyl-CoA produced in this reaction enters the TCA cycle, combining with oxaloacetate in the citrate synthase reaction. This irreversible step commits acetate units to oxidation and to energy production. Citrate synthase is allosterically inhibited by NADH and by succinyl-CoA. Citrate is then isomerized to isocitrate by aconitase in a two-step process involving aconitate as an intermediate. This reaction converts a tertiary alcohol - a poor substrate for further oxidation - to a secondary alcohol, which can be readily oxidized. Fluoroacetate blocks the TCA cycle *in vivo*, although it has no apparent effect on any single reaction of the TCA cycle. Citrate synthase converts fluoroacetate to fluorocitrate, which strongly inhibits aconitase.

Isocitrate is oxidatively decarboxylated by isocitrate dehydrogenase in a two-step reaction involving oxidation of the C-2 alcohol of isocitrate to form oxalosuccinate, followed by a β-decarboxylation reaction which expels the central carboxyl group as CO_2, leaving the product α-ketoglutarate. NADH produced in this reaction links the TCA cycle and the electron transport chain. Isocitrate dehydrogenase is allosterically inhibited by NADH and ATP and allosterically activated by ADP. A second oxidative decarboxylation occurs in the α-ketoglutarate dehydrogenase reaction, which involves a multienzyme complex similar in many respects to pyruvate dehydrogenase. The reaction produces NADH and succinyl-CoA - a thiol ester product. Succinyl-CoA is used in the next step to drive the synthesis of GTP (in animals) or ATP (in plants and bacteria). This reaction is catalyzed by succinyl CoA synthetase and is an example of substrate level phosphorylation, the only such reaction in the TCA cycle.

The TCA cycle is completed by converting succinate to oxaloacetate. The process involves an oxidation, a hydration reaction and a second oxidation. Succinate dehydrogenase, a membrane-bound enzyme and a part of the electron transport chain, catalyzes the oxidation of succinate to fumarate, linking this reaction to reduction of FAD. Fumarase then catalyzes the trans-hydration of fumarate to produce malate and malate dehydrogenase completes the cycle by oxidizing malate to oxaloacetate. The latter reaction is strongly endergonic, but the reaction is pulled forward by the favorable citrate synthase reaction. The net reaction accomplished by the TCA cycle produces two molecules of CO_2, one ATP and four reduced coenzymes per acetate group oxidized. Carbon introduced to the cycle in any given citrate synthase reaction is not expelled as CO_2 in that same turn of the cycle. The carbonyl carbon of any given acetyl-CoA survives the first turn intact, but is completely lost as CO_2 in the second turn. The methyl carbon survives two full turns, and then undergoes 50% loss in the third turn, 25% loss in the fourth turn, etc. The TCA cycle provides a variety of intermediates for biosynthetic pathways, including citrate, α-ketoglutarate, succinyl-CoA, fumarate and oxaloacetate. On the other hand, the cell also feeds some of these same molecules back into the TCA cycle from other reactions, replenishing TCA cycle intermediates in so-called anaplerotic reactions.

The main sites of regulation in the TCA cycle are pyruvate dehydrogenase, citrate synthase, isocitrate dehydrogenase and α-ketoglutarate dehydrogenase. All these enzymes are inhibited by NADH, and ATP inhibits both pyruvate dehydrogenase and isocitrate dehydrogenase. The cycle is turned on when ADP/ATP and NAD^+/NADH ratios are high. Acetyl-CoA activates pyruvate carboxylase and succinyl CoA inhibits both citrate synthase and α-ketoglutarate dehydrogenase.

Chapter 17

Electron Transport and Oxidative Phosphorylation

• •

Chapter Outline

- Oxidative phosphorylation driven by electron transport
  - Membrane associated processes
    - Plasma membrane of bacteria
    - Mitochondrial membrane of eukaryotes
      - Outer mitochondrial membrane: Permeable to $M_r < 10{,}000$ due to protein porin
      - Intermembrane space: Creatine kinase, adenylate kinase, cytochrome c
      - Inner mitochondrial membrane: Highly impermeable
        - High protein content
        - High content of unsaturated fatty acids
        - Cardiolipin and diphosphatidylglycerol: No cholesterol
        - Cristae: Folds that increase surface area
      - Matrix
        - TCA cycle enzymes (except succinate dehydrogenase)
        - Enzymes for catabolism of fatty acids
        - Circular DNA
        - Ribosomes, tRNAs
- Reduction potentials: Electrical potential generated when reactions involve movement of electrons
  - Standard reduction potentials: $\mathscr{E}_o'$
    - Measured using sample half-cell with oxidized and reduced forms at 1M
    - Reference half-cell 1M H^+ and H_2 gas at 1 atm
    - Electrons flow toward cell with larger positive reduction potential
    - Half-cell reactions written as reduction reaction: Electrons on reactant side
    - Positive values of $\mathscr{E}_o'$: Accept electrons: Oxidizing agents
    - Negative values of $\mathscr{E}_o'$: Donate electrons: Reducing agents
  - Standard free energy change related to standard reduction potential
    - $\Delta G°' = -n\mathscr{F}\Delta\mathscr{E}_o'$
      - n = number of electrons
      - $\mathscr{F}$ = Faraday's constant: 96,485 J/V·mol
      - $\Delta\mathscr{E}_o' = \mathscr{E}_o'(\text{final}) - \mathscr{E}_o'(\text{initial})$
- Electron transport chain
  - Electron mediators
    - Flavoproteins: Tightly bound FMN or FAD
    - Coenzyme Q (ubiquinone): 1 or 2 Electron transfers: Mobile within membrane
    - Cytochromes: Fe^{2+}/Fe^{3+}
      - Cytochrome a's: Isoprenoid (15-C) on modified vinyl and formyl in place of methyl
      - Cytochrome b's: Iron-protoporphyrin IX
      - Cytochrome c's: Iron-protoporphyrin IX linked to cysteine

- Iron-sulfur proteins: Fe^{2+}/Fe^{3+}: Several types
- Protein-bound copper: Cu^{+}/Cu^{2+}

➢ Electron transport complexes: Four complexes function independently

- Complex I: NADH-coenzyme Q reductase (NADH reductase)
  - Electron movement
    - [FMN] accepts electron pair from NADH
    - [$FMNH_2$] donates electrons to Fe-S
    - Fe-S donates electrons to coenzyme Q
  - Protons pumped from matrix to cytosol
  - Generally, protons pumped from N (negative) to P (positive) side of membrane
  - Supports 3 ATP when electrons flow from NADH to oxygen
- Complex II: Succinate-coenzyme Q reductase (succinate dehydrogenase)
  - Components
    - FAD
    - Fe-S centers
  - No protons pumped
  - Supports 2 ATP when electrons flow from succinate to oxygen
  - Other complexes that reduce coenzyme Q
    - Glycerolphosphate dehydrogenase: Reduces coenzyme Q: No protons pumped
    - Fatty acyl-CoA dehydrogenase: Reduces coenzyme Q: No protons pumped
- Complex III: Coenzyme Q-cytochrome c reductase
  - Components
    - Cytochromes b_L and b_H
    - Rieske protein: Fe-S protein
  - Q-cycle
  - Protons pumped
- Cytochrome c is mobile carrier connecting Complexes III and IV
- Complex IV: Cytochrome c oxidase
  - Components
    - Cytochrome a and Cu_A
    - Cytochrome a_3 and Cu_B
  - Protons pumped

❖ Mitchell's chemiosmotic hypothesis: Proton gradient used to drive ATP synthesis

➢ Electrochemical potential of protons across inner mitochondrial membrane

- pH gradient
- Electric potential
- Energy of proton electrochemical potential:

$$\Delta G = RT\ln\frac{[H^+]_{out}}{[H^+]_{in}} + \mathcal{F}\Delta\psi$$

➢ Protons per electron pair

- From succinate 6
- From NADH 10

➢ Four protons per ATP

- ADP uptake: 1 proton
- ATP synthesis: 3 protons

➢ One oxygen consumed per electron pair

➢ P/O

- From NADH: 2.5
- From succinate: 1.5

❖ ATP synthase: F_1F_0ATPase

➢ F_1: ATP synthesis: Spherical particles on inner membrane

➢ F_0: Proton channel in inner membrane

- F_1F_0ATPase is a rotary motor
- Inhibitors of oxidative phosphorylation
  - Complex I: Rotenone, ptericidin, amytal, mercurial compounds, Demerol
  - Complex II: 2-Thenoyltrifluoroacetone, carboxin
  - Complex III: Antimycin, myxothiazol
  - Complex IV: Cyanide, azide, carbon monoxide
  - ATP synthase: Oligomycin, DCCD
- Uncouplers: Stimulate electron transport: Short circuit proton gradient: Block ATP production
  - Proton ionophores: Lipid soluble substance with dissociable proton
  - Energy of proton gradient dissipated as heat
- Mitochondrial exchange and uptake
  - ATP/ADP translocate
    - ADP in: ATP out: 1 Proton in
  - Glycerolphosphate shuttle: Moves electrons from cytosol to mitochondria
    - Cytosolic glycerolphosphate dehydrogenase: NADH-dependent
    - Mitochondrial glycerolphosphate dehydrogenase: FAD-dependent
    - DHAP and glycerolphosphate exchanged
  - Malate-aspartate shuttle: Moves electrons from cytosol to mitochondria
    - Cytosolic and mitochondrial malate dehydrogenases both use NADH
    - Aspartate exchanged for glutamate
    - Malate exchanged for α-ketoglutarate
- Net ATP produced from glucose
  - Depends on shuttle used
  - ~30 ATP / glucose in eukaryotic cells
  - ~38 ATP / glucose in bacteria

Chapter Objectives

Oxidation/Reduction Reactions

Electron transport involves sequential oxidation/reduction reactions. For any electron carrier, we can think of the carrier as participating in a reaction in which electrons are either produced or consumed. If electrons are consumed, the carrier is reduced and if electrons are produced, the carrier is oxidized. But in reality, free electrons are not just present in solution waiting to react (electrons are not like a typical chemical substrate); rather electrons are exchanged between pairs of reacting molecules. The pairs are an oxidant or oxidizing agent and a reductant or reducing agent. An oxidant accepts electrons, is itself reduced, but oxidizes the reductant, the agent from which the electrons originated. A reductant donates electrons, is itself oxidized, but reduces an oxidant.

The tendency to donate or accept electrons is measured by standard reduction (redox) potentials. It should be clear how these measurements are made. A reference half-cell is connected to a sample half-cell by an agar salt bridge and a low-resistance pathway. (The agar salt bridge simply functions to complete the circuit between the half-cells connected by the low-resistance pathway.) The reference half-cell is H^+/H_2 at 1 M and 1 atmosphere. The reduction potential of this half-cell is 0 V by definition. (Clearly, no potential exists when the reference half-cell is connected to an identical sample cell.) If the reaction in the sample cell consumes electrons, electrons will flow from the reference cell to the sample cell unless a voltage is applied to prevent this flow. In this case a positive voltage is required. Thus, a positive standard redox potential indicates that the sample is an oxidant relative to the reference cell and a negative standard redox potential indicates that the sample is a reductant relative to the reference cell. The two key formulas to remember are: $\Delta G°' = -n\mathcal{F}\Delta\mathcal{E}_o'$ and $\mathcal{E} = \mathcal{E}_O' + \frac{RT}{nF} \times \ln\frac{[\text{oxidant}]}{[\text{reductant}]}$.

Electron Transport Chain

A simplified way of thinking about the electron transport chain is to divide the chain into two types of components, mobile electron carriers and membrane-bound protein complexes. The mobile electron carriers include: NAD^+/NADH, coenzyme Q or ubiquinone, cytochrome c, and oxygen. Substrates, such as malate (for malate dehydrogenase) or α-ketoglutarate (for α-ketoglutarate dehydrogenase) or succinate (for succinate dehydrogenase) can be thought of as mobile electron carriers as well; however, they are peripheral to the electron transport chain. There are four membrane-bound complexes that simply move electrons from one mobile electron carrier to another. The first complex is NADH-coenzyme Q reductase or NADH reductase, which moves electrons from NADH to CoQ. CoQ can also be reduced by succinate-coenzyme Q reductase also known as succinate dehydrogenase, which uses succinate as a source of electrons. Coenzyme Q-cytochrome c reductase moves electrons from $CoQH_2$ to cytochrome c. Finally, cytochrome oxidase moves electrons from reduced cytochrome c to molecular oxygen.

Proton Gradient Formation

Electron transport accomplishes two things: it regenerates reduced cofactors such as NAD^+ and [FAD], and it produces ATP. Understand how the energy of electron transport is used to form a proton gradient, which is used in turn to phosphorylate ADP to ATP. Protons are pumped out of the mitochondria by NADH-coenzyme Q reductase. Protons are also expelled in the Q cycle involving coenzyme Q-cytochrome c reductase. The essential point is that coenzyme Q carries electrons and protons whereas cytochrome c carries only electrons. Cytochrome c oxidase also moves protons but the details of how this is achieved are unknown.

ATP synthase

ADP phosphorylation and proton gradient dissipation are coupled by the ATP synthase or F_1F_0-ATPase. Understand how this protein complex is organized and situated in the inner mitochondrial membrane. The F_0 portion (o is for oligomycin) is a proton pore and the F_1 is an ATPase that functions in the reverse direction to produce ATP.

Inhibitors and Uncouplers

Appreciate the difference between an inhibitor and an uncoupler. Inhibitors block the action of some component of electron transport or ATP synthase. Uncouplers do not interfere with electron transport and in fact stimulate it. However, they provide an alternative pathway to dissipate the energy of electron transport.

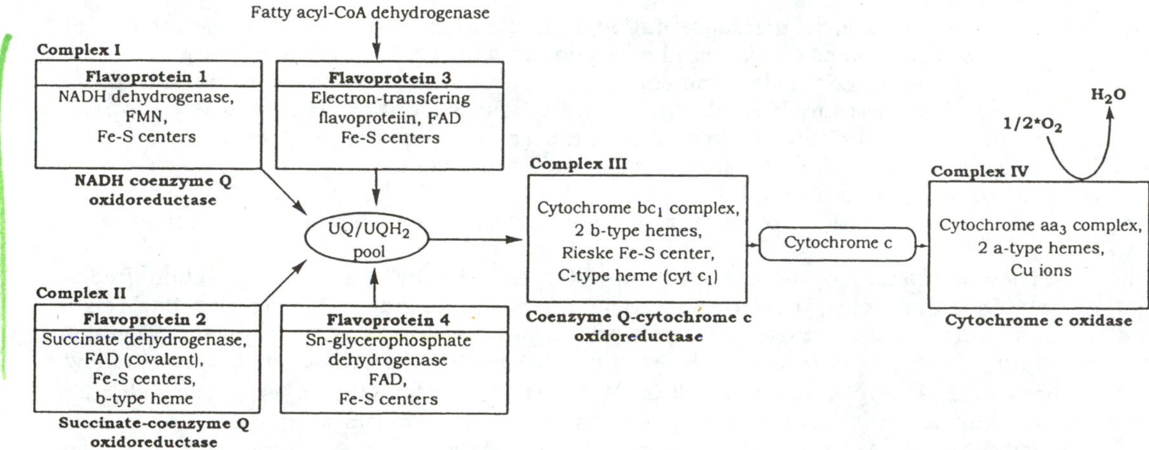

Figure 17.4. An overview of the complexes and pathways in the mitochondrial electron transport chain.

Problems and Solutions

1. ***For the following reaction,***

$$FAD + 2\ cyt\ c\ (Fe^{2+}) \rightleftharpoons FADH_2 + 2\ cyt\ c\ (Fe^{3+})$$

determine which of the redox couples is the electron acceptor and which is the electron donor under standard-state conditions, calculate the value of $\Delta\mathscr{E}_o'$, ***and determine the free energy change for the reaction.***

Answer: The reduction half-reactions and their standard reduction potentials for the reaction are (from Table 17.1):

$FAD + 2H^+ + 2e^- \rightarrow FADH_2$, $\mathscr{E}_o' = -0.219$ V*

And,

Cyto c (Fe^{3+}) + e^- → cyto c, (Fe^{2+}), $\mathscr{E}_o' = 0.254$ V.

The standard reduction potential is the voltage that is generated between a sample half-cell and reference half-cell (H^+/H_2). In effect, it is the voltage that must be applied to a circuit connecting a sample half-cell and the reference half-cell to prevent current from flowing. Using this convention we see that if the FAD half-cell is connected to the reference half-cell, a slightly negative voltage of -0.219 V must be applied to prevent electrons from flowing from the reference cell into the sample cell. Conversely, +0.254 V must be applied when the cytochrome c half-cell is connected to the reference cell. Thus, electrons have a greater tendency to flow from the reference cell to cytochrome c than to FAD. Therefore, electrons must move from $FADH_2$ to cytochrome c. Thus, $FADH_2$ is the electron donor and cytochrome c, Fe^{3+} is the electron acceptor.

$$\Delta\mathscr{E}_o' = \mathscr{E}_o'(\text{acceptor}) - \mathscr{E}_o'(\text{donor}) = \mathscr{E}_o'(\text{cyto c}) - \mathscr{E}_o'(\text{FAD})$$

$$\Delta\mathscr{E}_o' = +0.254 - (-0.219) = +0.473 \text{ V}$$

The free energy change for the reaction is given by :

$\Delta G^{\circ\prime} = -n\mathscr{F}\Delta\mathscr{E}_o'$, where n is the number of electrons, and

$$\mathscr{F} = \text{Faraday's constant} = 96.485\frac{\text{kJ}}{\text{V}\cdot\text{mol}}$$

$$\Delta G^{\circ\prime} = -2\times 96.485\frac{\text{kJ}}{\text{V}\cdot\text{mol}}\times 0.473\text{ V} = -91.3\frac{\text{kJ}}{\text{mol}}$$

* The value of -0.219 given for FAD in Table 17.1 is for free FAD. Protein-bound FAD has a range of standard reduction potential values. Using 0.02V, $\Delta G^{\circ\prime}$ = -45.2 kJ/mol.

2. *Calculate $\Delta\mathscr{E}_o'$ for the glyceraldehyde-3-phosphate dehydrogenase reaction, and calculate the free energy change for the reaction.*

Answer: Glyceraldehyde-3-phosphate dehydrogenase catalyzes the following reaction:

Glyceraldehyde-3-phosphate + P_i + NAD^+ ⇆ 1,3-bisphosphoglycerate + NADH + H^+

The relevant half reactions are (from Table 17.1):

$NAD^+ + 2H^+ + 2e^- \rightarrow NADH + H^+$, $\mathscr{E}_o' = -0.320$

and,

Glycerate-1,3-bisphosphate + 2 H^+ + 2 e^- → glyceraldehyde-3-phosphate + P_i, $\mathscr{E}_o' = -0.290$

Thus, NAD^+ is the electron acceptor and glyceraldehyde-3-phosphate is the electron donor.

$$\Delta\mathscr{E}_o' = \mathscr{E}_o'(\text{acceptor}) - \mathscr{E}_o'(\text{donor}) = \mathscr{E}_o'(\text{NAD}^+) - \mathscr{E}_o'(\text{G3P})$$

$$\Delta\mathscr{E}_o' = -0.320 - (-0.290) = -0.030 \text{ V}$$

The free energy change for the reaction is given by :

$\Delta G^{\circ\prime} = -n\mathscr{F}\Delta\mathscr{E}_o'$, where n is the number of electrons, and

$$\mathscr{F} = \text{Faraday's constant} = 96.485\frac{\text{kJ}}{\text{V}\cdot\text{mol}}$$

$$\Delta G^{\circ\prime} = -2\times 96.485\frac{\text{kJ}}{\text{V}\cdot\text{mol}}\times(-0.030\text{ V}) = 5.79\frac{\text{kJ}}{\text{mol}}$$

3. *For the following redox reaction,*

$NAD^+ + 2\,H^+ + 2\,e^- \rightarrow NADH + H^+$

suggest an equation (analogous to Equation 17.13) that predicts the pH dependence of this reaction, and calculate the reduction potential for this reaction at pH 8.

Answer: The NAD^+ reduction shown above may be rewritten as:

$NAD^+ + H^+ + 2\,e^- \rightarrow NADH$

However, a source of electrons is needed for the reaction to occur. Therefore, let us assume that the reaction is in aqueous solution with a general reductant of the form:

$$\text{Reductant} \rightarrow \text{Oxidant} + 2e^-$$

The overall reaction is:

$$NAD^+ + H^+ + \text{Reductant} \rightarrow NADH + \text{Oxidant}$$

$$\Delta G = \Delta G^{\circ\prime} + RT\ln\frac{[\text{Oxidant}][NAD^+]}{[\text{Reductant}][NADH][H^+]}$$

$$= \Delta G^{\circ\prime} + RT\ln\frac{[\text{Oxidant}][NAD^+]}{[\text{Reductant}][NADH]} + RT\ln\frac{1}{[H^+]}$$

$$= \Delta G^{\circ\prime} + RT\ln\frac{[\text{Oxidant}][NAD^+]}{[\text{Reductant}][NADH]} - RT\ln[H^+]$$

But, $\ln[H^+] = 2.303\log_{10}[H^+]$ and $pH \equiv -\log_{10}[H^+]$, so

$-\ln[H^+] = -2.303\log_{10}[H^+] = 2.303 \times pH$, thus

$$\Delta G = \Delta G^{\circ\prime} + RT\ln\frac{[\text{Oxidant}][NAD^+]}{[\text{Reductant}][NADH]} + 2.303 \times RT \times pH$$

In general, $\Delta G = -n\mathcal{F}\Delta\mathcal{E}$, or $\Delta\mathcal{E} = -\frac{\Delta G}{n\mathcal{F}}$ where n is the number of electrons, and

$\mathcal{F}$ = Faraday's constant = $96.485\frac{kJ}{V \cdot mol}$

$$\Delta\mathcal{E} = \Delta\mathcal{E}_o{}' - \frac{RT}{n\mathcal{F}}\ln\frac{[\text{Oxidant}][NAD^+]}{[\text{Reductant}][NADH]} - 2.303 \times \frac{RT}{n\mathcal{F}} \times pH$$

To calculate the reduction potential we assume that the reaction is being carried out under standard conditions. Thus, all reactants and products are at 1 M concentration. The above equation simplifies to:

$$\Delta\mathcal{E} = \Delta\mathcal{E}_o{}' - 2.303 \times \frac{RT}{n\mathcal{F}} \times pH$$

4. Sodium nitrite ($NaNO_2$) is used by emergency medical personnel as an antidote for cyanide poisoning (for this purpose, it must be administered immediately). Based on the discussion of cyanide poisoning in Section 17.10, suggest a mechanism for the life-saving effect of sodium nitrite.

Answer: Cytochrome c oxidase is the principle target of cyanide (CN^-) poisoning. Cyanide binds to the ferric (Fe^{3+}) or oxidized form of cytochrome a_3. Sodium nitrite may be administered intravenously in an attempt to combat cyanide poisoning. The intention of sodium nitrite treatment is to produce an alternate target for cyanide. Nitrite will oxidize the very abundant hemoglobin to methemoglobin (ferric hemoglobin) that will react with cyanide.

5. A wealthy investor has come to you for advice. She has been approached by a biochemist who seeks financial backing for a company that would market dinitrophenol and dicumarol as weight-loss medications. The biochemist has explained to her that these agents are uncouplers and that they would dissipate metabolic energy as heat. The investor wants to know if you think she should invest in the biochemist's company. How do you respond?

Answer: The structures of dicumarol and dinitrophenol are:

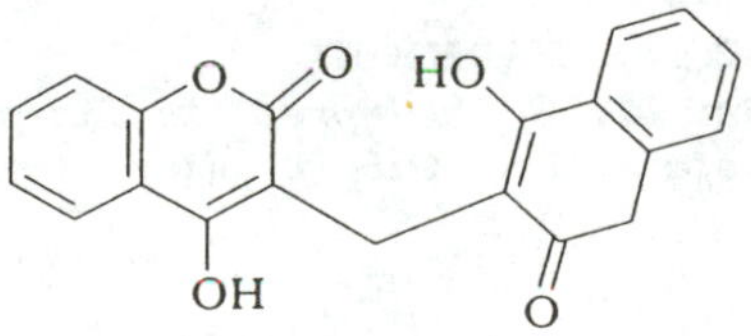

Dicumarol

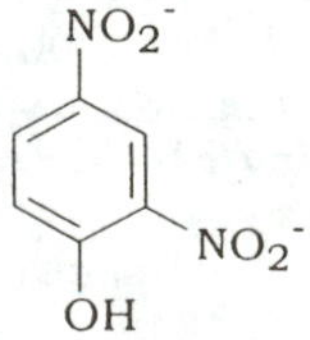

2,4-Dinitrophenol

The biochemistry of the suggestion is sound. (Beware: This is not an endorsement of the idea. Please read on.) Both compounds are uncoupling agents and act by dissipating the proton gradient across the inner mitochondrial membrane. Instead of being used to synthesize ATP, the energy of the proton gradient is dissipated as heat. As ATP levels decrease, electron transport will increase, and, glucose or fatty acids will be metabolized in an attempt to meet the false metabolic demand.

The compounds are both lipophilic molecules and are capable of dissolving in the inner mitochondrial membrane. Their hydroxyl groups have low pK_as because they are attached to conjugated ring systems. On the cytosolic (and higher pH) surface of the inner mitochondrial membrane the compounds will be protonated whereas on the matrix side they will be unprotonated.

So much for the theoretical biochemistry. In working with unfamiliar compounds, it is imperative to consult references to find out what is known about them. One good place to start is the MSDS (Material Safety Data Sheet). The MSDS is a summary of potential hazards of a compound provided by the manufacturer. Another good reference, and one found in virtually every biochemist's laboratory, is the Merck Index (Published by Merck & Co., Rahway, N.J.). The Merck Index informs us, in the section on human toxicity under 2,4-dinitrophenol, that this compound is highly toxic, produces an increase in metabolism (good for a diet), increased temperature, nausea, vomiting, collapse, and, death (a drastic weight loss indeed). Under dicumarol, human toxicity is not mentioned, however, under the subsection, therapeutic category, we are informed that the compound is used as an anticoagulant in humans. Dicumarol was first discovered as the agent responsible for hemorrhagic sweet clover disease in cattle. It is produced by microorganisms in spoiled silage and causes death by bleeding. It is used therapeutically as an anticoagulant but the therapy must be carefully monitored.

6. Assuming that 3 H^+ are transported per ATP synthesized in the mitochondrial matrix, the membrane potential difference is 0.18 V (negative inside), and the pH difference is 1 unit (acid outside, basic inside), calculate the largest ratio of [ATP]/[ADP][P_i] under which synthesis of ATP can occur.

Answer: The free energy difference per mole for protons across the inner mitochondrial membrane is given by Equation 17.25:

$$\Delta G = -2.303 \times RT \times (pH_{out} - pH_{in}) + \mathcal{F}\Delta\Psi$$

Where, pH_{out} is the pH outside the mitochondria; pH_{in} is the matrix pH; $\Delta\psi$ is the potential difference across the inner mitochondrial membrane, $V_{in} - V_{out}$; R is the gas constant, 8.314×10^{-3} kJ/mol K; $\mathcal{F}$ is Faraday's constant, 96.485 kJ/V mol; and T is temperature in K (°C + 273, assuming 25°C = 298).

$$\Delta G = -2.303 \times (8.314 \times 10^{-3})(298) \times (1) + 96.485 \times (-0.18)$$

$$\Delta G = -23.07 \frac{kJ}{mol}$$

For movement of 3 protons we have

$$\Delta G = 3 \times (-23.07 \frac{kJ}{mol}) = -69.2 kJ$$

What is the largest value of $\frac{[ATP]}{[ADP][P_i]}$ under which synthesis of ATP can occur?

For the reaction, ADP + P_i → ATP + H_2O , ΔG is given by:

$$\Delta G = \Delta G^{\circ\prime} + RT \ln \frac{[ATP]}{[ADP][P_i]}$$

$$\Delta G = +30.5 \frac{kJ}{mol} + (8.314 \times 10^{-3})(298) \ln \frac{[ATP]}{[ADP][P_i]}, \text{ and for 1 mole}$$

$$\Delta G = (+30.5 + 2.48 \ln \frac{[ATP]}{[ADP][P_i]}) kJ$$

For translocation of 3 protons coupled to synthesis of 1 ATP to be thermodynamically favorable the overall ΔG must be negative. Therefore,

$$\Delta G_{3H^+} + \Delta G_{ATP} < 0$$

$$-69.2+30.5+2.48\ln\frac{[\text{ATP}]}{[\text{ADP}][\text{P}_i]}<0$$

$$\text{Solving for } \frac{[\text{ATP}]}{[\text{ADP}][\text{P}_i]} \text{ we find :}$$

$$\frac{[\text{ATP}]}{[\text{ADP}][\text{P}_i]}<e^{\frac{69.2-30.5}{2.48}}=e^{15.6}=6\times10^{6}\ \text{M}^{-1}$$

7. Of the dehydrogenase reactions in glycolysis and the TCA cycle, all but one use NAD^+ as the electron acceptor. The lone exception is the succinate dehydrogenase reaction, which uses FAD, covalently bound to a flavoprotein, as the electron acceptor. The standard reduction potential for this bound FAD is in the range of 0.003 to 0.091 V (Table 17.1). Compared to the other dehydrogenase reactions of glycolysis and the TCA cycle, what is unique about succinate dehydrogenase? Why is bound FAD a more suitable electron acceptor in this case?

Answer: Succinate dehydrogenase converts succinate to fumarate, an oxidation of an alkane to an alkene. The other oxidation reactions in the TCA cycle and in glycolysis either convert alcohols to ketones or ketones to carboxyl groups. These oxidations are sufficiently energetic to reduce NAD^+. The standard redox potential (from Table 17.1) for reduction of fumarate to succinate is 0.031 V. In contrast, the redox potential for NAD^+ is -0.320 V. Thus, under standard conditions, if NAD^+ participated in the succinate/fumarate reaction, it would do so as a reductant (i.e., as NADH) passing electrons to fumarate to produce succinate, the exact opposite of what is accomplished in the TCA cycle. To remove electrons from succinate, an oxidant with a higher (more positive) redox potential is required. The covalently bound FAD of succinate dehydrogenase meets this requirement with a standard redox potential in the range of 0.003 to 0.091 V. (We will encounter another example of conversion of an alkane to an alkene with reduction of FAD in fatty acid metabolism or β-oxidation.)

8a. What is the standard free energy change ($\Delta G°'$), for the reduction of coenzyme Q by NADH as carried out by complex I (NADH-coenzyme Q reductase) of the electron transport pathway if $\mathcal{E}_o'(NAD^+/NADH+H^+)$ = -0.320 volts and $\mathcal{E}_o'(CoQ/CoQH_2)$ = +0.060 volts.
b. What is the equilibrium constant (Keq) for this reaction?
c. Assume that (1) the actual free energy release accompanying the NADH-coenzyme Q reductase reaction is equal to the amount released under standard conditions (as calculated above), (2) this energy can be converted into the synthesis of ATP with an efficiency = 0.75 (that is, 75% of the energy released upon NADH oxidation is captured in ATP synthesis), and (3) the oxidation of 1 equivalent of NADH by coenzyme Q leads to the phosphorylation of 1 equivalent of ATP.
Under these conditions, what is the maximum ratio of [ATP]/[ADP] attainable for oxidative phosphorylation when $[P_i]$ = 1 mM? (Assume $\Delta G°'$ for ATP synthesis = +30.5 kJ/mol.)

Answer: a. The relevant half reactions are (from Table 17.1):

$NAD^+ + 2H^+ + 2e^- \rightarrow NADH + H^+$, $\mathcal{E}_o' = -0.320$

and,

$CoQ\ (UQ) + 2\ H^+ + 2\ e^- \rightarrow CoQH_2\ (UQH_2)$, $\mathcal{E}_o' = +0.060$

Thus, CoQ is the electron acceptor and NADH is the electron donor.

$$\Delta\mathcal{E}_o' = \mathcal{E}_o'(\text{acceptor}) - \mathcal{E}_o'(\text{donor}) = \mathcal{E}_o'(\text{CoQ}) - \mathcal{E}_o'(\text{NAD}^+)$$

$$\Delta\mathcal{E}_o' = +0.060 - (-0.320) = +0.38\ \text{V}$$

The free energy change for the reaction is given by :

$\Delta G°' = -n\mathcal{F}\Delta\mathcal{E}_o'$, where n is the number of electrons, and

$$\mathcal{F} = \text{Faraday's constant} = 96.485\frac{\text{kJ}}{\text{V}\cdot\text{mol}}$$

$$\Delta G°' = -2\times 96.485\frac{\text{kJ}}{\text{V}\cdot\text{mol}}\times(0.38\ \text{V}) = -73.3\frac{\text{kJ}}{\text{mol}}$$

b. To calculate the equilibrium constant:

From $\Delta G^{\circ\prime} = -RT\ln K_{eq}$ we see that

$$K_{eq} = e^{\frac{-\Delta G^{\circ\prime}}{RT}}$$

The value of RT at 25°C is $(8.314\times10^{-3})(298)\frac{kJ}{mol} = 2.48\frac{kJ}{mol}$

Thus,

$$K_{eq} = e^{\frac{-\Delta G^{\circ\prime}}{RT}} = e^{\frac{-(-73.3)}{2.48}} = e^{29.6} = 7.14\times10^{12}$$

c. Assuming that ΔG = -73.3 kJ/mol, the amount of energy used to synthesize ATP is:

$$-73.3\frac{kJ}{mol}\times(0.75) = -55\frac{kJ}{mol}$$

Since $\Delta G = \Delta G^{\circ\prime} + RT\ln\frac{[ATP]}{[ADP][P_i]}$

$$\frac{[ATP]}{[ADP]} = [P_i]\times e^{\frac{\Delta G-\Delta G^{\circ\prime}}{RT}} = (1mM)\times e^{\frac{55-30.5}{2.48}} = (1\times10^{-3})\times e^{9.88}$$

$$\frac{[ATP]}{[ADP]} = 19.5$$

9. *Consider the oxidation of succinate by molecular oxygen as carried out via the electron transport pathway.*

$$succinate + \frac{1}{2}O_2 \rightarrow fumarate + H_2O$$

a. What is the standard free energy change (ΔG°′), for this reaction if $\mathscr{E}_o'$(fumarate/succinate) = +0.031 volts and $\mathscr{E}_o'(\frac{1}{2}O_2/H_2O)$ = +0.816 volts.

b. What is the equilibrium constant (K_{eq}) for this reaction?

c. Assume that (1) the actual free energy release accompanying succinate oxidation by the electron transport pathway is equal to the amount released under standard conditions (as calculated above), (2) this energy can be converted into the synthesis of ATP with an efficiency = 0.70 (that is, 70% of the energy released upon succinate oxidation is captured in ATP synthesis), and (3) the oxidation of 1 succinate leads to the phosphorylation of 2 equivalents of ATP.

Under these conditions, what is the maximum ratio of [ATP]/[ADP] attainable for oxidative phosphorylation when [P_i] = 1 mM? (Assume ΔG°′ for ATP synthesis = +30.5 kJ/mol.)

Answer: a. The relevant half reactions are (from Table 17.1):

Fumarate + $2H^+$ + $2e^-$ → succinate, $\mathscr{E}_o'$ = +0.031

and,

$\frac{1}{2}O_2$ + $2H^+$ + $2e^-$ → H_2O, $\mathscr{E}_o'$ = +0.816

Thus, O_2 is the electron acceptor and succinate is the electron donor.

$$\Delta\mathscr{E}_o' = \mathscr{E}_o'(acceptor) - \mathscr{E}_o'(donor) = \mathscr{E}_o'(O_2) - \mathscr{E}_o'(succinate)$$

$$\Delta\mathscr{E}_o' = +0.816 - (0.031) = +0.785\ V$$

The free energy change for the reaction is given by :

$\Delta G^{\circ\prime} = -n\mathscr{F}\Delta\mathscr{E}_o'$, where n is the number of electrons, and

$\mathscr{F}$ = Faraday's constant = $96.485\frac{kJ}{V\cdot mol}$

$$\Delta G^{\circ\prime} = -2\times96.485\frac{kJ}{V\cdot mol}\times(0.785\ V) = -151.5\frac{kJ}{mol}$$

b. To calculate the equilibrium constant:

From $\Delta G°' = -RT \ln K_{eq}$ we see that

$$K_{eq} = e^{\frac{-\Delta G°'}{RT}}$$

The value of RT at 25°C is $(8.314 \times 10^{-3})(298)\frac{kJ}{mol} = 2.48\frac{kJ}{mol}$

Thus,

$$K_{eq} = e^{\frac{-\Delta G°'}{RT}} = e^{\frac{-(-151.5)}{2.48}} = e^{61.1} = 3.39 \times 10^{26}$$

c. Assuming that ΔG = -151.3 kJ/mol, the amount of energy used to synthesize ATP is:

$$-151.5\frac{kJ}{mol} \times (0.70) = -106.1\frac{kJ}{mol} \text{ for 2 ATP synthesized}$$

Per ATP we have $53.05\frac{kJ}{mol}$

$$\text{Since } \Delta G = \Delta G°' + RT\ln\frac{[ATP]}{[ADP][P_i]}$$

$$\frac{[ATP]}{[ADP]} = [P_i] \times e^{\frac{\Delta G - \Delta G°'}{RT}} = (1mM) \times e^{\frac{53.05-30.5}{2.48}} = (1 \times 10^{-3}) \times e^{9.09}$$

$$\frac{[ATP]}{[ADP]} = 8.89$$

10. Consider the oxidation of NADH by molecular oxygen as carried out via the electron transport pathway

$$NADH + H^+ + \frac{1}{2}O_2 \rightarrow NAD^+ + H_2O$$

a. What is the standard free energy change ($\Delta G°'$), for this reaction if $\mathscr{E}_o'(NAD^+/NADH+H^+)$ = -0.320 volts and $\mathscr{E}_o'(\frac{1}{2}O_2/H_2O)$ = +0.816 volts.

b. What is the equilibrium constant (K_{eq}) for this reaction?

c. Assume that (1) the actual free energy release accompanying NADH oxidation by the electron transport pathway is equal to the amount released under standard conditions (as calculated above), (2) this energy can be converted into the synthesis of ATP with an efficiency = 0.75 (that is, 75% of the energy released upon NADH oxidation is captured in ATP synthesis), and (3) the oxidation of 1 NADH leads to the phosphorylation of 3 equivalents of ATP.
Under these conditions, what is the maximum ratio of [ATP]/[ADP] attainable for oxidative phosphorylation when [P_i] = 2 mM? (Assume $\Delta G°'$ for ATP synthesis = +30.5 kJ/mol.)

Answer

a. The relevant half reactions are (from Table 17.1):

$NAD^+ + 2H^+ + 2e^- \rightarrow NADH + H^+$, $\mathscr{E}_o'$ = -0.320

and,

$\frac{1}{2}O_2 + 2H^+ + 2e^- \rightarrow H_2O$, $\mathscr{E}_o'$ = +0.816

Thus, oxygen is the electron acceptor and NADH is the electron donor.

$$\Delta\mathscr{E}_o' = \mathscr{E}_o'(\text{acceptor}) - \mathscr{E}_o'(\text{donor}) = \mathscr{E}_o'(NAD^+) - \mathscr{E}_o'(G3P)$$

$$\Delta\mathscr{E}_o' = +0.816 - (-0.320) = +1.136 \text{ V}$$

The free energy change for the reaction is given by :

$\Delta G°' = -n\mathscr{F}\Delta\mathscr{E}_o'$, where n is the number of electrons, and

$$\mathscr{F} = \text{Faraday's constant} = 96.485\frac{kJ}{V \cdot mol}$$

$$\Delta G°' = -2 \times 96.485\frac{kJ}{V \cdot mol} \times (1.136 \text{ V}) = -219.2\frac{kJ}{mol}$$

b. To calculate the equilibrium constant:

From $\Delta G°' = -RT \ln K_{eq}$ we see that

$$K_{eq} = e^{\frac{-\Delta G°'}{RT}}$$

The value of RT at 25°C is $(8.314 \times 10^{-3})(298)\frac{kJ}{mol} = 2.48\frac{kJ}{mol}$

Thus,

$$K_{eq} = e^{\frac{-\Delta G°'}{RT}} = e^{\frac{-(-219.2)}{2.48}} = e^{88.4} = 2.46 \times 10^{38}$$

c. Assuming that ΔG = -219.2 kJ/mol, the amount of energy used to synthesize ATP is:

$$-219.2\frac{kJ}{mol} \times (0.75) = -164.4\frac{kJ}{mol} \text{ for 3 ATP synthesized}$$

Per ATP we have $54.8\frac{kJ}{mol}$

Since $\Delta G = \Delta G°' + RT \ln\frac{[ATP]}{[ADP][P_i]}$

$$\frac{[ATP]}{[ADP]} = [P_i] \times e^{\frac{\Delta G - \Delta G°'}{RT}} = (2mM) \times e^{\frac{54.8-30.5}{2.48}} = (2 \times 10^{-3}) \times e^{9.80}$$

$$\frac{[ATP]}{[ADP]} = 36$$

11. Write a balanced equation for the reduction of molecular oxygen by reduced cytochrome c as carried out by complex IV (cytochrome oxidase) of the electron transport pathway.

a. What is the standard free energy change ($\Delta G°'$), for this reaction if $\mathscr{E}_o'$(cyt c(Fe^{3+})/ cyt c(Fe^{2+})) = +0.254 volts and $\mathscr{E}_o'(\frac{1}{2}O_2/H_2O)$ = +0.816 volts.

b. What is the equilibrium constant (K_{eq}) for this reaction?

c. Assume that (1) the actual free energy release accompanying cytochrome c oxidation by the electron transport pathway is equal to the amount released under standard conditions (as calculated above), (2) this energy can be converted into the synthesis of ATP with an efficiency = 0.60 (that is, 60% of the energy released upon cytochrome c oxidation is captured in ATP synthesis), and (3) the reduction of 1 molecule of O_2 by reduced cytochrome c leads to the phosphorylation of 2 equivalents of ATP.

Under these conditions, what is the maximum ratio of [ATP]/[ADP] attainable for oxidative phosphorylation when [P_i] = 3 mM? (Assume $\Delta G°'$ for ATP synthesis = +30.5 kJ/mol.)

Answer: The balanced equation for transfer of electrons from cytochrome c to oxygen is:

$$4 \text{ Cytochrome c}(Fe^{2+}) + O_2 + 4H^+ \rightarrow 4 \text{ Cytochrome c}(Fe^{3+}) + 2H_2O$$

a. The relevant half reactions are (from Table 17.1):

$$\text{Cytochrome c}(Fe^{3+}) + e^- \rightarrow \text{Cytochrome c}(Fe^{2+}), \qquad \mathscr{E}_o' = +0.254$$

and,

$$\frac{1}{2}O_2 + 2H^+ + 2e^- \rightarrow H_2O \qquad \mathscr{E}_o' = +0.816$$

Thus, oxygen is the electron acceptor and cytochrome c is the electron donor.

$\Delta\mathcal{E}_o' = \mathcal{E}_o'(\text{acceptor}) - \mathcal{E}_o'(\text{donor}) = \mathcal{E}_o'(O_2) - \mathcal{E}_o'(\text{cytochrome c})$

$\Delta\mathcal{E}_o' = +0.816 - (+0.254) = +0.562\text{ V}$

The free energy change for the reaction is given by :

$\Delta G^{o\prime} = -n\mathcal{F}\Delta\mathcal{E}_o'$, where n is the number of electrons, and

$\mathcal{F}$ = Faraday's constant = $96.485\dfrac{\text{kJ}}{\text{V}\cdot\text{mol}}$

$$\Delta G^{o\prime} = -4\times 96.485\frac{\text{kJ}}{\text{V}\cdot\text{mol}}\times(0.562\text{ V}) = -217\frac{\text{kJ}}{\text{mol}}$$

b. To calculate the equilibrium constant:

From $\Delta G^{o\prime} = -RT\ln K_{eq}$ we see that

$$K_{eq} = e^{\frac{-\Delta G^{o\prime}}{RT}}$$

The value of RT at 25°C is $(8.314\times10^{-3})(298)\dfrac{\text{kJ}}{\text{mol}} = 2.48\dfrac{\text{kJ}}{\text{mol}}$

Thus,

$$K_{eq} = e^{\frac{-\Delta G^{o\prime}}{RT}} = e^{\frac{-(-217)}{2.48}} = e^{87.5} = 1.00\times10^{38}$$

c. Assuming that ΔG = -206.5 kJ/mol, the amount of energy used to synthesize ATP is:

$$-217\frac{\text{kJ}}{\text{mol}}\times(0.60) = -130.2\frac{\text{kJ}}{\text{mol}}\text{ for 2 ATP synthesized}$$

Per ATP we have $65.1\dfrac{\text{kJ}}{\text{mol}}$

Since $\Delta G = \Delta G^{o\prime} + RT\ln\dfrac{[\text{ATP}]}{[\text{ADP}][\text{P}_i]}$

$$\frac{[\text{ATP}]}{[\text{ADP}]} = [\text{P}_i]\times e^{\frac{\Delta G-\Delta G^{o\prime}}{RT}} = (3\text{mM})\times e^{\frac{65.1-30.5}{2.48}} = (3\times10^{-3})\times e^{14}$$

$$\frac{[\text{ATP}]}{[\text{ADP}]} = 3437$$

12. The standard reduction potential for (NAD^+/NADH) is -0.320 volts, and the standard reduction potential for (pyruvate/lactate) is -0.185 volts.

a. What is the standard free energy change, $\Delta G^{o\prime}$, for the lactate dehydrogenase reaction:

$$\text{NADH} + H^+ + \text{pyruvate} \rightarrow \text{lactate} + NAD^+$$

b. What is the equilibrium constant (K_{eq}) for this reaction?

c. If [pyruvate] = 0.05 mM and [lactate] = 2.9 mM and ΔG for the lactate dehydrogenase reaction = –15 kJ/mol in erythrocytes, what is the (NAD^+/NADH) ratio under these conditions?

Answer

a. The relevant half reactions are (from Table 17.1):

$NAD^+ + 2H^+ + 2e^- \rightarrow NADH + H^+$, $\mathcal{E}_o'$ = -0.320

and,

pyruvate + $2H^+ + 2e^- \rightarrow$ lactate, $\mathcal{E}_o'$ = -0.185

Thus, pyruvate is the electron acceptor and NADH is the electron donor.

$$\Delta\mathcal{E}_o' = \mathcal{E}_o'(\text{acceptor}) - \mathcal{E}_o'(\text{donor}) = \mathcal{E}_o'(\text{pyruvate}) - \mathcal{E}_o'(\text{NAD}^+)$$

$$\Delta\mathcal{E}_o' = -0.185 - (-0.320) = +0.135\ \text{V}$$

The free energy change for the reaction is given by :

$\Delta G^{\circ\prime} = -n\mathcal{F}\Delta\mathcal{E}_o'$, where n is the number of electrons, and

$$\mathcal{F} = \text{Faraday's constant} = 96.485\frac{\text{kJ}}{\text{V}\cdot\text{mol}}$$

$$\Delta G^{\circ\prime} = -2\times 96.485\frac{\text{kJ}}{\text{V}\cdot\text{mol}}\times(0.135\ \text{V}) = -26.05\frac{\text{kJ}}{\text{mol}}$$

b. To calculate the equilibrium constant:

From $\Delta G^{\circ\prime} = -RT\ln K_{eq}$ we see that

$$K_{eq} = e^{\frac{-\Delta G^{\circ\prime}}{RT}}$$

The value of RT at 25°C is $(8.314\times10^{-3})(298)\frac{\text{kJ}}{\text{mol}} = 2.48\frac{\text{kJ}}{\text{mol}}$

Thus,

$$K_{eq} = e^{\frac{-\Delta G^{\circ\prime}}{RT}} = e^{\frac{-(-26.05)}{2.48}} = e^{10.5} = 3.65\times10^4$$

c. The ratio of oxidized to reduced cofactor is calculated as follows:

$$\Delta G = \Delta G^{\circ\prime} + RT\ln\frac{[\text{lactate}][\text{NAD}^+]}{[\text{pyruvate}][\text{NADH}]},\ \text{or}$$

$$\frac{[\text{lactate}][\text{NAD}^+]}{[\text{pyruvate}][\text{NADH}]} = e^{\frac{\Delta G - \Delta G^{\circ\prime}}{RT}},\ \text{where } \Delta G = -15\frac{\text{kJ}}{\text{mol}},\ \Delta G^{\circ\prime} = -26.05\frac{\text{kJ}}{\text{mol}},\ RT = 2.48\frac{\text{kJ}}{\text{mol}}$$

$$\frac{[\text{NAD}^+]}{[\text{NADH}]} = \frac{[\text{pyruvate}]}{[\text{lactate}]}e^{\frac{-15-(-26.05)}{2.48}} = \frac{0.05\ \text{mM}}{2.9\text{mM}}e^{4.46} = 1.48$$

13. Assume that the free energy change, ΔG, associated with the movement of one mole of protons from the outside to the inside of a bacterial cell is -23 kJ/mol and 3 H+ must cross the bacterial plasma membrane per ATP formed by the F_1F_0ATP synthase. ATP synthesis thus takes place by the coupled process:

$$3\ H^+_{out} + ADP + P_i \leftrightarrows 3\ H^+_{in} + ATP + H_2O$$

a. If the overall free energy change ($\Delta G_{overall}$) associated with ATP synthesis in these cells by the coupled process is -21 kJ/mol, what is the equilibrium constant, K_{eq}, for the process?

b. What is $\Delta G_{synthesis}$, the free energy change for ATP synthesis, in these bacteria under these conditions?

c. The standard free energy change for ATP hydrolysis is $\Delta G^{\circ\prime}_{hydrolysis}$ is -30.5 kJ/mol. If [P_i] = 2 mM in these bacterial cells, what is the [ATP]/[ADP] ratio in these cells?

Answer

a.

$$\Delta G^{\circ\prime} = -RT\ln K_{eq},\ \text{or}$$

$$K_{eq} = e^{\frac{-\Delta G^{\circ}}{RT}} = e^{\frac{-21}{(8.314\times10^{-3})\times298}} = e^{8.48} = 4.80\times10^3\ M^{-1}$$

b. The overall free energy change for ATP synthesis accounts for proton movement and ATP synthesis. Because ΔG is a state property we can write:

$$\Delta G_{overall} = \Delta G_{synthesis} + \Delta G_{proton\ movement} \text{ or,}$$

$$\Delta G_{synthesis} = \Delta G_{overall} - \Delta G_{proton\ movement}$$

$$\Delta G_{synthesis} = -21\ \text{kJ} - 3\ \text{mol protons} \times (-23\frac{\text{kJ}}{\text{mol protons}})$$

$$\Delta G_{synthesis} = -21\ \text{kJ} + 69\ \text{kJ} = 48\ \text{kJ}$$

c.

$$\Delta G_{hydrolysis} = \Delta G^{\circ\prime}_{hydrolysis} + RT\ln\frac{[ADP][P_i]}{[ATP]} \text{ or}$$

$$\ln\frac{[ADP][P_i]}{[ATP]} = \frac{\Delta G_{hydrolysis} - \Delta G^{\circ\prime}_{hydrolysis}}{RT}$$

$$\ln\frac{[ADP][P_i]}{[ATP]} = \frac{(-48)-(-30.5)}{(8.314\times10^{-3})(298)} = -7.06$$

$$\frac{[ADP][P_i]}{[ATP]} = e^{-7.06} = 8.56\times10^{-4}$$

$$\frac{[ATP]}{[ADP][P_i]} = \frac{1}{8.56\times10^{-4}}$$

$$\frac{[ATP]}{[ADP]} = \frac{[P_i]}{8.56\times10^{-4}} = \frac{2\times10^{-3}}{8.56\times10^{-4}} = 2.34$$

Questions for Self Study

1. How does the formula $\Delta G^{\circ\prime} = -n\mathcal{F}\Delta\mathcal{E}_o'$ predict the direction of electron flow between two reduction half-reactions under standard conditions?

2. The electron transport chain is composed of four complexes. For the statements below indicate which complex or complexes fit the description. The complexes are: NADH-coenzyme Q reductase (complex I); succinate-coenzyme Q reductase (complex II); coenzyme Q-cytochrome c reductase (complex III); cytochrome c oxidase (complex IV).
   a. Contains two copper sites.
   b. Contains protein-bound FAD.
   c. Produces water upon electron transport.
   d. Is involved in the Q cycle.
   e. Moves electrons from substrate directly into the electron transport chain.
   f. Reduces Q.
   g. Is a citric acid cycle enzyme.
   h. Reduces a small protein localized on the outer surface of the inner mitochondrial membrane.
   i. Moves protons across the inner mitochondrial membrane.
   j. Moves electrons from a lipid-soluble mobile electron carrier to a water-soluble mobile electron carrier.
   k. Contains tightly bound FMN.
   l. Transfers electrons to oxygen.
   m. Moves electrons from NADH to CoQ.
   n. Moves electrons form cytochrome c to oxygen.
   o. Contains cytochromes.

3. The energy difference for protons across the inner mitochondrial membrane is given by:

$$\Delta G = 2.303RT\times\log\frac{H^+_{out}}{H^+_{in}} + Z\mathcal{F}\Delta\Psi$$

How would this equation be modified by the following conditions?
   a. If protons were uncharged species.
   b. If the inner mitochondrial membrane was freely permeable to ions other than protons.
   c. If the equation was written as a function of pH.

4. ATP synthase or F_1F_0-ATPase is composed of two complexes: a peripheral membrane complex and an integral membrane complex. What roles do the complexes play in oxidative phosphorylation?

5. What is the key difference between an electron transport inhibitor and an uncoupler? What are the consequences of each to electron transport, oxygen consumption, and ATP production?

Answers

1. In general, a reaction is spontaneous when $\Delta G < 0$. When the reaction is under standard conditions $\Delta G^{\circ\prime} < 0$. Thus, $\Delta\mathscr{E}_o' > 0$, $\mathscr{E}_o'{}_{final} - \mathscr{E}_o'{}_{initial} > 0$, or $\mathscr{E}_o'{}_{final} > \mathscr{E}_o'{}_{initial}$. Electrons will flow from the reduction half-reaction with the smaller standard reduction potential to the reduction half-reaction with the larger standard reduction potential.

2. a. IV; b. II; c. IV; d. III; e. II; f. I and II; g. II; h. III; i. I, III, and IV; j. III; k. I; l. IV; m. I; n. IV; o. III and IV.

3. a. $\Delta G = 2.303RT \times \log\frac{H^+_{out}}{H^+_{in}}$; b. $\Delta G = 2.303RT \times \log\frac{H^+_{out}}{H^+_{in}}$;
c. $\Delta G = -2.303RT \times (pH_{out} - pH_{in}) + Z\mathscr{F}\Delta\Psi$.

4. The peripheral membrane complex or F_1 unit is the site at which ADP and P_i are joined to form ATP. The integral membrane complex, F_0, is a proton channel.

5. An electron transport inhibitor blocks the movement of electrons through a component of the transport chain. Inhibitors will block electron transport for electrons entering the chain above the blockage point but not for those entering below the blockage point. Oxygen consumption and ATP production will, depending on the source of electrons, be diminished or stopped. Uncoupling agents provide an alternate pathway through which protons can reenter the mitochondrial matrix. Uncouplers stimulate electron transport and oxygen consumption but block ATP synthesis.

Additional Problems

1. Isolated, inner mitochondrial membranes do not transport electrons from NADH to O_2 unless cytochrome c is added. Why?

2. Describe one site in the electron transport chain responsible for pumping hydrogen ions out of the mitochondria during electron transport.

3. Nigericin is an ionophore whose structure is shown below.

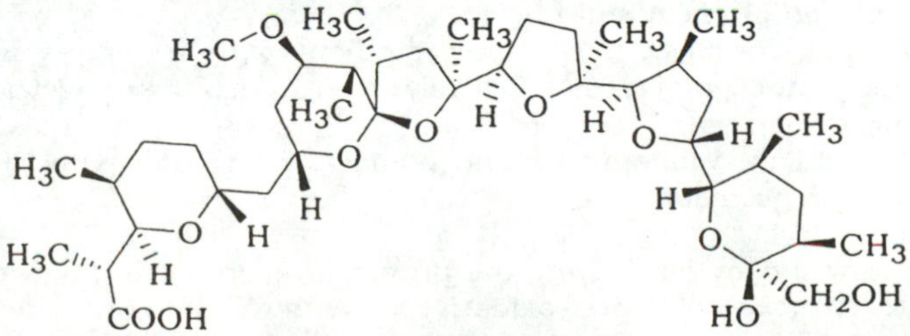

It is soluble in the inner mitochondrial membrane in either its acidic form with the carboxyl group protonated or as a potassium salt. Valinomycin is a cyclic compound composed of various L- and D- amino acids and lactic acid and functions as a potassium ionophore. The presence of nigericin, valinomycin, and potassium will effectively destroy the electrochemical potential of mitochondria. Why?

4. An uncoupling agent like dinitrophenol actually stimulates O_2 consumption. Why?

5. In the citric acid cycle, malate dehydrogenase catalyzes the following reaction:

$$\text{malate} + NAD^+ \rightarrow \text{oxaloacetate} + NADH$$

Given the following standard reduction potentials, calculate $\Delta G°'$

Reduction Half-Reaction	$\mathcal{E}_o'$ (V)
Oxaloacetate + $2H^+ + 2e^- \rightarrow$ malate	-0.166
$NAD^+ + H^+ + 2e^- \rightarrow NADH$	-0.320

In which direction will the reaction proceed under standard conditions. Which component functions as the oxidant? Reductant? Does this differ from the physiological direction?

Abbreviated Answers

1. Inner mitochondrial membranes contain all of the components necessary for electron transport except cytochrome c. Cytochrome c is a relatively small protein, M_r = 1,800, localized in the outer surface of the inner mitochondrial membrane and it is readily lost during purification procedures. Cytochrome c moves electrons between coenzyme Q-cytochrome c reductase and cytochrome c oxidase and must be present for electrons to flow from NADH to O_2.

2. There are three sites at which protons are pumped out of the mitochondria in response to electron transport: NADH dehydrogenase, coenzyme Q-cytochrome c reductase, and cytochrome c oxidase. The details of how protons are pumped by cytochrome c oxidase are unknown. Coenzyme Q-cytochrome c reductase is involved in the Q cycle. NADH dehydrogenase moves electrons from NADH to coenzyme Q using a flavoprotein and several Fe-S proteins to move electrons. In NADH dehydrogenase, the flavoprotein accepts electrons and protons from NADH on the matrix side of the inner mitochondrial membrane but donates electrons to a series of Fe-S proteins. CoQ is thought to play two roles in the movement of electrons through NADH dehydrogenase. It is the terminal electron acceptor for the complex. Additionally, CoQ is thought to participate at an intermediate stage of electron flow by accepting electrons from one Fe-S component and donating electrons to another Fe-S component. In the process, CoQ binds protons from the matrix side of the membrane and releases protons into the intermembrane space

3. Since valinomycin is specific for potassium, it will not affect the proton gradient by itself. In the presence of potassium, valinomycin will move potassium into the mitochondrial matrix until potassium is actually concentrated inside the mitochondria. This occurs because initially potassium moves down a potassium chemical gradient (assuming that the mitochondrial matrix has a low potassium concentration) until the potassium concentration is equal on both sides of the membrane. But, potassium continues to accumulate, being driven inside the cell by the electrical potential of the proton gradient. Net movement of potassium stops when the potassium concentration difference across the membrane is sufficient to produce a potassium electrical potential equal in magnitude and opposite in sign from the electrical potential of the proton gradient. However, the proton gradient still exists.

Nigericin alone will simply replace the proton gradient with a potassium gradient. The chemical potential of the proton gradient is effectively destroyed but an electrical potential, now due to potassium, is still present.

The combination of nigericin, valinomycin, and potassium will dissipate the proton gradient without creating a potassium gradient.

4. Uncoupling agents function by dissipating the proton gradient. Although electron transport continues to regenerate reduced cofactors, oxidative phosphorylation does not occur and ADP levels build up. This signals the cell to increase metabolic activity leading to an increase in oxygen consumption.

5. The relevant half reactions are (from Table 17.1):

$NAD^+ + 2H^+ + 2e^- \rightarrow NADH + H^+$, $\mathcal{E}_o'$ = -0.320

and,

oxaloacetate + $2H^+ + 2e^-$ → malate, $\mathscr{E}_o' = -0.166$

Thus, oxaloacetate is the electron acceptor and NADH is the electron donor.

$$\Delta\mathscr{E}_o' = \mathscr{E}_o'(\text{acceptor}) - \mathscr{E}_o'(\text{donor}) = \mathscr{E}_o'(\text{NAD}^+) - \mathscr{E}_o'(\text{oxaloacetate})$$

$$\Delta\mathscr{E}_o' = -0.320 - (-0.166) = -0.154\ \text{V}$$

The free energy change for the reaction is given by :

$\Delta G^{\circ\prime} = -n\mathscr{F}\Delta\mathscr{E}_o'$, where n is the number of electrons, and

$$\mathscr{F} = \text{Faraday's constant} = 96.485\frac{\text{kJ}}{\text{V}\cdot\text{mol}}$$

$$\Delta G^{\circ\prime} = -2\times 96.485\frac{\text{kJ}}{\text{V}\cdot\text{mol}}\times(-0.154\ \text{V}) = +29.7\frac{\text{kJ}}{\text{mol}}$$

A positive $\Delta G^{\circ\prime}$ indicates that the reaction, under standard conditions, is not favorable and will proceed in the reverse direction. Electrons will flow from NADH to oxaloacetate. Thus, NADH serves as a reductant and oxaloacetate serves as an oxidant. In the citric acid cycle, removal of oxaloacetate by citrate synthase lowers the concentration of oxaloacetate and drives the reaction toward NADH production.

Summary

Most of the metabolic energy obtained from sugars and similar metabolites entering glycolysis and the TCA cycle is funneled into NADH and [$FADH_2$] via oxidation/reduction reactions. Electrons stored in the form of reduced coenzymes and flavoproteins are then passed through an elaborate and highly organized chain of proteins and coenzymes, the so-called electron transport chain, finally reaching O_2, molecular oxygen, the terminal electron acceptor. In the course of electron transport, energy is stored in the form of a proton gradient across the inner mitochondrial membrane. This proton gradient then provides the energy to drive ATP synthesis in oxidative phosphorylation. Electron transport and oxidative phosphorylation are membrane-associated processes. Bacteria carry out these processes at (and across) the plasma membrane, and in eukaryotic cells they are carried out and localized in mitochondria. Mitochondria are surrounded by a simple outer membrane in addition to a more complex inner membrane. The proteins of the electron transport chain and oxidative phosphorylation are associated with the inner mitochondrial membrane.

The tendency of the components of the electron transport chain to transfer electrons to other species is characterized by an electron transfer potential or standard reduction potential, $\mathscr{E}_o'$. These potentials are determined by measuring the voltages generated in reaction half cells. Standard reduction potentials are measured relative to a reference half cell. The H^+/H_2 half cell normally serves as the reference half cell and is assigned a standard reduction potential of 0.0 V. The sign of $\mathscr{E}_o'$ for a redox couple can be used to indicate whether reduction or oxidation occurs for that couple in any real reaction (involving two redox couples). The redox couple with the more negative $\mathscr{E}_o'$ will act as an electron donor, and the redox couple with the more positive $\mathscr{E}_o'$ will act as an electron acceptor. For redox couples with a large positive standard reduction potential, the oxidized form of the couple has a strong tendency to be reduced.

Fractionation of the membranes containing the electron transport chain results in the isolation of four distinct complexes, including I) NADH-coenzyme-Q reductase, II) succinate-coenzyme-Q reductase, III) coenzyme-Q-cytochrome c reductase, and IV) cytochrome c oxidase. Oxidation of NADH and reduction of coenzyme-Q by complex I is accompanied by transport of protons from the matrix to the cytoplasmic side of the inner mitochondrial membrane. Complex II, succinate-coenzyme-Q reductase or succinate dehydrogenase, is an integral membrane protein, a TCA cycle enzyme and a flavoprotein. The net reaction carried out by this complex is oxidation of succinate and reduction of coenzyme-Q, with intermediate oxidation/reduction of [$FADH_2$]. In complex III, reduced coenzyme-Q passes its electrons to cytochrome c via a complex redox cycle involving three cytochromes, several FeS centers and proton transport across the inner mitochondrial membrane. Cytochrome c, a mobile electron carrier, passes electrons from complex III to complex IV, cytochrome c oxidase. Electron transfer in complex IV involves two hemes and two copper sites in cytochrome a and a_3 and results in proton transport across the inner mitochondrial membrane. In spite of their spatial proximity, the four major complexes of the electron transport chain operate and migrate independently in the inner mitochondrial membrane.

The elucidation of the mechanism, which couples electron transport and ATP synthesis, was one of the great challenges (and triumphs) of modern biochemistry. Peter Mitchell's chemiosmotic hypothesis postulates that a proton gradient generated by the electron transport chain is utilized to drive ATP synthesis. The mitochondrial complex, which carries out ATP synthesis, is called ATP synthase or F_1F_0-ATPase. The F_1 subunit of this complex is spherical in shape, consists of five major polypeptide chains and catalyzes ATP synthesis. The F_0 subunit forms the transmembrane pore or channel through which protons move to drive ATP synthesis. The essence of the Mitchell hypothesis was confirmed in a reconstitution experiment by E. Racker and W. Stoeckenius, using bacteriorhodopsin to generate a proton gradient and the ATP synthase to generate ATP. As shown by Paul Boyer, the energy provided by electron transport drives enzyme conformation changes, which regulate the binding and release of substrates on ATP synthase. The mechanism involves catalytic cooperativity between three interacting sites. Many of the details of electron transport and oxidative phosphorylation have been gained from studying the effects of specific inhibitors, including rotenone, which inhibits NADH-UQ reductase, and cyanide, which blocks complex IV, cytochrome c oxidase. Uncouplers dissipate the proton gradient across the inner mitochondrial membrane, which was created by electron transport. Molecular shuttle systems exist to carry the electrons of NADH into the mitochondrial matrix, because NADH and $FADH_2$ cannot be transported across the inner mitochondrial membrane. The two systems, which operate in this manner, are the glycerol phosphate shuttle and the malate-aspartate shuttle. ATP produced by oxidative phosphorylation is carried out of the mitochondria by an ATP/ADP translocase, which couples the exit of ATP with the entry of ADP.

The net yield of energy from glucose oxidation – either ~30 or ~32 ATP synthesized per glucose – depends upon the shuttle system used. The overall efficiency of metabolism (from glucose to the TCA cycle to electron transport and oxidative phosphorylation) is about 54%.

Chapter 18

Photosynthesis

Chapter Outline

- Photosynthesis: Two aspects
  - Light reactions: NADPH and ATP produced, oxygen evolved, water electron donor
  - Dark reactions: Carbon dioxide fixed
- Chloroplasts: Plastids
  - Outer membrane: Porous
  - Intermembrane space
  - Inner membrane
  - Stroma: Volume within inner membrane but outside thylakoids
  - Thylakoid: Paired folds
    - Lamellae
    - Thylakoid vesicles
    - Grana: Stacks of thylakoid vesicles
    - Internal thylakoid space (lumen)
    - Stroma lamellae: connects grana
- Photosynthetic pigments
  - Chlorophyll: Porphyrin-like molecule with Mg^{2+}
  - Secondary light-harvesting pigments: Broaden wavelength range
    - Carotenoids
    - Phycobilins
- Light energy
  - Absorption of energy
    - $E = h\nu$: h = Planck's constant, ν = frequency of light
    - $\nu = c/\lambda$: c = speed of light in a vacuum, λ = wavelength
    - Light photons: Packet of energy or quantum: $E = h\nu$
  - Energy excites π electrons
  - Release of energy
    - Heat
    - Fluorescence: Light but at longer wavelength: Lower frequency: Less energetic
    - Resonance energy transfer: Near neighbors exchange exciton: Förster resonance energy transfer
    - Energy transduction: Molecule with excited electron becomes better reductant
- Characteristics of photosynthesis
  - Membrane-localized
  - Pigments organized into units
    - Antennae molecules harvest light
    - Reaction center: Energy sink of unit
  - Eukaryotes: Two photosystems
    - PSI: 700nm: Excited electrons reduce $NADP^+$
    - PSII: 680nm: Splits water: Generates oxygen: Pumps protons: Reduces PSI

  - Prokaryotes: PSII-like but without oxygen evolution capacity
- Protein complexes: Two light harvesting complexes
  - PSI complex: Provides reducing power for production of NADPH (Used in Calvin cycle to fix carbon dioxide
  - PSII complex: Splits water producing oxygen (byproduct) and electrons: Electrons moved to PSI via electron transport chain
  - Cytochrome b_6 / cytochrome f complex
    - Moves electrons from PSII to PSI: Protons pumped (ATP produced): Noncyclic flow of electrons
    - Moves electrons from PSI to PSI: Protons pumped (ATP produced): Cyclic flow of electrons
- Z Scheme: Movement of electrons from water via PSII through cytochrome b_6/f complex through PSI to $NADP^+$
- PSII activities: H_2O/plastiquinone oxidoreductase
  - Electrons carried between H_2O and P680 by a specific tyrosyl residue
  - P680 to P680* to $P680^+$ by exciton absorption and electron release
  - Electron from P680 passes to pheophytin: Chlorophyll with 2 H^+ in place of Mg^{2+}
  - Pheophytin passes electron to plastoquinone
  - Plastoquinone (lipid mobile) passes electrons to the cytochrome b_6/f complex
  - $P680^+$ to P680: Electron from water through Mn complex and D (tyrosyl residue)
- Cytochrome b_6 / cytochrome *f* activities: Plastoquinol/plastocyanin oxidoreductase
  - Cytochrome *f*: Moves electrons from plastoquinone to plastocyanin
    - Plastocyanin: Copper atom involved in single-electron transfers
    - Diffuses along inside of thylakoid along membrane
  - Cytochrome b_6/f complex involved in cyclic electron transport: Cyclic photophosphorylation
- PSI activities: Plastocyanin/ferredoxin oxidoreductase
  - P700 to P700* to $P700^+$ by exciton absorption and electron release
  - Electron from P700 passes to quinone (A1)
  - Quinone (A1) passes electron to membrane-bound ferredoxins
  - Membrane-bound ferredoxins pass electrons to soluble ferredoxin
  - Soluble ferredoxin passes electrons to $NADP^+$: Ferredoxin/$NADP^+$ reductase
- The photosynthetic reaction center of the bacterium *Rhodopseudomonas viridis* serves as a model for PSI and PSII
- Quantum yields
  - Eight photons per O_2 evolved: Four per PSI and four for PSII
  - One proton per photon: Eight protons pumped
  - Three protons per ATP
  - Two NADPH produced
- Photophosphorylation
  - Proton translocations
    - Water to PSII
    - Q-cycle
    - 2 H^+/e^- transferred from H_2O to $NADP^+$
  - $\Delta p = \Delta\psi - 2.3RT \times \Delta pH/(\mathcal{F})$
  - CF_1CF_0ATP synthase couples proton release to ATP production
  - Noncyclic photophosphorylation: Electrons from water to $NADP^+$
  - Cyclic photophosphorylation: Electrons from PSI through the cytochrome b_6/f complex back to PSI
- Calvin-Benson cycle
  - Only known pathway of net CO_2 fixation

- ➢ Immediate product 3-phosphoglycerate: Reduced to glyceraldehyde-3-P using NADPH
- ➢ G-3-P used to produce 4-, 5-, 6-, and 7-C compounds
- ➢ Hexose principle product

❖ Rubisco: CO_2 + ribulose-1,5-BP → 3-PG
- ➢ Fixes carbon dioxide to carbon skeleton
- ➢ Most abundant enzyme in biosphere

❖ Calvin cycle enzymes regulated by photosynthetic input; indirectly by light

❖ Photorespiration: Rubisco: Ribulose bisphosphate carboxylase/oxygenase
- ➢ Oxygenase activity produces 3-PG + phosphoglycolate: O_2 consumed
- ➢ Phosphoglycolate converted to glyoxylate: O_2 consumed
- ➢ Glyoxylate to glycine to serine to 3-PB
- ➢ Glycine to serine: CO_2 produced
- ➢ Ratio of carboxylase to oxygenase activity is 3 or 4 to 1

❖ Hatch-Slack pathway: C-4 plants: CO_2 delivery system
- ➢ PEP + CO_2 to oxaloacetate to malate: Mesophyll cells
- ➢ Malate to pyruvate + CO_2: Bundle sheath cells
- ➢ Pyruvate and malate exchanged

❖ Crassulacean acid metabolism
- ➢ Stomata closed during the day: Malate decarboxylated to produce CO_2
- ➢ Stomata opened during the night: CO_2 + PEP to oxaloacetate and malate

Chapter Objectives

Photosynthesis

It is convenient to divide photosynthesis into light and dark reactions. Keep in mind that the light reactions harvest light energy and transduce it into the chemical energy of ATP and NADPH. ATP serves its usual role as energy currency. NADPH is a cellular reducing agent. In the dark reactions, CO_2 is fixed into organic compounds and reduced to a hydrated carbon. The objective is to produce carbohydrates from CO_2, the chemistry is carbon reduction, the biochemistry is the Calvin cycle.

Light Reactions

Light energy is used to split water in a reaction called photolysis producing O_2, H^+, and electrons. The electrons move along an electron transport chain leading to $NADP^+$. Analogous to the electron transport chain in mitochondria, electron flow in chloroplasts (and other photosynthetic systems) is accompanied by H^+ gradient formation, which is used, in turn, to synthesize ATP. Get the big picture down and then fill in the details (see Figure 18.12).

The abstraction of electrons from water is an energy-requiring process and in photosynthesis, the energy derives from sunlight. How is this energy harvested? Chlorophylls and accessory light-harvesting pigments are responsible. Production of O_2 from 2 H_2O generates 8 electrons. In the "Z" system, generation of each electron requires two quanta of light (only one quantum per electron is needed for the cyclic pathway of PSI). This explains why the pigment molecules are organized into photosynthetic units. The pigment molecules in a unit cooperate to collect light quanta and funnel the energy to a reaction center. Photosynthetic eukaryotes have two separate but interacting photosystems, PSI and PSII. PSII is responsible for O_2 production and has a reaction center P680. PSI reduces $NADP^+$ if coupled to PSII or by itself can move electrons in a cycle beginning and ending at its reaction center P700. The movement of electrons from water to $NADP^+$ requires excitation of PSII, electron transport from PSII to PSI, excitation of PSI, and electron transport from PSI to $NADP^+$. Electron transport from PSII to PSI is accompanied by proton-gradient formation. This gradient is subsequently used to produce ATP. When we get to the Calvin cycle we will see that there is a greater requirement for ATP than NADPH and the additional ATP can be generated by cyclic photophosphorylation. This process involves only PSI. An excited electron is fed from PSI into the electron transport chain leading from PSII to PSI and thus moves in a circle. However, cyclic electron flow is accompanied by proton-gradient formation and subsequent ATP production.

Dark Reactions

Rubisco is the key player in CO_2 fixation. Understand the carboxylase reaction leading to two phosphoglycerates. Rubisco also catalyzes an oxygenase reaction, which leads to photorespiration. The carboxylase reaction adds carbon at the oxidation state of a carboxylate group to ribulose-1,5-bisphosphate to produce phosphoglycerate. The carboxyl groups now must be reduced to aldehydes for carbohydrate synthesis to occur. The initial steps include phosphorylation to 1,3-BPG and reduction to glyceraldehyde-3-phosphate and P_i. (We saw these steps in gluconeogenesis and glycolysis.) The remainder of the Calvin cycle (Figure 18.21) converts glyceraldehyde to glucose, not an onerous task given our knowledge of gluconeogenesis, and regenerates ribulose-1,5-bisphosphate. This aspect of the cycle is reminiscent of the nonoxidative phase of the pentose phosphate pathway and uses isomerases, aldolases, transketolases, phosphatases, and a kinase to convert three-carbon sugars into five-carbon sugars.

Regulation

Calvin cycle activity is coordinated with the light reactions by three light-induced effects: light-induced pH changes, generation of reducing power, and Mg^{2+} efflux. Know the targets for these controls. Rubisco regulation by Mg^{2+} is dependent on CO_2 availability.

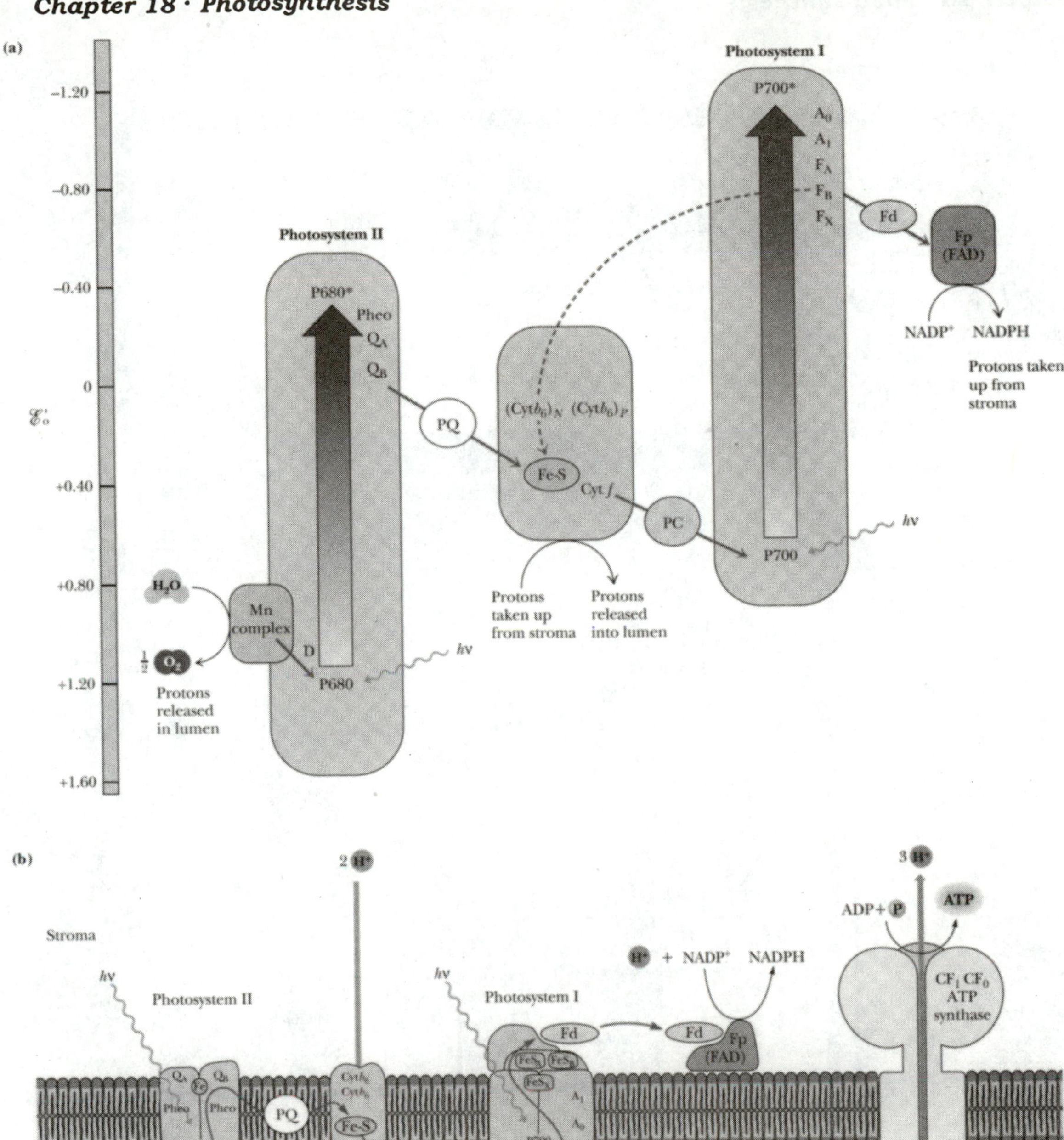

Figure 18.12 The Z-scheme of photosynthesis. The Z-scheme is a diagrammatic representation of photosynthetic electron flow from H_2O to NADP+.

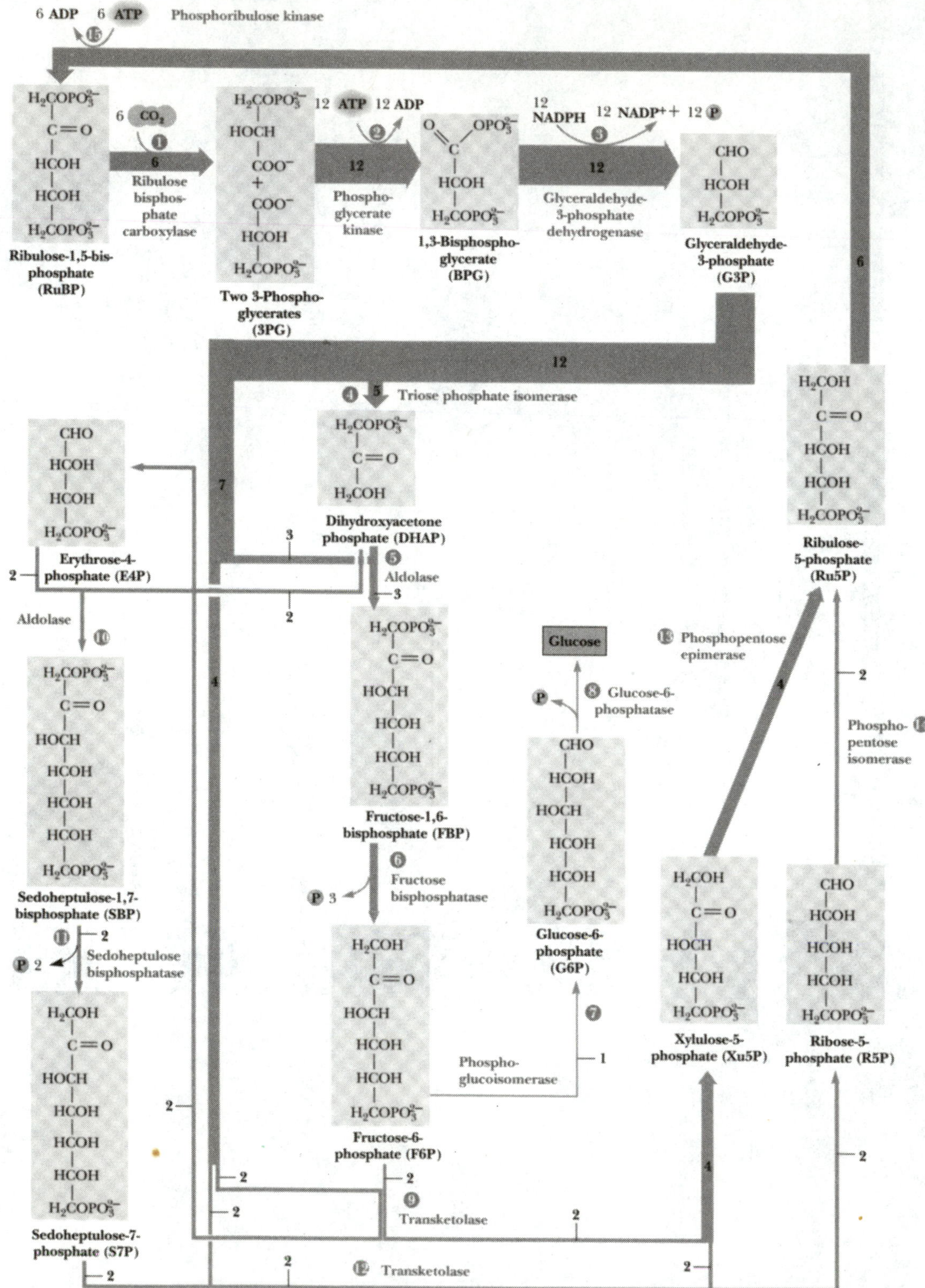

Figure 18.21 The Calvin-Benson cycle of reactions. The number of arrows at each step indicates the number of molecules reacting in a turn of the cycle that produces one molecule of glucose. Reactions are numbered as in Table 18.2 in the Garrett and Grisham text.

Problems and Solutions

1. In photosystem I, P700 in its ground state has an $\mathscr{E}_o'$ = +0.4 V. Excitation of P700 by a photon of 700-nm light alters the $\mathscr{E}_o'$ of P700* to -0.6 V. What is the efficiency of energy capture in this light reaction of P700?

Answer: The energy of a photon of light is given by:

$$E = \frac{hc}{\lambda},$$

$$h = \text{Planck's constant} = 6.626 \times 10^{-34}\,\text{J}\cdot\text{sec}$$

$$c = \text{speed of light in a vacuum} = 3 \times 10^{8}\,\frac{\text{m}}{\text{sec}}$$

λ = wavelength of light. In this case $\lambda = 700\text{ nm} = 700 \times 10^{-9}\text{m}$

$$\text{Thus, } E = \frac{(6.626 \times 10^{-34}\,\text{J}\cdot\text{sec}) \times (3 \times 10^{8}\,\frac{\text{m}}{\text{sec}})}{700 \times 10^{-9}\text{m}} = 2.84 \times 10^{-19}\,\text{J}$$

And, a mole of photons represents

$$(2.84 \times 10^{-19}\,\text{J}) \times 6.02 \times 10^{23} = 171\text{ kJ of energy}$$

Absorption of this amount of energy changes the redox potential of photosystem I from + 0.4 V to – 0.6 V. We can think of this as an oxidation-reduction reaction in which an electron is transferred from a donor with $\mathscr{E}_o'$ = + 0.4 V to an acceptor with $\mathscr{E}_o'$ = – 0.6 V. The $\Delta G^{\circ\prime}$ for this reaction can be calculated using:

$\Delta G^{\circ\prime} = -n\mathscr{F}\Delta\mathscr{E}_o'$, where n is the number of electrons, and

$$\mathscr{F} = \text{Faraday's constant} = 96.485\,\frac{\text{kJ}}{\text{V}\cdot\text{mol}}$$

$$\Delta\mathscr{E}_o' = \mathscr{E}_o'(\text{acceptor}) - \mathscr{E}_o'(\text{donor})$$

$$\Delta\mathscr{E}_o' = -0.6 - (+0.4) = -1.0\text{ V}$$

$$\Delta G^{\circ\prime} = -1 \times 96.485\,\frac{\text{kJ}}{\text{V}\cdot\text{mol}} \times (-1.0\text{ V}) = 96.5\,\frac{\text{kJ}}{\text{mol}}$$

$$\text{The efficiency} = \frac{96.5}{171} \times 100\% = 56.4\%$$

2. What is the $\mathscr{E}_o'$ for the light-generated primary oxidant of photosystem II if the light-induced oxidation of water (which leads to O_2 evolution) proceeds with a $\Delta G^{\circ\prime}$ of – 25 kJ/mol?

Answer:

From $\Delta G^{\circ\prime} = -n\mathscr{F}\Delta\mathscr{E}_o'$, where n is the number of electrons, and

$$\mathscr{F} = \text{Faraday's constant} = 96.485\,\frac{\text{kJ}}{\text{V}\cdot\text{mol}}$$

$$\Delta\mathscr{E}_o' = \mathscr{E}_o'(\text{acceptor}) - \mathscr{E}_o'(\text{donor})$$

We see that:

$$\mathscr{E}_o'(\text{acceptor}) - \mathscr{E}_o'(\text{donor}) = \frac{-\Delta G^{\circ\prime}}{n\mathscr{F}}, \text{ or}$$

$$\mathscr{E}_o'(\text{acceptor}) = -\frac{\Delta G^{\circ\prime}}{n\mathscr{F}} + \mathscr{E}_o'(\text{donor})$$

$$\mathscr{E}_o'(\text{acceptor}) = \frac{25\,\frac{\text{kJ}}{\text{mol}}}{4 \times 96.485\,\frac{\text{kJ}}{\text{V}\cdot\text{mol}}} + 0.816$$

$$\mathscr{E}_o'(\text{acceptor}) = 0.88\text{ V}$$

Generation of O_2 is a four-electron transfer, so n = 4.

3. Assuming that the concentrations of ATP, ADP, and P_i in chloroplasts are 3 mM, 0.1 mM, and 10 mM, respectively, what is the ΔG for ATP synthesis under these conditions? Photosynthetic electron transport establishes the proton-motive force driving photophosphorylation. What redox potential difference is necessary to achieve ATP synthesis under the foregoing conditions, assuming an electron pair is transferred per molecule of ATP generated?

Answer:

$$\Delta G = \Delta G^{\circ\prime} + RT \times \ln\frac{[ATP]}{[ADP][P_i]}$$

At T = 25°C (298 K),

$$\Delta G = 30.5\frac{kJ}{mol} + (8.314\times10^{-3}\frac{kJ}{mol\cdot K})(298K)\times\ln\frac{3\times10^{-3}}{(0.1\times10^{-3})(10\times10^{-3})}$$

$$\Delta G = 30.5\frac{kJ}{mol} + 19.8\frac{kJ}{mol}$$

$$\Delta G = 50.3\frac{kJ}{mol}$$

From $\Delta G = -n\mathcal{F}\Delta\mathcal{E}$, where n is the number of electrons, and

$$F = \text{Faraday's constant} = 96.485\frac{kJ}{V\cdot mol}$$

$$\Delta\mathcal{E} = -\frac{\Delta G}{n\mathcal{F}} = -\frac{50.3\frac{kJ}{mol}}{2\times96.485\frac{kJ}{V\cdot mol}}$$

$$\Delta\mathcal{E} = -0.26\text{ V}$$

4. Write a balanced equation for the synthesis of a glucose molecule from ribulose-1,5-bisphosphate and CO_2 that involves the first three reactions of the Calvin cycle and subsequent conversion of the two glyceraldehyde-3-P molecules into glucose by a reversal of glycolysis.

Answer:

CO_2 + H_2O + RuBP → ~~2 3-PG~~
~~2 3-PG~~ + ~~2 ATP~~ → ~~2 1,3-BPG~~ + ~~2 ADP~~
~~2 1,3-BPG~~ + **2 NADPH** → **2 NADP⁺ + 2 Pi** + ~~2 G-3-P~~
~~G-3-P~~ → ~~DHAP~~
~~G-3-P~~ + ~~DHAP~~ → ~~F-1,6-BP~~
~~F-1,6-BP~~ + ~~ADP~~ + **H⁺** → ~~F-6-P~~ + ~~ATP~~
~~F-6-P~~ → ~~G-6-P~~
~~G-6-P~~ + ~~ADP~~ + **H⁺** → **glucose** + ~~ATP~~

Net: **CO_2 + H_2O + 2 NADPH + 2 H^+ + RuBP → glucose + 2 $NADP^+$ + 2 P_i**

5. If noncyclic photosynthetic electron transport leads to the translocation of 3 H^+/e^- and cyclic photosynthetic electron transport leads to the translocation of 2 H^+/e^-, what is the relative photosynthetic efficiency of ATP synthesis (expressed as the number of photons absorbed per ATP synthesized) for noncyclic versus cyclic photophosphorylation? (Assume that the CF_1CF_0ATP synthase yields 1 ATP/3H^+).

Answer: In noncyclic photosynthetic electron transport, two photosystems operate in series to affect the transport of each electron from water to $NADP^+$. Thus two photons (commonly symbolized as *hν*, for the energy content of the photon) are required per electron transported. Accordingly, we have

$$\frac{2h\nu}{e^-}\times\frac{1e^-}{3H^+}\times\frac{3H^+}{ATP} = 2\ h\nu \text{ per ATP synthesized.}$$

In cyclic electron transport, only one photon is required to excite photosystem I. So,

$$\frac{1hv}{e^-} \times \frac{1e^-}{3H^+} \times \frac{3H^+}{ATP} = 1.5 \text{ hv per ATP synthesized.}$$

Thus, ATP synthesis by noncyclic electron transport is only 75% as efficient as ATP synthesis by cyclic electron transport.

It is noteworthy that this analysis ignores the fact that, unlike cyclic electron transport, noncyclic electron flow leads to reduction of $NADP^+$, which in itself represents a considerable capture of energy.

6. The overall equation for photosynthetic CO_2 fixation is

$$6\ CO_2 + 6\ H_2O \rightarrow C_6H_{12}O_6 + 6\ O_2$$

All the O atoms evolved as O_2 come from water; none comes from carbon dioxide. But 12 O atoms are evolved as 6 O_2, and only 6 O atoms as 6 H_2O in the equation. Also, 6 CO_2 have 12 O atoms, yet there are only 6 O atoms in $C_6H_{12}O_6$. How can you account for these discrepancies? (Hint: Consider the partial reactions of photosynthesis: ATP synthesis, NADP reduction, photolysis of water, and the overall reaction for hexose synthesis in the Calvin-Benson cycle.)

Answer: The net reaction for the Calvin Cycle given in Table 18.2 is:

$$6\ CO_2 + 18\ ATP + 12\ NADPH + 12\ H^+ + 12\ H_2O \rightarrow \text{glucose} + 18\ ADP + 18\ P_i + 12\ NADP^+$$

But, ATP was produced by phosphorylation of ADP as follows:

$$18\ ADP + 18\ P_i \rightarrow 18\ ATP + 18\ H_2O$$

NADPH was generated by

$$12\ H_2O + 12\ NADP^+ \rightarrow 12\ NADPH + 12\ H^+ + 6\ O_2$$

The net reaction is: $6\ CO_2 + 6\ H_2O \rightarrow \text{glucose} + 6\ O_2$.

What the overall reaction doesn't tell us is the source of the O_2. To generate 6 O_2, 12 H_2O's are split by the light reactions of photosynthesis. However, ATP used in the Calvin cycle accounts for production of 18 H_2O's.

Questions for Self Study

1. Match an item in the first column with an item in the second.

a. Plastid	1. Highly permeable
b. Thylakoid vesicle	2. Stack of thylakoid vesicles
c. Granum	3. Chloroplast
d. Inner membrane	4. Flattened sac
e. Outer membrane	5. Connects grana
f. Stroma lamellae	6. Highly impermeable

2. What are the four possible fates of the energy of a quantum of light after it is absorbed by photosynthetic pigments?

3. Fill in the blanks. Eukaryotic photosystems use light energy to move electrons from _____ to _____. During this process _____ is produced as a by-product in a reaction known as _____. Also, protons are moved from the _____ of the chloroplast into the _____ . This movement produces a proton gradient that is subsequently used to phosphorylate ADP to produce ATP in a process known as _____. The electrons follow a path known as the _____, which requires ____ quanta of light per electron to function. Light-dependent electron flow with the production of ATP can occur without by-product formation in a process known as _____ _____. This process involves only one of the photosystems, _____.

4. For the reactions, a and b, shown below, identify the reactants and products and name the enzymatic activities.

a. 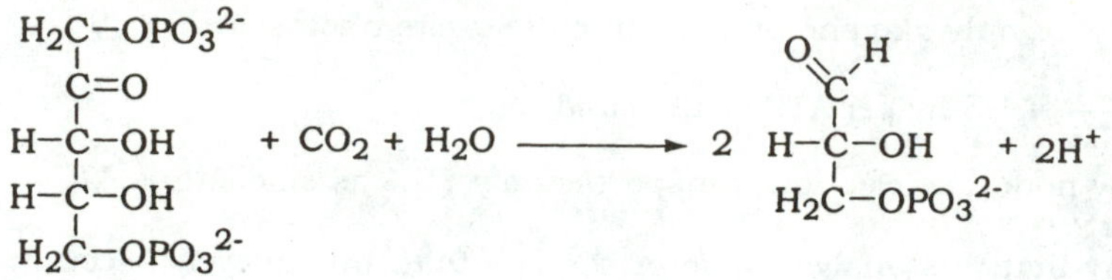
$$\begin{array}{c} H_2C{-}OPO_3^{2-} \\ | \\ C{=}O \\ | \\ H{-}C{-}OH \\ | \\ H{-}C{-}OH \\ | \\ H_2C{-}OPO_3^{2-} \end{array} + CO_2 + H_2O \longrightarrow 2\ \begin{array}{c} O\diagdown\ \diagup H \\ C \\ | \\ H{-}C{-}OH \\ | \\ H_2C{-}OPO_3^{2-} \end{array} + 2H^+$$

b. 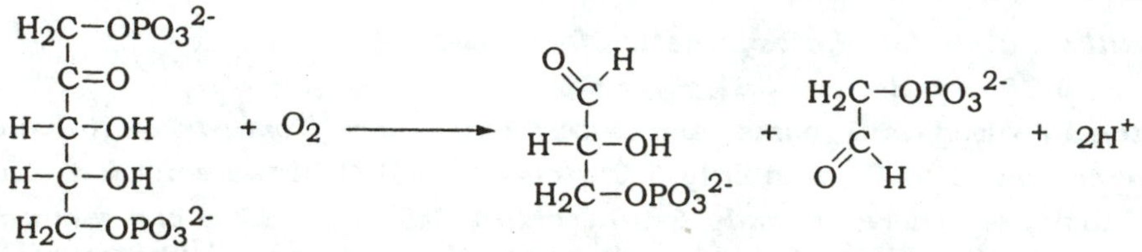
$$\begin{array}{c} H_2C{-}OPO_3^{2-} \\ | \\ C{=}O \\ | \\ H{-}C{-}OH \\ | \\ H{-}C{-}OH \\ | \\ H_2C{-}OPO_3^{2-} \end{array} + O_2 \longrightarrow \begin{array}{c} O\diagdown\ \diagup H \\ C \\ | \\ H{-}C{-}OH \\ | \\ H_2C{-}OPO_3^{2-} \end{array} + \begin{array}{c} H_2C{-}OPO_3^{2-} \\ | \\ C \\ O\diagup\ \diagdown H \end{array} + 2H^+$$

5. The Calvin cycle can be thought of as consisting of two phases, a reductive phase and a nonreductive phase. What are the inputs and outputs of each phase?

6. What is the role of malate production from PEP in C-4 plants ? In CAM plants?

Answers

1. a.3; b.4; c.2.; d.6; e.1; f.5.

2. Lost as heat, lost as light (fluorescence), resonance energy transfer, energy transduction.

3. Water; $NADP^+$; oxygen; photolysis; stroma; thylakoid lumen; photophosphorylation; Z scheme; 2; cyclic photophosphorylation; photosystem I (or P700).

4. a. Ribulose-1,5-bisphosphate, carbon dioxide and water react to form two molecules of 3-phosphoglycerate. The activity is the carboxylase activity of rubisco.
b. Ribulose-1,5-bisphosphate and oxygen combine to produce 3-phosphoglycerate and phosphoglycolate. The activity is oxygenase activity of rubisco.

5. In the reductive phase, ribulose-1,5-bisphosphate, carbon dioxide, water, ATP, and reducing equivalents in the form of NADPH are used to produce glyceraldehyde-3-phosphate. In the nonreductive phase, triose phosphates are used to produce glucose and to regenerate ribulose-1,5-bisphosphate consumed in the first phase.

6. In C-4 plants, malate serves as a carbon dioxide delivery system from mesophyll cells to bundle sheath cells where the Calvin cycle occurs. This allows carbon dioxide fixation to be carried out in a relatively oxygen-poor environment lessening the possibility of photorespiration. In CAM plants, malate production allows the plants to accumulate carbon dioxide at night when loss of water is minimized.

Additional Problems

1. Chlorophyll can be extracted from plant tissue with acetone. The acetone extract is a dark green solution. If an intense beam of visible light is directed at the solution, the solution will fluoresce with a reddish hue. However, this fluorescence is not observed for isolated chloroplasts even though the chlorophyll in chloroplasts is dissolved in a hydrophobic environment. Why does chlorophyll fluoresce in acetone but not in chloroplasts?

2. A researcher decided to demonstrate photosynthesis using isolated chloroplasts from spinach. The plan was to use an oxygen electrode to measure oxygen production as a function of light intensity. Chloroplasts were isolated by disrupting spinach leaves in a buffered solution, briefly centrifuging the sample to remove cell walls and other debris, and then centrifuging to pellet chloroplasts. The chloroplasts are resuspended in fresh buffer and repelleted by centrifugation.

Finally, the chloroplasts were resuspended in a small volume of buffer to make a concentrated solution of chloroplast fragments.

With great fanfare, the researcher loaded the chloroplasts into an oxygen electrode, illuminated the electrode with an intense source of light, and, before a packed audience of 250 excited undergraduates, found that the preparation was completely inactive. Can you offer a simple explanation?

3. In the scenario described in question 2 above, the researcher went back to the laboratory, got the photosynthesis experiment working perfectly, and scheduled another demonstration. Then, just before show time disaster struck, the oxygen electrode broke. Can you suggest an alternate method of measuring photosynthesis?

4. What do the so-called dark reactions of photosynthesis accomplish?

5. The Hatch-Slack pathway is involved in carbon dioxide fixation in C-4 plants. Describe this pathway (i.e. that portion of the pathway unique to C-4 plants) and identify a similar set of enzymes in animal cells that would accomplish this same process.

6. During electron transport in chloroplasts, hydrogen ions are pumped from one side of a membrane to another. Verbally or in a diagram describe where protons are concentrated in chloroplasts. How is this topologically analogous to proton pumping in mitochondria?

7. The metal ion in chlorophyll is magnesium. In hemoglobin and myoglobin and in the cytochromes, the metal ion is iron. In one sense the magnesium in chlorophyll behaves much like the iron in both hemoglobin and myoglobin, yet chlorophyll itself functions like the cytochrome porphyrins. Explain.

8. The membrane-bond ferredoxins found in the electron transport chain of photosystem I can pass electrons to either soluble ferredoxin or to cytochrome b_6/f complex. What are the outcomes of each path?

9. What experimental evidence supports the hypothesis that pigments are organized into photosynthetic units?

Abbreviated Answers

1. Fluorescence occurs when an electron is excited to a higher energy state, makes a transition to a slightly lower energy state, and then falls back to the unexcited ground state with loss of energy as a photon of light. Isolated chlorophyll fluoresces because the excited state has a long half life, making the transition to the slightly lower energy level more probable. In chloroplasts, the excited state is effectively quenched by resonance energy transfer so no florescence is observed.

2. For oxygen production to occur a terminal electron acceptor is necessary. Normally, $NADP^+$ serves this role and this small molecule cofactor was probably lost during chloroplast isolation.

3. One possibility is to use a pH meter to measure light-induced proton gradient formation. Oxygen production is accompanied by movement of protons into the thylakoid lumen. In a weakly buffered solution, this will cause a drop in pH that can be measured with a pH meter.

4. The dark reactions constitute the Calvin cycle in which CO_2 is fixed to ribulose-1,5-bisphosphate to produce 2 molecules of 3-phosphoglycerate, which are subsequently reduced by NADPH-dependent glyceraldehyde-3-phosphate dehydrogenase to glyceraldehyde-3-phosphate. Glyceraldehyde-3-phosphate is used to synthesize glucose and to resynthesize ribulose-1,5-bisphosphate.

5. In mesophyll cells, the enzyme pyruvate-phosphate dikinase catalyzes the conversion of pyruvate to PEP. PEP carboxylase produces oxaloacetate using PEP and bicarbonate (derived from CO_2) which is converted to either malate by an NADPH-specific malate dehydrogenase or to aspartate by transamination, and these four-carbon compounds are exported to bundle-sheath

cells. In the bundle-sheath cells, aspartate is converted to oxaloacetate by transamination and oxaloacetate is converted to pyruvate and CO_2 by a $NADP^+$-dependent oxidation catalyzed by malic enzyme. The CO_2 is subsequently used in the Calvin cycle whereas the pyruvate is transported back to the mesophyll cells for recycling.

In animal cells, the gluconeogenic enzyme pyruvate carboxylase produces oxaloacetate from pyruvate and CO_2 in the mitochondria. Oxaloacetate is converted to either malate or aspartate and exported to the cytosol. In the cytosol, malate and aspartate are converted back to oxaloacetate; however, PEP carboxykinase degrades oxaloacetate to PEP and CO_2.

6. During electron transport in chloroplasts, protons are moved from the stroma into the thylakoid lumen. The lumen pH decreases and the stromal pH increases. In mitochondria, protons are pumped from the matrix to the intramembrane space, which is connected to the cytosol. Thus, the cytosolic pH decreases and the mitochondrial matrix pH increases. It appears that the topology is quite different: the stroma of the chloroplast and the matrix of the mitochondria increase in pH upon proton pumping, and the intramembrane space of mitochondria and the thylakoid lumen of chloroplast decrease in pH. However, the biosynthesis of thylakoids is thought to occur by an invagination of the chloroplast outer membrane. Thus, the thylakoid lumen is topologically equivalent to the cytosol.

7. The irons in hemoglobin and myoglobin do not change oxidation state during normal physiological function and remain in the ferrous (2^+) oxidation state. The magnesium ion in chlorophyll behaves in a similar manner in that it does not change its oxidation state yet chlorophyll itself in oxidized and reduced during photosynthesis. The cytochromes are similarly oxidized and reduced during electron transport. However, the iron atom undergoes a change in oxidation state during the process.

8. Reduction of soluble ferredoxin leads to reduction of $NADP^+$, whereas reduction of cytochrome *bf* complex leads to cyclic flow of electrons, proton-gradient formation, and ATP synthesis.

9. By measuring O_2 evolved per flash of light at varying intensities of light flashes using the alga *Chlorella* the following data were obtained.

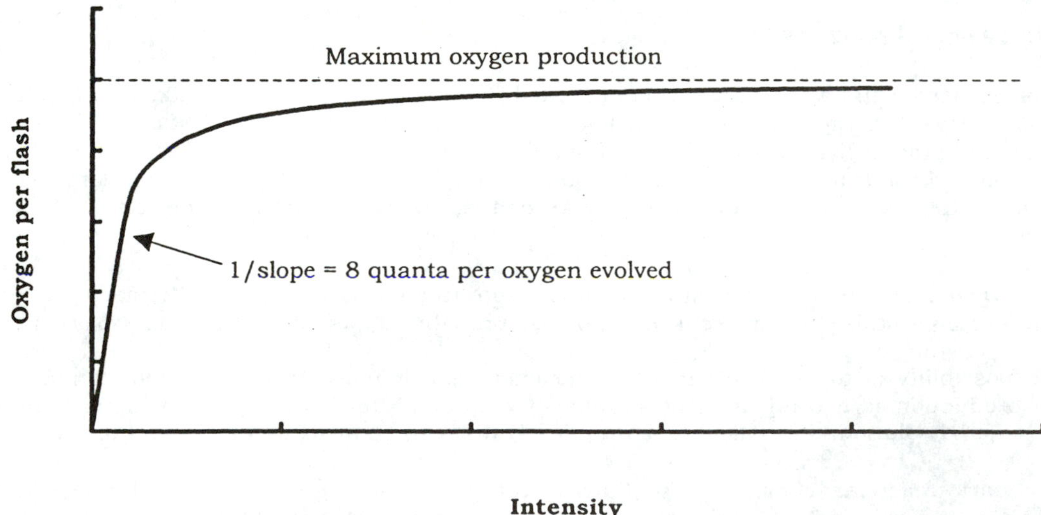

At saturating intensities of light, the maximum oxygen production corresponds to one oxygen evolved per 2400 chlorophylls in this organism. Thus, the chlorophylls must be acting in units of about this size.

Summary

Photosynthesis is the transformation of light energy into the chemical energy necessary to sustain life. Traditionally, synthesis of carbohydrate ($C_6H_{12}O_6$) has been viewed as the end product of the photosynthetic process, but, in reality, the primary products of photosynthesis are ATP and NADPH, the major energy sources for biosynthesis. These agents, as suppliers of phosphorylation potential and reducing potential, respectively, power many cellular processes, and therefore aptly reflect the role of photosynthesis as the ultimate and essentially universal source of biological energy.

The photosynthetic reactions are intimately associated with membranes, and in photosynthetic eukaryotes, a specialized organelle, the chloroplast, is the site of the photosynthetic process. Photosynthesis can be divided into the "light" reactions (the photochemical generation of energy) and the "dark" reactions (the utilization of chemical energy to drive biosynthesis of carbohydrate in the dark). These different reactions are localized in different chloroplast compartments: the light-dependent processes occur in membrane vesicles (the thylakoids), whereas the dark reactions occur in the stroma (the cytosol-like compartment of the chloroplast).

The transformation of light energy into chemical energy is a manifestation of the photoreactivity of chlorophyll. This π-electron-rich macrocycle readily absorbs photons (quanta of light energy), which promote π-electrons to higher energy orbitals, thereby rendering the Chl molecule a much more effective electron donor. Thus, excited-state chlorophyll (Chl*) initiates a series of electron transfer (oxidation-reduction) reactions whereby light energy is ultimately transduced into reducing power (NADPH) and phosphorylation potential (ATP).

The chlorophyll of photosynthetic organisms is organized into photosynthetic units in order to more efficiently harvest the incident light energy. Within such units, an array of several hundred Chl molecules serve as an antenna to collect photons and funnel quanta of energy to specialized Chl molecules, reaction centers, where the photochemical event occurs. All Chl is in specific Chl-protein complexes which reflect the roles of Chl in the photosynthetic process: a light harvesting complex (LHC), a photosystem I (PSI) complex functioning in $NADP^+$ reduction, and a photosystem II (PSII) complex involved in oxygen evolution. These three complexes are integral components of the thylakoid membrane. PSI and PSII contain unique reaction center chlorophylls, designated P700 in PSI and P680 in PSII.

PSII and PSI act in concert to mediate the light-driven transfer of electrons from H_2O to reduce $NADP^+$. The photoexcitation of P680 in PSII leads to the generation of a weak reductant, QH_2 (a reduced form of plastoquinone), and a strong oxidant, $P680^+$, which, via accepting electrons from water, leads to oxygen evolution. The reductant, QH_2, is coupled to PSI by an integral membrane complex containing Fe/S centers and cytochromes b_6 and f. Electrons transferred from Q via the cytochrome f complex serve to re-reduce the PSI reaction center Chl, $P700^+$. P700 upon photoexcitation generates a strong reductant, P700*, capable of reducing Fe/S centers whose $\mathcal{E}_o'$ are more negative than -0.6 V. The $P700^+$ thus formed has its electron hole filled by an electron from plastocyanin, the blue copper protein serving as e^- acceptor to the cytochrome b_6/f complex. The Fe/S centers on the reducing side of P700* provide electrons from water to reduce $NADP^+$. Their arrangement according to the redox potential scale and their organization into the unique PSI, PSII complexes bridged by the cytochrome b_6/f complex gives rise to a Z-shaped pathway, the so-called "Z" scheme of photosynthesis.

The primary events in photosynthesis are light-induced electron transfers which of necessity result in a situation where oxidized donor and reduced acceptor, for example, $P700^+$ and Chl a^-, are in close proximity. These neighboring, oppositely charged species invite charge recombination and consequent dissipation of the photochemical energy. To prevent this, the electron transfer reactions following photo-excitation of reactive centers must be very rapid so that the energized electron is quickly conducted away from the oxidized form of its original donor.

ATP produced by photosynthesis arises because photosynthetic electron transfer is accompanied by transmembrane proton translocations, which establish a proton-motive force across the membrane. This electrochemical H^+ gradient is tapped by the chloroplast ATP synthase, CF_1CF_0ATP synthase. Quantitative measurements indicate that 1 H^+ is translocated per photon entering the photosynthetic system and that the translocation of 3 H^+ generates the necessary electrochemical gradient to drive the synthesis of 1 ATP under conditions of noncyclic photophosphorylation. The other photophosphorylation pattern, cyclic photophosphorylation, involves a cyclic electron transfer reaction wherein light energy incident at P700 drives an electron through a series of acceptors back to refill $P700^+$. As a consequence of the e^- transfers, proton translocations ensue that can drive ATP formation.

Carbon dioxide fixation, the synthesis of carbohydrate from CO_2, while not uniquely a photosynthetic process, is nevertheless the major investment of the chemical energy derived from photosynthesis. Carbon dioxide is fixed in organic linkage through reaction with ribulose-1,5-bisphosphate to generate two molecules of 3-phosphoglycerate. The enzyme catalyzing this reaction is ribulose bisphosphate carboxylase or rubisco. Rubisco also has the capacity to use O_2 in place of CO_2 in the condensation with ribulose-1,5-bisphosphate, a reaction leading to a wasteful loss of this pentose. As an oxygen-consuming, CO_2-evolving reaction, this process is aptly termed photo-respiration and is viewed as a wasteful dissipation of RuBP, a crucial metabolic intermediate. The remaining reactions of Calvin cycle CO_2 fixation are similar to reactions in the pentose phosphate pathway and glycolysis. The aim of the Calvin cycle is to account for the synthesis of one hexose molecule ($C_6H_{12}O_6$) from 6 equivalents of carbon dioxide (6 CO_2). Key Calvin cycle enzymes are regulated by light-induced effects on chloroplast reducing power, pH and relative Mg^{2+} concentrations so that CO_2 fixation occurs only when light energy is available. An alternate means of CO_2 capture occurs in so-called C4 plants. These plants form C4 compounds (instead of C3-phosphoglycerate) as the initial products of CO_2 uptake. The C4 pathway is not an alternative process to the Calvin cycle but instead is an effective means for the transport and concentration of CO_2 within cells having high potential for Calvin cycle carbon dioxide fixation. C4 plants achieve greater photosynthetic efficiency by diminishing photorespiration. Typical C4 plants include tropical grasses such as sugar cane and maize.

Chapter 19

Gluconeogenesis, Glycogen Metabolism, and the Pentose Phosphate Pathway

• •

Chapter Outline

- Gluconeogenesis
  - Substrates: Pyruvate, lactate, glycerol, amino acids (except Lys and Leu)
  - Active in liver and kidney: Supplies glucose to brain and muscle
  - Seven of ten glycolytic steps in reverse
  - Three steps not reversal of glycolysis for two reasons
    - Energetics
    - Regulation
- Three unique gluconeogenesis steps
  - Pyruvate to PEP
    - Pyruvate carboxylase converts pyruvate to oxaloacetate
      - Biotin-dependent enzyme
      - ATP + bicarbonate to carbonylphosphate intermediate to carboxybiotin intermediate
      - Allosteric activation by acetyl-CoA
      - Localized in mitochondria
    - Oxaloacetate to malate in mitochondria and malate to oxaloacetate in cytosol
    - PEP carboxylase converts oxaloacetate to PEP
      - Decarboxylation drives reaction
      - GTP or ATP supplies P and drives reaction
  - Fructose-1,6-bisphosphate to fructose-6-phosphate
    - Hydrolysis
    - Fructose-1,6-bisphosphatase
      - Citrate stimulates
      - Fructose-2,6-bisphosphate inhibits
      - AMP inhibits
  - Glucose-6-phosphate to glucose
    - Hydrolysis
    - Glucose-6-phosphatase
      - Releases product into ER lumen
      - Enzyme expressed in liver and kidney (not in muscle or brain)
      - High Km for G6P: Substrate-level control
- Reciprocal regulation
  - Glucose to glucose-6-phosphate and back
    - Glucokinase high Km for glucose and not inhibited by G6P
    - Glucose-6-phosphatase high Km for G6P

- Fructose-6-phosphate to fructose-1,6-bisphosphate and back
  - Phosphofructokinase
    - AMP stimulates
    - Fructose-2,6-bisphosphate stimulates
    - Citrate inhibits
  - Fructose-1,6-bisphosphatase
    - AMP inhibits
    - Fructose-2,6-bisphosphate inhibits
    - Citrate stimulates
  - Fructose-2,6-bisphosphatase
    - Allosteric regulator of gluconeogenesis
    - Fructose-1,6-bisphosphate inhibits

- Dietary glycogen (and starch) catabolism
  - α-Amylase (saliva and pancreas) endoglycosidase
    - Hydrolyzes α(1→4) amylopectin and glycogen at random
    - Products: Maltose, maltotriose, oligosaccharides
  - Debranching enzyme: Two activities
    - Oligo (α1,4 → α1,4) glucantransferase: Moves three residues to new end
    - α(1 → 6) Glucosidase: Hydrolyzes branch
- Cellular glycogen
  - Glycogen phosphorylase is principal enzyme of glycogen catabolism
- Glycogen synthesis
  - Base: Glycogenin with glucose in acetal linkage to Tyr-OH
  - Glycogen synthase: α(1 → 4) linkage of glucose from UDP-glucose
  - UDP-glucose: UDP-glucose pyrophosphorylase: UTP + G-1-P to UDP-glucose (+ PP → 2P)
- Glycogen synthase regulation
  - D-form: Less active: Dependent on G6P: Phosphorylated by protein kinase
  - I-form: Active (independent of G6P) : Not phosphorylated (phosphate removed by phosphoprotein phosphatase)
- Hormonal regulation
  - Insulin: Released from β–cells islets of Langerhans of pancreas in response to high blood glucose
    - Increases uptake of glucose
    - Increases synthesis of glycolytic enzymes
    - Inhibits gluconeogenesis
  - Glucagon: α–cells islets of Langerhans of pancreas in response to low blood glucose
    - Stimulates liver to release glucose
    - Involved in long term glucose maintenance
  - Epinephrine (adrenaline) released by adrenal glands
    - Active on liver and muscle
    - Stimulates glycogen breakdown and glycolysis
    - Inhibits gluconeogenesis
- Pentose phosphate pathway (hexose monophosphate shunt, phosphogluconate pathway)
  - Supplies NADPH or ribose
  - Two phases
    - Oxidative phase: Glucose-6-P to ribulose-5-P
    - Nonoxidative phase: Ribulose-5-P to 3-, 4-, 5-, 6-, and 7-carbons ketoses and aldoses
- Glucose to ribulose-5-P: Oxidative
  - Glucose-6-phosphate dehydrogenase
    - NADPH produced
    - Gluconolactone produced
    - Enzyme inhibitors
      - NADPH
      - Intermediates of fatty acid biosynthesis
  - Gluconolactonase: Gluconolactone to 6-phosphogluconate
  - 6-Phosphogluconate dehydrogenase

- NADPH produced
- Ribulose-5-P and carbon dioxide produced

❖ Ribulose-5-P to various sugars: Nonreductive
- ➢ Phosphopentose isomerase: Ketose to aldose conversion
- ➢ Phosphopentose epimerase: Sugar interconversion
- ➢ Transketolase: 2-Carbon unit from keto donor to aldehyde acceptor
- ➢ Transaldolase: Ketose and aldose formation

Chapter Objectives

Gluconeogenesis

Although gluconeogenesis is not simply the reverse of glycolysis, the two pathways share many steps and so it would help to review glycolysis. Briefly, glucose is phosphorylated by hexokinase to glucose-6-phosphate, which is isomerized to fructose-6-phosphate. A second kinase reaction, catalyzed by phosphofructokinase, produces fructose-1,6-bisphosphate, which is cleaved by aldolase to produce dihydroxyacetone phosphate and glyceraldehyde-3-phosphate, two triose phosphates that are interconvertible. Glyceraldehyde-3-P is oxidized and phosphorylated to 1,3-bisphosphoglycerate, a high-energy compound used to support a substrate level phosphorylation of ADP leaving 3-phosphoglycerate, which is converted to 2-phosphoglycerate, dehydrated to phosphoenolpyruvate, and converted to pyruvate with accompanying ATP synthesis.

Gluconeogenesis starts with pyruvate and ends with glucose. Pyruvate can derive from lactate, amino acids, and citric acid cycle intermediates but not from acetyl-CoA units. Gluconeogenesis and glycolysis differ at three steps (see Figure 19.1), glucose to glucose-6-P, fructose-6-P to fructose-1,6-bisphosphate, and PEP to pyruvate. Understand how these three steps are bypassed and you will understand gluconeogenesis. The first two steps are catalyzed by phosphatases that hydrolyze phosphate from the substrates. The PEP to pyruvate step is a two-step process with pyruvate being converted to oxaloacetate by pyruvate carboxylase, a biotin-containing enzyme. Oxaloacetate is converted to PEP by PEP carboxykinase, which uses ATP or GTP hydrolysis, depending on the enzyme source, to drive the reaction.

Regulation

Glycolysis and gluconeogenesis are reciprocally regulated. You should remember the following regulatory pairs: glucose-6-phosphate inhibition of hexokinase and stimulation of glucose-6-phosphatase; fructose-2,6-bisphosphate stimulation of phosphofructokinase (PFK) and inhibition of fructose-1,6-bisphosphatase (FBPase); AMP stimulation of PFK and inhibition of FBPase; and, acetyl-CoA inhibition of pyruvate kinase and activation of pyruvate carboxylase. Additionally, ATP inhibits PFK and pyruvate kinase, citrate inhibits PFK, alanine inhibits pyruvate kinase, and cyclic-AMP dependent phosphorylation inhibits pyruvate kinase. Fructose-2,6-bisphosphate can be thought of as a signal to stimulate glycolysis.

Glycogen

Dietary glycogen and starch breakdown are catalyzed by α-amylase (β-amylase in plants), and debranching enzymes. Tissue glycogen is degraded by glycogen phosphorylase and synthesized by glycogen synthase. The latter enzyme is supplied with activated glucosyl residues in the form of UDP-glucose produced by UDP-glucose pyrophosphorylase. Understand the role of glycogenin, and branching enzyme in initiating glycogen synthesis and in introducing branches leading to multiple sites for rapid elongation or degradation of glycogen.

Regulation

Glycogen metabolism is carefully controlled by regulating glycogen phosphorylase and glycogen synthase. Glycogen phosphorylase is allosterically activated by AMP and inhibited by ATP and glucose-6-phosphate, whereas glycogen synthase is stimulated by glucose-6-phosphate. Both enzymes are subject to control by covalent modification. Glycogen phosphorylase exists in two forms, a less active *b* form and an active *a* form. Phosphorylase *a* is a phosphorylated form of phosphorylase *b*, catalyzed by a specific kinase, glycogen phosphorylase kinas, which is itself under covalent regulation by cAMP-dependent protein kinase. cAMP levels are regulated hormonally by hormone binding to surface receptors, leading to release of G-protein, and stimulation of adenylyl cyclase. Know this cascade. Glycogen synthase also exists in two forms that differ in both activity and in the state of phosphorylation. The less active phosphorylated glycogen synthase D depends on glucose-6-phosphate for activity, whereas the active, unphosphorylated, glycogen synthase I is independent of glucose-6-phosphate levels.

Pentose Phosphate Pathway

Admittedly, this is an impressive pathway but it may simplify matters to remember that it has an oxidative phase and a nonoxidative phase (see Figure 19.19). In the oxidative phase, glucose is converted to ribulose-5-phosphate and CO_2. Clearly, this must involve a phosphorylation (by hexokinase) and oxidations (by two NADP-dependent dehydrogenases). (With reflection, we might have anticipated that two oxidations are required. The CO_2 must derive from one of the alcoholic carbons of glucose via an aldehydic intermediate.)

The nonoxidative phase of the pentose phosphate pathway converts ribulose-5-phosphate into glucose-6-phosphate via 3-, 4-, 5-, 6- and 7- carbon sugar intermediates. Some of the key enzymes involved include the following: phosphopentose isomerase catalyzes a ketose to aldose conversion (We encountered similar biochemistry in glycolysis between glucose-6-P and fructose-6-P and between DHAP and glyceraldehyde-3-P.); phosphopentose epimerase (an epimerase changes the configuration about an asymmetric carbon; we encountered one in galactose catabolism.); transketolases (move two-carbon units between aldoses); and, a transaldolase (with a reaction mechanism similar to the aldolase encountered in glycolysis).

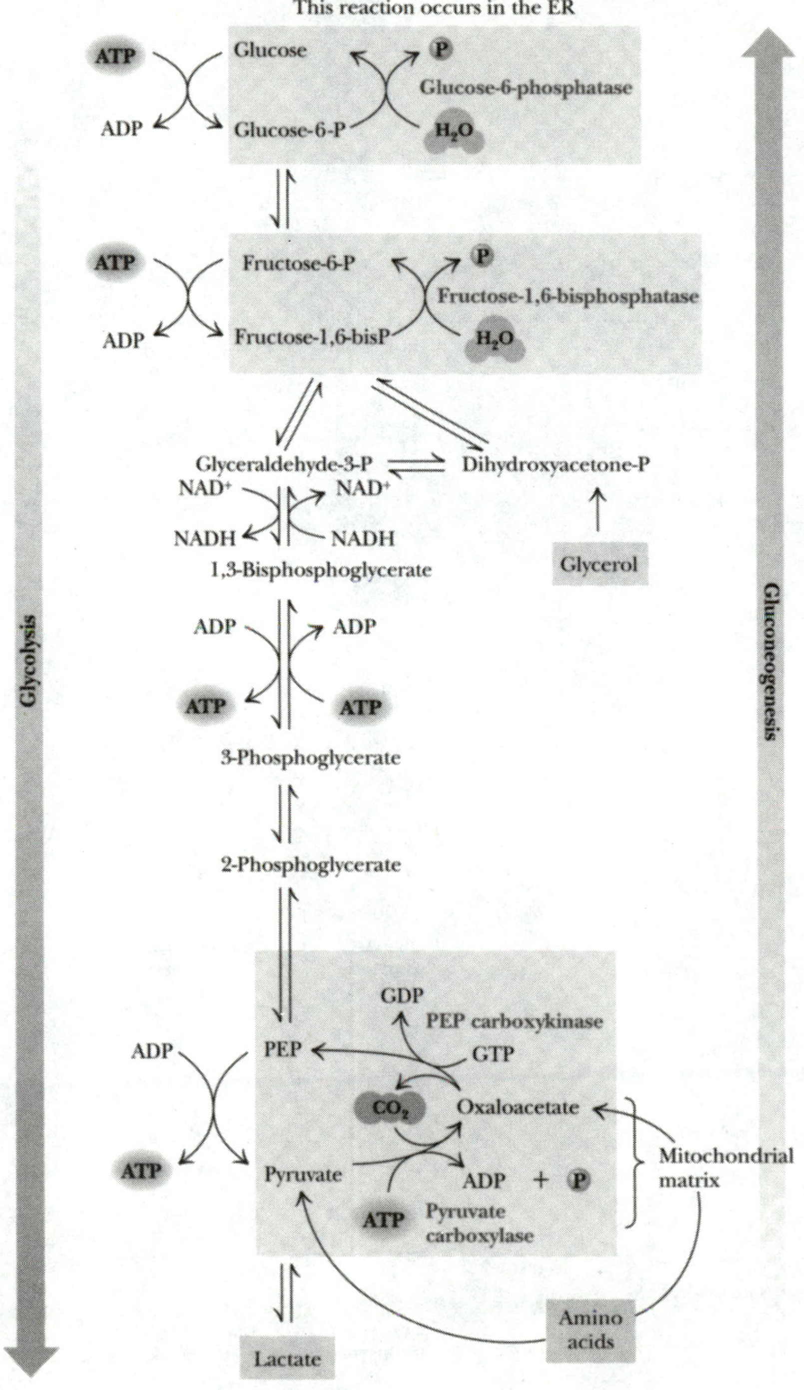

Figure 19.1 The pathways of gluconeogenesis and glycolysis. Species in the boxes indicate other entry points for gluconeogenesis (in addition to pyruvate).

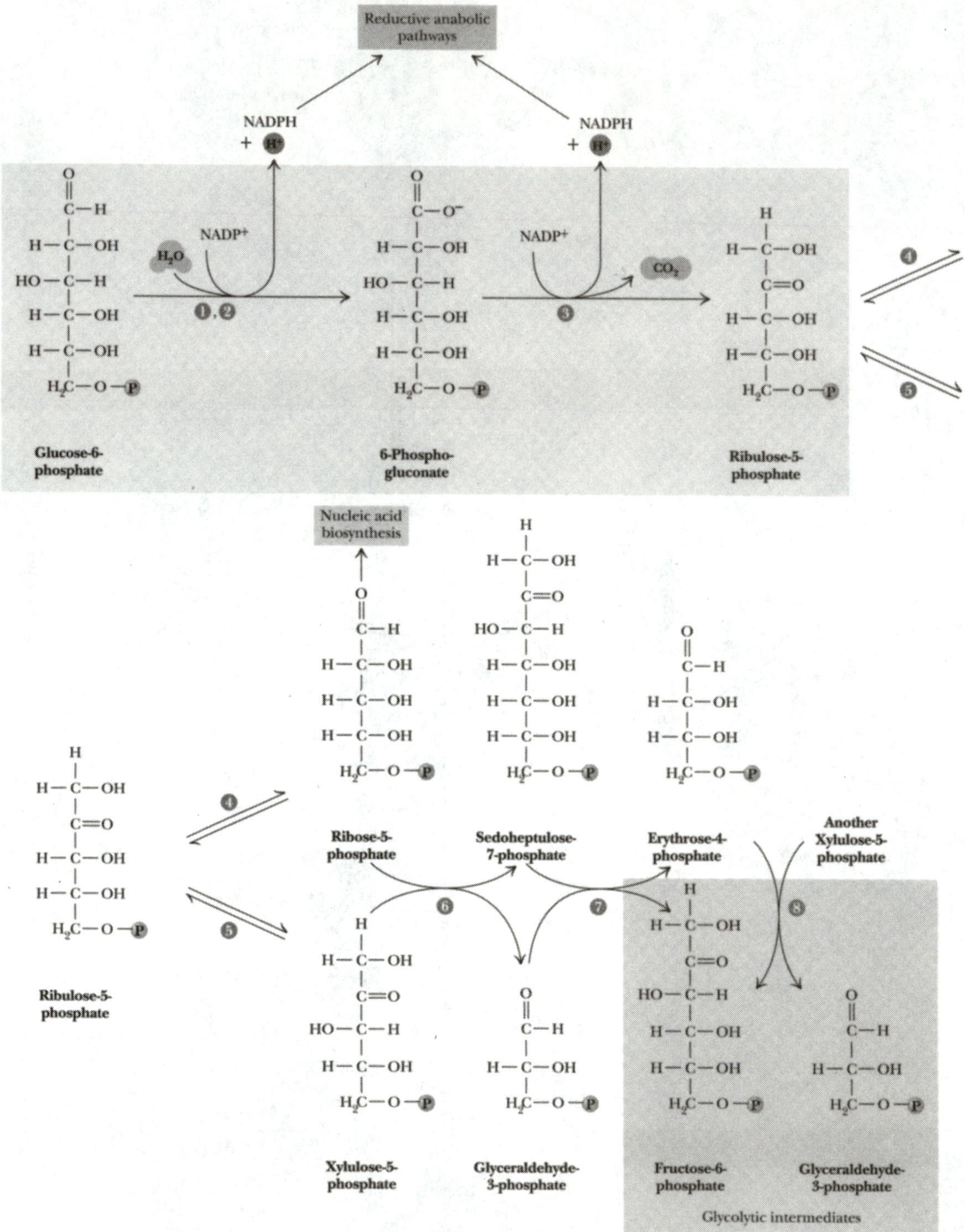

Figure 19.19 The pentose phosphate pathway. The numerals in the circles indicate steps discussed in the Garrett and Grisham text.

Problems and Solutions

1. Consider the balanced equation for gluconeogenesis in Section 19.1. Account for each of the components of this equation and the indicated stoichiometry.

Answer: The net reaction for the synthesis of glucose from pyruvate is:

$$2\ \text{Pyruvate} + 4\ \text{ATP} + 2\ \text{GTP} + 2\ \text{NADH} + 2\ \text{H}^+ + 2\ \text{H}_2\text{O} \rightarrow$$
$$\text{glucose} + 4\ \text{ADP} + 2\ \text{GDP} + 6\ \text{P}_i + 2\ \text{NAD}^+$$

The following reactions contribute to this equation:

$$2\ \text{Pyruvate} + 2\ \text{H}^+ + 2\ \text{H}_2\text{O} \rightarrow \text{glucose} + \text{O}_2$$
$$2\ \text{NADH} + 2\ \text{H}^+ + \text{O}_2 \rightarrow \text{NAD}^+ + \text{H}_2\text{O}$$
$$4\ \text{ATP} + 4\ \text{H}_2\text{O} \rightarrow 4\ \text{ADP} + 4\ \text{P}_i$$
$$2\ \text{GTP} + 2\ \text{H}_2\text{O} \rightarrow 2\ \text{GDP} + 2\ \text{P}_i$$

The six carbons of glucose derive from the 3-carbon compound pyruvate. In considering pyruvate, the keto carbon is oxidized relative to the average carbon of glucose and hence it must be reduced to form glucose. Reduction comes at the expense of NADH. (The carboxyl carbon is also oxidized; however, the methyl carbon is reduced relative to the average carbon of glucose. An internal oxidation-reduction during gluconeogenesis changes the oxidation state of these two carbons to match that of glucose carbons.)

The hydrolysis of high energy nucleoside triphosphates is accounted for as follows. In the first stage of gluconeogenesis, pyruvate is transformed to phosphoenolpyruvate in two steps. First, pyruvate is converted to oxaloacetate by pyruvate carboxylase, a biotin-containing enzyme. This reaction is driven by conversion of ATP to ADP and P_i. This is not a hydrolysis but rather a series of displacement reactions: ADP is displaced from ATP by bicarbonate; P_i is displaced from bicarbonate by biotin; biotin is displaced by pyruvate to produce oxaloacetate. Thus, hydrolysis does not occur and water is not consumed but two ATPs produce two ADPs and two P_is. Next, oxaloacetate is converted to PEP by PEP carboxykinase, a GTP-dependent enzyme, which accounts for two GTPs and two GDPs. The two remaining ATPs are consumed when 1,3-bisphosphoglycerate and ADP are produced from 3-phosphoglycerate and ATP by phosphoglycerate kinase.

We must still account for two waters, and four P_is. Reduction of two moles of 1,3-bisphosphoglycerate to two moles of glyceraldehyde-3-phosphate releases two P_i. Finally, two hydrolysis reactions catalyzed by fructose-1,6-bisphosphatase and glucose-6-phosphatase account for two H_2O consumed and two P_i produced.

Pyruvate derives from a number of sources including citric acid cycle intermediates, amino acids, and lactate. NADH may be supplied indirectly by the mitochondria. The inner mitochondrial membrane is impermeable to oxaloacetate so it is converted to malate which is transported to the cytosol. In the cytosol, malate dehydrogenase converts malate back to oxaloacetate with reduction of NAD^+ to NADH.

2. Calculate $\Delta G^{\circ\prime}$ and ΔG for gluconeogenesis in the erythrocyte using the data in Table 15.2 (assuming $NAD^+/NADH = 20$, [GTP] = [ATP], and [GDP] = [ADP]). See how closely your values match those in section 19.1.

Answer: To calculate $\Delta G^{\circ\prime}$ we can use the information presented in Table 15.1 for glycolytic intermediates in erythrocytes for all of the reactions of gluconeogenesis with the exception of the three unique steps. We can then use the concentrations given in Table 15.2 to calculate ΔG values for the same reactions. We will handle the three unique reactions below.

Conversion of PEP to pyruvate has a $\Delta G^{\circ\prime}$ of -31.7 kJ/mol. In addition to the ATP utilized (produced in glycolysis) an additional high energy phosphate, this time GTP, was hydrolyzed to drive the reaction. Assuming that GTP and ATP hydrolysis are equivalent, the overall $\Delta G^{\circ\prime}$ for the reaction is given by (+31.7 - 30.5)kJ/mol = **+1.2** kJ/mol.

$$\Delta G \text{ is calculated using :}$$

$$\Delta G = \Delta G'^{o} + RT \times \ln\frac{[PEP][ADP]^2[P_i]}{[pyruvate][ATP]^2}$$

$$\Delta G = +1.2\frac{kJ}{mol} + 8.314\times10^{-3}\times310\times\ln\frac{[(0.023\times10^{-3})(0.14\times10^{-3})^2(1\times10^{-3})}{(0.051\times10^{-3})(1.85\times10^{-3})^2}$$

$$\Delta G = -32.0\frac{kJ}{mol}$$

Note that the temperature used in these calculations is 37°C, the temperature at which eurythrocytes operate, and the temperature used for the ΔG calculations in Table 15.1.

For the conversion of fructose-1,6-bisphosphate to fructose-6-phosphate, $\Delta G°'$ = **-16.3** kJ/mol (given in Equation 15.6).

$$\Delta G \text{ is calculated using :}$$

$$\Delta G = \Delta G'^{o} + RT \times \ln\frac{[F6P][P_i]}{[F1,6BP]}$$

$$\Delta G = -16.3\frac{kJ}{mol} + 8.314\times10^{-3}\times310\times\ln\frac{[(0.014\times10^{-3})(1\times10^{-3})}{(0.031\times10^{-3})}$$

$$\Delta G = -36.2\frac{kJ}{mol}$$

For conversion of glucose-6-phosphate to glucose, using the value of $\Delta G°'$ = **-13.9** kJ/mol given in Table 3.3, we have:

$$\Delta G \text{ is calculated using :}$$

$$\Delta G = \Delta G'^{o} + RT \times \ln\frac{[glucose][P_i]}{[G6P]}$$

$$\Delta G = -13.9\frac{kJ}{mol} + 2.58\times\ln\frac{[(5.0\times10^{-3})(1\times10^{-3})}{(0.083\times10^{-3})}$$

$$\Delta G = -21.1\frac{kJ}{mol}$$

The overall ΔG is calculated by summing the ΔG values given in Table 19.1 (with signs reversed) for all the reactions with the exception of hexokinase, phosphofructokinase, and pyruvate kinase. These are replaced by the ΔG values calculated above. (The enolase, phosphoglycerate mutase and kinase, and glyceraldehyde phosphate dehydrogenase reactions are all multiplied by 2, as is the ΔG calculated for conversion of pyruvate to PEP.) Thus,

$$\Delta G_{gluconeogenesis} = -21.1 + 2.92 - 36.2 + 18.8 + 0.\,23 + 2(-2.41 + 1.29 - 0.1 - 0.83 - 32.0)$$

$$= -103.4\frac{kJ}{mol}$$

$$\Delta G°'_{gluconeogenesis} = -13.9 - 1.67 - 16.3 + 14.2 - 23.9 + 2(-7.56 - 6.3 + 18.9 - 4.4 + 1.2)$$

$$= -37.9\frac{kJ}{mol}$$

3. Use the data of Figure 19.11 to calculate the percent inhibition of fructose-1,6-bisphosphatase by 25 μM fructose-2,6-bisphosphate when fructose-1,6-bisphosphate is (a) 25 μM and (b) 100 μM.

Answer: From Figure 19.11 (a) we can estimate that at 25 μM fructose-1,6-bisphosphate the activity of fructose-1,6-bisphosphatase in the absence of fructose-2,6-bisphosphate is approximately 15 units. In the presence of 25 μM fructose-2,6-bisphosphate, the activity is only 1 unit. The percent inhibition is given by

$$\frac{15-1}{15}\times100\% = 93.3\%$$

At 100 μM fructose-1,6-bisphosphate the activities of fructose-1,6-bisphosphatase without and with fructose-2,6-bisphosphate are 14 and 8. The percent inhibition in this case is

$$\frac{14-8}{14}\times 100\% = 42.9\%$$

4. *Suggest an explanation for the exergonic nature of the glycogen synthase reaction (ΔG°' = -13.3 kJ/mol). Consult Chapter 3 to review the energetics of high-energy phosphate compounds if necessary.*

Answer: Glycogen synthase adds glucose units in α(1→4) linkage to glycogen using UDP-glucose as a substrate and releasing UDP. In comparing UDP-glucose to UDP, we expect very little difference in destabilization due to electrostatic repulsion and no significant entropy factors. The presence of a glucose moiety does interfere with phosphorous resonance states of UDP. The standard free energy of hydrolysis of UDP-glucose is -31.9 kJ/mol whereas the ΔG°' for glycogen synthase reaction is -13.3 kJ/mol.

5. *Using the values in Table 20.1 for body glycogen content and the data in part b of the illustration for HUMAN BIOCHEMISTRY - Carbohydrate Utilization in Exrcise, calculate the rate of energy consumption by muscles in heavy exercise (in J/sec). Use the data for fast-twitch muscle.*

Answer: From Table 20.1 we are informed that a 70-kg person has about 120 g (dry weight) of glycogen, equivalent to 1,920 kJ of available energy. From *HUMAN BIOCHEMISTRY* we see glycogen supplies are exhausted after approximately 60 minutes of heavy exercise. The rate of energy consumption is given by:

$$\frac{1{,}920\ \text{kJ}}{60\ \text{min}} = 32\frac{\text{kJ}}{\text{min}}$$

$$32\frac{\text{kJ}}{\text{min}} = 32\frac{\text{kJ}}{\text{min}} \times \frac{1000\ \text{J}}{\text{kJ}} \times \frac{1\,\text{min}}{60\,\text{sec}} = 533\frac{\text{J}}{\text{sec}}$$

6. *What would be the distribution of carbon from positions 1,3, and 6 of glucose after one pass through the pentose phosphate pathway if the primary need of the organism is for ribulose-5-phosphate and the oxidative steps are bypassed?*

Answer: The oxidative steps in the pentose phosphate pathway occur during conversion of glucose-6-phosphate to ribulose-5-phosphate at two steps. Since these oxidative steps are to be bypassed, the carbons of glucose must enter the pathway by another route. There are two possibilities, fructose-6-phosphate and glyceraldehyde-3-phosphate. C-1, C-3, and C-6 of glucose become C-1, C-3, and C-6 of fructose-6-P. C-1 and C-6 of glucose become C-1 of glyceraldehyde-3-P and C-3 of glucose becomes C-1 of glyceraldehyde. Using transketolase, fructose-6-phosphate and glyceraldehyde-3-phosphate may be converted to xylulose-5-phosphate and erythrose-4-phosphate in the following reaction:

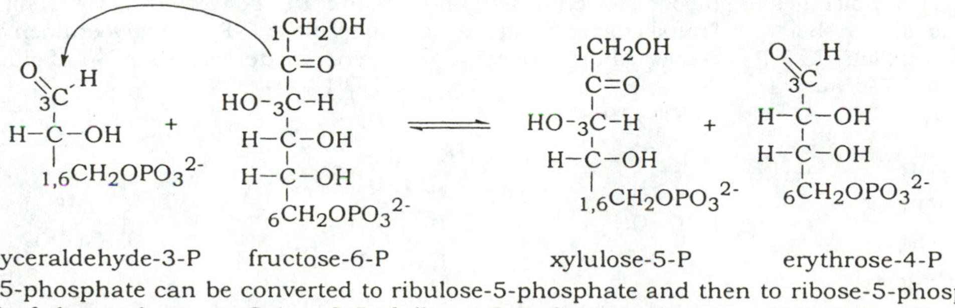

```
                          1CH2OH
  O  H                     |              1CH2OH         O  H
   \\/                     C=O              |             \\/
   3C                      |               C=O            3C
   |                    HO-3C-H             |              |
 H-C-OH      +             |          HO-3C-H      +    H-C-OH
   |                     H-C-OH  <==>       |              |
 1,6CH2OPO3(2-)            |             H-C-OH          H-C-OH
                         H-C-OH             |              |
                           |          1,6CH2OPO3(2-)   6CH2OPO3(2-)
                        6CH2OPO3(2-)
```

Glyceraldehyde-3-P   fructose-6-P   xylulose-5-P   erythrose-4-P

Xylulose-5-phosphate can be converted to ribulose-5-phosphate and then to ribose-5-phosphate. Thus, C-1 of glucose becomes C-1 or C-5 of ribose, C-3 of glucose becomes C-3 of ribose, and C-6 of glucose becomes C-5 of ribose.

Erythrose-4-phosphate, a 4-carbon compound, reacts with another fructose-6-phosphate to produce glyceraldehyde-3-phosphate and sedoheptulose-7-phosphate in a reaction catalyzed by transaldolase.

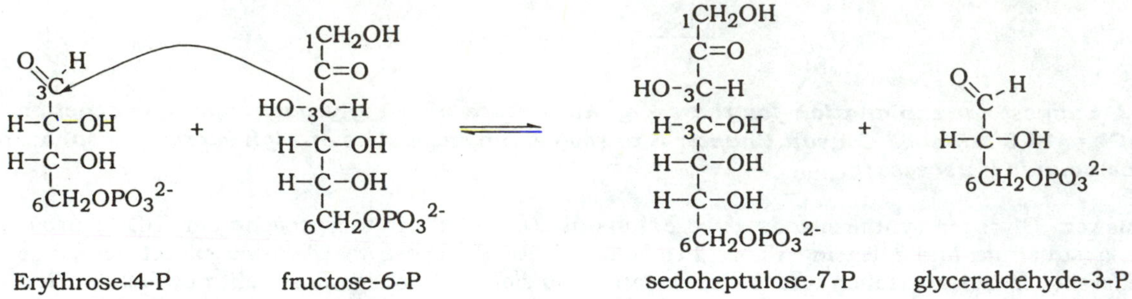

Erythrose-4-P: $O{=}{}_3C{-}H$, H–C–OH, H–C–OH, ${}_6CH_2OPO_3^{2-}$

\+ fructose-6-P: ${}_1CH_2OH$, C=O, HO–${}_3$C–H, H–C–OH, H–C–OH, ${}_6CH_2OPO_3^{2-}$

⇌ sedoheptulose-7-P: ${}_1CH_2OH$, C=O, HO–${}_3$C–H, H–${}_3$C–OH, H–C–OH, H–C–OH, ${}_6CH_2OPO_3^{2-}$

\+ glyceraldehyde-3-P: $O{=}C{-}H$, H–C–OH, ${}_6CH_2OPO_3^{2-}$

Transketolase can convert these two products to two 5-carbon sugars, namely xylulose-5-phosphate and ribose-5-phosphate.

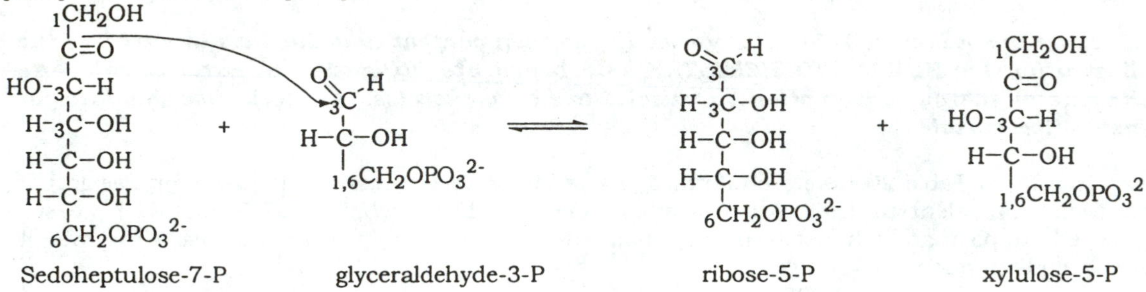

Sedoheptulose-7-P: ${}_1CH_2OH$, C=O, HO–${}_3$C–H, H–${}_3$C–OH, H–C–OH, H–C–OH, ${}_6CH_2OPO_3^{2-}$

\+ glyceraldehyde-3-P: $O{=}{}_3C{-}H$, H–C–OH, ${}_{1,6}CH_2OPO_3^{2-}$

⇌ ribose-5-P: $O{=}{}_3C{-}H$, H–${}_3$C–OH, H–C–OH, H–C–OH, ${}_6CH_2OPO_3^{2-}$

\+ xylulose-5-P: ${}_1CH_2OH$, C=O, HO–${}_3$C–H, H–C–OH, ${}_{1,6}CH_2OPO_3^{2}$

Ribose-5-P and xylulose-5-P are converted to ribulose. Glucose labeled at C-6 only shows up as label in C-5 of ribulose. Carbon 1 of glucose is converted to C-1 and C-5 of ribulose. Carbon 3 of glucose is converted to C-1, C-2, and C-3 of ribulose as shown above.

7. What is the fate of carbon from positions 2 and 4 of glucose-6-P after one pass through the scheme shown in Figure 19.19?

Answer: In order to meet high demands for ATP and NADPH, the oxidative reactions of the pentose phosphate pathway can be used to produce NADPH and ribulose-5-phosphate, which is converted to glyceraldehyde-3-phosphate and can be used in glycolysis to produce pyruvate and ATP. In order to accomplish this, ribulose-5-phosphate must be used to produce either glyceraldehyde-3-phosphate or fructose-6-phosphate. This is achieved using a transketolase reaction, a transaldolase reaction, and finally a transketolase reaction in a series that is exactly opposite to the one outlined in Problem 6.

The oxidative phase of the pathway will convert glucose-6-P, labeled at either carbon 2 or 4, to ribulose, labeled at either carbon 1 or 3. Let us examine the reactions outlined in Problem 6 in reverse. Two molecules of glucose labeled at C-2 and C-4 are first converted to 5-carbon sugars labeled as shown below. Transketolase produces sedoheptulose-7-P and glyceraldehyde-3-P. When glyceraldehyde-3-P is converted to pyruvate, C-1 of pyruvate derives from C-4 of glucose.

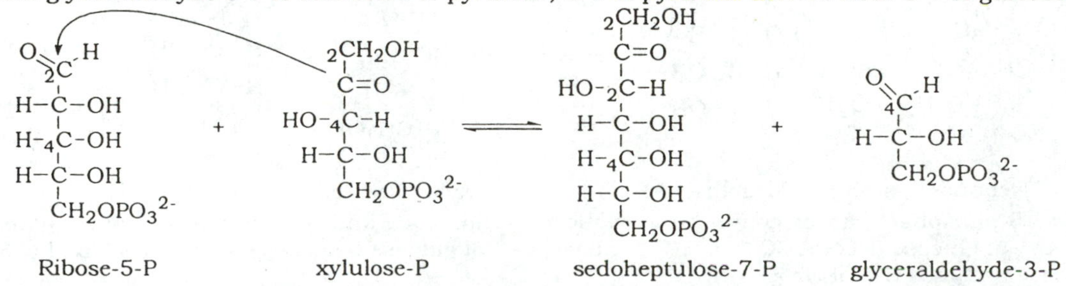

Ribose-5-P: $O{=}{}_2C{-}H$, H–C–OH, H–${}_4$C–OH, H–C–OH, $CH_2OPO_3^{2-}$

\+ xylulose-P: ${}_2CH_2OH$, C=O, HO–${}_4$C–H, H–C–OH, $CH_2OPO_3^{2-}$

⇌ sedoheptulose-7-P: ${}_2CH_2OH$, C=O, HO–${}_2$C–H, H–C–OH, H–${}_4$C–OH, H–C–OH, $CH_2OPO_3^{2-}$

\+ glyceraldehyde-3-P: $O{=}{}_4C{-}H$, H–C–OH, $CH_2OPO_3^{2-}$

In the next reaction sedoheptulose-7-P reacts with glyceraldehyde to form fructose-6-P and erythrose-4-P. As shown below, fructose is labeled at C-1, C-3, and C-4. C-1 and C-4 become carbon 1 in pyruvate. Thus, C-1 of pyruvate derives label from C-2 and C-4 of glucose whereas C-3 of pyruvate derives from C-2 of glucose.

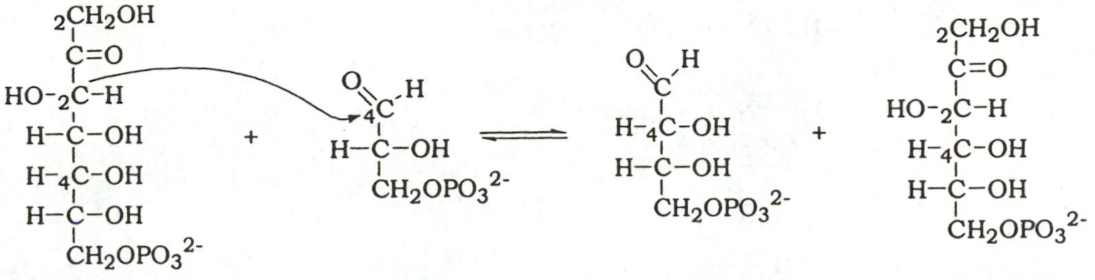

```
  2CH2OH                                                    2CH2OH
   C=O                                     O   H             C=O
HO-2C-H                  O   H              \\C/          HO-2C-H
 H-C-OH        +          \\4C               |              H-4C-OH
 H-4C-OH               H-C-OH     <==>    H-4C-OH    +     H-C-OH
 H-C-OH                 CH2OPO3^2-        H-C-OH            CH2OPO3^2-
  CH2OPO3^2-                               CH2OPO3^2-
```

Sedoheptulose-7-P glyceraldehyde-3-P erythrose-4-P fructose-6-P

In the next reaction erythrose reacts with xylulose, derived from labeled glucose, to produce glyceraldehyde and fructose as shown below. Again we see that conversion of glyceraldehyde to pyruvate results in labeled C-1 derived from C-4 of glucose. Conversion of fructose into two pyruvates results in C-1 label derived from C-4 of glucose and C-3 label derived from C-2 of glucose.

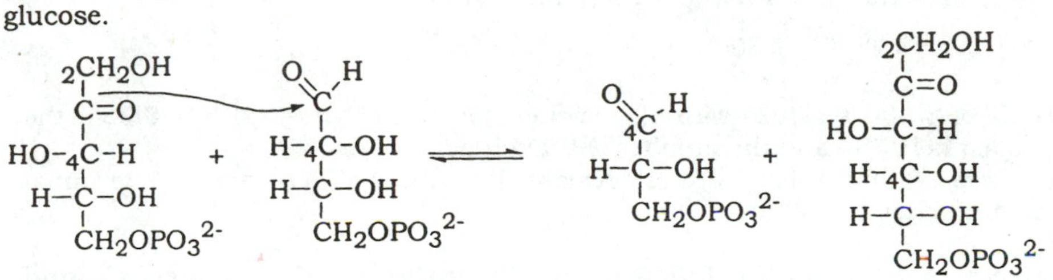

```
                                                          2CH2OH
  2CH2OH          O   H                                    C=O
   C=O             \\C            O   H                 HO-C-H
HO-4C-H    +    H-4C-OH   <==>     \\4C          +       H-4C-OH
 H-C-OH          H-C-OH          H-C-OH                  H-C-OH
  CH2OPO3^2-      CH2OPO3^2-      CH2OPO3^2-              CH2OPO3^2-
```

Xylulose-5-P erythrose-4-P glyceraldehyde-3-P fructose-6-P

8. Imagine a glycogen molecule with 8000 glucose residues. If branches occur every eight residues, how many reducing ends does the molecule have? If branches occur every 12 residues, how many reducing ends does it have? How many nonreducing ends does it have in each of these cases?

Answer: The reducing end of a polysaccharide contains a free aldehyde or keto group. In glycogen, the linkage between glucose residues is α(1→4) for nonbranch residues and α(1→6) for branch residues. Thus, both branch residues and nonbranch residues have their anomeric carbons, C-1, in glycosidic bonds. Further, C-1 of the first glucose residue is tied up in linkage to glycogenin at the core so there are no free reducing ends, regardless of the size of the glycogen molecule.

The number of nonreducing ends is very much dependent on the frequency of branches. In the diagram below, let n = number of residues in a branch, b = a branch residue, and e = nonreducing end.

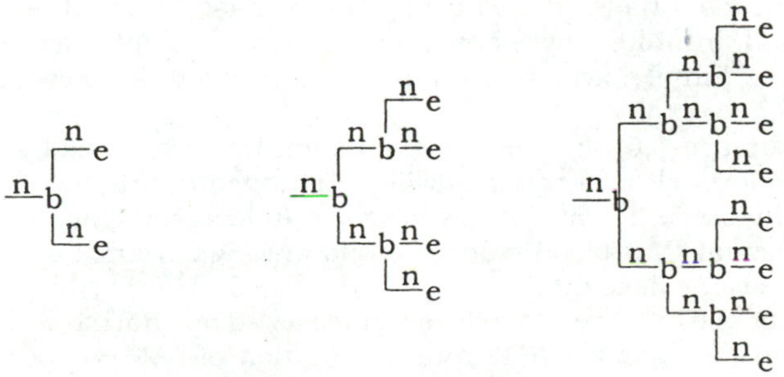

```
                                                   n
                                                  ┌─e
                                              n  n
                                               ┌b─e
                       n                    n  n  n
                      ┌─e               ┌─b──b──e
         n          n  n               │     └n
        ┌─e        ┌─b─e               │       e
   n              n                 n  │       n
   ─b             ─b                ───b      ┌─e
    └n             └n  n               │  n  n  n
      e             ─b─e               └─b──b──e
                     └n                     └n  n
                       e                      b─e
                                               └n
                                                 e
```

From this diagram the following table may be constructed.

Number of residues	number of branch points (b)	number of ends (e)
3n	1	2
7n	3	4
15n	7	8
.	.	.
$[(2^x - 1) + 2^x]n$	$2^x - 1$	2^x

For a glycogen molecule of 8,000 residues with 8 residues per branch (n = 8) we have:

$$[(2^x - 1) + 2^x]n = 8{,}000 \text{ residues}$$

For n = 8.

$$[(2^x - 1) + 2^x] \times 8 = 8{,}000 \text{ or}$$

$$2^x + 2^x = \frac{8{,}000}{8} + 1$$

$$2 \times 2^x = 1001$$

$$2^x = \frac{1001}{2}$$

$$x = \frac{\log\frac{1001}{2}}{\log 2} = \frac{2.699}{0.301} = 8.967$$

The number of nonreducing ends is given by

$$2^x = 2^{8.967} = 500$$

For a glycogen molecule of 8,000 residues with 12 residues per branch (n = 12), x = 8.364, the number of nonreducing ends is 334 and the number of branch points is 333.
Note: 8,000 residues forms 8,000 ÷ 12 = 666 12-residue branches. There are 334 terminal branches, 333 internal branches.

9. Explain the effects of each of the following on the rates of gluconeogenesis and glycogen metabolism:

a. Increasing the concentration of tissue fructose-1,6-bisphosphate
b. Increasing the concentration of blood glucose
c. Increasing the concentration of blood insulin
d. Increasing the amount of blood glucagon
e. Decreasing levels of tissue ATP
f. Increasing the concentration of tissue AMP
g. Decreasing the concentration of fructose-6-phosphate.

Answer: Gluconeogenesis and glycolysis are reciprocally controlled. In glycolysis, control is exerted at the strongly exergonic steps catalyzed by hexokinase, phosphofructokinase, and pyruvate kinase. In gluconeogenesis, these steps are bypassed with different reactions catalyzed by glucose-6-phosphatase, fructose-1,6-bisphosphatase, and pyruvate carboxylase and PEP carboxykinase. Glucose-6-phosphatase is under substrate-level control by glucose-6-phosphate. Fructose-1,6-bisphosphatase is allosterically inhibited by AMP and activated by citrate. This enzyme is also inhibited by fructose-2,6-bisphosphate, which is produced from fructose-6-phosphate by phosphofructokinase 2. Phosphofructokinase 2 is allosterically activated by fructose-6-phosphate and inhibited by phosphorylation of a single serine residue by cAMP-dependent protein kinase. Finally, acetyl-CoA stimulates pyruvate carboxylase.

a. Increasing fructose-1,6-bisphosphate will stimulate glycolysis by stimulating pyruvate kinase and, therefore, gluconeogenesis will be inhibited indirectly.
b. High blood glucose will stimulate glycolysis and decrease gluconeogenesis.
c. Insulin is a signal that blood glucose levels are high. Since additional glucose is not required, gluconeogenesis is inhibited.
d. Glucagon signals the liver to release glucose. This hormone sets off a cascade of signals leading to increase in cAMP and stimulation of cAMP-dependent protein kinase. Pyruvate kinase is inhibited by phosphorylation leading to inhibition of glycolysis. Phosphofructokinase 2 is also inactivated by phosphorylation leading to a drop in the level of fructose-2,6-bisphosphate which causes relief of inhibition of fructose-1,6-bisphosphatase and stimulation of gluconeogenesis.
e. A decrease in ATP will stimulate glycolysis and inhibit gluconeogenesis.
f. AMP inhibits fructose-1,6-bisphosphatase and inhibits gluconeogenesis.
g. Fructose-6-phosphate is not a regulatory molecule but it is rapidly equilibrated with glucose-6-phosphate, a substrate for glucose-6-phosphatase. Decreased levels of glucose-6-phosphate will lead to inhibition of gluconeogenesis.

Glycogen metabolism is regulated as follows. Glycogen phosphorylase is allosterically activated by AMP and is inhibited by ATP and glucose-6-phosphate. Further, phosphorylation of serine-14 activates glycogen phosphorylase. Glycogen synthase is reciprocally regulated by phosphorylation: It is inactive when phosphorylated. The synthase is also allosterically activated by glucose-6-phosphate.

a. Fructose-1,6-bisphosphate will inhibit glycogen catabolism.
b. High blood glucose leads to an increase in glycogen anabolism.
c. Insulin stimulates glycogen anabolism by increasing glucose intracellularly, which leads to an increase in glucose-6-phosphate, an allosteric activator of glycogen synthase.
d. Glucagon will release glucose from glycogen by setting off a signal cascade ending with phosphorylation, and stimulation, of glycogen phosphorylase and inhibition of glycogen synthase.
e. ATP inhibits glycogen phosphorylase and inhibits glycogen catabolism. A drop in ATP will stimulate glycogen breakdown.
f. AMP activates glycogen phosphorylase leading to stimulation of glycogen catabolism.
g. Fructose-6-phosphate, through glucose-6-phosphate, will activate glycogen synthase leading to stimulation of glycogen formation.

10. The free-energy change of the glycogen phosphorylase reaction is $\Delta G^{\circ\prime} = +3.1$ kJ/mol. If $[P_i] = 1$ mM, what is the concentration of glucose-1-P when this reaction is at equilibrium.

Answer: For the reaction: $\text{glycogen}_n + P_i \leftrightarrows \text{glycogen}_{n-1} + \text{glucose-1-P}$

$$\Delta G^{\circ\prime} = -RT \times \ln K_{eq} = -RT \times \ln \frac{[\text{glucose-1-P}][\text{glycogen}_{n-1}]}{[P_i][\text{glycogen}_n]}$$

Since $[\text{glycogen}_n] = [\text{glycogen}_{n-1}]$ this simplifies to:

$$\Delta G^{\circ\prime} = -RT \times \ln \frac{[\text{glucose-1-P}]}{[P_i]}$$

Solving for [glucose - 1 - P] we find

$$[\text{glucose-1-P}] = [P_i] \times e^{\frac{-\Delta G^{\circ\prime}}{RT}} = [(1 \times 10^{-3}) \times e^{\frac{-3.1}{8.314 \times 10^{-3} \times 298}}$$

$$[\text{glucose-1-P}] = 0.286\text{mM}$$

Questions for Self Study

1. Fill in the blanks. Gluconeogenesis is the production of _____ from pyruvate and other 3-carbon and 4-carbon compounds. In animals it occurs mainly in two organs, the ____ and _____. This pathway utilized several of the steps of glycolysis but three key glycolytic steps are bypassed. They are the reactions catalyzed by _____, _____, and _____. The conversion of pyruvate to PEP is a two step process in which pyruvate is first converted to _____, a citric acid cycle intermediate with the addition of ____. The enzyme catalyzing this step uses the vitamin _____. Production of PEP from pyruvate results in the hydrolysis of _____ high-energy phosphate bonds. The other two steps unique to gluconeogenesis are catalyzed by _____ and _____. These enzymes carry out _____ reactions.

2. The Cori cycle is an interplay of glycolysis and gluconeogenesis between two tissues. Answer true or false.
   a. The liver is responsible for gluconeogenesis. _____
   b. Pyruvate is transported by the circulatory system. _____
   c. Production of lactate by muscle allows ATP production by fermentation. _____
   d. Glucose-6-phosphatase is localized in the endoplasmic reticulum of muscle. _____
   e. High levels of NADH in the muscle favor reduction of pyruvate. _____

3. Name three important control molecules for glycolysis and gluconeogenesis.

4. What is remarkable about phosphofructokinase-2 and fructose-2,6-bisphosphatase enzymatic activities?

5. There are two enzymes necessary to digest dietary glycogen, α-amylase and debranching enzyme. What are the functions of each?

6. Glycogen phosphorylase cleaves glucose units from tissue glycogen. This enzyme is carefully regulated by allosteric controls and by covalent modification. In the diagram presented below place the following: phosphorylase kinase, AMP, ATP, glucose-6-phosphate, glucose, phosphoprotein phosphatase 1, phosphorylase a, phosphorylase b and the active and inactive states.

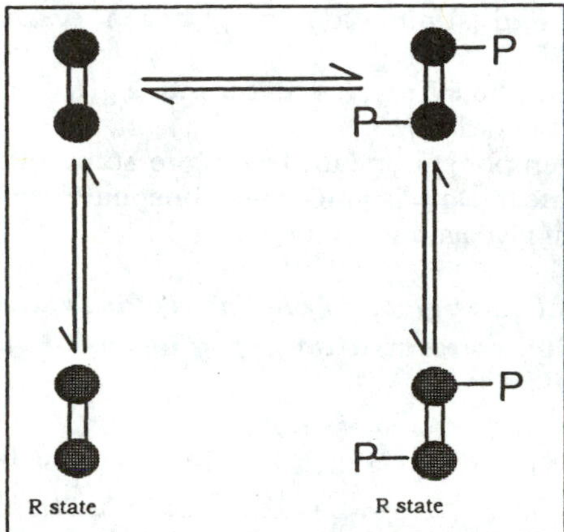

7. Describe the cascade of events from hormone stimulation to inactivation of glycogen synthase and activation of glycogen phosphorylase.

8. In the oxidative steps of the pentose phosphate pathway, glucose is converted to ribulose-5-phosphate and carbon dioxide in three steps. This short metabolic sequence serves two immediate purposes. What are they?

9. In the nonoxidative steps of the pentose phosphate pathway, 5-carbon sugars are metabolized back to glucose. The enzymes involved include an isomerase (I), an epimerase (E), transketolases (TK) and a transaldolase (TA). Which enzyme best fits each of the statements below?

a. Thiamine pyrophosphate-dependent enzyme. _____
b. Causes conversion of configuration about asymmetric carbon. _____
c. Ketose-aldose conversion. _____
d. Forms Schiff base intermediate. _____
e. Interconverts ribulose-5-phosphate and ribose-5-phosphate. _____
f. Forms enediolate intermediate, reduction of which may occur on either side of the carbon. _____
g. Reduction of adduct by sodium borohydride leads to inactivation. _____
h. Moves two carbons from a ketose to an aldose. _____

Answers

1. Glucose; liver; kidney; hexokinase (or glucokinase); phosphofructokinase; pyruvate kinase; oxaloacetate; carbon dioxide; biotin; two; fructose-1,6-bisphosphatase; glucose-6-phosphatase; hydrolysis.

2. a. T; b. F; c. T; d. F; e. T.

3. Fructose-2,6-bisphosphate, AMP, ATP, citrate, acetyl-CoA, alanine, and glucose-6-phosphate.

4. Both activities are properties of the same protein molecule, a bifunctional or tandem enzyme.

5. α-Amylase is an endoglycosidase that hydrolyzes α-(1→4) linkages to produce monosaccharides, disaccharides, and short oligosaccharides in addition to limit dextrins, highly branched polysaccharides. Debranching enzyme degrades limit dextrins by removing branches.

6.

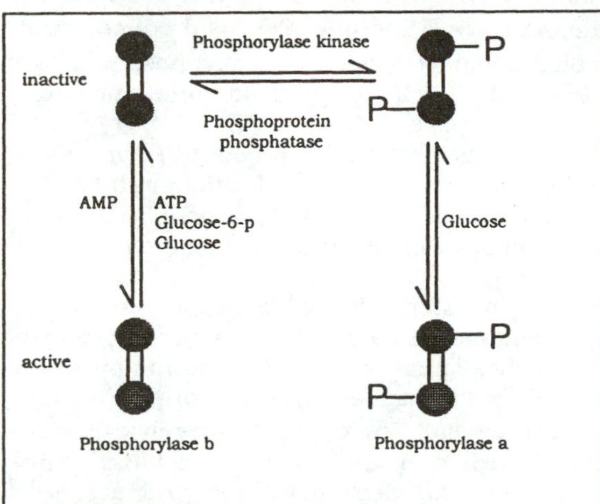

7. Hormone binding to surface receptors leads to stimulation of adenylyl cyclase, which raises the levels of cellular cAMP. This second messenger binds to and activates protein kinase, which phosphorylates and inactivates glycogen synthase. cAMP-dependent protein kinase also phosphorylates glycogen phosphorylase kinase leading to activation. Finally, phosphorylation of glycogen phosphorylase by glycogen phosphorylase kinase leads to activation.

8. The oxidative steps produce 5-carbon sugars and reducing equivalents in the form of NADPH.

9. a. TK; b. E; c. I; d. TA; e. I; f. E; g. TA; h. TK.

Additional Problems

1. What are the three potential substrate cycles in glycolysis/gluconeogenesis and how is each regulated?

2. Pyruvate carboxylase reacts bicarbonate and pyruvate to produce oxaloacetate. It is a vitamin-dependent enzyme. Which vitamin does it use? Describe the reaction the enzyme catalyzes.

3. Alanine inhibits pyruvate kinase. What relationship does alanine have to this enzyme?

4. How are glycogen phosphorylase and glycogen synthase reciprocally regulated?

5. The oxidative phase of the pentose phosphate pathway is the following sequence: glucose → glucose-6-phosphate → 6-phosphogluconolactone → 6-phosphogluconate → ribulose-5-phosphate. Describe the biochemistry of this sequence.

6. Can triacylglycerol (a.k.a. fat) be used in gluconeogenesis?

Abbreviated Answers

1. The conversion of glucose and glucose-6-phosphate catalyzed by glucokinase (in liver) and glucose-6-phosphatase is a substrate cycle confined to cells expressing glucose-6-phosphatase, such as kidney and liver cells. A substrate cycle exists between fructose-6-phosphate and fructose-1,6-bisphosphate catalyzed by phosphofructokinase and fructose-1,6-bisphosphatase. This cycle is regulated by reciprocal control of the two enzymes by fructose-2,6-bisphosphate and AMP. The conversion of PEP to pyruvate in glycolysis and of pyruvate to oxaloacetate to PEP in

gluconeogenesis is the final substrate cycle. Acetyl-CoA levels play a key role in regulating this cycle.

2. Pyruvate carboxylase is a biotin-dependent reaction. Biotin participates in one-carbon reactions in which the carbon is fully oxidized. The enzyme catalyzes a two-step reaction. In the first step, carboxyphosphate, an activated form of CO_2, is produced by nucleophilic attack on the γ-phosphate of ATP by a bicarbonate oxygen. Carboxyphosphate is used to transfer CO_2 to biotin, which carboxylates the methyl carbon of pyruvate to produce oxaloacetate.

3. Deamination of alanine produces pyruvate, a product of pyruvate kinase. The presence of alanine is a signal that an alternative carbon source for the production of pyruvate is present. In general, proteins are not stored to the extent that carbohydrates are, and they are therefore broken down into amino acids, which are subsequently metabolized.

4. Glycogen phosphorylase is allosterically activated by AMP and inhibited by glucose-6-phosphate, whereas glycogen synthase is stimulated by glucose-6-phosphate. Both enzymes are also regulated by covalent modification. There are two forms of glycogen phosphorylase. The active form, phosphorylase *a*, differs from the less active form, phosphorylase *b*, by a phosphate group on Ser^{14}. Ser^{14} is phosphorylated by glycogen phosphorylase kinase, a kinase that also exists in an active, phosphorylated form, and an inactive form. Phosphorylation of glycogen phosphorylase kinase is a result of cAMP-dependent protein kinase. Glycogen synthase also has two forms, an active, glucose-6-phosphate-independent form (glycogen synthase I) and a less active, glucose-6-phosphate-dependent form (glycogen synthase D). Glycogen synthase I is converted to glycogen synthase D by phosphorylation at a number of sites catalyzed by several kinases.

5. The first step is a phosphorylation catalyzed by hexokinase. The next step is an oxidation of glucose-6-phosphate by NADP-dependent glucose-6-phosphate dehydrogenase. 6-Phosphogluconolactone is converted to phosphogluconate by hydrolysis catalyzed by gluconolactonase. Phosphogluconate is oxidatively decarboxylated to ribulose-5-phosphate and CO_2 by NADP-dependent 6-phosphogluconate dehydrogenase. So, the steps are phosphorylation, oxidation, hydrolysis, and oxidative decarboxylation.

6. A majority of the carbons in triacylglycerol is unavailable for incorporation into glucose because they are converted into acetyl-CoA units. However, the glycerol backbone can be used as a carbon source for gluconeogenesis. (For triacylglycerols with fatty acids with an odd number of carbons, a three-carbon unit is metabolized into succinate, a citric acid cycle intermediate and therefore a substrate for gluconeogenesis.

Summary

In order to maintain normal levels of metabolic activity, cells must sustain an appropriate level of sugars, especially glucose. Glucose is synthesized from noncarbohydrate precursors by a process known as gluconeogenesis, and it can also be supplied via breakdown of glycogen (in animals) or starch (in plants or in the diet of animals). The pentose phosphate pathway is the primary source of NADPH, the reduced coenzyme essential to most reductive biosynthetic processes, including fatty acid and amino acid biosynthesis. The pentose phosphate pathway also produces ribose-5-P, an important component of ATP, $NAD(P)^+$, FAD, coenzyme A, DNA and RNA.

Substrates for gluconeogenesis include pyruvate, lactate, most of the amino acids (except leucine and lysine), glycerol and all the TCA cycle intermediates. Because fatty acids are not substrates for gluconeogenesis, animals cannot carry out net synthesis of sugars from acetyl-CoA. Major sites of gluconeogenesis include the liver (90%) and kidneys (10%). Glucose produced by these organs is utilized in brain, heart, muscle and red blood cells to meet their metabolic needs. Gluconeogenesis is not merely the reverse of glycolysis. In order that the net pathway may be exergonic and that the pathway may be suitably regulated, three key enzymatic reactions replace hexokinase (glucokinase), phosphofructokinase, and pyruvate kinase from glycolysis. These reactions are catalyzed by glucose-6-phosphatase, fructose-1,6-bisphosphatase, and the pair of pyruvate carboxylase/PEP carboxykinase. The overall conversion of pyruvate to glucose is made exergonic by these substitutions: the last three steps of gluconeogenesis combine for a $\Delta G°'$ of -30.5 kJ/mole.

The pyruvate carboxylase reaction, which occurs in the mitochondrial matrix, is biotin-dependent and utilizes ATP and bicarbonate as substrates. Pyruvate carboxylase is allosterically activated by acyl coenzyme A. When ATP and acetyl-CoA are high, gluconeogenesis is favored, but when ATP and acetyl-CoA are low, acetyl-CoA is consumed by the TCA cycle for the net production of energy and ATP. Whereas pyruvate carboxylase consumes an ATP to drive a carboxylation, the PEP carboxykinase uses decarboxylation to facilitate formation of PEP. AMP inhibits fructose-1,6-bisphosphatase, and the inhibition by AMP is enhanced by fructose-2,6-bisphosphate. Glucose-6-phosphatase, the final step in gluconeogenesis, is located in the membrane of the endoplasmic reticulum. Glucose formed in liver ER by this reaction is transported to the plasma membrane and secreted directly into the bloodstream for distribution to other organs and cells.

The gluconeogenic pathway is driven by hydrolysis of ATP and GTP - the net free energy change for the pathway from pyruvate to glucose is -37.7 kJ/mole. The pathway is controlled by substrate cycles involving pairs of reactions such as phosphofructokinase (glycolysis) and fructose-1,6-bisphosphatase (gluconeogenesis). Moreover, reciprocal regulatory control ensures that glycolysis is active when gluconeogenesis is inactive and *vice versa*. Acetyl-CoA is a potent allosteric effector of these two pathways, inhibiting pyruvate kinase (glycolysis) and activating pyruvate carboxylase. Hydrolysis of fructose-1,6-bisphosphate by fructose-1,6-bisphosphatase is allosterically inhibited by fructose-2,6-bisphosphate, which also activates phosphofructokinase in another example of reciprocal control. Cellular levels of fructose-2,6-bisphosphate are controlled by phosphofructokinase-2 and fructose-2,6-bisphosphatase, two activities of a bifunctional enzyme which are regulated by phosphorylation and dephosphorylation of a single Ser residue of the 49,000 Dalton subunit.

Adult human beings typically metabolize about 160 g of carbohydrates daily, either from digestion of dietary carbohydrates or from breakdown of tissue glycogen. Dietary carbohydrates are degraded by α-amylase. Digestion is a highly efficient (and unregulated) process in which 100% of ingested food is absorbed and metabolized. By contrast, tissue glycogen is metabolized in a highly regulated process. Glycogen breakdown is stimulated by glucagon and epinephrine, whereas glycogen synthesis is activated by insulin. In addition to stimulating glycogen synthesis in liver and muscle, insulin also stimulates transport of glucose across cell membranes and activates several key glycolytic enzymes. Binding of glucagon to liver cell membranes activates intracellular adenylyl cyclase, producing cyclic AMP, a second messenger which activates a protein kinase. This kinase in turn phosphorylates and activates phosphorylase kinase, which phosphorylates phosphorylase, the enzyme which catalyzes the first step in glycogen breakdown. Such kinase cascades facilitate the amplification of hormonal signals.

Glycogen synthesis involves the transfer of activated glucose units (in the form of sugar nucleotides) to a growing chain in the glycogen synthase reaction. The glycogen chain is initiated on a protein core, glycogenin, with the first glucose residue linked covalently to a Tyr-OH. Glycogen synthase catalyzes the transfer of glucosyl units from UDP-glucose to the C-4 hydroxyl groups of non-reducing ends of glycogen strands. Glycogen branching occurs by transfer of six- or seven-residue segments from the nonreducing end of a linear chain at least eleven residues in length to the C-6 hydroxyl of a glucose residue further up the chain.

The pentose phosphate pathway converts glucose-6-phosphate in two successive oxidations and five non-oxidative reactions to a variety of carbohydrate intermediates, producing NADPH and ribose-5-phosphate as well. Oxidation of glucose-6-phosphate by glucose-6-phosphate dehydrogenase, followed by ring opening and further oxidation by 6-phosphogluconate dehydrogenase yields ribulose-5-phosphate, whereas phosphopentose epimerase produces xylulose-5-phosphate. Further rearrangement to produce other carbohydrate intermediates is carried out by transketolase and transaldolase. Transketolase is a thiamine pyrophosphate-dependent enzyme, and transaldolase functions via formation of an active site Schiff base with a lysine residue on the protein. Utilization of glucose-6-phosphate in glycolysis and the pentose phosphate pathway depends on the cell's needs for ATP, NADPH and ribose-5-P. Depending on the cell's needs for these metabolites, three different metabolic routes produce needed NADPH or ribose-5-P, either selectively or in relatively balanced amounts.

Chapter 20

Lipid Metabolism

• •

Chapter Outline

- Triacylglycerol as energy storage molecule
  - Highly reduced carbon source
  - Stored without large amounts of water
  - Stored in specialized cells: Adipocytes (adipose or fat cells)
- Hormonal regulation
  - Adrenaline, glucagon, ACTH mobilize fatty acids from adipocytes
  - Hormone binding to surface receptors leads to stimulation of cAMP-dependent protein kinase
  - Triacylglycerol lipase (hormone-sensitive lipase) activated by phosphorylation
    - Triacylglycerol lipase produces fatty acid and diacylglycerol
    - Diacylglycerol lipase produces fatty acid and monoacylglycerol
    - Monoacylglycerol lipases produces fatty acid and glycerol
      - Fatty acids metabolized by β-oxidation
      - Glycerol metabolized by glycolysis
- Dietary triacylglycerols
  - Pancreatic lipases secreted into duodenum release fatty acids and glycerol
    - Lipase activity depends on bile salts functioning as detergents
    - Short-chain fatty acids directly absorbed
    - Long-chain fatty acids recondensed onto glycerol in epithelial cells
      - Resynthesized triacylglycerol complexed to proteins
      - Chylomicrons released into lymphatic system
- β-Oxidation
  - Oxidation of β-carbon, cleavage of C_{α}- C_{β} bond
  - Occurs in mitochondria
  - Two-carbon acetate units split off as acetyl-CoA
- Fatty acid activation and transport into mitochondria
  - Short-chain fatty acids
    - Diffuse into mitochondria
    - Acyl-CoA synthetase (Acyl-CoA ligase or fatty acid thiokinase) produces acyl-CoA derivatives
    - ATP to AMP + PP_i ($PP_i \rightarrow 2P_i$) drives synthesis
  - Long-chain fatty acids: Activated in cytosol
    - Acyl-CoA synthetase produces acyl-CoA derivatives
    - Carnitine acyltransferase I produces acylcarnitine
      - Outside surface of inner mitochondrial membrane
    - Carnitine acyltransferase II reforms acyl-CoA
      - Matrix surface of inner mitochondrial membrane
- Enzymology of β-oxidation: Four steps
  - Acyl-CoA dehydrogenase
    - Produces C-C double bond
    - Noncovalent (but tightly bound) FAD reduced

- $FADH_2$ passes electrons through electron transfer protein to coenzyme Q
- Three enzymes: Chain-length specific
- Enoyl-CoA hydrase (crotonase)
  - Adds water across double bond
  - Three kinds of enzymes
    - trans-enoyl-CoA derivatives to L-β-hydroxyacyl-CoA
- L-Hydroxyacyl-CoA dehydrogenase
  - Oxidizes hydroxyl group
  - NAD^+ reduced to NADH
- Thiolase (β-ketothiolase)
  - Cleaves off acetyl-CoA unit
  - Acyl-CoA two carbons shorter
  - Last cleavage releases
    - Two acetyl-CoA for even carbon numbered chains
    - One propionyl-CoA and one acetyl-CoA for odd carbon numbered chains
- Complete oxidation of one palmitic acid yields 106 molecules of ATP: 33 % efficiency
- Fatty acids (triacylglycerols) supply: Physiological importance
  - Energy reserves for some migratory birds
  - Water for animals that cannot drink: 130 water molecules per palmitate
- Odd-chain fatty acids: Triacylglycerols in plant and marine organism
  - Propionyl-CoA converted to succinyl-CoA
    - Propionyl-CoA also from Met, Val, Ile degradation
    - Propionyl-CoA to succinyl-CoA: Three steps
      - Propionyl-CoA carboxylase produces D-methylmalonyl-CoA
        - Biotin-dependent reaction
        - ATP hydrolysis drives reaction
      - Methylmalonyl-CoA epimerse
        - D-methylmalonyl CoA to L-methylmalonyl-CoA
        - Epimerization not racemization
      - Methylmalonyl-CoA mutase
        - Carbonyl-CoA moiety moved between carbons
        - B_{12}-dependent reaction
    - Succinyl-CoA TCA cycle intermediate: To metabolize completely
      - Succinyl-CoA to malate
      - Malate transported to cytosol
      - Cytosolic malate to pyruvate via malic enzyme
- Unsaturated fatty acids: β-Oxidation and additional reactions
  - Monounsaturated fatty acids: Enoyl-CoA isomerase: cis-Δ^3 to trans-Δ^2 isomerization of double bond
  - Polyunsaturated fatty acids: 2,4-Dienoyl-CoA reductase
- Peroxisomal β-oxidation
  - Acyl-CoA oxidase: Oxygen accepts electrons: H_2O_2 produced
  - operates on chains with > 8 carbons
  - glyoxysomes have a similar set of enzymes
- α-Oxidation: Branched-chain fatty acids: Phytol: Chlorophyll breakdown product
- ω-Oxidation (ω, last letter of Greek alphabet)
  - Dicarboxylic acid produced
  - Cytochrome P450 requires NADPH and oxygen
  - ω-Carbon oxidized to carboxylic acid
- Ketone bodies: Acetoacetate, β-hydroxybutyrate, acetone
  - Synthesis occurs in liver
  - Converts acetyl units into ketone bodies
  - Ketone bodies serve as fuel for brain, heart, muscle

- Ketogenesis: Four reactions
  - Thiolase
    - Reversal of last step in β-oxidation
    - Acetoacetyl-CoA produced from two acetyl-CoA
  - HMG-CoA synthase: β-Hydroxy-β-methylglutaryl-CoA
    - This and thiolase reaction in mitochondria leads to ketone body formation
    - In cytosol, HMG-CoA production fuels cholesterol synthesis
  - HMG-CoA lyase: Produces acetoacetate
  - β-Hydroxybutyrate dehydrogenase: Produces β-hydroxybutyrate
- Fatty acid biosynthesis
  - Biosynthesis localized in cytosol: Fatty acid degradation in mitochondria
  - Intermediates held on acyl carrier protein (ACP): Phosphopantetheine group attached to serine: CoA in degradation
  - Fatty acid synthase: Multienzyme complex
  - Carbons derived from acetyl units
    - Acetyl CoA to malonyl CoA by carboxylation
    - Acetyl unit added to fatty acid with decarboxylation of malonyl CoA
  - Carbonyl carbons of acetyl units reduced using NADPH
  - Two carbon units added until chain reaches 16 carbons
- Source of acetyl units
  - Amino acids, glucose
  - Acetyl CoA used to produce citrate in mitochondria
  - Citrate exported to cytosol: ATP-citrate lyase forms acetyl-CoA and oxaloacetate
- Source of NADPH
  - Oxaloacetate utilization
    - Oxaloacetate (from citrate) to malate: NADH dependent reaction
    - Malate to pyruvate: Malic enzyme: NADPH produced
  - Pentose phosphate pathway
- Malonyl-CoA production: Acetyl-CoA carboxylase
  - Biotin-dependent enzyme
  - ATP drives carboxylation
  - Enzyme regulation: Filamentous polymeric form active
    - Citrate favors active polymer
    - Palmitoyl-CoA favors inactive protomer (polymer's monomer or building block molecule)
    - Citrate/palmitoyl-CoA effects depend on state of phosphorylation of protein
      - Unphosphorylated protein binds citrate with high affinity: Activation
      - Phosphorylated protein binds palmitoyl with high affinity: Inactivation
- Acetyl transacetylase: Acetylates acyl carrier protein (ACP): Destined to become methyl end of fatty acid
- Malonyl transacetylase: Malonylates ACP
- β-Ketoacyl-ACP synthase (acyl-malonyl ACP condensing enzyme): Accepts acetyl group: Transfers acyl group to malonyl-ACP
  - Malonyl carboxyl group released: Decarboxylation drives synthesis
  - Malonyl-ACP converted to acetoacetyl-ACP
- β-Ketoacyl-ACP reductase
  - Carbonyl carbon reduced to alcohol
  - NADPH provides electrons
- β-Hydroxyacyl-ACP dehydratase: Elements of water removed: Double bond created
- 2,3-trans-Enoyl-ACP reductase
  - Double bond reduced
  - NADPH provides electrons
- Subsequent cycles: C-16: Palmitoyl-CoA
- Additional modifications
  - Elongation: In mitochondria and endoplasmic reticulum
    - Mitochondrial-based system uses reversal of β-oxidation
    - Endoplasmic reticulum-based system uses malonyl-CoA

- Monounsaturation: One double bond
  - Bacteria: Oxygen-independent pathway: Chemistry performed on carbonyl carbon
  - Eukaryotes: Oxygen-dependent pathway
- Polyunsaturation
  - Plants can add double bonds between C-9 and methyl end
  - Animals
    - Add double bonds between C-9 and carboxyl end
    - Require essential fatty acids to have double bonds closer to methyl end

- Regulation
  - Malonyl-CoA inhibition of carnitine-acyl transferase: Blocks fatty acid uptake
  - Citrate/palmitoyl regulation of acetyl-CoA carboxylase
- Complex lipids
  - Glycerolipids: Glycerol backbone
    - Glycerophospholipids
    - Triacylglycerols
  - Sphingolipids: Sphingosine backbone
  - Phospholipids
    - Sphingolipids
    - Glycerophospholipids
- Glycerolipid biosynthesis
  - Phosphatidic acid is precursor
    - Glycerokinase produces glycerol-3-P
    - Glycerol-3-phosphate acyltransferase acylates C-1 with saturated fatty acid: Monoacylglycerol phosphate
    - Eukaryotes can produce monoacylglycerol phosphate using DHAP
    - Acyldihydroxyacetone phosphate reduced by NAPDH to monoacylglycerol phosphate
    - Acyltransferase acylates C-2: Phosphatidic acid
  - Phosphatidic acid used to synthesize two precursors of complex lipids
    - Diacylglycerol: Precursor of
      - Triacylglycerol: Diacylglycerol acyltransferase
      - Phosphatidylethanolamine, phosphatidylcholine
        - Ethanolamine phosphorylated
        - CTP and phosphoethanolamine produce CDP-ethanolamine
        - Transferase moves phosphoethanolamine onto diacylglycerol
        - Dietary choline: As per ethanolamine
        - Phosphatidylethanolamine to phosphatidylcholine by methylation
      - Phosphatidylserine: Serine for ethanolamine exchange
    - CDP-diacylglycerol
      - Phosphatidate cytidylyltransferase produces CDP-diacylglycerol
      - CDP-diacylglycerol used to produce phosphaphatidyl inositol, phosphaphatidyl glycerol, and cardiolipin
  - Plasmalogens: α,β-Unsaturated ether-linked chain at C-1
    - DHAP acetylated
    - Acyl group exchanged for alcohol
    - Keto group on DHAP reduced to alcohol and acylated
    - Head group attached
    - Desaturase produces double bonds
- Sphingolipid biosynthesis
  - Serine and palmitoyl-CoA condensed with decarboxylation to produce 3-ketosphinganine
  - Reduction forms sphinganine
  - Sphinganine acylated to form N-acyl sphinganine
  - Desaturation produces ceramide, the building block for all other sphingolipids
- Eicosanoids: Derived from 20-C fatty acids: Arachidonate is precursor
  - Local hormones: Prostaglandins, thromboxanes, leukotrienes, hydroxyeicosanoic acids
  - Prostaglandins
    - Cyclopentanoic acid formed from arachidonate by prostaglandin endoperoxidase synthase

        - Asprin inhibits enzyme
- Cholesterol biosynthesis: In liver
    - Mevalonate biosynthesis
        - Thiolase condenses two acetyl-CoA to produce acetoacetyl-CoA
        - HMG-CoA synthase produces HMG-CoA
        - HMG-CoA reductase produces mevalonate
            - Rate limiting step in cholesterol biosynthesis
            - Regulation
                - Inactivated by cAMP-dependent protein kinase
                - Short half life of enzyme when cholesterol levels high
                - Gene expression regulated
            - Pharmacological target for blood cholesterol regulation
    - Isopentenyl pyrophosphate and dimethylallyl pyrophosphate from mevalonate
    - Squalene to lanosterol to cholesterol
        - 20 steps from lanosterol to cholesterol
        - All steps in the endoplasmic reticulum
- Bile salts
    - Glycocholic acid
    - Taurocholic acid
- Steroid hormones
    - Cholesterol to pregnenolone
    - Pregnenone to progesterone, the precursor of all steroid hormones
        - Sex hormones are androgens and estrogens
        - Corticosteroids are glucocorticoids and mineralocorticoids (e.g., aldosterone)
    - Steroid hormones modulate transcription in the nucleus

Chapter Objectives

β-Oxidation

β-Oxidation is a simple series of four steps leading to degradation of fatty acyl-coenzyme A derivatives. You should know how fatty acids are supplied to β-oxidation. Lipases hydrolyze triacylglycerols. The fatty acids released are converted to acyl-CoA derivatives by synthetases that use two high-energy phosphoanhydride bonds to drive synthesis. β-Oxidation is reminiscent of the steps in the citric acid cycle leading from succinate to oxaloacetate. You might recall that they included oxidation by an FAD-dependent enzyme to produce a double bond, hydration of the double bond, and reduction of the alcoholic carbon to a ketone by an NAD^+-dependent enzyme. In β-oxidation (Figure 20.7), a similar series operates but with an acyl-CoA as substrate and thiolysis of the product to release acetyl-CoA and a fatty acyl-CoA, two carbons shorter. The FAD-dependent enzyme, acyl-CoA dehydrogenase, will move electrons into the electron transport chain at the level of coenzyme Q and each electron pair will support production of 1.5 ATP. Odd-carbon fatty acids are metabolized by β-oxidation to yield several acetyl-CoAs and one propionyl-CoA, a three-carbon acyl-CoA. Propionyl-CoA is metabolized by a vitamin B_{12}-dependent pathway to succinyl-CoA, a citric acid cycle intermediate. Oxidation of unsaturated fatty acids (Figure 20.15) requires additional enzymes. Keep this in mind by remembering that typical double bonds in fatty acids are in *cis* configuration whereas the double-bond formed during β-oxidation is *trans*. Double bonds starting at an odd-numbered carbon are simply isomerized to the *trans* configuration and moved, by one carbon, closer to the carboxyl end. Double bonds at even-numbered carbons are ultimately reduced.

Ketone Bodies

Acetyl units can be joined to form the four-carbon compound, acetoacetate, which together with β-hydroxybutyrate, a reduced form of acetoacetate, are used as a source of fuel for certain organs of the body and in times of glucose shortage. You should know the metabolic sequence leading to acetoacetate production. An intermediate in the pathway, β-hydroxy-β-methylglutaryl-CoA (HMG-CoA) is an important intermediate in cholesterol synthesis.

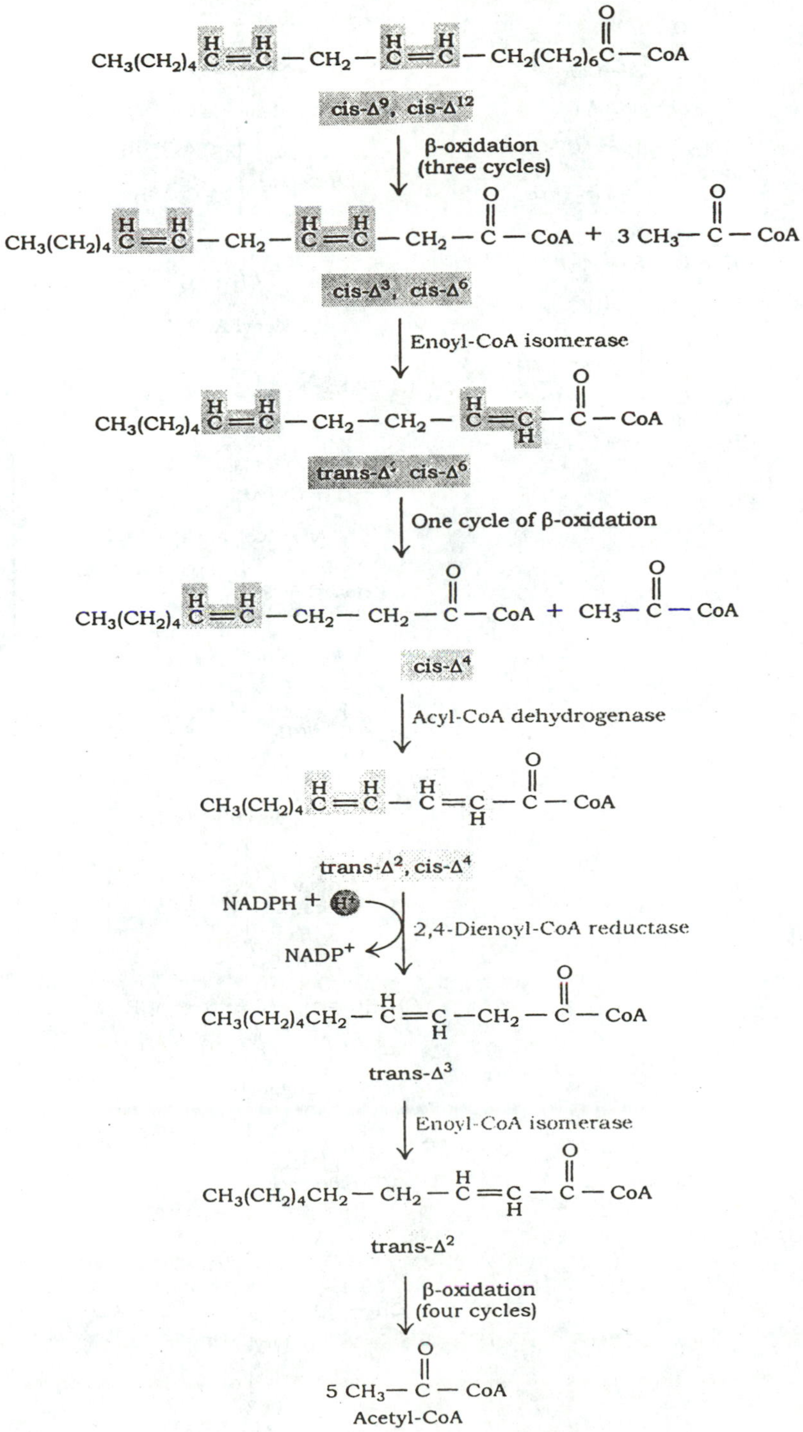

$$CH_3(CH_2)_4\overset{H}{C}=\overset{H}{C}-CH_2-\overset{H}{C}=\overset{H}{C}-CH_2(CH_2)_6\overset{O}{\overset{\|}{C}}-CoA$$

cis-Δ^9, cis-Δ^{12}

β-oxidation
(three cycles)

$$CH_3(CH_2)_4\overset{H}{C}=\overset{H}{C}-CH_2-\overset{H}{C}=\overset{H}{C}-CH_2-\overset{O}{\overset{\|}{C}}-CoA + 3\ CH_3-\overset{O}{\overset{\|}{C}}-CoA$$

cis-Δ^3, cis-Δ^6

Enoyl-CoA isomerase

$$CH_3(CH_2)_4\overset{H}{C}=\overset{H}{C}-CH_2-CH_2-\overset{H}{C}=\underset{H}{C}-\overset{O}{\overset{\|}{C}}-CoA$$

trans-Δ^2, cis-Δ^6

One cycle of β-oxidation

$$CH_3(CH_2)_4\overset{H}{C}=\overset{H}{C}-CH_2-CH_2-\overset{O}{\overset{\|}{C}}-CoA + CH_3-\overset{O}{\overset{\|}{C}}-CoA$$

cis-Δ^4

Acyl-CoA dehydrogenase

$$CH_3(CH_2)_4\overset{H}{C}=\overset{H}{C}-\overset{H}{C}=\underset{H}{C}-\overset{O}{\overset{\|}{C}}-CoA$$

trans-Δ^2, cis-Δ^4

NADPH + H^+

NADP$^+$

2,4-Dienoyl-CoA reductase

$$CH_3(CH_2)_4CH_2-\overset{H}{C}=\underset{H}{C}-CH_2-\overset{O}{\overset{\|}{C}}-CoA$$

trans-Δ^3

Enoyl-CoA isomerase

$$CH_3(CH_2)_4CH_2-CH_2-\overset{H}{C}=\underset{H}{C}-\overset{O}{\overset{\|}{C}}-CoA$$

trans-Δ^2

β-oxidation
(four cycles)

$$5\ CH_3-\overset{O}{\overset{\|}{C}}-CoA$$

Acetyl-CoA

Figure 20.15 The oxidation pathway for polyunsaturated fatty acids, illustrated for linoleic acid. Three cycles of β-oxidation of linoleoyl-CoA yield the cis-Δ^3,cis-Δ^6 intermediate, which is converted to a trans-Δ^2, cis-Δ^6 intermediate. An additional round of β-oxidation gives cis-Δ^4 enoyl-CoA, which is oxidized to the trans-Δ^2, cis-Δ^4 species by acyl-CoA dehydrogenase. The subsequent action of 2,4-dienoly-CoA reductase yields the trans-Δ^3 product, which is converted by enoyl-CoA isomerase to the trans-Δ^2 form. Normal β-oxidation then produces five molecules of acetyl-CoA.

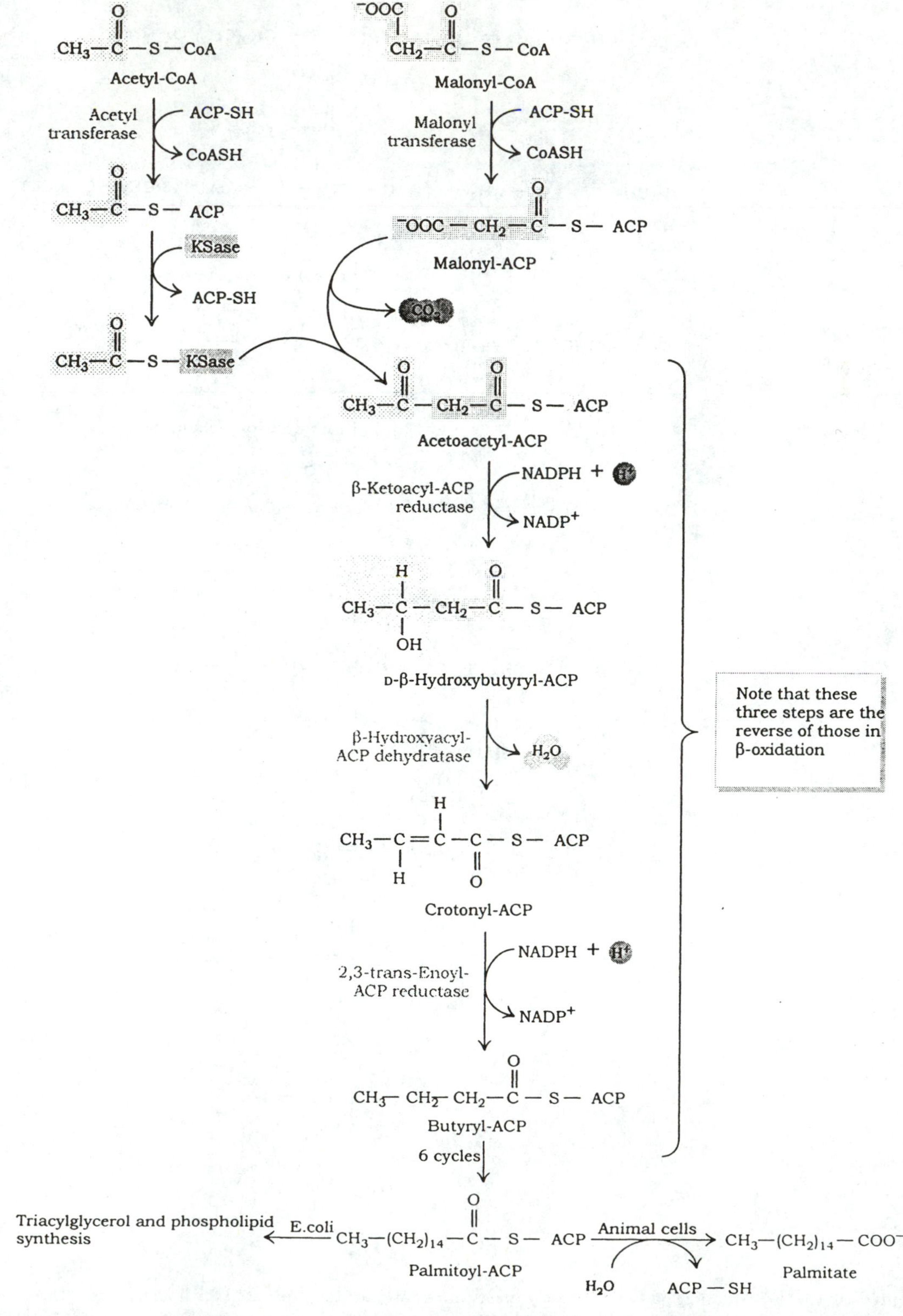

Figure 20.25 The pathway of palmitate synthesis from acetyl-CoA and malonyl-CoA. Acetyl and malonyl building blocks are introduced as acyl carrier protein conjugates. Decarboxylation drives the β-ketoacyl-ACP synthase and results in the addition of two-carbon units to the growing chain. Concentrations of free fatty acids are extremely low in most cells, and newly synthesized fatty acids exist primarily as acyl-CoA esters.

6?1-695-9292

Fatty Acid Biosynthesis

The steps of fatty acid biosynthesis (Figure 20.25) are similar in chemistry to the reverse of β-oxidation. Two-carbon acetyl units are used to build a fatty acid chain. The carbonyl carbon is reduced to a methylene carbon in three steps: reduction to an alcoholic carbon, dehydration to a carbon-carbon double bond intermediate, and reduction of the double bond. The two reduction steps utilize NADPH as reductant. Two-carbon acetyl units are moved out of the mitochondria as citrate and activated by carboxylation to malonyl-CoA. We have already seen similar carboxylation reactions and should remember that biotin is involved when carbons are added at the oxidation level of a carboxyl group. The enzyme, acetyl-CoA carboxylase, is regulated by polymerization/depolymerization with the filamentous polymeric state being active. You should understand the regulatory effects of citrate (favors polymer formation), palmitoyl-CoA (depolymerizes), and covalent phosphorylation (blocks citrate binding) on acetyl-CoA carboxylase activity.

In β-oxidation, we saw that the phosphopantetheine group of coenzyme A functioned as a molecular chauffeur for two-carbon acetyl units. In synthesis, phosphopantetheine, attached to the acyl carrier protein, functions as a molecular chaperone by guiding the growth of fatty acid chains.

In plants and bacteria, the steps of fatty acid biosynthesis are catalyzed by individual proteins whereas in animals a large multifunctional protein is involved. Synthesis starts with formation of acetyl-ACP and malonyl-ACP by specific transferases. The carboxyl group of malonyl-ACP departs, leaving a carbanion that attacks the acetyl group of acetyl-ACP to produce a four-carbon β-ketoacyl intermediate, which is subsequently reduced by an NADPH-dependent reductase, dehydrated, and reduced a second time by another NADPH-dependent reductase. To continue the cycle, ACP is recharged with malonyl group, malonyl group decarboxylates and attacks the acyl-ACP. The original acetyl group is the methyl-end of the fatty acid, whereas the malonyl groups are added at the carboxyl end. NADPH is supplied by the pentose phosphate pathway and by malic enzyme, which converts the oxaloacetate skeleton, used to transport acetyl groups out of the mitochondria as citrate, into pyruvate and CO_2 with $NADP^+$ reduction.

Additional elongation and introduction of double bonds can occur after synthesis of a C_{16} fatty acid. Elongation can occur in the endoplasmic reticulum where malonyl CoA is utilized or in the mitochondria where acetyl-CoA is used. Introduction of double bonds occurs via oxygen-independent mechanisms in bacteria and oxygen-dependent mechanisms in eukaryotes. Be familiar with the reaction catalyzed by stearoyl-CoA desaturase, involving stearoyl-CoA and oxygen as substrates and oleoyl-CoA and water as products.

Complex Lipids

The glycerolipids, including glycerophospholipids and triacylglycerols, are synthesized from glycerol, fatty acids, and head groups. Synthesis starts with the formation of phosphatidic acid from glycerol-3-phosphate and fatty acyl-CoA. C-1 is esterified usually with a saturated fatty acid. Phosphatidic acid may be converted to diacylglycerol and then to triacylglycerol. Alternately, diacylglycerol can be used to synthesize phosphatidylethanolamine and phosphatidylcholine with CDP-derivatized head groups serving as substrates. Phosphatidylserine is produced by exchange of the ethanol head-group from PE with serine. Phosphatidylinositol, phosphatidylglycerol, and cardiolipin (two diacylglycerols linked together by glycerol) are synthesized using CDP-diacylglycerol as an intermediate. Plasmalogens are synthesized from acylated DHAP. The acyl group is exchanged for a long-chain alcohol followed by reduction of the keto carbon of DHAP, acyl group transfer from acyl-CoA to C-2, head group transfer from CDP-ethanolamine and formation of a *cis* double bond between C-1 and C-2 of the long-chain alcohol.

The sphingolipids all derive from ceramide, whose synthesis starts with bond formation between palmitic acid and the α-carbon of serine (with loss of the serine carboxyl carbon as bicarbonate). After a few steps a second fatty acid is attached to serine in amide linkage. Subsequent sugar additions lead to cerebrosides and gangliosides.

Prostaglandins

The prostaglandins are produced from arachidonic acid released by phospholipase A_2 action on phospholipids. Production of these local hormones is blocked by aspirin, and nonsteroid anti-inflammatory agents such as ibuprofen and phenylbutazone.

Cholesterol

Cholesterol derives from HMG-CoA, a product we already encountered in ketone body formation. You might recall that ketone bodies are produced from acetyl-CoA units. HMG-CoA is a six-carbon CoA derivative produced from three acetyl units. The rate limiting step in cholesterol synthesis is formation of 3*R*-mevalonate from HMG-CoA by HMG-CoA reductase, which catalyzes two NADPH-dependent reductions. This enzyme is carefully regulated by 1) phosphorylation leading to inactivation, 2) degradation, and 3) gene expression. Mevalonate, a six-carbon intermediate, is converted to isopentenyl pyrophosphate, which is used to synthesize cholesterol. Cholesterol is the precursor of bile salts and the steroid hormones. You should understand how lipoproteins are responsible for movement of cholesterol and other lipids in the body.

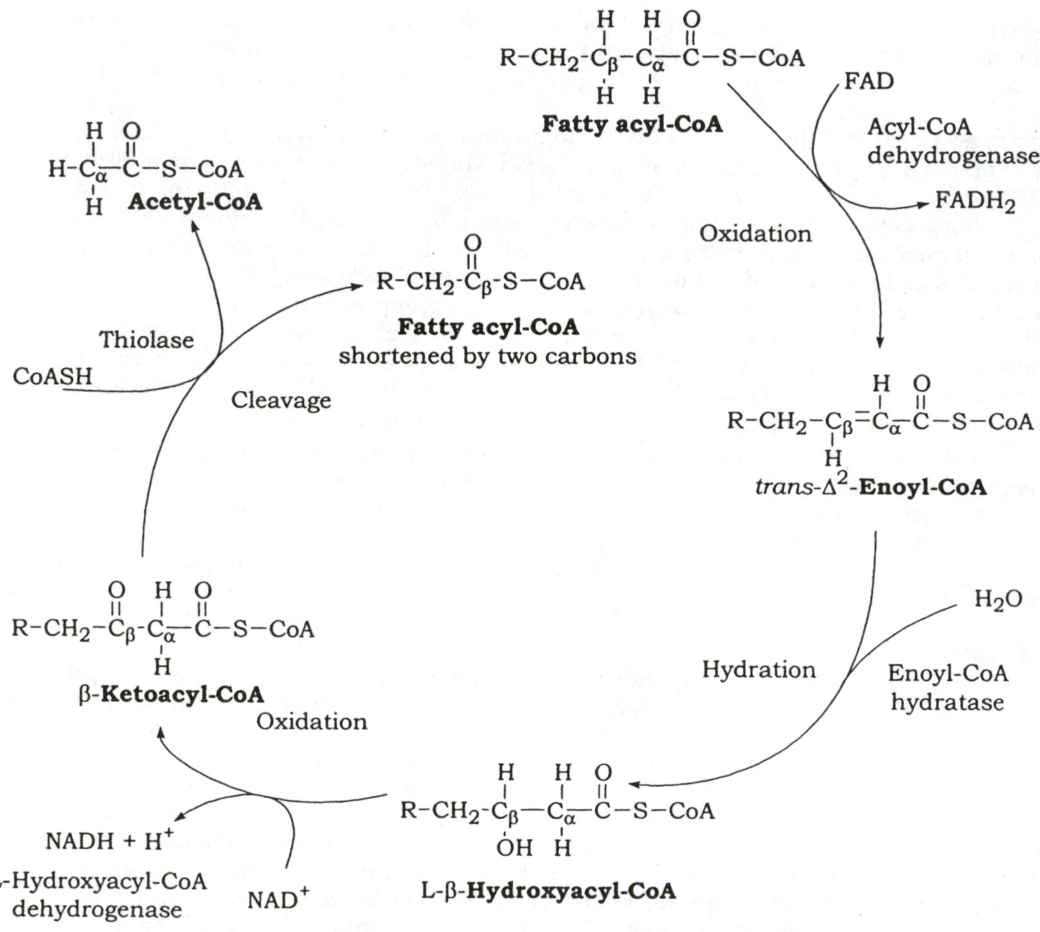

Figure 20.7 The β-oxidation of saturated fatty acids involves a cycle of four enzyme-catalyzed reactions. Each cycle produces single molecules of $FADH_2$, NADH, and acetyl-CoA and yields a fatty acid shortened by two carbons. (The delta [Δ] symbol connotes a double bond, and its superscript indicates the lowest-numbered carbon involved.)

Problems and Solutions

1. Calculate the volume of metabolic water available to a camel through fatty acid oxidation if it carries 30 lb of triacylglycerol in its hump.

Answer: If we assume that the fatty acid chains in the triacylglycerol are all palmitic acid, the fatty acid content of a triacylglycerol as a percent of the total molecular weight is calculated as follows:

$$\% \text{ Palmitic acid} = \frac{3 \times M_{r,\text{palmitic acid}}}{3 \times M_{r,\text{palmitic acid}} + M_{r,\text{glycerol}} - 3 \times M_{r,\text{water}}} \times 100\%$$

$$M_{r,\text{palmitic acid}} = 256.4\,(C_{16}H_{32}O_2),\ M_{r,\text{glycerol}} = 92.09\,(C_3H_8O_3),\ M_{r,\text{water}} = 18$$

$$\% \text{ Palmitic acid} = \frac{3 \times 256.4}{3 \times 256.4 + 92.09 - 3 \times 18} \times 100\% = 95.3\%$$

Thirty pounds of triacylglycerol corresponds to:

$$30\ \text{lb} \times \frac{453.6\ \text{g}}{\text{lb}} = 13{,}608\ g \text{ of triacylglycerol}$$

If 95.3% of this is composed of palmitic acid we have:

$$0.953 \times 13{,}608\ g = 13{,}608\ g \text{ of palmitic acid which corresponds to:}$$

$$\frac{13{,}608\ \text{g}}{256.4 \frac{\text{g}}{\text{mol}}} = 50.6 \text{ mol of palmitic acid}$$

β-oxidation of palmitate produces 130 moles of H_2O per mole of palmitate. Therefore, 50.6 moles of palmitate metabolized gives:

$$50.6\ \text{mol} \times 130 \frac{\text{mol water}}{\text{mol}} = 6{,}578 \text{ mol water}$$

At $18 \frac{\text{g}}{\text{mol}}$ this represents $6{,}578\ \text{mol} \times 18 \frac{\text{g}}{\text{mol}} = 118{,}404$ g water or 118.4 liters

2. Calculate the approximate number of ATP molecules that can be obtained from the oxidation of cis-11-heptadecenoic acid to CO_2 and water.

Answer: *cis*-11-Heptadecenoic acid is a 17-carbon fatty acid with a double bond between carbons 11 and 12. β-oxidation, ignoring for the moment the presence of the double bond, will produce 7 acetyl-CoA units and one propionyl-CoA. The 7 acetyl-CoA units are metabolized in the citric acid cycle with the following stoichiometry:

$$7 \text{ Acetyl-CoA} + 14\ O_2 + 70 \text{ ADP} + 70\ P_i \rightarrow 7 \text{ CoA} + 77\ H_2O + 14\ CO_2 + 70 \text{ ATP}$$

In producing the 7 acetyl-CoAs, 7 β-carbons had to be oxidized, and only 6 of these by the complete β-oxidation pathway. Thus, 6 FAD → 6 $FADH_2$, and 6 NAD^+ → 6 NADH. And, if electrons from $FADH_2$ produce 1.5 ATP, whereas electrons from NADH produce 2.5 ATPs, we have 9 ATPs from $FADH_2$ and 15 ATPs from NADH. Now we will deal with the double bond. The presence of a double bond means that a carbon is already partially reduced, so, the FAD-dependent reduction step is bypassed and only NADH is generated. Thus, one acetyl-CoA unit is generated along with only 2.5 ATPs. Finally, we consider the propionyl-CoA unit which is metabolized into succinyl CoA. In this pathway, propionyl-CoA is converted to methylmalonyl-CoA, a process driven in part by ATP hydrolysis. Succinyl-CoA is a citric acid cycle intermediate that is metabolized to oxaloacetate. This process yields 1 GTP, one FAD-dependent oxidation, and one NAD^+-dependent reaction. Thus a net of (1 + 1.5 + 2.5 - 1) 4 ATPs are produced. Succinyl-CoA cannot be consumed by the citric acid cycle and we have only converted it to oxaloacetate. One way of oxidizing succinate completely is to remove the carbons from the mitochondria as malate, and convert them to CO_2 and pyruvate using malic enzyme. This reaction is an oxidative-decarboxylation; $NADP^+$ is reduced in the cytosol. There is a slight reduction in ATP production potential for electrons on NADPH in the cytosol but we will ignore this and assume that the reduction is equivalent energetically to reduction of malate to oxaloacetate (which we have already taken into account). Pyruvate is metabolized to acetyl-CoA with production of NADH which supports 2.5 ATPs synthesized. The acetyl-CoA unit results in 10 ATP. To summarize: Oxidation of 7 β-carbons produce 24 ATPs; oxidation of an additional β-carbon (the one involved in a double bond) produces 2.5 ATPs; oxidation of 7 acetyl-CoAs contributes 70 ATPs; oxidation of an additional acetyl-CoA (derived from propionyl-CoA) yields 16.5 ATPs. The total is 113.

3. Phytanic acid, the product of chlorophyll that causes problems for individuals with Refsum's disease, is 3,7,11,15-tetramethylhexadecanoic acid. Suggest a route for its oxidation that is consistent with what you have learned in this chapter. (Hint: The

methyl group at C-3 effectively blocks hydroxylation and normal β-oxidation. You may wish to initiate breakdown in some other way.)

Answer: The structure of phytanic acid is:

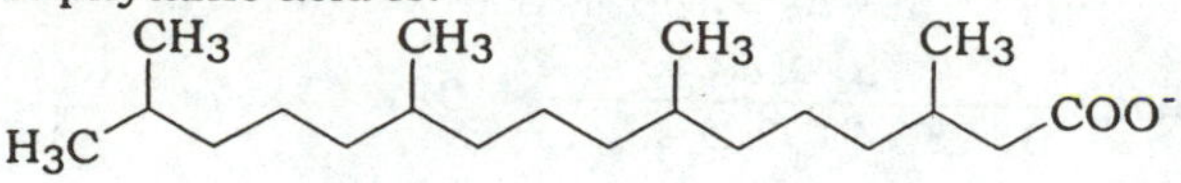

$$H_3C-CH(CH_3)-(CH_2)_3-CH(CH_3)-(CH_2)_3-CH(CH_3)-(CH_2)_3-CH(CH_3)-CH_2-COO^-$$

Normally it is metabolized by α-oxidation. The enzyme phytanic acid α-oxidase hydroxylates the α-carbon to produce hydroxyphytanic acid, which is decarboxylated by phytanate α-oxidase to produce pristanic acid. Pristanic acid can form a coenzyme A ester which is metabolized by β-oxidation to yield 3 propionyl-CoAs, 3 acetyl-CoAs and 2-methyl-propionyl-CoA. The reaction sequence is shown below.

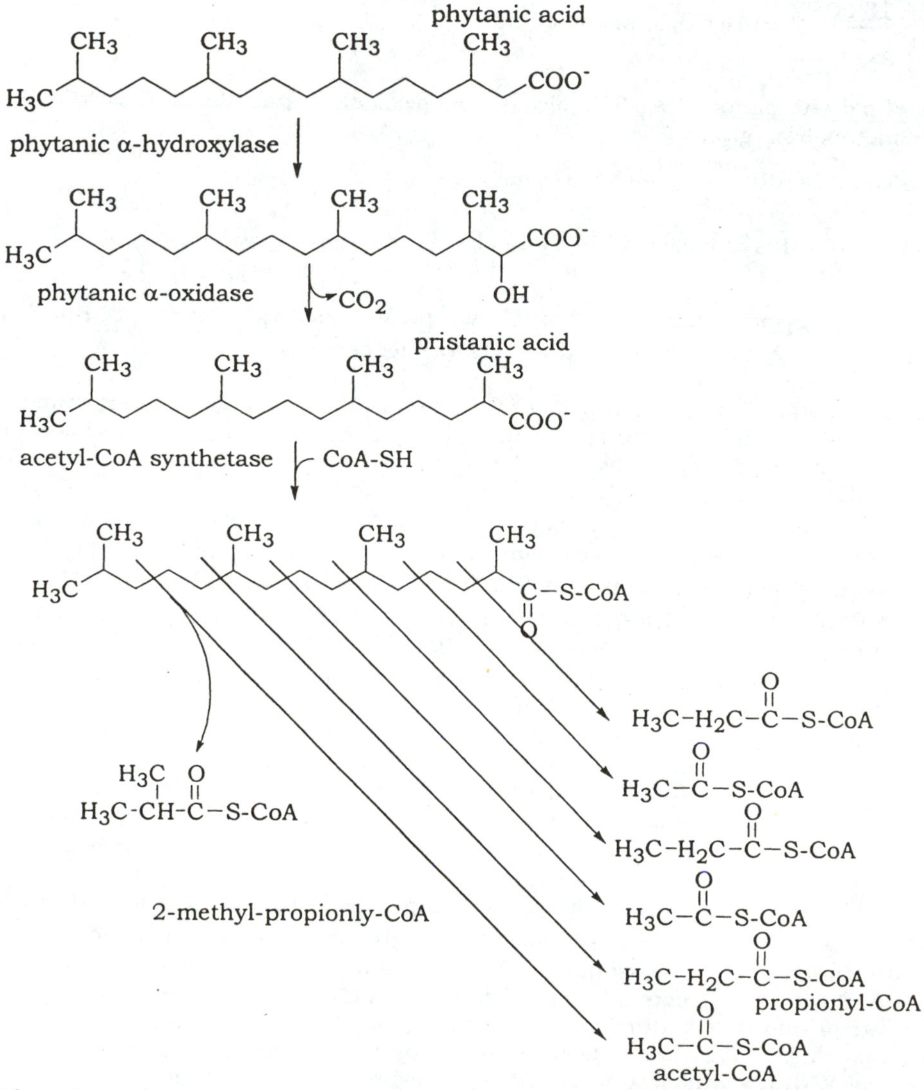

phytanic acid

$$H_3C-CH(CH_3)-(CH_2)_3-CH(CH_3)-(CH_2)_3-CH(CH_3)-(CH_2)_3-CH(CH_3)-CH_2-COO^-$$

phytanic α-hydroxylase ↓

$$H_3C-CH(CH_3)-(CH_2)_3-CH(CH_3)-(CH_2)_3-CH(CH_3)-(CH_2)_3-CH(CH_3)-CH(OH)-COO^-$$

phytanic α-oxidase ↓ → CO_2

pristanic acid

$$H_3C-CH(CH_3)-(CH_2)_3-CH(CH_3)-(CH_2)_3-CH(CH_3)-(CH_2)_3-CH(CH_3)-COO^-$$

acetyl-CoA synthetase ↓ ← CoA-SH

$$H_3C-CH(CH_3)-(CH_2)_3-CH(CH_3)-(CH_2)_3-CH(CH_3)-(CH_2)_3-CH(CH_3)-C(=O)-S\text{-}CoA$$

$H_3C-H_2C-C(=O)-S\text{-}CoA$

$H_3C-C(=O)-S\text{-}CoA$

$H_3C-H_2C-C(=O)-S\text{-}CoA$

$H_3C-C(=O)-S\text{-}CoA$

$H_3C-H_2C-C(=O)-S\text{-}CoA$ propionyl-CoA

$H_3C-C(=O)-S\text{-}CoA$ acetyl-CoA

$H_3C-CH(CH_3)-C(=O)-S\text{-}CoA$

2-methyl-propionly-CoA

4. Even though acetate units, such as those obtained from fatty acid oxidation, cannot be used for net synthesis of carbohydrate in animals, labeled carbon from ¹⁴C-labeled acetate can be found in newly synthesized glucose (for example in liver glycogen) in animal tracer studies. Explain how this can be. Which carbons of glucose would you expect to be the first to be labeled by ¹⁴C-labeled acetate?

Answer: Acetate, as acetyl-CoA, enters the citric acid cycle where its two carbons show up as carbons 1 and 2 or carbons 3 and 4 of oxaloacetate after one turn of the cycle. C-1 and C-4 label derive from acetate labeled at the carboxyl carbon, whereas C-2 and C-3 label derive from label at the methyl carbon of acetate. Oxaloacetate is converted to PEP with carbons 1, 2, and 3 becoming carbons 1, 2, and 3 of PEP, and so label is expected at carbon 1 if carboxy-labeled acetate is used and at carbons 2 or 3 if methyl-labeled acetate is used. Conversion of PEP to glyceraldehyde-3-phosphate results in label either at carbon 1 or at carbon 2 or 3. Isomerization to dihydroxyacetone phosphate labels the same carbons. Carbons 1 of DHAP and glyceraldehyde-3-phosphate become carbons 3 and 4 of fructose-1,6-bisphosphate so aldolase will label fructose-1,6-bisphosphate at carbons 3 and 4 from carboxy-labeled acetate and carbons 1, 2, 5, and 6 from methyl-labeled acetate.

5. Overweight individuals who diet to lose weight often view fat in negative ways because adipose tissue is the repository of excess caloric intake. However, the "weighty" consequences might be even worse if excess calories were stored in other forms. Consider a person who is 10 lb "overweight," and estimate how much more he or she would weigh if excess energy were stored in the form of carbohydrate instead of fat.

Answer: There are two problems to consider: carbohydrates are a more oxidized form of fuel than fats; and, storage of carbohydrates requires large amounts of water which will add greatly to its weight. From Table 20.1 we see that the energy content of fat and carbohydrate are 37 kJ/g and 16 kJ/g, respectively. Fat has 2.3 times (37 kJ/g ÷ 16 kJ/g) the energy content of carbohydrate on a per weight basis. Therefore, ten pounds of fat is equivalent to 23 pounds of carbohydrate.

6. What would be the consequences of a deficiency of vitamin B_{12} for fatty acid oxidation? What metabolic intermediates might accumulate?

Answer: Vitamin B_{12} is a coenzyme prosthetic group for methylmalonyl-CoA mutase. This enzyme is active in metabolism of odd-chain fatty acids. The last round of β–oxidation releases acetyl-CoA and propionyl-CoA from odd-chain fatty acids. Propionyl-CoA is further metabolized by being converted to (S)-methylmalonyl-CoA by propionyl-CoA carboxylase, a biotin-containing enzyme. (S)-Methylmalonyl-CoA racemase then converts (S)-methylmalonyl-CoA to (R)-methylmalonyl-CoA, which is metabolized to succinyl CoA by methylmalonyl-CoA mutase. This enzyme is a vitamin B_{12}-dependent enzyme, and a deficiency in vitamin B_{12} will cause a build-up of (R)-methylmalonyl-CoA. Depending on the amount of odd chain fatty acids (and propionic acid) in the diet, a deficiency of B_{12} is expected to act as a coenzyme A trap, slowly accumulating coenzyme A as methylmalonyl-CoA.

7. Write a properly balanced chemical equation for the oxidation to CO_2 and water of (a) myristic acid, (b) stearic acid, (c) α-linolenic acid, and (d) arachidonic acid.

Answer: a. Myristic acid is a saturated 14:0 fatty acid, $CH_3(CH_2)_{12}COOH$. To metabolize myristic acid, it is first activated to a coenzyme A derivative in the following reaction:

$$CH_3(CH_2)_{12}COOH + ATP + CoA\text{-}SH \rightarrow CH_3(CH_2)_{12}CO\text{-}S\text{-}CoA + AMP + PP_i$$

Six cycles of β-oxidation convert myristoyl-CoA to 7 acetyl-CoA units. The balanced equation for β-oxidation is:

$$CH_3(CH_2)_{12}CO\text{-}S\text{-}CoA + 6\ CoA\text{-}SH + 6\ H_2O + 6\ NAD^+ + 6\ FAD \rightarrow 7\ CH_3CO\text{-}S\text{-}CoA + 6\ NADH + 6\ H^+ + 6\ FADH_2$$

The citric acid cycle metabolizes acetyl-CoA units as follows:

$$7\ CH_3CO\text{-}S\text{-}CoA + 21\ H_2O + 21\ NAD^+ + 7\ FAD \rightarrow 14\ CO_2 + 7\ CoA\text{-}SH + 21\ NADH + 21\ H^+ + 7\ FADH_2$$

This is accompanied by substrate-level GDP phosphorylation which is equivalent to:

$$7\ ADP + 7\ P_i \rightarrow 7\ ATP + 7\ H_2O$$

Electron transport recycles NAD^+ as follows:

$$27\ NADH + 27\ H^+ + 13.5\ O_2 \rightarrow 27\ NAD^+ + 27\ H_2O$$

which supports the production of $2.5 \times 27\ (ADP + P_i \rightarrow ATP + H_2O)$ or

$$67.5\ ADP + 67.5\ P_i \rightarrow 67.5\ ATP + 67.5\ H_2O$$

FAD is recycled by:

$$13\ FADH_2 + 6.5\ O_2 \rightarrow 13\ FAD + 13\ H_2O$$

which supports the production of $1.5 \times 13\ (ADP + P_i \rightarrow ATP + H_2O)$ or

$$19.5\ ADP + 19.5\ P_i \rightarrow 19.5\ ATP + 19.5\ H_2O$$

The AMP and PP_i produced in the very first reaction can be metabolized by hydrolysis of PP_i and phosphorylation of AMP to ADP using ATP. Thus, $PP_i + H_2O \rightarrow 2\ P_i$ and $AMP + ATP \rightarrow 2\ ADP$:

$$AMP + ATP + PP_i + H_2O \rightarrow 2\ ADP + 2\ P_i$$

If we sum these equations we find:

$$CH_3(CH_2)_{12}COOH + 92\ ADP + 92\ P_i + 20\ O_2 \rightarrow 92\ ATP + 14\ CO_2 + 106\ H_2O$$

b. Stearic acid or octadecanoic acid is 18:0 and its oxidation is given by:

$$CH_3(CH_2)_{16}COOH + 120\ ADP + 120\ P_i + 26\ O_2 \rightarrow 120\ ATP + 18\ CO_2 + 138\ H_2O$$

c. α-Linolenic acid is a polyunsaturated 18-carbon fatty acid with double bonds at carbons 9, 12, and 15. In two out of a total of 8 rounds of β-oxidation, reduction by FAD is bypassed. Therefore 2 fewer moles of $FADH_2$ are reduced in the electron transport chain than for a fully saturated fatty acid and 1.0 fewer moles of O_2 are consumed. The amount of ATP is reduced by $1.5 \times 2 = 3.0$. In addition, a NADH was actually consumed after the fifth cycle of β-oxidation to resolve a conjugated double bond. This accounts for 2.5 few ATP and 0.5 fewer O_2 than in b. The balanced equation is:

$$CH_3CH_2(CH{=}CH\ CH_2)_3(CH_2)_6COOH + 114.5\ ADP + 114.5\ P_i + 24.5\ O_2 \rightarrow 114.5\ ATP + 18\ CO_2 + 129.5\ H_2O$$

d. Arachidonic acid is 5,8,11,14-eicosatetraenoic acid, a 20-carbon fatty acid with four double bonds. It undergoes a total of 9 cycles of β–oxidation, with 2 cycles not producing $FADH_2$ and consumption of two NADH to resolve two cases of conjugated double bonds. The balanced equation is:

$$CH_3(CH_2)_4(CH{=}CH\ CH_2)_4(CH_2)_2COOH + 126\ ADP + 126\ P_i + 27\ O_2 \rightarrow 126\ ADP + 20\ CO_2 + 142\ H_2O$$

8. How many tritium atoms are incorporated in acetate if a molecule of palmitic acid is oxidized in 100% tritiated water?

Answer: For palmitic acid to be converted to acetate, it must pass through β-oxidation, each cycle, except the last, will result in incorporation of tritium at the α-carbon, which is destined to become carbon 2 of acetate. You might recall that the four steps of β-oxidation are dehydrogenation (oxidation), hydration, dehydrogenation (oxidation), and thiolysis. During the hydration step, the elements of water are added across a carbon-carbon double bond. In the subsequent dehydrogenation, the hydroxyacyl group, is oxidized to a keto group, with loss of two hydrogens, one of which is derived from water. The other water hydrogen remains on the α-carbon. In the thiolase reaction, a proton is abstracted to form a carbanion that is subsequently reprotonated. The proton is donated by a group on the enzyme but it is likely to be an exchangeable proton and thus be tritiated. Thus, a C-2 carbon will have two tritium atoms, one from the hydration step and a second from the thiolase reaction.

9. What would be the consequences of a carnitine deficiency for fatty acid oxidation?

Answer: Carnitine functions as a carrier of activated fatty acids across the inner mitochondrial membrane. A deficiency of carnitine is expected to effectively block fatty acid metabolism by preventing mitochondrial uptake of fatty acids.

10. Use the relationships shown in Figure 20.21 to determine which carbons of glucose will be incorporated into palmitic acid. Consider the cases of both citrate that is immediately exported to the cytosol following its synthesis and citrate that enters the TCA cycle.

Answer: The six carbons of glucose are converted into two molecules each of CO_2 and acetyl units of acetyl-coenzyme A. Carbons 1, 2, and 3 of glyceraldehyde derive from carbons 3 and 4, 2 and 5, and 1 and 6 of glucose respectively. Carbon 1 of glyceraldehyde is lost as CO_2 in conversion to acetyl-CoA, so we expect no label in palmitic acid from glucose labeled only at carbons 3 and 4. The carbonyl carbon and the methyl carbon of the acetyl group of acetyl-CoA derive from carbons 2 and 5, and carbons 1 and 6 of glucose, respectively. The methyl carbon is incorporated into palmitoyl-CoA at every even-numbered carbon, whereas the carbonyl carbon is incorporated at every odd-numbered carbon.

Acetyl-CoA is produced in the mitochondria and exported to the cytosol for fatty acid biosynthesis by being converted to citrate. The cytosolic enzyme, citrate lyase, converts citrate to acetyl-CoA and oxaloacetate. When newly synthesized citrate is immediately exported to the cytosol, the labelling pattern described above will result. However, where citrate is instead metabolized in the citric acid cycle, back to oxaloacetate, label derived from acetyl-CoA shows up at carbons 1, 2, 3 and 4 of oxaloacetate. These carbons do not get incorporated into palmitoyl-CoA.

11. Based on the information presented in the text and in Figures 20.23, suggest a model for the regulation of acetyl-CoA carboxylase. Consider the possible roles of subunit interaction, phosphorylation, and conformation changes in your model.

Answer: Acetyl-CoA carboxylase catalyzes the formation of malonyl-CoA, the committed step in synthesis of fatty acids. This enzyme is a polymeric protein composed of protomers, or subunits, of 230 kD. In the polymeric form, the enzyme is active whereas in the protomeric form the enzyme is inactive. Polymerization is regulated by citrate and palmitoyl-CoA such that citrate, a metabolic signal for excess acetyl units, favors the polymeric and, therefore, active form of the enzyme whereas palmitoyl-CoA shifts the equilibrium to the inactive form. The activity of acetyl-CoA carboxylase is also under hormonal regulation. Glucagon and epinephrine stimulate cyclic AMP-dependent protein kinase that will phosphorylate a large number of sites on the enzyme. The phosphorylated form of the enzyme binds citrate poorly and citrate binding occurs only at high citrate levels. Citrate is a tricarboxylic acid with three negative charges and its binding site on the enzyme is likely to be composed of positively-charged residues. Phosphorylation introduces negative charges which may be responsible for the decrease in citrate binding.

In the phosphorylated form, low levels of palmitoyl-CoA will inhibit the enzyme. Thus, the enzyme is sensitive to palmitoyl-CoA binding and to depolymerization in the phosphorylated form. If we assume that the palmitoyl-CoA binding site is located at a subunit-subunit interface, and that phosphorylated, and hence negatively charged subunits interact with lower affinity than do unphosphorylated subunits, we see that it is easier for palmitoyl-CoA to bind to the enzyme.

12. Consider the role of the pantothenic acid groups in animal fatty acyl synthase and the size of the pantothenic acid group itself, and estimate a maximal separation between the malonyl transferase and the ketoacyl-ACP synthase active sites.

Answer: In fatty acyl synthase, pantothenic acid is attached to a serine residues as shown below.

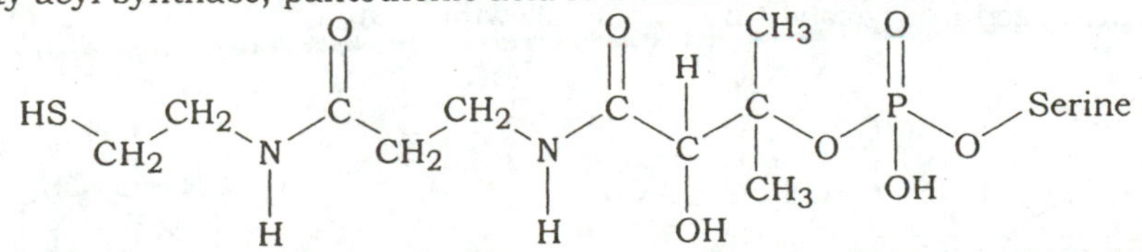

HS–CH$_2$–CH$_2$–N(H)–C(=O)–CH$_2$–CH$_2$–N(H)–C(=O)–C(H)(OH)–C(CH$_3$)(CH$_3$)–O–P(=O)(OH)–O–Serine

The approximate distance from the pantothenic group to the α-carbon of serine is calculated as follows. For carbon-carbon single bonds the bond length is approximately 0.15 nm. The distance between carbon atoms is calculated as follows.

109°
0.15 nm
35.5°
35.5°
d

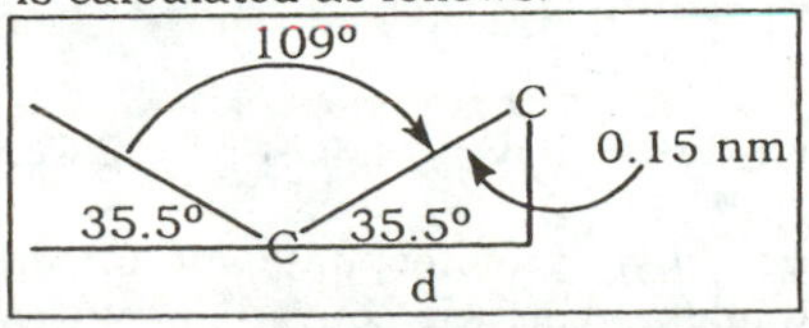

$$d = 0.15\ \text{nm} \times \cos(35.5^\circ) = 0.12\ \text{nm}$$

Let us use this length for carbon-carbon single bonds, carbon-oxygen bonds, oxygen-phosphorous bonds, and carbon-nitrogen bonds exclusive of the amide bond. For the amide bond we will use a distance of 0.132 nm. The overall length is approximately 1.85 nm from the α-carbon of serine to the sulfur. The maximal separation between malonyl transferase and ketoacyl-ACP is about twice this distance or approximately 3.7 nm. The actual distance between these sites is smaller than this upper limit.

13. Write a balanced, stoichiometric reaction for the synthesis of phosphatidylethanolamine from glycerol, fatty acyl-CoA, and ethanolamine. Make an estimate of the $\Delta G°'$ for the overall process.

Answer: The synthesis of phosphatidylethanolamine involves the convergence of two separate pathways: A diacylglycerol backbone is synthesized from glycerol and fatty acids; ethanolamine is phosphorylated and activated by transfer to CTP to produce CDP-ethanolamine. CDP-ethanolamine: 1,2-diacylglycerol phosophoethanolamine transferase then catalyzes the formation of phosphatidylethanolamine from diacyl-glycerol and CDP-ethanolamine.

Starting from glycerol, production of diacylglycerol involves the following reactions:

$$\text{Glycerol} + ATP^{4-} \rightarrow \text{glycerol-3-phosphate} + ADP^{3-} + H^+$$
$$\text{Glycerol-3-phosphate} + \text{fatty acyl-CoA} \rightarrow \text{lysophosphatidic acid} + \text{CoA-SH}$$
$$\text{Lysophosphatidic acid} + \text{fatty acyl-CoA} \rightarrow \text{phosphatidic acid} + \text{CoA-SH}$$
$$\text{Phosphatidic acid} + H_2O \rightarrow \text{diacylglycerol} + P_i^{2-}$$

We have:

$$\text{Glycerol} + 2\text{ fatty acyl-CoA} + H_2O + ATP^{4-} \rightarrow \text{diacylglycerol} + 2\text{ CoA-SH}^+\ ADP^{3-} + P_i^{2-} + H^+$$

Production of CDP-ethanolamine involves the following:

$$\text{Ethanolamine} + ATP^{4-} \rightarrow \text{phosphoethanolamine} + ADP^{3-} + H^+$$
$$\text{Phosphoethanolamine} + CTP^{4-} \rightarrow \text{CDP-ethanolamine} + PP_i^{\,4-}$$
$$PP_i^{\,4-} + H_2O \rightarrow 2\ P_i^{2-}$$

Or,

$$\text{Ethanolamine} + ATP^{4-} + CTP^{4-} + H_2O \rightarrow \text{CDP-ethanolamine} + ADP^{3-} + 2\ P_i^{2-} + H^+$$

Finally, for the reaction catalyzed by CDP-ethanolamine: 1,2-diacylglycerol phosophoethanolamine transferase we have:

$$\text{Diacylglycerol} + \text{CDP-ethanolamine} \rightarrow \text{phosphatidylethanolamine} + CMP^{2-} + H^+$$

The balanced, stoichiometric reaction is:

$$\text{Glycerol} + \text{ethanolamine} + 2\text{ fatty acyl-CoA} + 2\ ATP^{4-} + CTP^{4-} + 2\ H_2O \rightarrow$$
$$\text{Phosphatidylethanolamine} + 2\text{ CoA-SH} + 2\ ADP^{3-} + 2\ H^+ + 3\ P_i^{2-} + CMP^{2-}$$

14. Write a balanced, stoichiometric reaction for the synthesis of cholesterol from acetyl-CoA.

Answer: The immediate precursors of cholesterol are isopentenyl pyrophosphate and dimethylallyl pyrophosphate, both of which derive from hydroxymethylglutaryl-CoA (HMG-CoA). HMG-CoA is synthesized from acetyl-CoA by the following route:

$$H_3C{-}\overset{O}{\overset{\|}{C}}{-}S\text{-CoA} + H_3C{-}\overset{O}{\overset{\|}{C}}{-}S\text{-CoA} \xrightarrow{\;-\text{CoA-SH}\;} H_3C{-}\overset{O}{\overset{\|}{C}}{-}CH_2{-}\overset{O}{\overset{\|}{C}}{-}S\text{-CoA} \xrightarrow[-\text{CoA-SH}]{H_2O\ +\ H_3C{-}\overset{O}{\overset{\|}{C}}{-}S\text{-CoA}} {}^-OOC{-}CH_2{-}\overset{OH}{\underset{CH_3}{C}}{-}CH_2{-}\overset{O}{\overset{\|}{C}}{-}S\text{-CoA}$$

The reaction is:

$$3\text{ Acetyl-CoA} + H_2O \rightarrow \text{HMG-CoA} + 2\text{ CoA-SH}$$

HMG-CoA is anabolized into isopentenyl pyrophosphate and dimethylallyl pyrophosphate, both of which are used to synthesize squalene which is converted by way of lanosterol into cholesterol. (The next question asks us to trace carbons from mevalonate to cholesterol, so it is worthwhile now to look at these reactions in detail).

Synthesis of isopentenyl pyrophosphate from HMG-CoA is as follows:

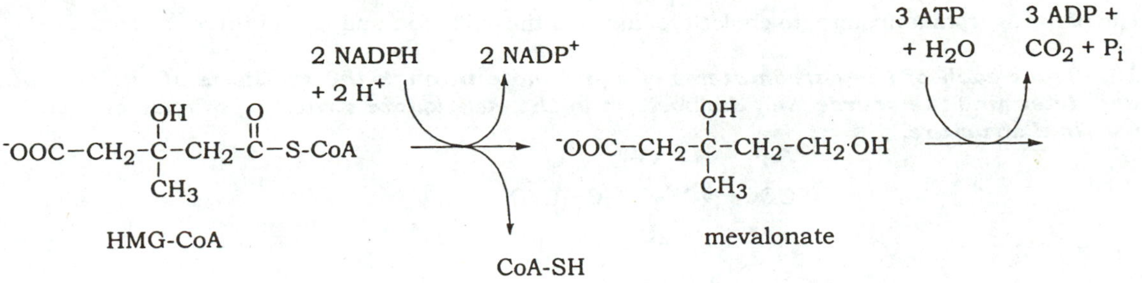

$$^{-}OOC{-}CH_2{-}C(OH)(CH_3){-}CH_2{-}C(=O){-}S{-}CoA \xrightarrow[\text{CoA-SH}]{2\,NADPH + 2\,H^+ \rightarrow 2\,NADP^+} {}^{-}OOC{-}CH_2{-}C(OH)(CH_3){-}CH_2{-}CH_2OH \xrightarrow{3\,ATP + H_2O \rightarrow 3\,ADP + CO_2 + P_i}$$

HMG-CoA   mevalonate

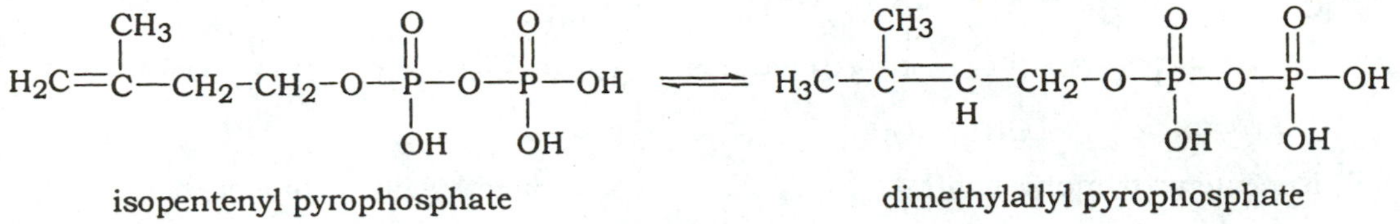

$$H_2C{=}C(CH_3){-}CH_2{-}CH_2{-}O{-}P(=O)(OH){-}O{-}P(=O)(OH){-}OH \rightleftharpoons H_3C{-}C(CH_3){=}C(H){-}CH_2{-}O{-}P(=O)(OH){-}O{-}P(=O)(OH){-}OH$$

isopentenyl pyrophosphate   dimethylallyl pyrophosphate

Overall, the reaction is:

HMG-CoA + 2 NAPDH + 2 H^+ + H_2O + 3 ATP → isopentenyl pyrophosphate (or dimethylallyl pyrophosphate) + CoA-SH + 2 $NADP^+$ + 3 ADP + P_i + CO_2

Production of squalene using isopentenyl pyrophosphate and dimethylallyl pyrophosphate proceeds as follows. Two farnesyl pyrophosphates are produced from two dimethylallyl pyrophosphates and four isopentenyl pyrophosphates. The farnesyl pyrophosphates are reacted to produce squalene as follows:

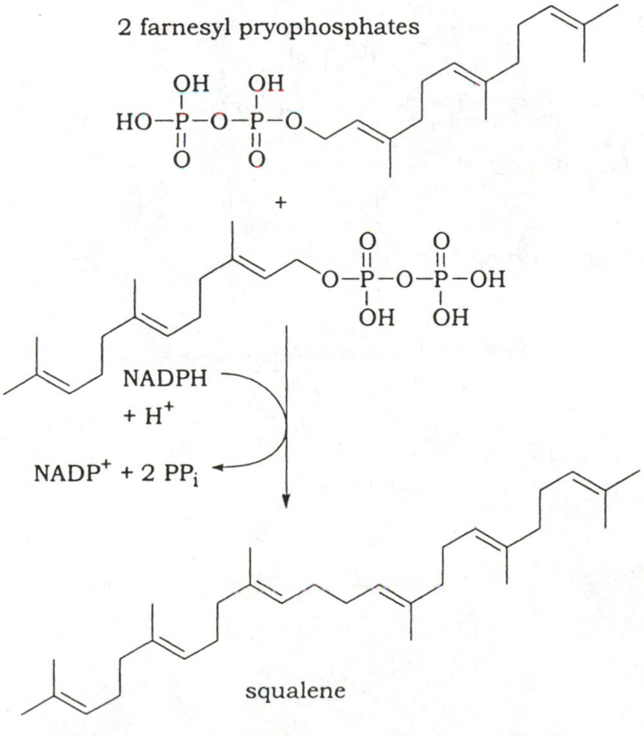

2 farnesyl pryophosphates

HO–P(O)(OH)–O–P(O)(OH)–O–(farnesyl)

+

(farnesyl)–O–P(O)(OH)–O–P(O)(OH)–OH

NADPH + H^+ → $NADP^+$ + 2 PP_i

squalene

Squalene is converted to lanosterol in two steps catalyzed by squalene epoxidase and squalene oxidocyclase. Squalene + 0.5 O_2 + NADPH → lanosterol

The overall equation for acetyl-CoA to lanosterol is:

$$18\ \text{Acetyl-CoA} + 13\ NADPH + 13\ H^+ + 18\ ATP + 0.5\ O_2 \rightarrow$$
$$\text{Lanosterol} + 18\ \text{CoA-SH} + 13\ NADP^+ + 18\ ADP + 6\ P_i + 6\ PP_i + 6\ CO_2$$

The pathway from lanosterol to cholesterol involves the oxidation and loss of three carbons.

15. Trace each of the carbon atoms of mevalonate through the synthesis of cholesterol, and determine the source (i.e., the position in the mevalonate structure) of each carbon in the final structure.

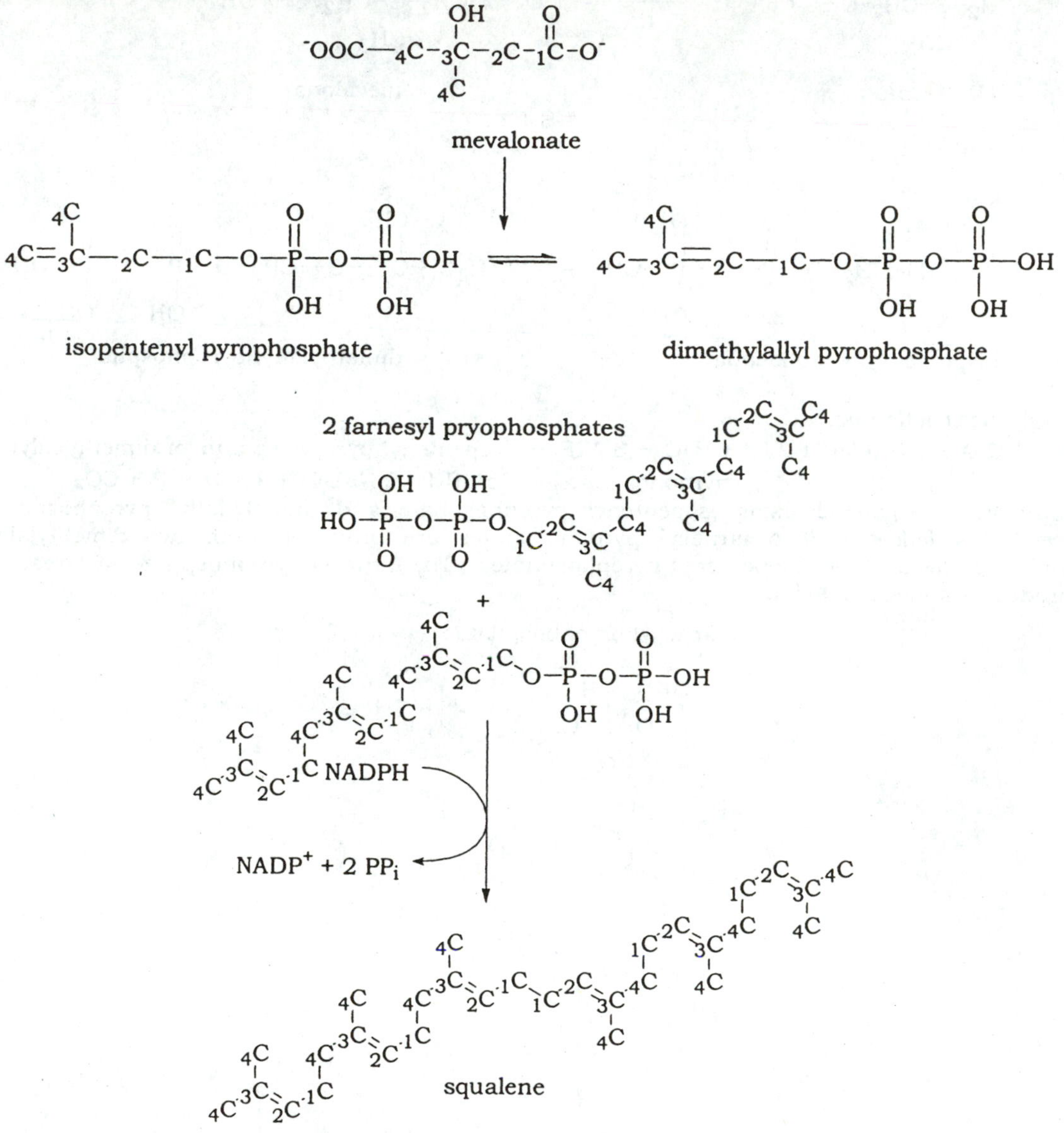

$$^{-}OOC{-}{}_4C{-}{}_3C(OH)({}_4C){-}{}_2C{-}{}_1C(=O){-}O^{-}$$

mevalonate

$${}_4C{=}{}_3C({}_4C){-}{}_2C{-}{}_1C{-}O{-}P(=O)(OH){-}O{-}P(=O)(OH){-}OH \rightleftharpoons {}_4C{-}{}_3C({}_4C){=}{}_2C{-}{}_1C{-}O{-}P(=O)(OH){-}O{-}P(=O)(OH){-}OH$$

isopentenyl pyrophosphate | dimethylallyl pyrophosphate

2 farnesyl pryophosphates

$$HO{-}P(=O)(OH){-}O{-}P(=O)(OH){-}O{-}{}_1C{-}{}_2C{=}{}_3C(C_4){-}C_4{-}{}_1C{-}{}_2C{=}{}_3C(C_4){-}C_4{-}{}_1C{-}{}_2C{=}{}_3C(C_4){-}C_4$$

+

$${}_4C{-}{}_3C({}_4C){=}{}_2C{-}{}_1C{-}{}_4C{-}{}_3C({}_4C){=}{}_2C{-}{}_1C{-}{}_4C{-}{}_3C({}_4C){=}{}_2C{-}{}_1C{-}O{-}P(=O)(OH){-}O{-}P(=O)(OH){-}OH$$

NADPH → NADP^{+} + 2 PP$_i$

$${}_4C{-}{}_3C({}_4C){=}{}_2C{-}{}_1C{-}{}_4C{-}{}_3C({}_4C){=}{}_2C{-}{}_1C{-}{}_4C{-}{}_3C({}_4C){=}{}_2C{-}{}_1C{-}{}_1C{-}{}_2C{=}{}_3C({}_4C){-}{}_4C{-}{}_1C{-}{}_2C{=}{}_3C({}_4C){-}{}_4C{-}{}_1C{-}{}_2C{=}{}_3C({}_4C){-}{}_4C$$

squalene

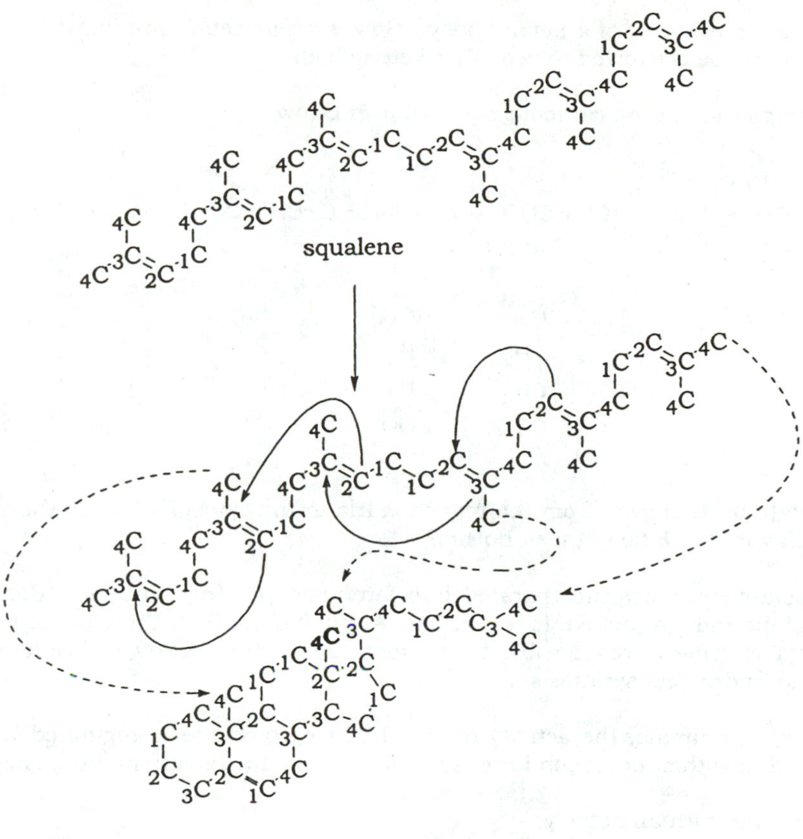

squalene

cholesterol

16. Identify the lipid synthetic pathways that would be affected by abnormally low levels of CTP.

Answer: Phosphatidylethanolamine and phosphatidylcholine synthesis depend on the formation of CDP-ethanolamine and CDP-choline respectively. Phosphatidyl-inositol and phosphatidylglycerol biosynthesis utilize CDP-diacylglycerol. The synthetic pathways of all of these compounds may be affected if the cell experiences low levels of CTP.

Questions for Self Study

1. What is the difference between initiation of catabolism of short-chain fatty acids and long-chain fatty acids?

2. β-Oxidation consists of cycles of a sequence of oxidation, hydration, oxidation and cleavage. The first three steps in this sequence are reminiscent of the metabolism of which steps in the citric acid cycle?

3. During β-oxidation a double bond is transiently produced in the fatty acid chain. How is this double bond different than double bonds found in mono- and polyunsaturated fatty acids?

4. Fill in the blanks. Odd-carbon fatty acids are metabolized by β-oxidation until the last three carbons are released as _____. This compound is converted to _____, a citric acid cycle intermediate in a series of three steps. In the first step, a carboxyl group is added by propionyl-CoA carboxylase. This enzyme uses the water soluble vitamin _____ during catalysis. The third reaction is catalyzed by methylmalonyl-CoA mutase, a vitamin ____ dependent enzyme.

5. Acetoacetate is an example of a ketone body. How is acetoacetate produced from acetyl-CoA? How can acetoacetate be converted to two other ketone bodies?

6. Label the compounds in the catabolic series shown below.

$$H_3C-\underset{OH}{\overset{H}{C}}-CH_2-COO^- \xrightarrow{NAD^+} H_3C-\overset{O}{\overset{\|}{C}}-CH_2-COO^- \longrightarrow H_3C-\overset{O}{\overset{\|}{C}}-CH_2-\overset{O}{\overset{\|}{C}}-S\text{-}CoA \xrightarrow{CoA\text{-}SH} 2\ H_3C-\overset{O}{\overset{\|}{C}}-S\text{-}CoA$$

(second step: $O{=}C(S\text{-}CoA)-CH_2-CH_2-COO^-$ → $COO^--CH_2-CH_2-COO^-$)

7. How are acetate units moved from the mitochondria to the cytosol? What other role does the acetate carrier play in regulation of metabolism?

8. Although acetate units are incorporated into fatty acids during synthesis, they derive from three-carbon compounds attached to coenzyme A. What is this three-carbon coenzyme A derivative? What enzyme is responsible for its formation? How many high-energy phosphate bonds are cleaved to drive its synthesis?

9. Describe how, in animals, the activity of acetyl-CoA carboxylase is regulated by citrate and palmitoyl-CoA and how this regulation is sensitive to covalent modification of the enzyme.

10. Match an enzyme with an activity.

a. Acetyltransferase	1. Keto carbon converted to alcoholic carbon.
b. Dehydrase	2. Attaches malonyl group to fatty acid synthase.
c. Malonyltransferase	3. Attaches acetyl group to acyl carrier protein.
d. Enoyl reductase	4. Carbon-carbon double bond reduced.
e. β-Ketoacyl reductase	5. Condensation of acetyl group and malonyl group.
f. β-Ketoacyl synthase	6. Enoyl intermediate formed.

11. From the following list of compounds identify those that are cholesterol derivatives and appropriately identify each cholesterol derivative as a hormone (H), bile salt (B), or vitamin (V).

a. Prostaglandin D_2
b. Glycocholic acid.
c. Squalene
d. Testosterone
e. Arachidonic Acid.
f. Cortisol
g. Cholecalciferol
h. Progesterone
i. Thromboxanes
j. Taurocholic acid
k. Aldosterone

12. What is the rate-limiting step in cholesterol biosynthesis?

13. In eukaryotes, glycerolipids are all derived from phosphatidic acid. Draw the structure of phosphatidic acid and outline its biosynthesis from dihydroxyacetone-phosphate and from glycerol-3-phosphate.

14. What is the role of cytidine in lipid biosynthesis?

15. Fill in the blanks. The prostaglandins are _____ that function locally and at very low concentrations. They are synthesized from _____, a 20-carbon polyunsaturated fatty acid.

Mammals can produce this fatty acid from ____ ($18{:}2^{\Delta 9,12}$) but must acquire this polyunsaturated fatty acid from their diet.

Answers

1. Short-chain fatty acids are transported into the mitochondria as free fatty acids where they are converted to acyl-CoA derivatives. Long-chain fatty acids are attached to carnitine and moved across the inner membrane as O-acylcarnitine. They are then transferred to coenzyme A.

2. Succinate to fumarate to malate to oxaloacetate.

3. The double bond transiently produced during fatty acid catabolism is *trans* whereas double bonds found in naturally occurring fatty acids are usually *cis*. Also, double bonds in naturally occurring fatty acids may start at even-numbered carbons as well as odd-numbered carbons in the fatty acid chain.

4. Propionyl-CoA; succinyl-CoA; biotin, B_{12}.

5. Acetoacetate is produced in a three-step metabolic sequence from acetyl-CoA. Acetoacetyl-CoA is produced in the first step and subsequently converted to β-hydroxy-β-methylglutaryl-CoA (HMG-CoA). HMG-CoA is then metabolized to acetoacetate and acetyl-CoA. Decarboxylation of acetoacetate leads to production of acetone. Reduction of acetoacetate leads to production of β-hydroxybutyrate.

6. β-Hydroxybutyrate, acetoacetate, acetoacetyl-CoA, and acetyl-CoA are across the top line. Succinyl-CoA and succinate are shown on the bottom.

7. Acetate units on acetyl-CoA are used to produce citrate in the mitochondria. Citrate is exported to the cytosol where it is converted to acetyl-CoA and oxaloacetate. Citrate inhibits phosphofructokinase and thus serves as a regulator of glycolysis. It also stimulates fatty acid synthesis.

8. Malonyl-CoA is produced by acetyl-CoA carboxylase at the expense of one high-energy phosphate bond.

9. Acetyl-CoA carboxylase is active in a polymeric state. The equilibrium between active polymer and inactive protomers is affected by citrate and palmitoyl-CoA. Citrate is an allosteric activator of the enzyme and shifts the equilibrium to the polymer. Palmitoyl-CoA shifts the equilibrium to the inactive, protomeric state. The enzyme is phosphorylated by a number of kinases and the phosphorylated state has a low affinity for citrate and a high affinity for palmitate.

10. a. 3; b. 6; c. 2; d. 4; e. 1; f. 5.

11. b. B; d. H; f. H; g. V; h. H; j. B; k. H.

12. The production 3*R*-mevalonate from HMG-CoA catalyzed by HMG-CoA reductase.

13. Dihydroxyacetone phosphate is converted to 1-acyldihydroxyacetone-phosphate by an acyltransferase reaction and reduced to 1-acylglycerol-3-phosphate by a reductase. This compound can also be synthesized from glycerol-3-phosphate by acyltransferase. Transfer of a second acyl group to C-2 produces phosphatidic acid whose structure is shown below.

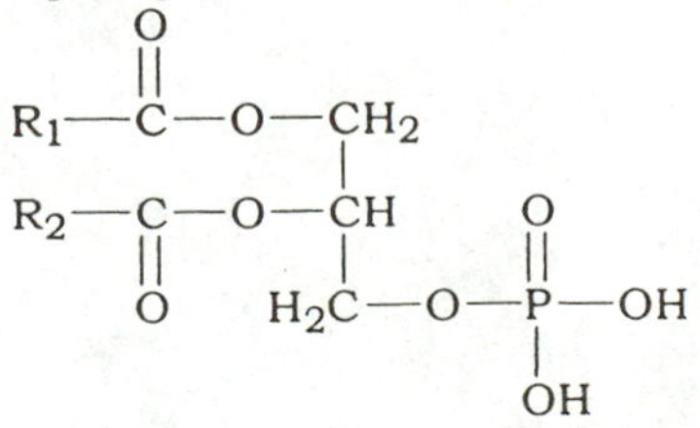

```
         O
         ||
R1—C—O—CH2
              |
R2—C—O—CH       O
         ||       |        ||
         O     H2C—O—P—OH
                           |
                          OH
```

14. The head groups of phosphatidylethanolamine and phosphatidylcholine derive from cytidine diphosphate derivatives. CDP-diacylglycerol is a precursor of phosphatidylinositol, phosphatidylglycerol, and cardiolipin.

15. Hormones; arachidonic acid; linoleic acid.

Additional Problems

1. It has been suggested that conversion of gasoline-powered automobiles to ethanol will require larger fuel tanks to keep the same effective range. Can you suggest a simple reason why this is the case?

2. Production of free fatty acids from fat deposits involves three separate lipases, triacylglycerol lipase, diacylglycerol lipase, and monoacylglycerol lipase. Fatty acid production is also under hormonal regulation but only one of the lipases is sensitive. Which lipase is hormonally controlled, how is regulation achieved, and why does it make good metabolic sense to regulate this particular lipase?

3. β-Oxidation begins with conversion of a free fatty acid to an acyl-coenzyme A derivative, a reaction catalyzed by acyl-CoA synthetase. Explain why this enzyme is named synthetase.

4. Describe the chemistry involved in the four steps of β-oxidation.

5. Metabolism of fatty acids with *cis* double bonds requires one or two additional enzymes. Explain.

6. In insulin-dependent diabetes mellitus, insulin deficiency can lead to shock as a result of metabolic acidosis. A person in shock may have a detectable odor of acetone on their breath. Explain the source of acetone and the cause of acidosis.

7. What are the sources of carbons for fatty acid biosynthesis? What is the role of the citrate-malate-pyruvate shuttle in making carbon compounds available for fatty acid biosynthesis?

8. Movement of citrate out of the mitochondria coordinates glycolysis and fatty acid biosynthesis. Explain.

9. Name the three water soluble vitamins that are crucial to fatty acid synthesis and briefly describe the roles they play in this process.

10. Why do mammals require certain essential fatty acids in their diet?

11. Outline the synthesis of glycerophospholipid.

12. Lovastatin lowers serum cholesterol by interfering with HMG-CoA reductase, the enzyme that catalyzes the rate limiting step in cholesterol synthesis. The drug is administered as an inactive lactone that is activated by hydrolysis to mevinolinic acid, a competitive inhibitor of HMG-CoA reductase. Can you recall another lactone hydrolysis reaction encountered in an earlier chapter?

13. Synthesis of the steroid hormones from cholesterol starts with the reaction catalyzed by desmolase shown below. Why is this a critical reaction for formation of steroid hormones?

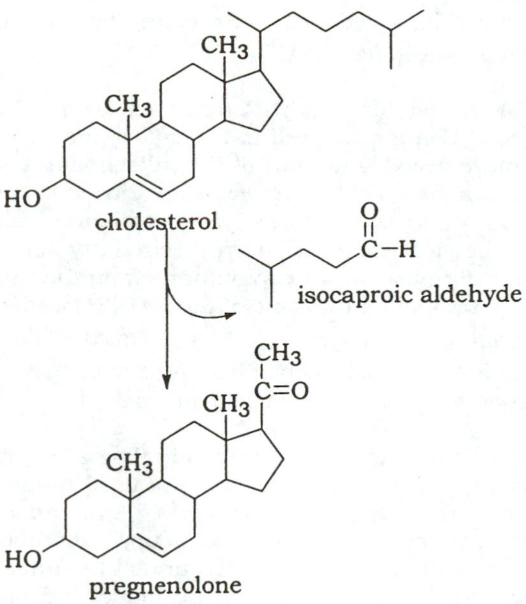

CH3
CH3
HO
cholesterol
isocaproic aldehyde
CH3
C=O
CH3
CH3
HO
pregnenolone

Abbreviated Answers

1. We are informed in Table 19.1 that fat has over twice the energy content of glycogen on a per gram basis. The reason is that the average carbon in fat is less oxidized than the average carbon in glycogen and so the amount of energy released in oxidation of carbon to CO_2 is greater. The same is true in comparing gasoline and ethanol. Gasoline is composed primarily of hydrocarbons whereas the average carbon in ethanol, C_2H_6O, is slightly more oxidized.

2. Adrenaline, glucagon, and adrenocorticotropic hormone mobilize fatty acids from adipocytes by stimulation of triacylglycerol lipase or hormone-sensitive lipase. The hormones bind to surface receptors, which results in the stimulation of adenylyl cyclase and an increase in intracellular cyclic AMP levels. cAMP activates protein kinase, which in turn phosphorylates triacylglycerol lipase leading to activation of this enzyme. Diacylglycerol lipase is active against the product of triacylglycerol lipase activity and monoacylglycerol lipase requires the product of diacylglycerol lipase activity. Thus, only the first enzyme in this series of enzymes needs to be regulated.

3. Synthetases use cleavage of high energy phosphoanhydride bonds to drive product formation. Acyl-CoA synthetase uses ATP to AMP + PP_i conversion to produce acyl-CoA.

4. Starting from acyl-CoA the α- and β-carbons are reduced to form *trans*-Δ^2enoyl-CoA and hydrated to L-β-hydroxyacyl-CoA. The β-carbon is oxidized to β-ketoacyl-CoA and cleaved to produce acetyl-CoA and an acyl-CoA two carbons shorter.

5. For *cis* double bonds starting at an odd carbon, β-oxidation proceeds normally until the double bond is between the β and γ carbons. An enoyl-CoA isomerase is used to rearrange the *cis*-Δ^3 bond to a *trans*-Δ^2 bond, which is hydrated, oxidized and cleaved normally. For *cis* double bonds starting at an even carbon, β-oxidation cycles eventually produce a *cis*-Δ^4 intermediate, which is converted to a *trans*-Δ^2,*cis*-Δ^4 intermediate by acyl-CoA dehydrogenase. Next, a 2,4-dienoyl-CoA reductase, in mammals, produces a *trans*-Δ^3 enoyl-CoA, which is converted to a *trans*-Δ^2 enoyl-CoA isomer by enoyl-CoA isomerase, a normal β-oxidation cycle intermediate. In *E. coli,* the double bound between carbons 4 and 5 is simply reduced by a 2,4-enoyl-CoA reductase to give *trans*-Δ^2 enoyl-CoA.

6. Acidosis and acetone are a consequence of ketone body formation. In liver mitochondria, acetoacetate and β-hydroxybutyrate are produced and excreted into the blood to be transported to other tissues to be used as fuels. These compounds are acidic with pK_a = 3.5 and in high

concentrations causes a drop in blood pH. Acetoacetate, an unstable compound, will nonenzymatically breakdown into acetone and CO_2.

7. The immediate source of carbons are acetyl-CoAs which are produced from carbohydrates, amino acids, and lipids. Acetyl-CoA is produced in the mitochondria but fatty acid biosynthesis occurs in the cytosol. To move acetyl units out of the mitochondria, they are condensed onto oxaloacetate to form citrate, in a citric acid cycle reaction. Citrate is then transported out of the mitochondria to the cytosol, where ATP-citrate lyase catabolizes citrate to acetyl-CoA and oxaloacetate. This cytosolic acetyl-CoA is used to synthesize fatty acids. So, the citrate-malate-pyruvate shuttle is responsible for moving two-carbon units from the mitochondria to the cytosol. However, it has another purpose: it supplies some of the NADPH needed for fatty acid synthesis. Cytosolic oxaloacetate is reduced to malate and then oxidatively decarboxylated to CO_2 and pyruvate by malic enzyme, in an $NADP^+$-dependent reaction. The NADPH thus formed is consumed during the reduction steps of fatty acid biosynthesis.

8. When glycolysis was covered, it was pointed out that phosphofructokinase activity is inhibited by citrate. In this chapter, we saw how citrate is used to move two-carbon units from the mitochondria to the cytosol for fatty acid biosynthesis. An increase in the concentration of citrate is a signal that the citric acid cycle is backing up, either because energy stores are satisfactory or because there is an abundance of acetyl units. In either case, there is not reason to continue glycolysis. Movement of citrate out of the mitochondria shifts acetyl units from degradation via the citric acid to storage via cytosolic fatty acid synthesis and serves to turn down glycolysis at phosphofructokinase.

9. Biotin is a component of acetyl-CoA carboxylase. Nicotinamide is found in NADPH. Phosphopantetheine is covalently attached to acyl carrier protein. Biotin functions as an intermediate carrier of activated carboxyl groups in malonyl-CoA biosynthesis by acetyl-CoA carboxylase. NADPH is required at two reduction steps in each round of chain elongation in fatty acid biosynthesis. Phosphopantotheine serves as a carrier of the growing fatty acid. This group is covalently attached to a serine residue in acyl carrier protein and serves to carry acetyl groups, malonyl groups, and acyl groups during various stages of fatty acid biosynthesis.

10. Mammals cannot introduce a double bond beyond C-9 in a given fatty acid. The prostaglandins are synthesized from linoleic acid, $\Delta^{9,12}$-octadecadienoic acid, which cannot be produced by mammals and is therefore an essential fatty acid.

11. The components of glycerophospholipids are glycerol, phosphate, fatty acids, and an alcoholic head group. Synthesis starts with either glycerol (via reduction of glyceraldehyde) or DHAP being converted to glycerol-3-phosphate by glycerokinase or glycerol-3-phosphate dehydrogenase, respectively. Glycerol-3-phosphate is converted to 1-acylglycerol-3-P and then to phosphatidic acid (1,2-diacylglycerol-3-P) by two acyltransferase reactions. Phosphatidic acid serves as the precursor for triacylglycerol and the glycerophospholipids phosphatidylethanolamine (PE), phosphatidylcholine (PC), phosphatidylserine (PS), phosphatidylinositol (PI), phosphatidylglycerol (PG), and cardiolipin (diphosphatidylglycerol). Phosphatidic acid is converted to diacylglycerol, which is converted to PE or PC by transferases using CDP-derivatized ethanolamine or choline. Alternatively, phosphatidic acid can be converted to its CDP derivative, CDP-diacylglycerol, which is metabolized to PI or PG. PS is produced from PE using serine to displace ethanolamine.

12. In the pentose phosphate pathway, conversion of 6-phosphogluconolactone to 6-phosphogluconate involves hydrolysis of a lactone.

13. The steroid hormones are transported in the blood to target tissues and must therefore be slightly more soluble than cholesterol. The reaction catalyzed by desmolase removes the hydrocarbon tail of cholesterol, making the product more soluble.

Summary

Fatty acids represent the principal form of energy storage for many organisms. Fatty acids are advantageous forms of stored energy, since 1) the carbon in fatty acids is highly reduced (thus yielding more energy in oxidation than other forms of carbon), and 2) fatty acids are not hydrated and can pack densely in storage tissues. Fatty acids are acquired readily in the diet and can also be synthesized from carbohydrates and amino acids. Whether dietary or synthesized, triacylglycerols are a major source of fatty acids. Release of fatty acids from triacylglycerols is triggered by hormones such as adrenaline and glucagon, which bind to plasma membrane receptors and activate (via cyclic AMP) the lipases, which hydrolyze triacylglycerols to fatty acids and glycerol. A large fraction of dietary triacylglycerols is absorbed in the intestines and passes into the circulatory system as chylomicrons.

Following Knoop's discovery that fatty acids were degraded to two-carbon units, Lynen and Reichart showed that these two-carbon units were acetyl-CoA. The process of β-oxidation of fatty acids begins with the formation of an acyl-CoA derivative by acyl-CoA synthetase, but transport of the acyl chain into the mitochondrial matrix occurs via acyl-carnitine derivatives, which are converted back to acyl-CoA derivatives in the matrix. β-Oxidation involves 1) formation of a C_{α}-C_{β} double bond, 2) hydration across the double bond and 3) oxidation of the β-carbon, followed by 4) a thiolase cleavage, leaving acetyl-CoA and the CoA ester of the fatty acid chain, shorter by two carbons. (A metabolite of hypoglycin, found in unripened akee fruit, inhibits the first of these reactions, resulting in vomiting, convulsions and death.) Repetition of this cycle of reactions degrades fatty acids with even numbers of carbons completely to acetyl-CoA. β-Oxidation of fatty acids with odd numbers of carbon atoms yields acetyl-CoA plus a single molecule of propionyl-CoA, which is subsequently converted to succinyl-CoA in a sequence of three reactions. Complete β-oxidation of a molecule of palmitic acid generates 129 high energy phosphate bonds, i.e., formation of 129 ATP from ADP and P_i. This massive output of energy from β-oxidation permits migratory birds to travel long distances without stopping to eat. The large amounts of water formed in β-oxidation allow certain desert and marine organisms to thrive without frequent ingestion of water.

Unsaturated fatty acids are also catabolized by β-oxidation, but two additional mitochondrial enzymes - an isomerase and a novel reductase - are required to handle the *cis*-double bonds of naturally occurring fatty acids. Degradation of polyunsaturated fatty acids also requires the activity of 2,4-dienoyl-CoA reductase. Although β-oxidation in mitochondria is the principal pathway of fatty acid catabolism, a similar pathway also exists in peroxisomes and glyoxysomes. Fatty acids with branches at odd carbons in the chain are degraded by α-oxidation. Phytol, an isoprene breakdown product of chlorophyll, is degraded via α-oxidation. Defects in this pathway lead to Refsum's disease, which is characterized by poor night vision, tremors and neurologic disorders and which is caused by accumulation in the body of phytanic acid. In the ER of eukaryotic cells, ω-oxidation leads to the synthesis of small amounts of dicarboxylic acids. This is accomplished in part by cytochrome P-450, which hydroxylates the terminal carbon atom of the fatty acid chain. Ketone bodies are a significant source of fuel and energy for certain tissues. Acetoacetate and 3-hydroxybutyrate are the preferred substrates for kidney cortex and heart muscle, and ketone bodies are a major energy source for the brain during periods of starvation. Ketone bodies represent easily transportable forms of fatty acids which move through the circulatory system without the need for complexation with serum albumin and other fatty acid binding proteins.

The biosynthesis of lipid molecules proceeds via mechanisms and pathways which are different from those of their degradation. In the synthesis of fatty acids, for example, 1) intermediates are linked covalently to the -SH groups of acyl carrier proteins instead of coenzyme A, 2) synthesis occurs in the cytosol instead of the mitochondria, 3) the nicotinamide coenzyme used is NADPH instead of NADH, and 4) in eukaryotes, the enzymes of fatty acid synthesis are associated in one large polypeptide chain instead of being separate enzymes. Fatty acids are synthesized by the addition of two carbon acetate units which have been activated by the formation of malonyl-CoA, decarboxylation of which drives the reaction forward. Once the growing fatty acid chain reaches 16 carbons in length, it dissociates from the fatty acid synthase and is subject to the introduction of unsaturations or additional elongation. Acetyl-CoA needed for fatty acid synthesis is provided in the cytosol by citrate, which is transported across the mitochondrial membrane and converted to acetyl-CoA and oxaloacetate by ATP-citrate lyase. Formation of malonyl-CoA by acetyl-CoA carboxylase (ACC), a biotin-dependent enzyme, commits acetate units to fatty acid synthesis. In animals, ACC is a multifunctional protein which forms long, filamentous polymers. It is allosterically activated (and polymerized) by citrate and inhibited (and depolymerized) by palmitoyl-CoA. Affinities for both these regulators are decreased by phosphorylation of the enzyme at up to 8 to 10 separate sites. The fatty acid synthesis reactions involve formation of O-acetyl and O-malonyl enzyme intermediates, followed by transfer of the acetyl group to the -SH of an acyl carrier protein (ACP) and then to the β-ketoacyl-ACP synthase. Transfer of the malonyl group to the ACP is followed by decarboxylation of the malonyl group and condensation of the remaining two-carbon unit with the carbonyl carbon of the acetate group on the synthase. This is followed by reduction of the β-carbonyl to an alcohol, dehydration to yield a *trans*-α,β double bond and reduction to yield a saturated bond. Introduction of unsaturations in the nascent chain occurs by O_2-dependent and O_2-independent pathways and may be followed by further chain elongation. Several mechanisms are utilized to introduce multiple unsaturations in a fatty acid chain. Regulation of fatty acid synthesis is related to regulation of fatty acid breakdown and the activity of the TCA cycle, because of the importance of acetyl-CoA in all these processes. Malonyl-CoA inhibits carnitine transport, blocking fatty acid oxidation. Citrate activates ACC and palmitoyl-CoA inhibits, both in chain-length-dependent fashion. The enzymes of fatty acid synthesis are also under hormonal control.

Glycerolipid synthesis is built around the synthesis of phosphatidic acid from glycerol-3-phosphate or dihydroxyacetone phosphate. Specific acyltransferases add acyl chains to these glycerol derivatives. Other glycerolipids, such as phosphatidylcholine (PC) and phosphatidylethanolamine (PE) are synthesized from phosphatidic acid via CDP-diacylglycerol and diacylglycerol. Base exchange converts phosphatidylethanolamine to phosphatidylserine. Other phospholipids, such as phosphatidylinositol, phosphatidylglycerol and cardiolipin are synthesized from CDP-diacylglycerol. Dihydroxyacetone phosphate is a precursor to the plasmalogens. Platelet activating factor (PAF), an ether lipid, dilates blood vessels, reduces blood pressure and aggregates platelets. Sphingolipids are produced via condensation of serine and palmitoyl-CoA by 3-ketosphinganine synthase and reduction of the ketone product to form sphinganine. Acylation followed by desaturation yields ceramide, the precursor to other sphingolipids and cerebrosides.

Eicosanoids, derived from arachidonic acid by oxidation and cyclization, are ubiquitous local hormones. They include the prostaglandins, thromboxanes, leukotrienes and other hydroxyeicosanoic acids. A variety of stimuli, including histamine, epinephrine, bradykinin, proteases and other agents associated with tissue injury and inflammation, can stimulate the release of eicosanoids, which have short half-lives and are rapidly degraded. Aspirin acetylates endoperoxide synthase on its cyclooxygenase subunit, irreversibly inhibiting the synthesis of prostaglandins.

Cholesterol biosynthesis begins with mevalonic acid, which is formed from acetyl-CoA by thiolase, HMG-CoA synthase and HMG-CoA reductase. The HMG-CoA reductase reaction is the rate-limiting step in cholesterol biosynthesis. Inhibition of this enzyme by lovastatin blocks cholesterol biosynthesis and can significantly lower serum cholesterol. Mevalonate is converted to squalene via isopentenyl pyrophosphate and dimethylallyl pyrophosphate, which join to yield farnesyl pyrophosphate and then squalene. Squalene is cyclized in two steps and converted to lanosterol. The conversion of lanosterol to cholesterol requires another 20 steps.

Steroids such as the bile acids and steroid hormones are synthesized from cholesterol via key intermediates such as pregnenolone and progesterone. The male hormone testosterone is a precursor to the female hormones including estradiol. Steroid hormones modulate transcription of DNA to RNA in the cell nucleus. The corticosteroids, including glucocorticoids and mineralocorticoids, synthesized by the adrenal glands, are important physiological regulators.

Chapter 21

Amino Acid and Nucleotide Metabolism

Chapter Outline

- Nitrogen cycle
  - Nitrogen moved into biosphere
    - N_2 in atmosphere changed to ammonium
      - Nitrogen fixation: Plants, fungi, bacteria
    - NO_3^- (nitrate) in soil and oceans changed to ammonium
      - Nitrate assimilation: Bacteria
  - Animals need reduced dietary N and excrete N as ammonium or organic nitrogen
    - Ammonium to NO_3^- (nitrate): Nitrifying bacteria
    - NO_3^- (nitrate) to N_2: Denitrifying bacteria
- Ammonium entry into organic linkage
  - Carbamoyl-phosphate synthetase I
    - Urea cycle
    - $NH_4^+ + HCO_3^- + 8\ ATP \rightarrow$ carbamoyl-phosphate $+ 2\ ADP + P_i + 2\ H_2$
    - N-Acetylglutamate activator
  - Glutamate dehydrogenase (GDH)
    - $NH_4^+ + \alpha$-ketoglutarate + NADPH + $H^+ \rightarrow$ glutamate + $NADP^+ + 2\ H_2O$
      - Produces glutamate using NADPH
  - Glutamine synthetase (GS)
    - Glu + $NH_4^+ \rightarrow$ Gln: ATP driven
  - GDH and GS responsible for most of the NH_4^+ assimilated into organic compounds
- Glutamine Synthetase regulation: Dodecameric complex
  - Feedback inhibition
    - Gly, Ala, Ser inhibit by competing with Glu
    - AMP competes with ATP
    - His, Trp, CTP, carbamoyl-P, glucosamine-6-P inhibit
  - Covalent modification
    - Adenylylation of Tyr inactivates
    - Degree of inactivation proportional to number of subunits adenylylated
    - Adenylyl transferase (AT): Adenylylates and deadenylylates
      - AT·P_{IIA} adenylylates
      - AT·P_{IID} deadenylylates
        - Gln activates P_{IIA} inhibits P_{IID}
        - α-Ketoglutarate inhibits P_{IIA} activates P_{IID}
  - Gene expression: GS subunit encoded by Gln A

        - Transcriptional enhancer: NR_I: Must be phosphorylated
            - NR_{II}: NR_I kinase
            - NR_{II}/P_{IID}-complex: NR_I-phosphatase
- Amino acid biosynthesis: Transamination of appropriate α-keto acid
    - Essential amino acids: Required in diet
    - Nonessential amino acids: Synthesized by organism
    - Transamination by aminotransferases
        - Glutamate aminotransferase most common
- Carbon skeletons
    - α-Ketoglutarate family
        - Glu, Gln
        - Pro from Glu
        - Arg from Glu via ornithine
            - Ornithine's metabolic fates
                - Arg precursor
                - Urea cycle intermediate
                - Arg breakdown product
    - Aspartate family
        - Asp, Asn
        - Thr, Met, Lys
    - Pyruvate family
        - Ala, Val, Leu
    - 3-Phosphoglycerate family
        - Ser, Gly, Cys
    - Aromatic amino acids
        - Shikimate path produces chorismate
        - Chorismate branch point
            - Phe, Tyr
            - Trp
- Amino acid degradation: 20 Amino acids converted to 7 metabolic intermediates
    - Pyruvate
        - Ala, Ser, Cys, Gly, Trp (carboxyl-, α–, β-carbons), Thr
    - Oxaloacetate
        - Asp, Asn
    - α-Ketoglutarate
        - Glu, Gln, Pro, Arg, His
    - Succinyl-CoA
        - Val, Ile, Met
    - Acetoacetate
        - Leu (+ acetyl-CoA)
        - Lys, Trp ring
        - Phe, Tyr (+ fumarate)
    - Acetyl-CoA
        - Leu (+ fumarate)
    - Fumarate
        - Phe, Tyr (+ acetoacetate)
    - Ketogenic vs glucogenic
    - Excess dietary nitrogen is secreted
- Nucleotides
    - NTPs: RNA biosynthesis
        - ATP: Energy currency
        - UTP: Carbohydrate metabolism
        - CTP: Phospholipid metabolism
        - GTP: Protein synthesis and signal transduction
    - dNTPs: DNA biosynthesis

- Formation: Two options
  - De novo synthesis
  - Salvage pathway
- Purine biosynthesis: IMP biosynthesis – 11 steps
  - Synthesized on ribose-5-P
  - Ribose-5-phosphate pyrophosphokinase reaction rate limiting: PRPP formed
    - ADP, GDP inhibitors
  - Glutamine phosphoribosyl pyrophosphate amidotransferase
    - Committed step
    - GMP, GDP, GTP inhibitory site
    - AMP, ADP, ATP inhibitory site
    - Azaserine inhibition: Antitumor agent
  - GAR synthetase
  - GAR transformylase: Uses formyl-THF
    - Methotrexate inhibits
  - FGAR amidotransferase (FGAM synthetase)
    - Inactivated by azaserine
  - AIR synthetase
    - Formation of imidazole ring
    - Activation of formyl group for second ring closure
  - AIR carboxylase
  - SAICAR synthetase
    - This and next enzyme on bifunctional polypeptide
  - Adenylosuccinase (adenylosuccinate lyase) produces AICAR
    - AICAR: Histidine biosynthesis
  - AICAR transformylase: Uses formyl-THF: Produces FAICAR
    - Methotrexate inhibits
  - IMP cyclohydrolase: Dehydration and ring closure
    - This and AICAR transformylase on bifunctional polypeptide
  - IMP biosynthesis consumes 7 ATP equivalents
- IMP precursor of AMP and GMP
  - AMP production
    - Adenylosuccinate synthetase: GTP-dependent reaction
    - Adenylosuccinase: AMP and fumarate
  - GMP production
    - IMP dehydrogenase
    - GMP synthetase: ATP-dependent reaction
- NTP production
  - Adenylate kinase: (d)AMP + ATP $\rightarrow$ (d)ADP + ADP
  - Guanylate kinase (d)GMP + ATP $\rightarrow$ (d)GDP + ADP
  - Nucleoside diphosphate kinase: (d)NDP + ATP $\rightarrow$ (d)NTP + ADP
- Purine salvage
  - Adenosine phosphoribosyltransferase: APRT
  - Hypoxanthine-guanine phosphoribosyltransferase: HGPRT
- Purine nucleotide degradation
  - Dietary nucleotides
    - Nucleotidases and phosphatases produce nucleosides
    - Nucleosidases or nucleoside phosphorylases produce bases
  - Intracellular nucleotides
    - Nucleotidase produces nucleosides
    - Purine nucleoside phosphorylase releases base
      - A and dA converted to inosine by adenosine deaminase
        - Adenosine deaminase deficiency causes severe combined immunodeficiency syndrome
        - Gene therapy attempted to supply functional enzyme
    - Xanthine

                - Hypoxanthine (from inosine) to xanthine by xanthine oxidase
                - Guanine to xanthine by guanine deaminase
            - Uric acid: From xanthine by xanthine oxidase
- Anaplerotic cycle: Fumarate production
    - AMP to IMP via adenosine deaminase: Degradation
    - IMP to AMP via adenylosuccinate synthetase and adenylosuccinase: Fumarate produced
- Uric acid: Metabolism/excretion
    - Primates: End product (but urea production accounts for nitrogen excretion also)
        - Excess levels leads to crystallization and gout
        - Allopurinol blocks xanthine oxidase
- Pyrimidine biosynthesis: Precursors are carbamoyl-P and aspartate
    - Carbamoyl-P synthetase II
        - In mammals : CPS-II and following two enzymes on multifunctional polyprotein
            - Metabolic channeling
        - UDP/UTP inhibitory, PRPP and ATP stimulatory
    - ATCase: Carbamoyl-P condensed with aspartate
        - Regulation
            - *E. coli* enzyme inhibited by CTP, activated by ATP
    - Dihydroorotase: Ring closure
    - Dihydroorotate dehydrogenase
        - NAD^+-linked in bacteria
        - Quinone-linked in eukaryotes
    - Orotate phosphoribosyltransferase
        - Transferase and decarboxylase (below) on UMP synthase: In mammals single polypeptide
    - OMP decarboxylase to form UMP
- UTP and CTP formation
    - Nucleoside monophosphate kinase: UMP + ATP $\rightarrow$ UDP + ADP
    - Nucleoside diphosphate kinase: UDP + ATP $\rightarrow$ UTP + ADP
    - CTP synthase: Animation of UTP to give CTP
- Pyrimidine degradation
    - U and C: β-Alanine, ammonium, carbon dioxide
    - T: β-Aminoisobutyric acid
- Deoxyribonucleotide biosynthesis
    - Ribonucleotide reductase
        - $(R1)_2(R2)_2$ subunit structure
            - R1: Two regulatory sites
                - Overall activity site: ATP activates, dATP inhibits
                - Substrate specificity site
                    - Bound ATP specifies UDP or CDP
                    - Bound dTTP specifies GDP
                    - Bound dGTP specifies ADP
        - NADPH source of electrons
            - Electrons moved to thioredoxin by thioredoxin reductase
- Thymine nucleotides
    - dUDP to dUTP to dUMP: dUTPase catalyzes last step; Or, dCMP to dUMP by dCMP deaminase
    - dUMP to dTMP by thymidylate synthase
        - 5-Fluorouracil converted to inhibitor of thymidylate synthase

Chapter Objectives

Nitrogen Acquisition

All organisms produce important compounds containing reduced nitrogens. However, the principal forms of nitrogen in the environment, N_2 (dinitrogen) and NO_3^- (nitrate), contain nitrogen in an oxidized state. Nitrogen is abstracted from the environment and converted to NH_4^+

(ammonium) by two pathways: nitrogen assimilation and nitrogen fixation. Know the basic chemistry involved and the important enzymes in these two processes.

Ammonium Utilization

Ammonium is incorporated into organic compounds in the following ways.

1. $NH_4^+ + HCO_3^- + 2\ ATP \rightarrow H_2N{-}\overset{O}{\overset{\|}{C}}{-}O{-}PO_3^{2-} + 2\ ADP + P_i + H^+$
catalyzed by carbamoyl-phosphate synthetase I, a urea cycle enzyme.
2. NH_4^+ + α-ketoglutarate + NADPH → glutamate + $NADP^+$ + H_2O catalyzed by glutamate dehydrogenase.
3. NH_4^+ + glutamate + ATP → glutamine + ADP + P_i catalyzed by glutamine synthetase.

Glutamate dehydrogenase has a higher K_m for ammonium than does glutamine synthetase and in conditions in which ammonium is not abundant, reaction 3 above predominates and glutamate is consumed. To replenish glutamate, glutamate synthase or GOGAT (glutamate:oxo-glutarate amino-transferase) catalyzes

NADPH + α-ketoglutarate + glutamine → 2 glutamate + $NADP^+$

Regulation of *E. coli* Glutamine Synthetase

Regulation of this key enzyme in nitrogen metabolism is achieved by three mechanisms. The enzyme is allosterically regulated by nine inhibitors including Gly, Ala, Ser, His, Trp, CTP, AMP, carbamoyl-phosphate, and glucosamine-6-phosphate. Covalent modification of glutamine synthetase by adenylylation of a tyrosine residue leads to inactivation. Gene expression is regulated to control enzyme levels.

The level of adenylylation of glutamine synthetase is regulated by adenylyl transferase. This enzyme both adds and removes adenylyl groups but activity in general is dependent on a complex between adenylyl transferase and a regulatory protein P_{II}. P_{II} is itself a target for covalent modification by uridylyl transferase. Uridylylated P_{II} complexed with adenylyl transferase results in deadenylylation activity, deadenylylation of glutamine synthetase, and glutamine synthetase activity. Unuridylylated P_{II} complexed to adenylyl transferase stimulates deadenylylation activity leading to inhibition of glutamine synthetase. Uridylyl transferase activity is dependent on glutamine.

Amino Acid Biosynthesis

The key point to remember about amino acid biosynthesis is that an α-keto acid is first produced and then an amino group is introduced by transamination from glutamate, catalyzed by an aminotransferase. Some α-keto acids important in this regard include α–ketoglutarate, oxaloacetate, and pyruvate. α–Ketoglutarate, a citric acid cycle intermediate, is used to produce glutamic acid, glutamine, proline, and ornithine, an amino acid intermediate in the urea cycle, which leads to arginine synthesis. Oxaloacetate, another citric acid cycle intermediate, is the precursor of aspartic acid, asparagine, threonine, methionine and lysine. Four of the carbons of isoleucine derive from oxaloacetate via threonine, with two carbons coming from pyruvate. Pyruvate is used to form alanine, valine, and leucine. The glycolytic intermediate, 3-phosphoglyc-erate, is used to synthesize serine, glycine, and cysteine. (You might recall a relationship between serine and glycine in photorespiration.) The aromatic amino acids, phenylalanine, tyrosine, and tryptophan, all have chorismate as a common intermediate. Chorismate is produced from erythrose-4-phosphate, a pentose phosphate intermediate, and PEP via shikimate

Amino Acid Degradation

The α-amino group is removed from amino acids by transamination and excess nitrogen is excreted in one of three ways depending on the organism. It may be released as free ammonia, as urea (Figure 21.12), or as uric acid. The carbon skeletons are metabolized into seven metabolic intermediates including pyruvate, oxaloacetate, α–ketoglutarate, acetyl-CoA, fumarate, and acetoacetate. Alanine, serine, glycine, cysteine, and parts of threonine and tryptophan are converted to pyruvate. Aspartate and asparagine produce oxaloacetate. Glutamate, glutamine, proline, arginine, and histidine produce α-ketoglutarate. Valine, isoleucine, and methionine are degraded to succinyl-CoA. Leucine is converted to acetyl-CoA and acetoacetate, a ketone body. Lysine is decarboxylated twice to yield acetoacetate. Phenylalanine is converted to tyrosine by hydroxylation and tyrosine is degraded into acetoacetate and fumarate. Amino acids that can support gluconeogenesis are termed glucogenic amino acids whereas amino acids producing only acetyl-CoA or acetoactate are ketogenic.

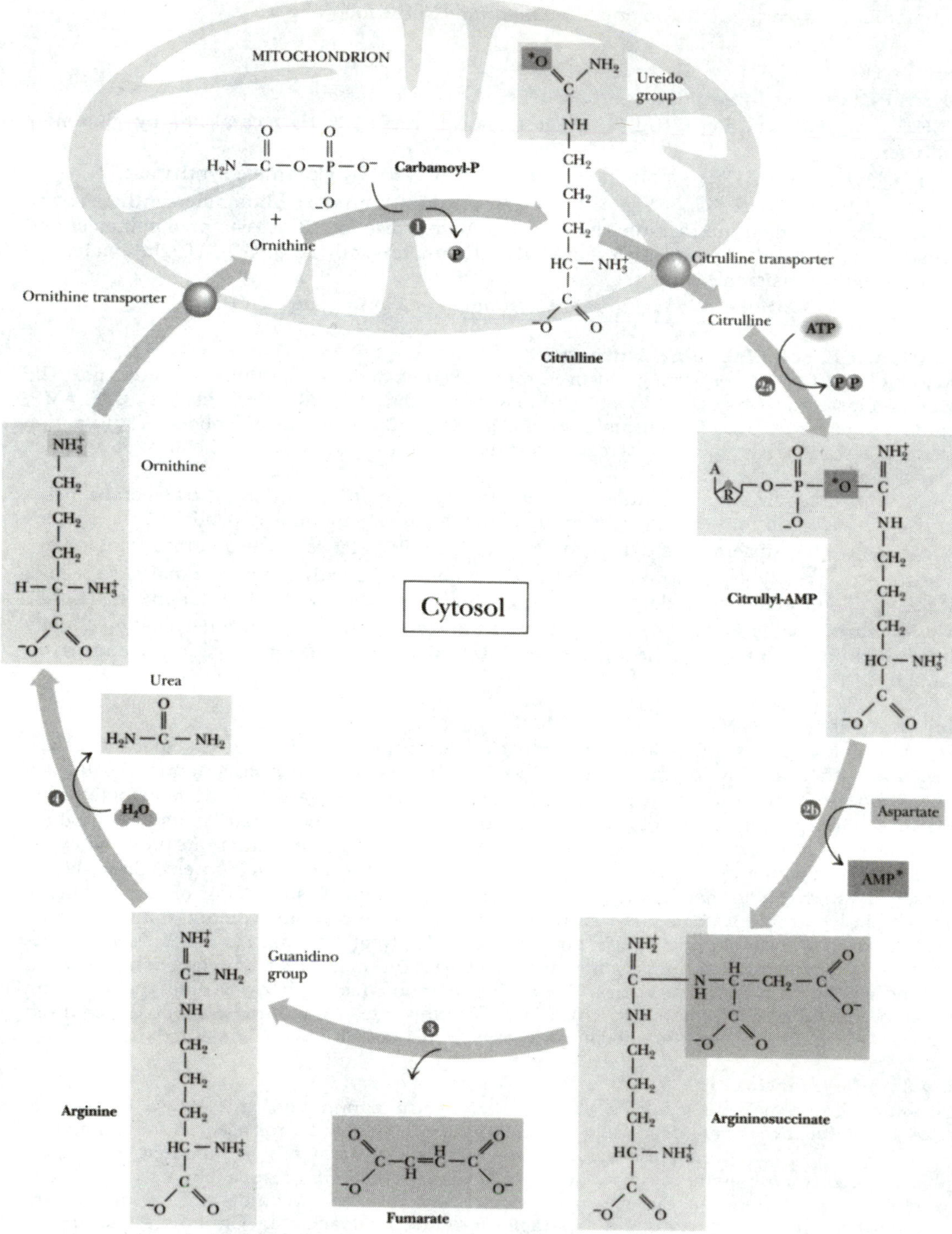

Figure 21.12 The urea cycle.

Purine Biosynthesis

Purine biosynthesis (Figure 21.26) starts with 5-phosphoribosyl pyrophosphate and the complete pathway involves formation of nucleotide derivatives. N-9 is the first component of the purine ring to be added from the side chain amino group of glutamine. The five-membered, imidazole ring components include N-9 and a formylglycine residue introduced by amide linkage of glycine to N-9 followed by formylation. Before imidazole ring formation is completed, N-3 is incorporated from glutamine. Imidazole ring formation produces an amino derivative of imidazole ribotide or AIR. To complete synthesis of the purine inosine monophosphate, the precursor of ATP and GTP, two carbons and one nitrogen are needed. The first carbon is introduced as a carboxyl group, the remaining nitrogen derives from the α-amino group of aspartic acid. Aspartic acid is attached by an amide linkage to carboxylated AIR and the carbon skeleton is eliminated as fumarate. (A similar reaction occurs in the urea cycle.) Formylation of N-3, incorporated prior to imidazole-ring closure, and subsequent ring closure produces inosine monophosphate. Amination of IMP using the α-amino group of aspartate and releasing the carbon skeleton of aspartate as fumarate produces AMP. GMP is produced by oxidation of C-2 followed by amination using the side chain nitrogen of glutamine.

Purine Degradation

The common pathway for purine degradation is to produce free bases, which are metabolized to xanthine, then uric acid (Figure 21.31), and depending on the organism, allantoin, allantoic acid, urea, or CO_2, H_2O and ammonia.

Purine Points to Consider

Know why methotrexate inhibits purine formation and why the sulfonamides (structural analogs of p-aminobenzoic acid (PABA, an ingredient in sunscreen)) function as folic acid antagonists. Understand the role of phosphoribosyltransferases in purine salvage pathways and the consequence of HGPRT deficiency (Lesch-Nyhan syndrome). Gout is a consequence of excess uric acid production. Know why allopurinol is used to treat gout.

Pyrimidine Biosynthesis

In contrast to purine synthesis, pyrimidine ring formation (Figure 21.36) is completed before being attached to ribose. Synthesis starts with production of carbamoyl-phosphate from bicarbonate and the side chain amino group of glutamine. (We encountered carbamoyl-phosphate synthesis in the urea cycle where it was produced in the mitochondria. Carbamoyl-P synthesis leading to pyrimidine biosynthesis is catalyzed by a second synthetase located in the cytosol.) The nitrogen and carbon of carbamoyl-phosphate account for two of the ring components of the six-membered pyrimidine ring. The remaining components derive from aspartate. Aspartic acid is carbamoylated by amide linkage of a carbamoyl group to the α-amino group of aspartic acid. This is followed by ring closure and oxidation to produce orotate (uracil-6-carboxylic acid, with the carboxyl group from the α-carboxyl group of aspartic acid). Orotate is attached to a phosphoribosyl group and decarboxylated to form UMP.

Deoxyribonucleotide Biosynthesis

Ribonucleotide reductase catalyzes the conversion of NDP to dNDP without regard to base (see problem 6 for details on regulation of this important enzyme). The reducing power derives from thioredoxin, a small protein with reactive sulfhydryl groups that undergo oxidation-reduction cycles. The reducing power of thioredoxin is replenished by NADPH-dependent thioredoxin reductase. Thymine nucleotides are synthesized from dUMP by methylation with methylene-tetrahydrofolate as methyl-group donor.

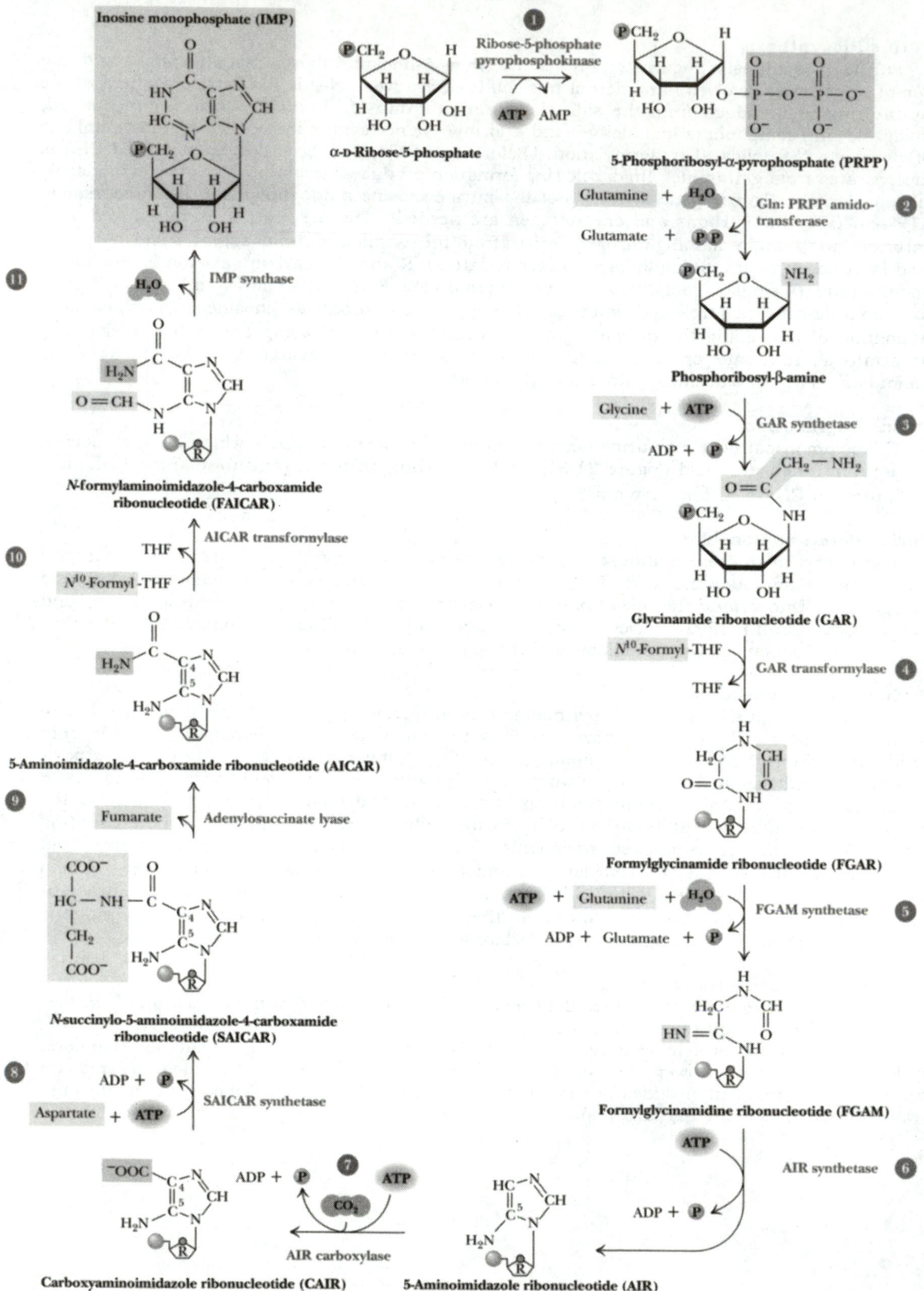

Figure 21.26 The *de novo* pathway for purine synthesis. The first purine product of this pathway, IMP, serves as a precursor to AMP and GMP.

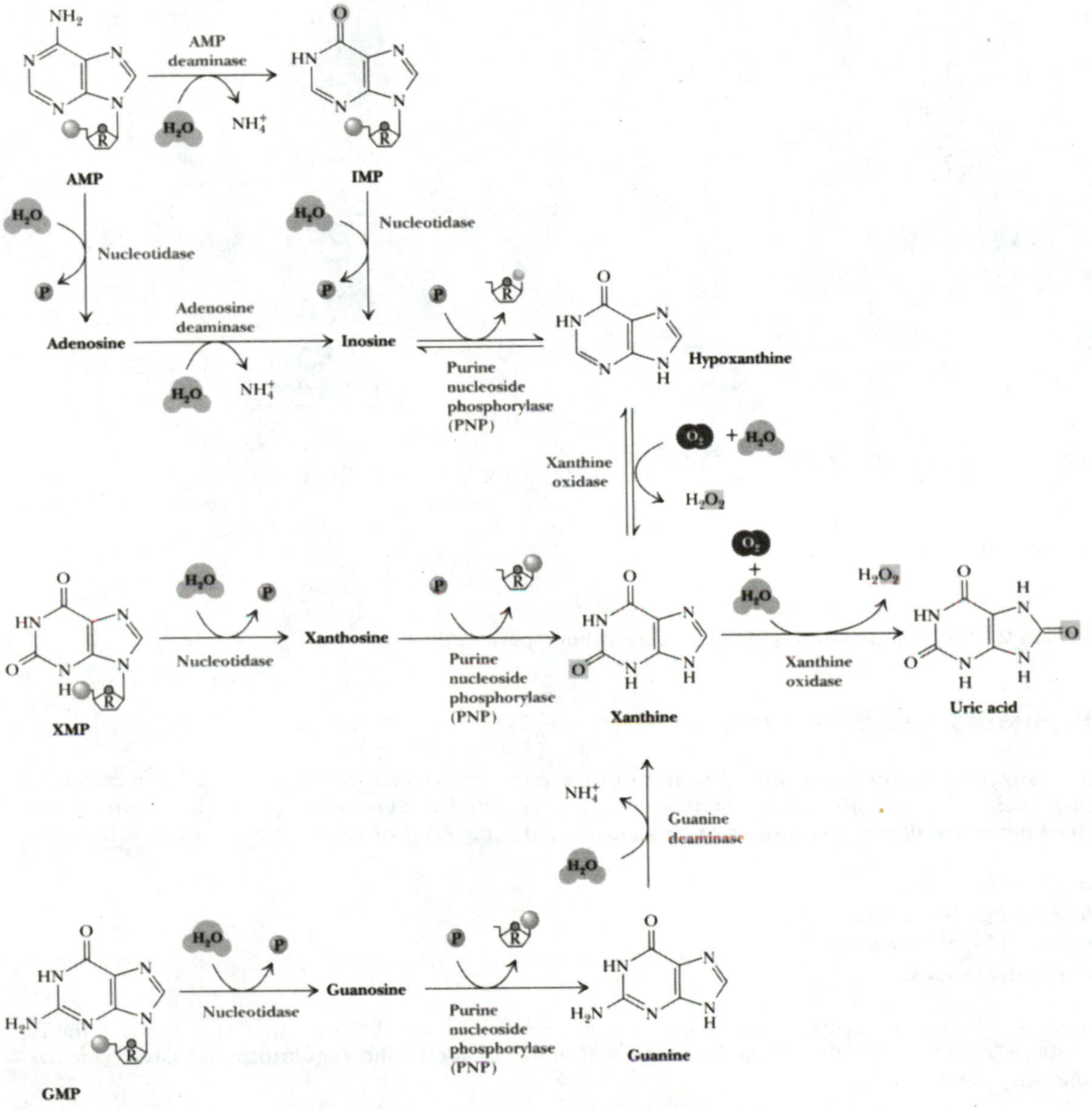

Figure 21.31 The major pathways for purine catabolism in animals. Catabolism of the different purine nucleotides converges in the formation of uric acid.

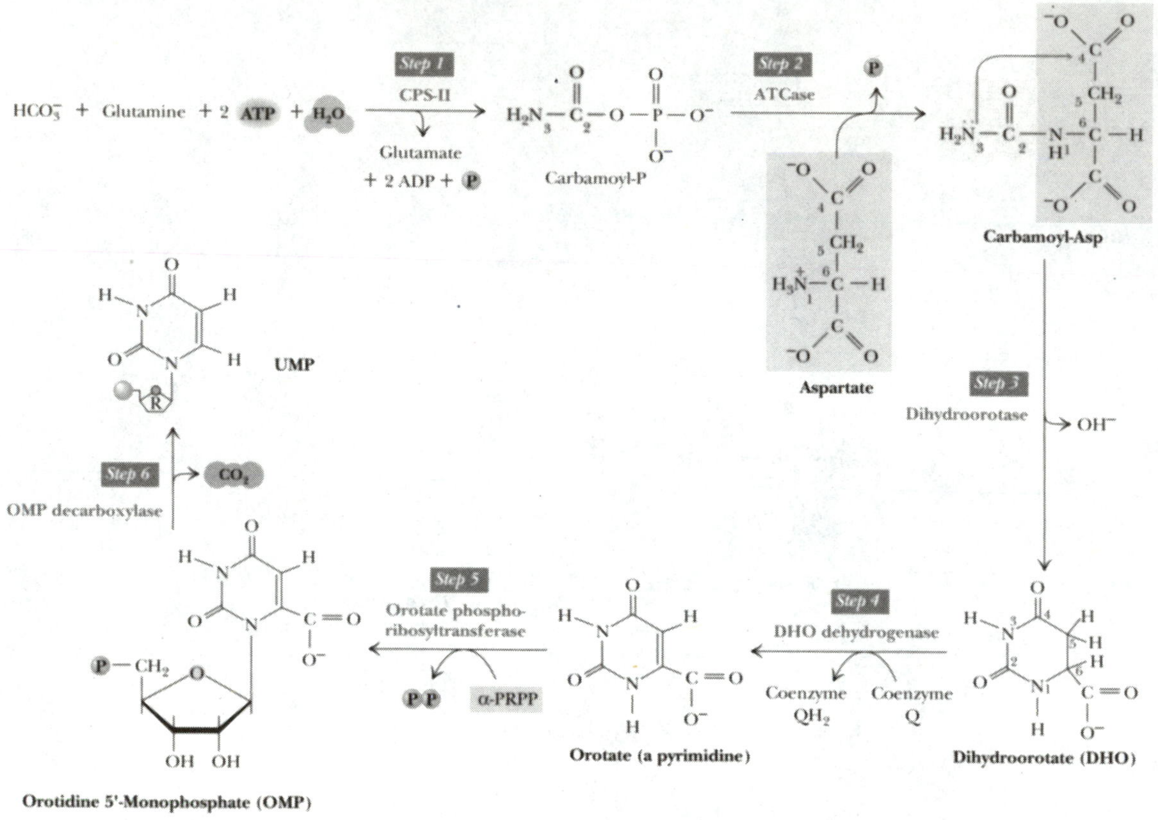

Figure 21.36 The *de novo* pyrimidine biosynthetic pathway.

Problems and Solutions

1. Suppose at certain specific metabolite concentrations in vivo the cyclic cascade regulating E. coli glutamine synthetase has reached a dynamic equilibrium where the average state of GS adenylylation is poised at n = 6. Predict what change in n will occur if:

a. [ATP] increase.

b. P_{IID}/ P_{IIA} increases.

c. [α-KG]/[Gln] increases.

d. [P_i] decreases.

Answer: Glutamine synthetase is a key enzyme in nitrogen metabolism. In *E. coli* the enzyme is a 600-kD dodecamer that is sensitive to a number of metabolic regulators. It catalyzes the following reaction

$$\text{glutamate} + \text{ATP} + NH_3 \rightarrow \text{glutamine} + \text{ADP} + P_i$$

The activity of the enzyme is further regulated by covalent modification; specifically the enzyme is inhibited by adenylylation catalyzed by ATP:GS:adenylyl transferase or AT. Thus, the activity of AT determines *n*, the average state of GS adenylylation. AT is itself regulated by interaction with a regulatory protein, P_{II}, which also can exist it two states, P_{IIA} and P_{IID}. Interaction with P_{IIA} results in adenylyl transferase activity whereas interaction with P_{IID} results in deadenylylation activity. The interconversion of P_{II} between its two states is a result of uridylylation/deuridylylation catalyzed by a converter enzyme, UT. Uridylylated P_{II} is in the P_{IID} state whereas deuridylylated P_{II} is in the P_{IIA} state. UT catalyzes both uridylylation and deuridylylation, the exact activity dependent on the ratio of [Gln]/[α-ketoglutarate]. Excess glutamine is a signal that additional glutamine is not required. Therefore, GS must be inactivated by adenylylation, a reaction itself dependent on P_{IIA} which is formed by deuridylylation of P_{II}.

a. An increase in [ATP] is expected to favor adenylylation of GS causing n to increase leading to increased inhibition.

b. An increase in P_{IID} levels, leading to deadenylylation, will cause a decrease in n and decreased inhibition.

c. An increase in the ratio of [α-KG]/[Gln] is a signal that Gln is needed. Thus, GS activity is increased by deadenylylation and n will decrease.

d. A decrease in [P_i], the substrate of the deadenylylation reaction, will lead to a decrease in deadenylylation, and an increase in n.

2. *How many ATP equivalents are consumed in the production of one equivalent of urea by the urea cycle?*

Answer: In the urea cycle, the amino acid ornithine serves as a skeleton to synthesize arginine which is hydrolyzed to form urea and ornithine. The two nitrogens in urea derive immediately from ammonia and the amino group of aspartic acid. The nitrogen from ammonia is introduced during the synthesis of citrulline from ornithine and carbamoyl phosphate. Carbamoyl phosphate is produced from ammonia and bicarbonate by carbamoyl phosphate synthetase, in a reaction driven by hydrolysis of two 2 ATPs. Citrulline and aspartate are combined to form argininosuccinate in a reaction catalyzed by argininosuccinate synthetase and driven by hydrolysis of ATP to AMP and PP_i. This step is equivalent to hydrolysis of two high-energy phosphate bonds because PP_i is hydrolyzed by pyrophosphatase into 2 P_i. So far we have accounted for 4 ATPs consumed to form one urea and one fumarate from one CO_2, one NH_3, and one aspartate.

However, there are some hidden energetic benefits to the urea cycle. Arginino-succinate is resolved into arginine and fumarate by argininosuccinase. In order to convert fumarate back into aspartate it is hydrated to malate and oxidized to oxaloacetate (two citric acid cycle steps). Malate to oxaloacetate is accompanied by NADH production, which is equivalent to approximately 2.5 ATPs. (An aminotransferase reaction converts oxaloacetate to aspartate.) Finally, ammonia is generated by glutamate dehydrogenase in an oxidative deamination of glutamate to α-ketoglutarate with accompanying reduction of NAD^+ to NADH accounting for another 2.5 ATP equivalents. Therefore, if we account for ammonia and aspartate, we see that as long as there is a source of amino acids, the urea cycle actually accounts for the net production of 1 ATP.

3. *Why are persons on a high-protein diet advised to drink lots of water?*

Answer: Humans do not store excess proteins or amino acids, and a dietary excess of either will result in degradation of amino acids and production of urea by the urea cycle. The urea cycle is carried out by liver cells, which release urea to the bloodstream to be carried to the kidney where it is excreted in urine. Excess water will be required to meet this increase in urea synthesis and urine production.

4. *If PEP labeled with ^{14}C in the 2-position serves as precursor to chorismate synthesis, which C atom in chorismate is radioactive?*

Answer: Chorismate, which is synthesized from PEP and erythrose-4-phosphate, is a precursor to the aromatic amino acids, tyrosine, phenylalanine, and tryptophan. Its synthesis is as follows

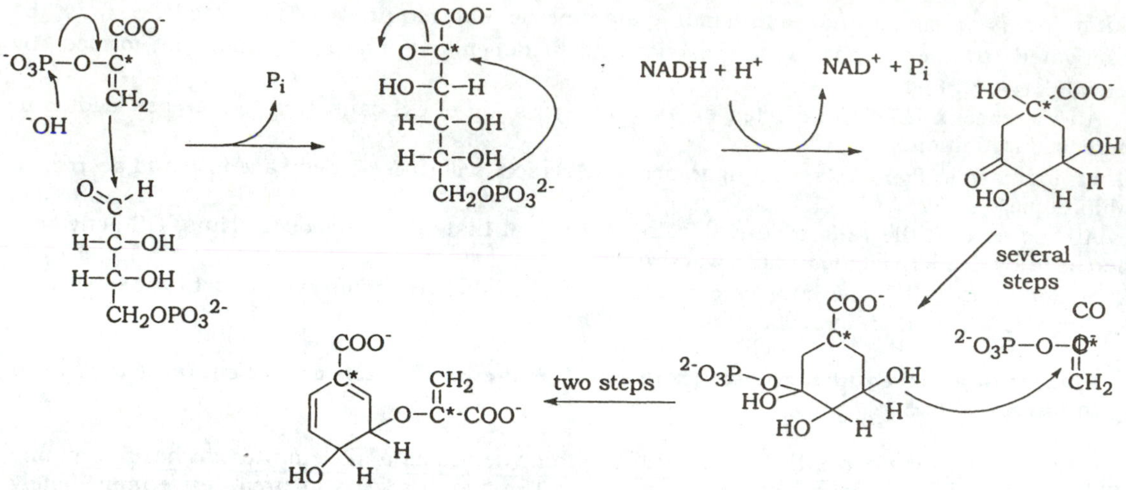

P_i

$NADH + H^+$ $NAD^+ + P_i$

several steps

two steps

5. *Write a balanced equation for the synthesis of glucose (by gluconeogenesis) from aspartate.*

Answer: In gluconeogenesis (chapter 19), pyruvate is converted to glucose with the following stoichiometry

$$2\text{ pyruvate} + 2\text{ NADH} + 2\text{ H}^+ + 4\text{ ATP} + 2\text{ GTP} + 6\text{ H}_2\text{O}$$
$$\downarrow$$
$$\text{glucose} + 2\text{ NAD}^+ + 4\text{ ADP} + 2\text{ GDP} + 6\text{ P}_i$$

ATP is consumed at two steps: Conversion of pyruvate to oxaloacetate and production of 1,3-bisphosphoglycerate from 3-phosphoglycerate. In addition, GTP is consumed by PEP carboxykinase in producing PEP from oxaloacetate.

With aspartate as a source of carbons for gluconeogenesis, oxaloacetate is produced by transamination of aspartate to α-ketoglutarate.

$$2\text{ aspartate} + 2\ \alpha\text{-ketoglutarate} \rightarrow 2\text{ oxaloacetate} + 2\text{ glutamate}$$

Thus, one energy utilizing step in gluconeogenesis starting from pyruvate is bypassed when aspartate is metabolized and CO_2 shows up as a product.

$$2\text{ oxaloacetate} + 2\text{ ADP} + 2\text{ P}_i \rightarrow 2\text{ pyruvate} + 2\text{ CO}_2 + 2\text{ H}_2\text{O} + 2\text{ ATP}$$

The balanced equation is

$$2\text{ aspartate} + 2\ \alpha\text{-ketoglutarate} + 2\text{ NADH} + 2\text{ H}^+ + 2\text{ ATP} + 2\text{ GTP} + 4\text{ H}_2\text{O}$$
$$\downarrow$$
$$\text{glucose} + 2\text{ glutamate} + 2\text{ NAD}^+ + 2\text{ ADP} + 2\text{ GDP} + 4\text{ P}_i + 2\text{ CO}_2$$

We may wish to metabolize glutamate with glutamate dehydrogenase as follows

$$2\text{ glutamate} + 2\text{ NAD(P)}^+ + 2\text{ H}_2\text{O} \rightarrow 2\ \alpha\text{-ketoglutarate} + 2\text{ NH}_3 + 2\text{ NAD(P)H} + 2\text{ H}^+$$

If we then consider this reaction, the balanced equation is

$$2\text{ Asp} + 2\text{ H}^+ + 2\text{ ATP} + 2\text{ GTP} + 6\text{ H}_2\text{O} \rightarrow \text{glucose} + 2\text{ ADP} + 2\text{ GDP} + 4\text{ P}_i + 2\text{ CO}_2 + 2\text{ NH}_3$$

6. *Draw the purine and pyrimidine ring structures, indicate the metabolic source of each atom in the rings.*

Answer: For purine biosynthesis

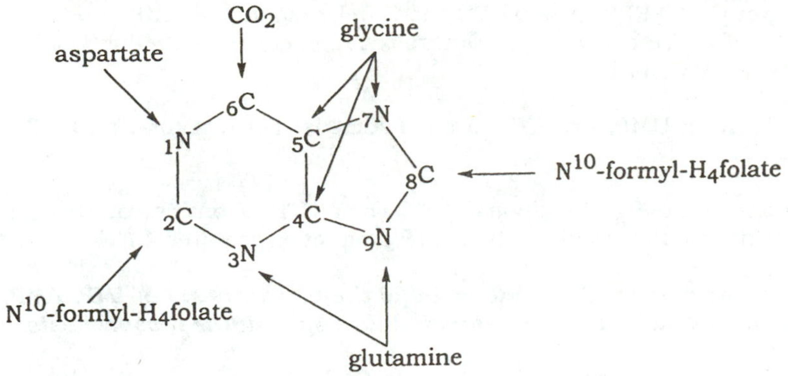

CO_2 glycine aspartate
$_6$C $_7$N $_1$N $_5$C $_8$C ← N^{10}-formyl-H_4folate
$_2$C $_4$C $_9$N $_3$N
N^{10}-formyl-H_4folate
glutamine

For pyrimidine biosynthesis

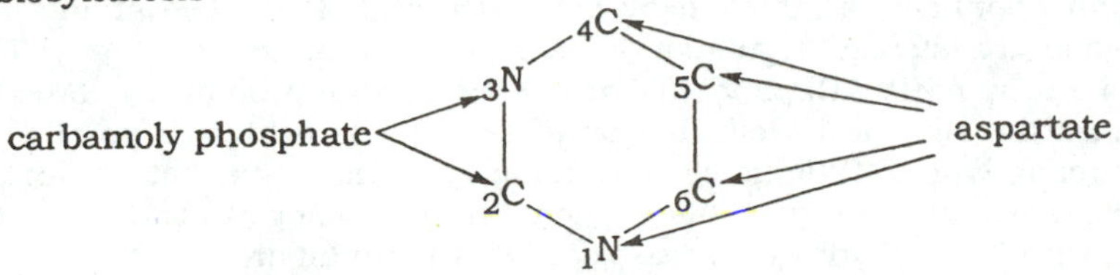

$_4$C $_3$N $_5$C
carbamoly phosphate aspartate
$_2$C $_6$C $_1$N

7. Starting from glutamine, aspartate, glycine, CO_2, and N^{10}-formyl THF, how many ATP equivalents are expended in the synthesis of (a) ATP, (b) GTP, (c) UTP, and (d) CTP?

Answer: Synthesis of purines begins with pyrophosphorylation of ribose-5-phosphate using ATP to produce 5-phosphoribosyl-α-pyrophosphate (PRPP) and AMP which means that two ATP equivalents have been expended. PRPP is then converted to glycinamide ribotide (GAR) in two steps, the first of which incorporates N-9 from glutamine with release of PP_i and the second adds glycine (see answer to problem 1) with hydrolysis of ATP driving the reaction. Thus, these three steps account for 2 ATP AMP + ADP + PP_i + P_i or, 3 ATP 3 ADP + 3 P_i. GAR is converted to formylglycinamide ribotide (FGAR) which in two steps is converted to 5-aminoimidazole ribotide (AIR), with each step driven by ATP hydrolysis. So far we have accounted for 5 ATPs. AIR is then carboxylated and subsequently reacted with aspartate to produce 5-aminoimidazole-4-(N-succinylocarboxamide) ribotide (SACAIR). Both reactions are dependent on ATP hydrolysis, accounting for the sixth and seventh ATPs utilized. SACAIR is three steps away from inosine monophosphate (IMP), the precursor of both AMP and GMP.

a. To produce ATP from IMP, the amino group at C-6 is incorporated from aspartate with GTP hydrolysis driving the reaction and subsequent elimination of fumarate to produce AMP. Finally, AMP is phosphorylated to ATP at the expense of two high energy phosphate bonds. So, to produce ATP a total of 10 high energy phosphate bonds are hydrolyzed, 7 for production of IMP, 1 to produce AMP, and 2 to produce ATP. (If we start from ribose instead of ribose-5-phosphate we must account for an additional ATP.)

b. Production of GTP from IMP requires an oxidation step to convert IMP to xanthosine monophosphate (XMP), the direct precursor of AMP. This step generates reduced NAD^+ which is equivalent to production of 3.0 high energy phosphate bonds. Next, XMP is converted to GMP with ATP hydrolysis to AMP and PP_i driving the reaction. This step is equivalent to hydrolysis of 2 ATPs. Finally GMP is converted to GTP with 2 more ATPs used. The overall synthesis of GTP consumes 8 ATP.

Pyrimidine synthesis begins with formation of carbamoyl phosphate from bicarbonate and the amino group of glutamine with hydrolysis of 2 ATP. Carbamoyl phosphate is reacted with aspartate to produce carbamoyl aspartate, which is subsequently converted to dihydroorate. Dihydroorotate is oxidized to orotate with accompanying reduction of NAD^+, a step equivalent to +3.0 ATP. Orotate is converted to a nucleotide monophosphate, orotidine monophosphate (OMP). The sugar moiety derives from 5-phospho-α-D-ribosyl-1-pyrophosphate (PRPP). The pyrophosphate group of PRPP is from ATP; it is released upon OMP production. This then accounts for a net of 2 ATP equivalents. Decarboxylation converts OMP to UMP. The total ATPs consumed to produce UMP is 1.

c. To produce UTP from UMP, two ATPs are consumed. Thus, a total of 3 ATP are needed to synthesize UTP.

d. CTP formation requires an additional ATP when UTP is aminated, using glutamine and hydrolysis of ATP to drive the reaction Thus, CTP production requires 4 ATP.

8. Illustrate the key points of regulation in (a) the biosynthesis of IMP, AMP, and GMP; (b) E. coli pyrimidine biosynthesis; and (c) mammalian pyrimidine biosynthesis.

Answer: a. IMP is the precursor of both AMP and GMP and its production is regulated by both purine nucleotides. Regulation occurs in the first two steps catalyzed by ribose-5-phosphate pyrophosphorylase and glutamine phosphoribosyl-pyrophosphate amidotransferase. Ribose-5-phosphate pyrophosphorylase is inhibited by both ADP and GDP. Glutamine phosphoribosyl-pyrophosphate amidotransferase regulation is similar, but more complex. This enzyme is sensitive to inhibition by AMP, ADP, and ATP at one site, and inhibition by GMP, GDP, and GTP at another site, and the enzyme is activated by PRPP. IMP is a branch point to either AMP or GMP. AMP production from IMP initiates with adenylosuccinate synthetase, an enzyme that is inhibited by AMP, the final product of the branch. In an analogous fashion, GMP production from IMP initiates with IMP dehydrogenase subject to GMP inhibition.

b. The first step in pyrimidine biosynthesis is the formation of carbamoyl phosphate in a reaction catalyzed by aspartate transcarbamoylase. In *E. coli*, ATCase is feedback inhibited by CTP and activated by ATP.

c. In contrast, regulation of pyrimidine biosynthesis in mammals occurs at the level of carbamoyl phosphate synthesis. Both UTP and CTP inhibit carbamoyl phosphate synthetase II, whereas PRPP and ATP are allosteric activators.

9. Indicate which reactions of purine or pyrimidine metabolism are affected by the inhibitors (a) azaserine, (b) methotrexate, (c) sulfonamides, (d) allopurinol, and (e) 5-fluorouracil.

Answer: a. Azaserine (O-diazoacetyl-L-serine) and a related compound DON (6-diaxo-5-oxo-L-norleucine) are glutamine analogs. Their structures are shown below.

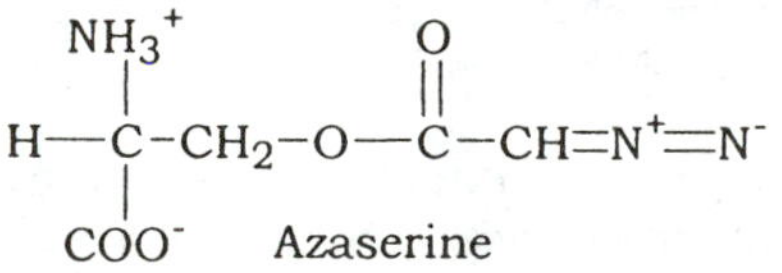

```
   NH3+          O
   |             ||
H—C-CH2-O—C-CH=N+=N-
   |
   COO-     Azaserine

   NH3+             O
   |                ||
H—C-CH2-CH2-C—CH=N+=N-
   |
   COO-        DON

   NH3+
   |               O
   |               ||
H—C-CH2CH2-C—NH2
   |
   COO-    Glutamine
```

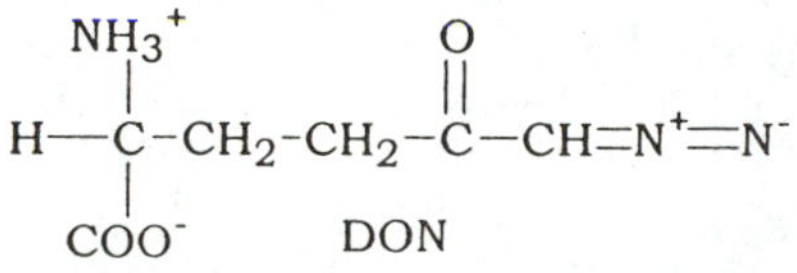

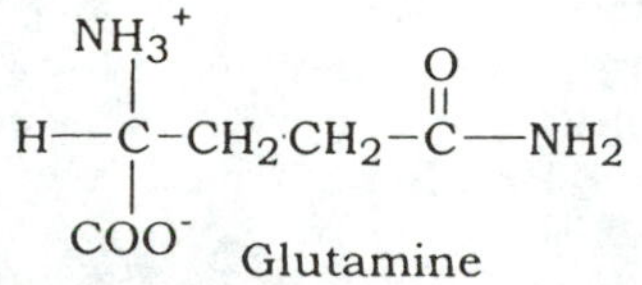

Azaserine and DON bind to glutamine-binding proteins and are capable of reacting with nucleophiles, leading to covalent modification and inactivation. The following reactions in purine synthesis are sensitive to these inhibitors

1. PRPP + glutamine + H_2O → β-5-phosphoribosylamine + glutamate + PP_i
catalyzed by glutamine:PRPP amidophosphoribosyl transferase.
2. FGAR + ATP + glutamine + H_2O → FGAM + ADP + glutamate + P_i

catalyzed by FGAM synthetase.

3. XMP + glutamine + ATP + H_2O → GMP + glutamate + AMP + PP_i
catalyzed by GMP synthetase.

For pyrimidine synthesis, there are two reactions sensitive to these inhibitors.

1. 2 ATP + HCO_3^- + glutamine + H_2O → 2 ADP + glutamate + carbamoyl phosphate + P_i
catalyzed by carbamoyl phosphate synthetase II.
2. UTP + glutamine + ATP + H_2O → CTP + glutamate + ADP + P_i
catalyzed by CTP synthetase.

b. Methotrexate is an analog of dihydrofolate and will compete, with extremely high affinity, for folate binding sites on enzymes. The following reactions in purine biosynthesis will be affected.

1. GAR + N^{10}-formyl-THF → FGAR + THF
catalyzed by GAR transformylase.
2. AICAR + N^{10}-formyl-THF → FAICAR + THF
catalyzed by AICAR transformylase.

In pyrimidine synthesis, methotrexate will block the formation of dTMP produced in the following reaction

1. dUMP + N^5,N^{10}-methylene-THF → dTMP + DHF

catalyzed by thymidylate synthase.

c. Sulfonamides are structural analogs of para-aminobenzoic acid and they competitively inhibit production of folic acid in bacteria. Folic acid is a precursor of THF and the same steps indicated above for methotrexate inhibition will be affected by sulfonamides. However, inhibition will occur because of an inability of bacteria to produce folic acid; inhibition is a result of a lack of substrate as opposed to the presence of an inhibitor of the enzymes for these steps. Since folic acid is a dietary requirement for animals, sulfonamides will not have an effect on nucleotide biosynthesis in animals.

d. Allopurinol is a suicide inhibitor of xanthine oxidase. Xanthine oxidase will hydroxylate allopurinol to form alloxanthine, which remains tightly bound to the enzyme and thus inactivates it. This enzyme plays a key role in degradation of purines. Both adenosine and guanosine are metabolized to xanthine, which is subsequently converted to uric acid by xanthine oxidase. Uric acid formation is blocked.

e. 5-Fluorouracil in and of itself is not an important inhibitor of nucleotide metabolism but it is capable of being converted to 5-fluorodeoxyuridylate (FdUMP), which is a potent inhibitor of thymidylate synthase.

10. Since dUTP is not a normal component of DNA, why do you suppose ribonucleotide reductase has the capacity to convert UDP to dUDP?

Answer: Thymidine, a component of DNA, is produced by methylation of dUMP by the enzyme thymidylate synthase. dUMP is derived from dUTP through the action of dUTP diphosphohydrolase in the following reaction:

dUTP + H_2O → dUMP + PP_i

Thus, to produce dUMP a source of dUTP is required. Ribonucleotide reductase converts NDP to dNDP without regard to base. Thus, dUDP is produced and can be phosphorylated to dUTP. Ultimately UDP is the precursor to dTTP, which is one of the deoxynucleotide triphosphates required for DNA synthesis.

11. Describe the underlying rationale for the regulatory effects exerted on ribonucleotide reductase by ATP, dATP, dUDP, dCDP, and dGDP.

Answer: The diagram below shows the various levels of nucleotide regulation of ribonucleotide reductase. By starting with ATP bound to the overall specificity site and following the arrows we see that the enzyme first produces dCDP and dUTP which leads to increased dTTP which binds to the substrate specificity site by competing with ATP. dTTP leads to production of dGTP which in turn leads to dATP and enzyme inactivation.

ATP thus sets off this cascade of changing activity.

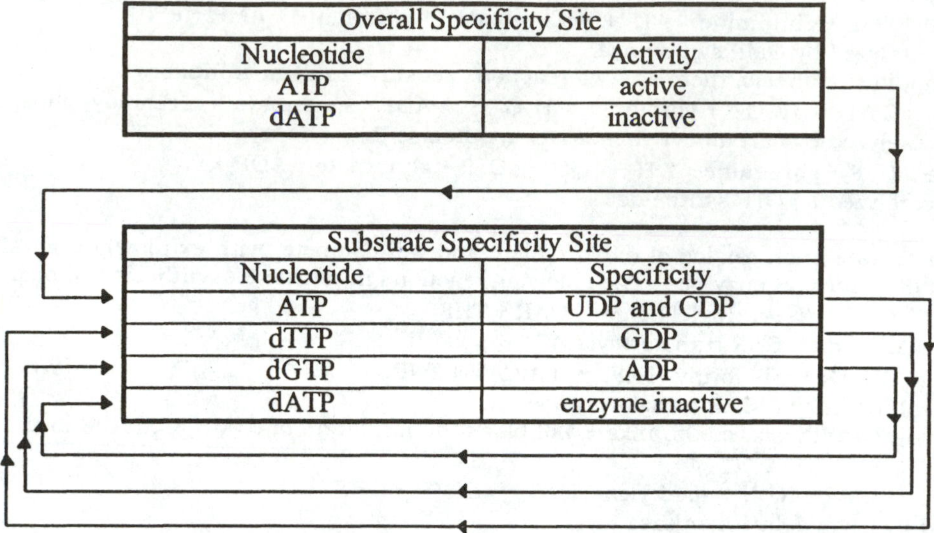

Overall Specificity Site	
Nucleotide	Activity
ATP	active
dATP	inactive

Substrate Specificity Site	
Nucleotide	Specificity
ATP	UDP and CDP
dTTP	GDP
dGTP	ADP
dATP	enzyme inactive

12. By what pathway(s) does the ribose released upon nucleotide degradation enter intermediary metabolism and become converted to cellular energy? How many ATP equivalents can be recovered from one equivalent of ribose?

Answer: Ribose is metabolized in the pentose phosphate pathway, glycolysis, and the citric acid cycle. To enter the pentose phosphate pathway, ribose is phosphorylated at the expense of ATP and converted to ribulose-5-phosphate, which is metabolized as follows: Ribulose-5-phosphate is used to produce triose phosphate equivalents that are fed through glycolysis into the citric acid cycle. Thus, 5 triose phosphates must follow this route to account for 3 ribose units. Starting with 3 ribose (or a total of 15 carbons), the nonoxidative phase of the pentose phosphate pathway is used to produce a total of 5 triose phosphates. For example, using ribulose-5-phosphate isomerase, ribulose-5-phosphate epimerase, transketolase, and transaldolase, three ribose-5-phosphates are converted to two fructose-6-phosphate and glyceraldehyde-3-phosphate. To convert fructose-6-phosphate to glyceraldehyde-3-phosphate, ATP is consumed. Thus, to this stage, a total of 5 ATP are consumed; three to phosphorylate ribose initially* and two additional ATP to metabolize fructose-6-phosphate. Net to this stage: -5 ATPs

The 5 glyceraldehyde-3-phosphates are metabolized to pyruvate with production of 5 NADH, equivalent to 15 ATP, and 10 ATP directly for a total of 25 ATP. Pyruvate metabolism accounts for 1 NADH to form acetyl CoA, and 1 GTP, 3 NADH, and 1 $FADH_2$ in the citric acid cycle for a total of 15 ATP per pyruvate. So, 5 pyruvates yield 75 ATP.

The total is -5 + 25 + 75 = 95 ATP per 3 ribose or 31.7 ATP per ribose.

* If nucleoside phosphorylase is used to degrade nucleosides, ribose-1-phosphate is produced so the net would be -2 ATP instead of -5 ATP. The extra expense of 3 ATPs is avoided by rearranging ribose-1-phosphate to ribose-5-phosphate.

Questions for Self Study

1. Ammonium is incorporated into organic compounds as a result of action of which three enzymes?

2. Write a two step reaction sequence leading from the citric acid cycle intermediate, α-ketoglutarate, to the amino acid glutamine and name the two enzymes involved.

3. Complete the following reaction, name the general enzyme type that catalyzes this kind of reaction, and identify the products.

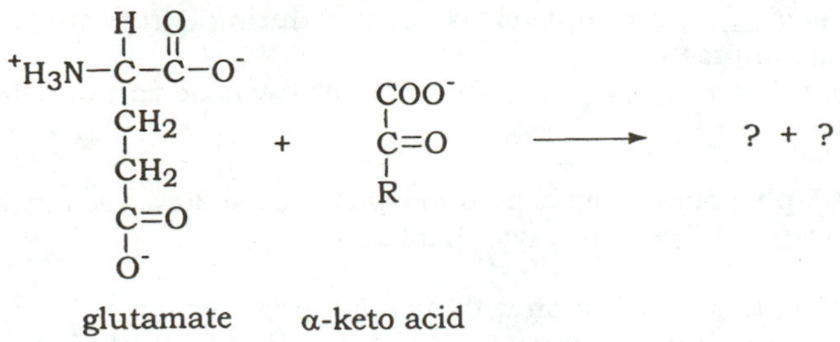

$$^{+}H_3N-\underset{CH_2-CH_2-C(=O)O^-}{\overset{H}{C}}-\overset{O}{\overset{\|}{C}}-O^- \; + \; \underset{R}{\overset{COO^-}{C}}=O \longrightarrow \; ? + ?$$

glutamate α-keto acid

4. Match the nonessential amino acids with their precursor molecules.

a. Glutamate	1. serine and H_2S
b. Glutamine	2. serine (one step)
c. Aspartate	3. phenylalanine
d. Asparagine	4. glutamate (one step)
e. Alanine	5. glutamate (several steps)
f. Serine	6. 3-phosphoglycerate
g. Glycine	7. pyruvate
h. Proline	8. aspartate
i. Tyrosine	9. α-ketoglutarate
j. Cysteine	10. oxaloacetate

5. Fill in the blanks. The 20 common amino acids are degraded into 7 metabolic intermediates. These include the citric acid cycle fuel _____, four citric acid cycle intermediates _____, _____, _____, and _____, a ketone body, _____, and the end product of glycolysis, _____.

6. Identify the compound drawn below. Its synthesis and subsequent degradation convert an amino acid into fumarate and one of the 20 common amino acids. Name the amino acid consumed and produced. This compound is part of which metabolic cycle?

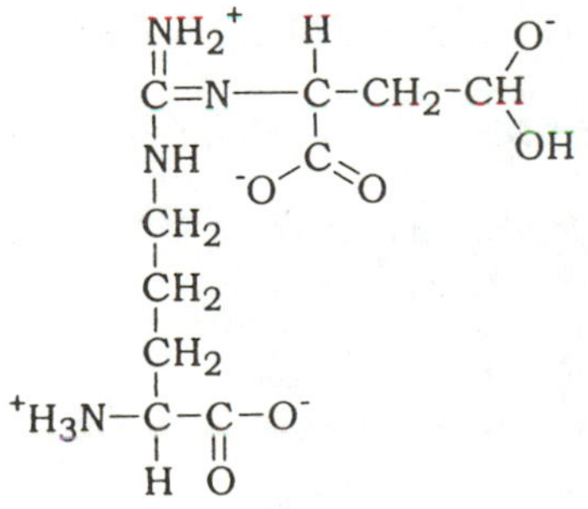

NH2+ H O-
C=N—C-CH2-CH
NH -O-C=O OH
CH2
CH2
CH2
+H3N-C-C-O-
H O

7. Chorismate is a key metabolic intermediate that is derived from erythrose-4-phosphate and two PEP in the shikimate pathway. Given that the pathway does not contain oxidation steps, indicate the carbons in chorismate that derive from PEP and from erythrose-4-phosphate.

COO-
CH2
O-C-COO-
H
HO H

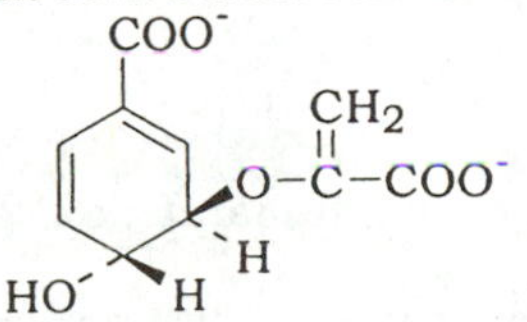

8. List four compounds that derive from chorismate.

9. The de novo synthesis of AMP and GMP start with IMP.

a. Synthesis of this compound starts with _____.

b. Two of the carbons of the purine ring of IMP are supplied by the one-carbon pool as formyl groups attached to _____.

c. Two of the nitrogens derive from the side chain of _____, a common amino acid.

d. One carbon derives from _____ in a reaction that interestingly does not require biotin (nor ATP hydrolysis).

e. The amino acid _____ is completely consumed during IMP synthesis accounting for two carbons and one nitrogen.

f. One nitrogen derives from _____, an amino acid whose carbon skeleton is released as fumarate.

10. The intermediate 5-phosphoribosyl-α-pyrophosphate plays key roles in both de novo purine biosynthesis and in purine salvage pathways. Explain.

11. Match the disease or clinical condition with its appropriate enzyme.

a. Lesch-Nyhan Syndrome	1. Xanthine Oxidase
b. Severe Combined Immunodeficiency Syndrome (SCID)	2. Hypoxanthine-guanine phosphoribosyltransferase (HGPRT)
c. Gout	3. Adenosine deaminase

12. In mammals there are two enzymes that synthesize carbamoyl phosphate. The enzymes are localized in different compartments of the cell and participate in different metabolic pathways. Draw the structure of carbamoyl phosphate. Name the two enzymes that synthesize it, identify their cellular locations, and state in which metabolic pathways they function.

13. Fill in the blanks.
The enzyme, _____, produces deoxyribonucleotide diphosphates from ribonucleotide diphosphates. The reducing equivalents for this reaction come from NADPH but are supplied to the enzyme through _____, a small protein with two cysteine residues that undergo reversible oxidation-reduction cycles between cysteine sulfhydryls and _____ disulfide.

Answers

1. Carbamoyl-phosphate synthetase I, glutamate dehydrogenase, glutamine synthetase.

2. Glutamate dehydrogenase:

$$NH_4^+ + \alpha\text{-ketoglutarate} + NADPH \rightarrow \text{glutamate} + NADP^+$$

Glutamine synthetase :

$$\text{Glutamate} + NH_4^+ + ATP \rightarrow \text{glutamine} + ADP + P_i$$

3. Aminotransferases catalyze the following reaction.

$$^{+}H_3N-CH(CH_2-CH_2-COO^-)-COO^- + R-C(=O)-COO^- \longrightarrow {}^-OOC-C(=O)-CH_2-CH_2-COO^- + {}^{+}H_3N-CH(R)-COO^-$$

glutamate · α-keto acid · α-ketogutarate · amino acid

4. a. 9; b. 4; c. 10; d. 8; e. 7; f. 6; g. 2; h. 5; I. 3; j. 1.

5. Acetyl-CoA; α-ketoglutarate; succinyl-CoA; fumarate; oxaloacetate; acetoacetate; pyruvate.

6. The compound is arginosuccinate, a urea cycle molecule. Its formation and degradation convert aspartate to fumarate and produce the amino acid arginine.

7. The unlabelled carbons derive from erythrose.

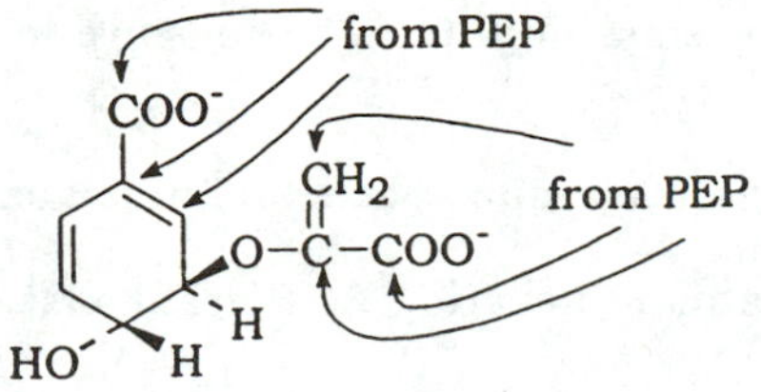

from PEP

COO^-

CH_2 from PEP

$O-\overset{\|}{C}-COO^-$

H

HO H

8. Any four of tryptophan, phenylalanine, tyrosine, Vitamins E and K, Coenzyme Q, plastoquinone, and folic acid.

9. a. ribose-5-phosphate; b. tetrahydrofolate; c. glutamine; d. carbon dioxide; e. glycine; f. aspartate.

10. The purine ring of IMP is synthesized on a ribose moiety and the first reaction involves 5-phosphoribosyl-α-pyrophosphate. This derivatized sugar is used as a substrate for adenine phosphoribosyltransferase and hypoxanthine-guanine phosphoribosyltransferase to salvage free bases.

11. a. 2; b. 3; c. 1

12. Carbamoyl phosphate synthetase II is a cytosolic enzyme involved in pyrimidine biosynthesis. Carbamoyl phosphate synthetase I is a mitochondrial enzyme of the urea cycle (and arginine biosynthesis). The structure of carbamoyl phosphate is shown.

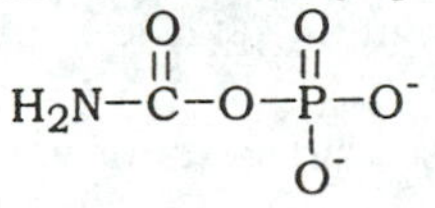

$$H_2N-\overset{O}{\overset{\|}{C}}-O-\underset{O^-}{\overset{O}{\overset{\|}{P}}}-O^-$$

13. Ribonucleotide reductase; thioredoxin; cystine.

Additional Problems

1. Draw the structure of the citric acid cycle intermediates that can be converted to amino acids by transamination. Which amino acids do they produce?

2. The two enzymes glutamate-pyruvate aminotransferase and glutamate-oxaloacetate aminotransferase were formerly used to diagnose damage to heart tissue after an infarction. By comparing the levels of these enzymes in serum drawn at various times after an incident, an estimate of the extent of heart muscle damage could be made. Can you suggest a simple coupled assays to measure the activity of these two enzymes?

3. The polyamine, spermidine, is produced from two amino acids. Given the structure of spermidine, can you suggest which two amino acids are involved?

$$^{+}H_3N\text{/\\/\\/}\underset{H}{N}\text{\\/\\/\\/}NH_3^{+}$$

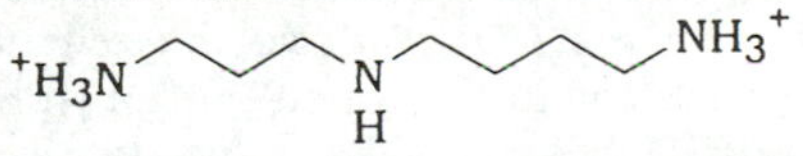

4. Describe the reaction catalyzed by glutamine synthetase, suggest a reaction mechanism, and explain why this reaction is favorable even in low concentrations of ammonium.

5. The essential amino acids in mammals include phenylalanine but not tyrosine. Draw the side chains of these two amino acids to convince yourself that tyrosine is in fact more complicated than phenylalanine in structure and explain why phenylalanine is required but tyrosine is not.

6. What amino acids are involved in the urea cycle?

7. Foods containing aspartame have a warning to phenylketonurics. Why?

8. Excrement from birds makes an excellent fertilizer. Why?

9. Would you expect aminotransferase activity to be sensitive to sodium borohydride ($NaBH_4$)? Explain.

10. Why are the sulfonamides effective against bacterial but not animal cells?

11. Describe the anaplerotic pathway involving AMP deaminase, adenylosuccinate synthetase, and adenylosuccinase.

12. Some patients with gout suffer from a defect in PRPP amidotransferase. What reactions does this enzyme catalyze? Can you suggest a defect in the enzyme that might lead to gout?

13. How does purine biosynthesis differ fundamentally from pyrimidine biosynthesis?

Abbreviated Answers

1. Oxaloacetate is converted to aspartate and α–ketoglutarate is converted to glutamate. The structures of the α-keto acids are shown below.

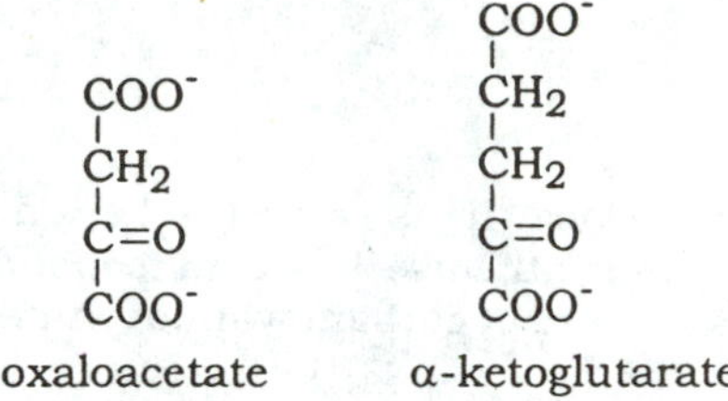

2. Glutamate-pyruvate aminotransferase and glutamate-oxaloacetate aminotransferase catalyze the following reactions.

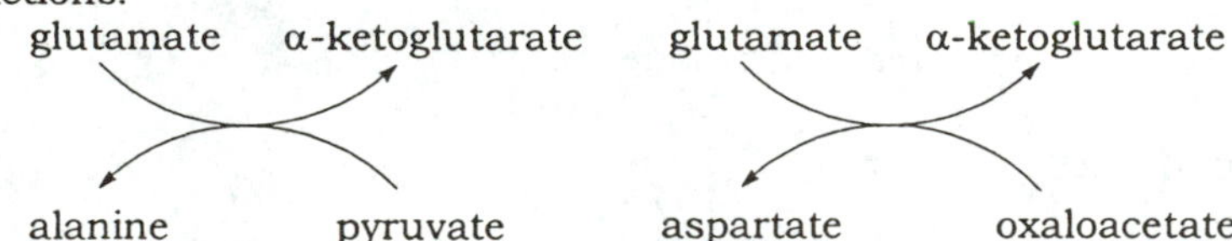

For glutamate-pyruvate aminotransferase, lactate dehydrogenase can be used to produce lactate from pyruvate with oxidation of NADH, which can be followed spectrophotometrically at 340 nm. The aminotransferase reaction can be forced toward pyruvate production by adding excess glutamate and alanine. In a similar manner, glutamate-oxaloacetate aminotransferase activity can be measured using malate dehydrogenase.

3. The butanediamine group, known as putrescine, is produced by decarboxylation of ornithine, a urea cycle intermediate. The aminopropane group derives from methionine via S-adenosylmethionine by decarboxylation and loss of 5'-methylthioadenosine.

4. Glutamine synthetase converts glutamate and ammonium to glutamine. The name synthetase implies that the enzyme uses ATP to drive the reaction. The reaction mechanism involves nucleophilic attack of a γ-carboxylate oxyanion on the γ-phosphorus of ATP to produce γ-glutamyl-phosphate intermediate. The phosphate group is subsequently displaced by ammonium to produce glutamine.

5. Phenylalanine can be converted to tyrosine by phenylalanine hydroxylase in an irreversible oxidation reaction involving molecular oxygen. The side chains of phenylalanine and tyrosine are

OH

CH_2 CH_2

6. Arginine, ornithine, citrulline, and aspartate are involved in the urea cycle. Only arginine and aspartate are common amino acids found in proteins. Ornithine and citrulline are not incorporated into proteins during protein synthesis. (Citrulline was first isolated from watermelon, *Citrullus vulgaris*.)

7. Aspartame is a low-calorie sweetener composed of aspartic acid in amide linkage to phenylalanine methyl ester. The compound is metabolized to aspartic acid and phenylalanine, which cannot be metabolized by phenylketonurics.

8. Birds excrete excess ammonia as uric acid in a thick, pasty substance. The high nitrogen content of uric acid makes it an ideal natural fertilizer.

9. The mechanism of action of aminotransferases involves a Schiff base intermediate on pyridoxal phosphate or Vitamin B_6. The incoming amino acid is deaminated with the amino group being transferred to an enzyme-bound pyridoxal phosphate. This step involves Schiff base formation and sodium borohydride can reduce the Schiff base and inactivate the enzyme.

10. Sulfonamides are structural analogs of p-aminobenzoic acid, a component of folate. Animals do not synthesize folate. However, bacteria synthesize it *de novo* or using precursors including p-aminobenzoic acid. Thus, sulfonamides will compete with p-aminobenzoic acid and lead to inhibition of folic acid biosynthesis.

11. AMP deaminase converts AMP to IMP. Adenylosuccinate synthetase attaches aspartate to IMP to form adenylosuccinate, which is degraded into AMP and fumarate, a citric acid cycle intermediate.

12. PRPP amido transferase produces 5-phosphoribosylamine from 5-phosphoribosyl pyrophosphate and glutamine. It is the first step in purine biosynthesis. The defect in this enzyme cannot lead to enzyme inactivation as this would have the effect of lowering uric acid levels (and perhaps being lethal). The enzyme is subject to feedback inhibition by guanine nucleotides and adenine nucleotides. A defect in allosteric inhibitor binding might lead to a loss of feedback inhibition of purine production leading to increased uric acid production and gout.

13. Purines are synthesized as nucleotide derivatives whereas pyrimidines are produced as the free base orotate, which is subsequently converted to a nucleotide.

Summary

Nitrogen is a crucial macronutrient. In the common inorganic forms of nitrogen prevalent in the environment, nitrogen is in an oxidized state. Its assimilation into organic compounds requires reduction to the level of NH^+_4. Two metabolic pathways accomplish this end: nitrate assimilation and nitrogen fixation. Virtually all higher plants and algae and many fungi and bacteria are capable of nitrate assimilation. Nitrogen fixation is an exclusively prokaryotic process, but nitrogen-fixing bacteria do occur in symbiotic association with selected plants and animals. Animals themselves lack both of these pathways and thus are dependent on plants and microorganisms for a dietary source of reduced-N, principally in the form of protein.

The incorporation of significant quantities of ammonium into organic linkage occurs via only three enzymatic reactions, carbamoyl phosphate synthetase I, glutamate dehydrogenase (GDH) and glutamine synthetase (GS). The latter two of these three reactions are by far the most important. Two pathways of ammonium assimilation are prevalent. If $[NH^+_4]$ levels are adequate, the GDH/GS pathway operates so that one equivalent of α-ketoglutarate picks up two equivalents of NH^+_4 and glutamine is formed. *E. coli* GS is regulated at three levels: it is sensitive to cumulative feedback inhibition by a collection of end products of nitrogen metabolism, its activity is controlled by a bi-cyclic cascade system of covalent modification, and its synthesis is regulated at the level of gene expression.

Amino acid biosynthesis is basically a matter of synthesizing the appropriate α-keto acid carbon skeletons for the various amino acids, followed by transamination of the α-keto acid by a glutamate-dependent aminotransferase. Plants and most micro-organisms can synthesize all of the 20 amino acids commonly found in proteins, but mammals lack the metabolic pathways for synthesis of 10 of these. It is a generality of nature that the longer the metabolic pathway, the more likely it is that the pathway has been lost over evolutionary time. ("Longer" here means "the greater the number of unique reactions intervening between central metabolic intermediates and the final end product"). The amino acids that mammals cannot synthesize are called "essential amino acids" to denote that it is essential for the mammal to obtain these amino acids in their diet. Amino acid biosynthetic pathways can be grouped according to shared precursors, as in the "α-ketoglutarate family of amino acids" (Glu, Gln, Pro, Arg and in some organisms, Lys) or "the 3-phosphoglycerate family" (Ser, Gly and Cys). The aromatic amino acids (Phe, Tyr and Trp) are synthesized from chorismate, a metabolic intermediate central to the biosynthesis of practically all common aromatic compounds. In addition, the histidine and the aromatic amino acid biosynthetic pathways share intermediates in common with the pathways for synthesis of purine and pyrimidine nucleotides.

Nearly all organisms have the ability to synthesize purine and pyrimidine nucleotides *de novo.* Since these nucleotides are the direct precursors for RNA and DNA synthesis, their formation is a particularly active process in rapidly proliferating cells, as seen in bacterial infections or cancerous malignancies. Consequently, enzymes of nucleotide biosynthesis are often the targets of drugs designed to halt uncontrolled cell division.

The nine atoms of the purine ring are contributed by five separate precursors: CO_2, the amino acids glycine, aspartate and glutamine, and one-carbon units contributed by THF derivatives. The purine ring is constructed on a ribose-5-phosphate scaffold. PRPP, the "active form" of ribose-5-P, is the limiting substance in purine biosynthesis. The *de novo* purine synthetic pathway has as its first purine product, IMP. IMP synthesis is regulated at the first two steps: 1) Ribose-5-P pyrophosphokinase is allosterically inhibited by ADP and GDP; 2) Glutamine-PRPP amidotransferase is subject to cumulative feedback inhibition by adenine and guanine nucleotides. This latter enzyme has two sites for allosteric inhibition, one where AMP, ADP and ATP act, and a second where GMP, GDP and GTP act. The two prominent purines, AMP and GMP, are formed from IMP via a bifurcated pathway with two reactions in each branch. The first enzyme in the AMP branch, adenylosuccinate synthetase, has GTP as a substrate and is feedback inhibited by AMP. The initial step in the GMP branch, IMP dehydrogenase, is feedback inhibited by GMP, while the second step, GMP synthetase, uses ATP as its energy source. This product inhibition and reciprocity in energy donor provides an effective regulatory mechanism for balancing the formation of AMP and GMP to meet cellular needs.

Purine salvage pathways furnish a means to recover free purine bases and nucleosides released as a consequence of nucleic acid degradation and turnover. Though these salvage enzymes seem peripheral to the major utilization, deficiencies in them, as in the absence of HGPRT (hypoxanthine-guanine phosphoribosyltransferase) in Lesch-Nyhan Syndrome, have tragic consequences. Further, aberrations in purine catabolism, as in ADA (adenosine deaminase) deficiency, the underlying cause of Severe Combined Immunodeficiency Syndrome or "SCID", can also be devastating. The end product of purine catabolism is uric acid. An excess accumulation of uric acid leads to the clinical disorder known as gout. Allopurinol, an analog of hypoxanthine, is an effective treatment. Allopurinol binds tightly to xanthine oxidase, inhibiting its activity and preventing uric acid build-up. Purine metabolism even fulfills an anaplerotic role. In skeletal muscle, a cyclic series of purine transformations known as the purine nucleoside cycle has the net effect of converting aspartate to fumarate to replenish the level of citric acid cycle intermediates.

De novo pyrimidine biosynthesis first yields the six-membered pyrimidine ring; only then is the ribose-5-moiety added to create a nucleotide. Just two precursors, carbamoyl-P and aspartate, provide the requisite six atoms of the pyrimidine. In *E. coli*, pyrimidine synthesis is regulated at aspartate transcarbamoylase (ATCase), the step where carbamoyl-P and aspartate condense to form carbamoyl-aspartate. ATCase is the paradigm of allosteric enzymes (Chapter 12); it is feedback inhibited by the ultimate ribonucleotide end product of the pyrimidine pathway, CTP, and is activated by ATP. In animals, carbamoyl phosphate synthetase II (CPS-II) is the committed step in pyrimidine biosynthesis. Animal CPS-II is feedback inhibited by UDP and UTP, while ATP and PRPP are allosteric activators. Pyrimidine synthesis in mammals provides an example of "metabolic channeling". The first three reactions are catalyzed by a multi-functional polypeptide, which contains the active sites for CPS-II, ATCase and DHOase. A second multi-functional polypeptide carries both OPRTase and OMP decarboxylase activities (reactions 5 and 6 of the pathway). Such multi-functional proteins achieve metabolic channeling: the direct transfer of substrate from one active site to the next, avoiding dissociation from the protein followed by diffusion (and dilution) in free solution to the next enzyme.

Deoxyribonucleotides have only one metabolic purpose - to serve as building blocks in DNA synthesis. Ribonucleoside diphosphates (NDPs) are the substrates for deoxyribonucleotide formation; NDP reduction at the 2'-position forms the corresponding dNDP. The enzyme mediating this reaction is ribonucleotide reductase, and NADPH is the ultimate source of reducing power for dNDP synthesis. Ribonucleotide reductase has two distinct allosteric sites in addition to its active site, and its activity is regulated by an elegant feedback control circuit that balances the supply of dNTPs. One of the active sites determines the overall activity of the enzyme depending on whether ATP or dATP is bound. ATP activates; dATP inhibits. The second allosteric site is the substrate specificity site. Occupation of this site by its effectors, ATP, dTTP, dGTP or dATP, determines which NDP substrate (CDP, UDP, GDP or ADP) is bound and reduced in the active site.

Synthesis of thymine deoxynucleotides begins with the deoxyribonucleotides, dUDP and dCDP, both of which can lead to dUMP: dUDP→dUTP→dUMP or dCDP→dCMP→dUMP. The conversion of dCDP to dUMP is catalyzed by dCMP deaminase, an allosteric enzyme that provides a control point for regulation of dTTP formation. Thymidylate synthase catalyzes the synthesis of dTMP from dUMP and N^5, N^{10}-methylene THF. Because of its pivotal role in the pathway of dTTP formation, thymidylate synthase has become an attractive target for chemotherapeutic agents designed to selectively inhibit cell division through denial of adequate amounts of DNA precursors.

Chapter 22

Metabolic Integration and Organ Specialization

Chapter Outline

- Metabolism: Three functional blocks
  - Catabolic activities
    - Foods oxidized to carbon dioxide and water
    - ATP and NADPH produced
  - Anabolic activities
    - Metabolic intermediates from catabolism converted to variety of molecules
    - ATP and NADPH consumed
  - Macromolecular synthesis
    - Anabolic products used to synthesize biopolymers
    - ATP principle source of energy
      - GTP: Protein synthesis
      - CTP: Phospholipid synthesis
      - UTP: Polysaccharide synthesis
- Phototrophs
  - Photochemical activities
    - Light energy used to produce ATP and NADPH
  - Carbon dioxide fixation
    - ATP and NADPH used to fix carbon dioxide and convert to intermediate
- Ten key intermediates in catabolism serve as raw material for most of anabolism
  - Carbohydrates
    - Triose-P
    - Tetrose-P
    - Pentose-P
    - Hexose-P
  - α-Keto acids
    - Pyruvate
    - Oxaloacetate

      α-Ketoglutarate
  - CoA derivatives
    - Acetyl-CoA

      Succinyl-CoA
  - PEP
- ADP/ATP and $NADP^+$/NADPH couple catabolism to anabolism
- Stoichiometries of ATP utilization
  - Reaction stoichiometry: Chemical stoichiometry: Reactants and products balanced
  - Obligate coupling stoichiometry: Initial substrate of one pathway and final product of second pathway determine stoichiometry
  - Evolved coupling stoichiometry: Stoichiometry of ATP coupled to pathway, not fixed
- Thermodynamic role of ATP

- ATP coupling allows reaction sequences to be thermodynamically favorable
- Kinetic controls ensure [ATP]/[ADP][P_i] ratio remains high

- Metabolic roles of ATP
  - Stoichiometry establishes large Keq: Unidirectional process
  - ATP (AMP and ADP) allosteric effectors
  - Adenylate kinase interconverts ATP, ADP, and AMP
  - Energy charge: $\text{E.C.} = \frac{1}{2} \times \left(\frac{2 \times [\text{ATP}] + [\text{ADP}]}{[\text{ATP}] + [\text{ADP}] + [\text{AMP}]} \right)$; E.C. varies from 0 to 1
    - R response to E.C.
      - Active when E.C. low: Activity decreases as E.C. approaches 1
      - Enzymes in catabolic pathways show R response
    - U response
      - Active when E.C. close to 1: Activity decreases as E.C. decreases
      - Enzymes in anabolic pathways show U response
  - E.C. oscillates around 0.85 to 0.88 in healthy cells
- Human metabolism
  - Fuel stores
    - Glycogen: Liver and muscle
    - Triacylglycerol: Adipose tissue
    - Protein: Muscle
  - Fuel utilization preferences: Glycogen > triacylglycerol > protein
  - Organ metabolism and interplay
    - Brain
      - High respiratory metabolism
      - No fuel reserves
      - Glucose preferred fuel: From diet or from liver via gluconeogenesis
      - β-Hydroxybutyrate during starvation: From liver
    - Muscle
      - Fatty acid, glucose, and ketone body metabolism at rest
      - P-creatine, and glycogen utilization during intense activity
      - Fatigue caused by decrease in pH, not by depletion of reserves
      - Fasting muscle utilizes amino acids from protein
    - Heart
      - Preferred fuel is fatty acids
      - Minimal reserves: Fatty acids, glucose, and ketone bodies must be supplied
    - Adipose tissue
      - Glucose converted to acetyl-CoA and fatty acids
      - Fatty acids also supplied by liver
      - Triacylglycerol production relies on glycerol-3-P from glucose
      - Glucose plays pivotal role
        - Source of glycerol-3-P
        - Fuel for pentose phosphate pathway
      - Brown fat
        - Thermogenin – uncoupling protein
        - Proton gradient converted to heat, not ATP
    - Liver
      - Buffers blood glucose levels
      - Fatty acid metabolism and ketogenesis
      - Glucose-6-P plays key role
        - Glycogen metabolism
        - Gluconeogenesis/glycolysis
        - Production of NADPH from pentose phosphate
        - Production of acetyl-CoA for ATP via oxidative phosphorylation

Chapter Objectives

Metabolic Integration

So far we have developed a complicated picture of intermediary metabolism and it is time to attempt to simplify and unify. There are a small number of intermediates that serve crucial roles in intermediary metabolism. These include sugar phosphates, pyruvate, oxaloacetate, α-ketoglutarate, acetyl-CoA, succinyl-CoA, and PEP. The sugar phosphates are found in glycolysis, gluconeogenesis, the pentose phosphate pathway, and the Calvin cycle. Pyruvate, oxaloacetate and α-ketoglutarate are keto acids. Pyruvate derives from a number of sources including glycolysis and amino acids and is the port of entry into the citric acid cycle for glucose derived carbons. Oxaloacetate and α-ketoglutarate are citric acid cycle intermediates and both can be produced from amino acids by deamination. Acetyl-CoA is consumed in the citric acid cycle and is a common denominator between fatty acids, sugars, and amino acid. Succinyl-CoA, a citric cycle intermediate, is the place of entry of propionate from dietary sources and odd-chain fatty acid catabolism, is a product of amino acid catabolism, and is used in heme biosynthesis.

ATP and NADPH serve critical roles in coupling catabolism and anabolism. Catabolism is largely oxidative in nature, leading to reduction of cofactors NAD^+ and FAD. Anabolic pathways are reductive with NADPH usually serving as the immediate source of electrons. This coenzyme is reduced in the pentose phosphate pathway. Additionally, cycles exist to move electrons from NADH to $NADP^+$. Catabolic pathways are exergonic and lead to synthesis of ATP. ATP is then consumed in anabolic, energy requiring pathways. The coupling of ATP production to a complex metabolic pathway such as aerobic oxidation of glucose to CO_2 and H_2O has a stoichiometry that is not defined by simple chemical considerations. Rather ATP coupling stoichiometry is an evolved quantity. Nevertheless, under physiological conditions, the complete oxidation of glucose gives high yields of ATP. Furthermore, the process is always far from equilibrium making regulation by kinetic controls possible.

Thermodynamic Role of ATP in Metabolism

Diametric pathways leading to synthesis and degradation of the same intermediate are always characterized by different ATP coupling coefficients. The coupling of ATP to pathways insures that they are thermodynamically favorable and, thus, must be regulated kinetically. ATP serves an additional role in kinetic regulation by functioning as an allosteric effector in many key reactions.

Energy Charge

The energy charge, E.C., is defined as

$$\text{E.C.} = \frac{1}{2}\frac{(2[\text{ATP}]+[\text{ADP}])}{([\text{ATP}]+[\text{ADP}]+[\text{AMP}])}$$

and can have values between 0 and 1.0. In healthy cells, the E.C. is maintained between 0.85 and 0.88.

The regulatory enzymes of metabolism respond to the E.C. in reciprocal manner – enzymes regulating catabolism are active at low E.C. and inactive at high E.C., whereas the regulatory enzymes of anabolism are inactive at low and active at high E.C. This inactivity/activity enzymatic profile causes the E.C. to oscillate around a value of 0.85 to 0.88 in healthy cells.

Problems and Solutions

1. Assume the following intracellular concentrations in muscle tissue: ATP = 8 mM, ADP = 0.9 mM, AMP = 0.04 mM, P_i = 8 mM. What is the energy charge in muscle?

Answer: The energy charge is given by:

$$\text{Energy Charge} = \frac{1}{2}\times\frac{(2\times[\text{ATP}]+[\text{ADP}])}{[\text{ATP}]+[\text{ADP}]+[\text{AMP}]}$$

$$\text{Energy Charge} = \frac{1}{2}\times\frac{(2\times 8\times 10^{-3}+0.9\times 10^{-3})}{8\times 10^{-3}+0.9\times 10^{-3}+0.04\times 10^{-3}}$$

$$\text{Energy Charge} = 0.945$$

2. *Strenuous muscle exertion (as in the 100-meter dash) rapidly depletes ATP levels. How long will 8 mM ATP last if 1 gram of muscle consumes 300 μmol of ATP per minute? (Assume muscle is 70% water). Muscle contains phosphocreatine as a reserve of phosphorylation potential. Assuming [phosphocreatine] = 40 mM, [creatine] = 4 mM, and $\Delta G^{\circ\prime}$ (phosphocreatine + H_2O → creatine + P_i) = -43.3 kJ/mol, how low must [ATP] become before it can be replenished by the reaction phosphocreatine + ADP → ATP + creatine. [Remember, $\Delta G^{\circ\prime}$ (ATP hydrolysis) = -30.5 kJ/mol.]*

Answer: One gram of muscle contains approximately 0.7 g of H_2O, or 0.7 mL. If the [ATP] = 8 mM, 0.7 mL contains

$$0.7\times10^{-3}\text{L}\times8\times10^{-3}\frac{\text{mol}}{\text{L}}=5.60\times10^{-6}\text{ mol or }5.6\ \mu\text{mol ATP}$$

If ATP is consumed at the rate of 300 μmol per min it will last

$$\frac{5.6\ \mu\text{mol}}{300\ \frac{\mu\text{mol}}{\text{min}}}=0.019\text{ min or }1.12\text{ sec}$$

Phosphocreatine and ATP are coupled by the following reaction

phosphocreatine + ADP → ATP + creatine

Which is the sum of two reactions

phosphocreatine + H_2O → creatine + P_i $\quad\Delta G^{\circ\prime}$= -43.3 kJ/mol

and $\quad P_i$ + ADP → ATP + H_2O $\quad\Delta G^{\circ\prime}$= 30.5 kJ/mol

Thus, the overall $\Delta G^{\circ\prime}$= -12.8 kJ/mol. For the reaction to be favorable, ΔG must be less than zero or

$$\Delta G=\Delta G^{\circ\prime}+RT\ln\frac{[\text{creatine}][\text{ATP}]}{[\text{phosphocreatine}][\text{ADP}]}<0\text{, or}$$

$$\Delta G^{\circ\prime}+RT\ln\frac{[\text{creatine}][\text{ATP}]}{[\text{phosphocreatine}][\text{ADP}]}<0$$

$$RT\ln\frac{[\text{creatine}][\text{ATP}]}{[\text{phosphocreatine}][\text{ADP}]}<-\Delta G^{\circ\prime}$$

$$\frac{[\text{ATP}]}{[\text{ADP}]}<\frac{[\text{phosphocreatine}]}{[\text{creatine}]}\times e^{\frac{-\Delta G^{\circ\prime}}{RT}}$$

$$\frac{[\text{ATP}]}{[\text{ADP}]}<\frac{40\text{ mM}}{4\text{ mM}}\times e^{\frac{-(-12.8)}{2.58}}$$

$$\frac{[\text{ATP}]}{[\text{ADP}]}<1{,}435$$

Here we have assumed T=37°C, and that the $\Delta G^{\circ\prime}$ values for creatine and ADP phosphorylation are the same as at 25°C. Under these conditions and assumptions, the reaction is favorable and ADP is phosphorylated at the expense of phosphocreatine when [ATP]< 1,435×[ADP] and [Cr-P] = 40 mM and [Cr] = 4 mM.

3. *The standard reduction potentials for the (NAD^+/NADH) and the ($NADP^+$/NADPH) couples are identical, namely -320 mV. Assuming the in vivo concentration ratios NAD^+/NADH = 20 and $NADP^+$/NADPH = 0.1, what is ΔG for the following reaction?*

NADPH + NAD^+ → $NADP^+$ + NADH

Calculate how many ATP equivalents can be formed from ADP + P_i by the energy released in this reaction.

Answer: From $\Delta G = -n\mathcal{F}\Delta\mathcal{E}_o$, where n is the number of electrons transferred, $\mathcal{F}$ is Faraday's constant (96,494 J/V· mol), and $\Delta\mathcal{E}_o'$ is the change in redox potential. We can calculate ΔG given that:

$$\Delta\mathscr{E} = \Delta\mathscr{E}^{o\prime} - \frac{RT}{n\mathscr{F}}\ln\frac{[NADH][NADP^+]}{[NAD^+][NADPH]}$$

$$\Delta\mathscr{E}^{o\prime} = \mathscr{E}^{o\prime}_{acceptor} - \mathscr{E}^{o\prime}_{donor} = -320\text{ mV} - (-320\text{ mV}) = 0$$

$$\Delta\mathscr{E} = 0 - \frac{8.314\times10^{-3}\times298}{2\times96{,}494}\ln\frac{0.1}{20}$$

$$\Delta\mathscr{E} = 6.80\times10^{-5}\text{V}$$

$$\Delta G = -n\mathscr{F}\Delta\mathscr{E} = -2\times96{,}494\times6.80\times10^{-5}\text{V}$$

$$\Delta G = -13.1\frac{\text{kJ}}{\text{mol}}$$

Under standard conditions and under physiological conditions hydrolysis of ATP is a very favorable reaction with large negative ΔG values (on the order of -30 kJ/mol and -50 kJ/mol respectively). The ΔG for the reaction of electron transfer from NADPH to NAD^+ will not support ATP synthesis under these conditions.

4. Assume the total intracellular pool of adenylates (ATP + ADP + AMP) = 8 mM, 90% of which is ATP. What are [ADP] and [AMP] if the adenylate kinase reaction is at equilibrium? Suppose [ATP] drops suddenly by 10%. What are the concentrations now for ADP and AMP, assuming adenylate kinase reaction is at equilibrium? By what factor has the AMP concentration changed?

Answer: Adenylate kinase catalyzes the following reaction

$$\text{ATP + AMP} \rightarrow \text{ADP + ADP}$$

The equilibrium constant for the reaction K_{eq} = 1.2 (given in Figure 22.2).

$$K_{eq} = \frac{[ADP]^2}{[ATP][AMP]}$$

Let T = [ATP] + [ADP] + [AMP], the total concentration of adenylates

If [ATP] = 90% × T = 0.9 × 8 mM = 7.2 mM

[ADP] + [AMP] = 8 mM - 7.2 mM = 0.8 mM, or

[AMP] = 0.8 mM - [ADP]

Substituting this expession and the value of 7.2 mM for [ATP] into the equilibrium equation we find:

$$K_{eq} = \frac{[ADP]^2}{7.2\text{ mM}\times(0.8\text{ mM} - [ADP])}\text{, or}$$

$$[ADP]^2 + 7.2\text{ mM}\times K_{eq}\times[ADP] - 7.2\text{ mM}\times0.8\text{ mM}\times K_{eq} = 0$$

By substitution K_{eq} = 1.2 we find :

$$[ADP]^2 + 8.64\times10^{-3}\times[ADP] - 6.91\times10^{-6} = 0$$

A quadratic equation whose solution is [ADP] = 0.737 mM

[AMP] = 0.8 mM - [ADP] = 0.8 mM - 0.737 mM = 0.063 mM

Thus, we have :

[AMP] = 0.063 mM, [ADP] = 0.737 mM, and [ATP] = 7.2 mM

If the [ATP] suddenly falls by 10%, we have:

$[ATP] = 7.2\ mM - 10\% \times 7.2\ mM = 6.48\ mM$
$[ADP] + [AMP] = 8\ mM - 6.48\ mM = 1.52\ mM$, or
$[AMP] = 1.52\ mM - [ADP]$
Substituting this expession and the value of 6.48 mM for [ATP]
into the equilibrium equation we find :

$$K_{eq} = \frac{[ADP]^2}{6.48\ mM \times (1.52\ mM - [ADP])}, \text{or}$$

$$[ADP]^2 + 6.48\ mM \times K_{eq} \times [ADP] - 6.48\ mM \times 1.52\ mM \times K_{eq} = 0$$

By substitution $K_{eq} = 1.2$ we find :

$$[ADP]^2 + 7.78 \times 10^{-3} \times [ADP] - 1.18 \times 10^{-5} = 0$$

A quadratic equation whose solution is [ADP] = 1.30 mM
[AMP] = 1.52 mM, [ADP] = 0.22 mM, [ATP] = 6.48 mM

This relatively modest change in [ATP], causes almost a doubling of [ADP] levels $(\frac{1.30\ mM}{0.737\ mM} = 1.76)$ and a 3.5-fold increase in [AMP] $(\frac{0.22\ mM}{0.063\ mM} = 3.5)$.

5. Leptin not only induces synthesis of fatty acid oxidation enzymes and uncoupler protein-2 in adipocytes, but it also causes inhibition of acetyl-CoA carboxylase, resulting in a decline in fatty acid biosynthesis. This effect on acetyl-CoA carboxylase, as an additional consequence, enhances fatty acid oxidation. Explain how leptin-induced inhibition of acetyl-CoA carboxylase might promote fatty acid oxidation.

Answer: Acetyl-CoA carboxylase catalyzes the production of malonyl-CoA and inhibition of this enzyme will immediately inhibit fatty acid biosynthesis because malonyl-CoA is a substrate for fatty acid synthase. Malonyl has another regulatory role in fatty acid metabolism: it inhibits carnitine acyltransferase,the enzyme responsible for fatty acid uptake by the mitochondria. As malonyl-CoA levels fall, carnitine acyltransferase will cause an increased uptake of fatty acids into the mitochondria where they are metabolized by β-oxidation

Questions for Self Study

1. Describe the interplay between β-oxidation, a catabolic pathway, and fatty acid biosynthesis, an anabolic pathway. How are the two connected? Does degradation of a fatty acid provide sufficient resources to support resynthesis of a fatty acid of the same length?

2. What two important compounds couple anabolism and catabolism?

3. Metabolism in a typical aerobic heterotropic cell (a chemoheterotroph) can be described as an interaction of three functional blocks: catabolism, anabolism, and macromolecular synthesis and growth. How is this picture altered for a chemoautotroph? A photoheterotroph? A photoautotroph?

4. The balanced equation for oxidation of glucose via glycolysis, the citric acid cycle, and electron transport is $C_6H_{12}O_6 + 6\ O_2 + 38\ ADP + 38\ P_i \rightarrow 6\ CO_2 + 38\ ATP + 44\ H_2O$. Of the ATPs produced, how many derive from simple chemical stoichiometry and how many from evolved coupling stoichiometry?

5. Describe the reaction catalyzed by adenylate kinase. What important allosteric regulator is produced by this reaction?

6. Table 22.1 is reproduced below with the last three columns randomly ordered. Rearrange the entries in these columns to correct Table 22.1

Organ	Energy Reservoir	Preferred Substrate	Energy Source Exported
Brain	Glycogen, triacylglycerol	Fatty acids	Fatty acids, glycerol
Skeletal muscle (resting)	Triacylglycerol	Glucose(ketone bodies during starvation)	Fatty acids, glucose, ketone bodies
Skeletal muscle (prolonged exercise)	Glycogen	Fatty acids	None
Heart muscle	None	Glucose	Lactate
Adipose tissue	Glycogen	Amino acids, glucose, fatty acids	None
Liver	None	Fatty acids	None

Answers

1. In β-oxidation, fatty acids are converted into acetyl-CoA units. During the process, coenzymes are reduced. Fatty acid biosynthesis involves joining of acetyl units and oxidation of reduced coenzymes. Per acetyl unit, the number of reduced coenzymes produced in β-oxidation balances the number of reduced coenzymes consumed during fatty acid synthesis. However, fatty acid synthesis is driven by hydrolysis of high-energy phosphate bonds whereas β-oxidation, once primed, is thermodynamically spontaneous. To support anabolism of fatty acids from a catabolic series, some of the acetyl units of the catabolic series would have to be consumed in the citric acid cycle to provide energy for fatty acid biosynthesis.

2. ATP and NADPH.

3. For a photoautotroph an additional block is added in which light energy is converted to ATP and reducing equivalents, which are used, in turn, to fix carbon dioxide. A photoheterotroph can also harvest light but cannot fix carbon dioxide. Thus, its block lacks the ability to produce reducing equivalents using light energy and to fix carbon dioxide. The chemoautotroph can fix carbon dioxide with reducing equivalents derived from inorganic compounds.

4. The conversion of glucose to two pyruvates produces two ATP. Two additional ATPs are generated by substrate level phosphorylation in the citric acid cycle in which two pyruvates are metabolized to carbon dioxide and water. Thus, the simple chemical reaction stoichiometry is 4 ATP.

5. The reaction catalyzed by adenylate kinase is ATP + AMP → 2 ADP. In the reverse reaction, AMP, an important allosteric regulator, is produced from ADP.

6. Table 22.1

Organ	Energy Reservoir	Preferred Substrate	Energy Source Exported
Brain	None	Glucose(ketone bodies during starvation)	None
Skeletal muscle (resting)	Glycogen	Fatty acids	None
Skeletal muscle (prolonged exercise)	None	Glucose	Lactate
Heart muscle	Glycogen	Fatty acids	None
Adipose tissue	Triacylglycerol	Fatty acids	Fatty acids, glycerol
Liver	Glycogen, triacylglycerol	Amino acids, glucose, fatty acids	Fatty acids, glucose, ketone bodies

Additional Problems

1. A plot of the rate of running versus distance for recent world records of several races is shown below. It appears that the data can be divided into three linear regions. For each region, describe the source of energy being utilized by muscle.

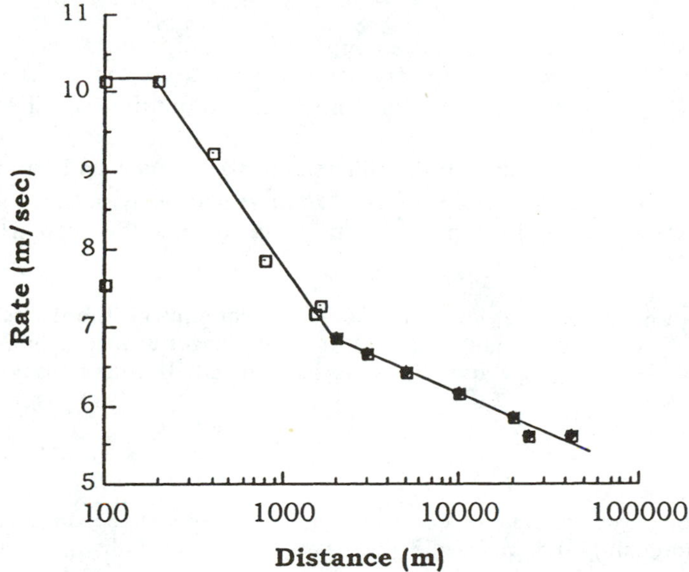

2. In brown fat, the protein thermogenin functions in a substrate cycle. Describe this cycle and state its purpose.

3. The energy of ATP hydrolysis is a highly exergonic process with a large negative free energy change. We have discussed in this chapter how coupling of ATP hydrolysis is used to drive pathways. Give an example of a reaction driven by ATP hydrolysis and describe how ATP hydrolysis is coupled to the reaction.

4. Explain why NADPH and NADH have different energy equivalents?

5. Why is there a difference in energy between cytosolic NADH and mitochondrial NADH?

6. Carl and Gerti Cori were the first to describe the metabolic interplay between muscle and liver involving lactic acid formation in muscle and glucose formation in liver. This metabolic couple is known as the Cori cycle. Describe it.

Abbreviated Answers

1. The first two points on the graph represent the winning performances for the 100- and 200-meter dash. These races are run almost completely on high-energy phosphate stores of ATP and phosphocreatine, which limit the muscle to about 20 sec of intense activity. The group of races constituting the line of high slope is using a combination of aerobic and anaerobic respiration. At the end of these races high-energy phosphate stores are exhausted, lactic acid levels have risen, and aerobic respiration is functioning maximally. The long-distance races rely completely on aerobic respiration. (It is of interest to note that the 2000-meter race, a grueling contest, is at the break-point between the long-distance races and the anaerobic/aerobic-dependent races.)

2. Thermogenin functions as a proton channel in the inner mitochondrial membrane. The substrate cycle involves protons as substrates. Protons are pumped out of the mitochondria by electron transport but are allowed back in via thermogenin. Thus, protons move from one side of the membrane to the other. The purpose is to generate heat.

3. Any one of the biotin containing enzymes is a good example. Biotin is involved in metabolism of one-carbon units at the oxidation level of a carboxylate. The reaction mechanism involves formation of a phosphorylated bicarbonate intermediate by nucleophilic attack of bicarbonate on the γ-phosphate of ATP, releasing ADP. This activated carboxylate is then transferred to enzyme-bound biotin and from there to substrate to produce the carboxylated product. Carboxylation of biotin thus depends on ATP hydrolysis.

4. The oxidized and reduced forms of these dinucleotides are maintained by the cell at very different levels. [NADPH] is used as a primary source of electrons for metabolic pathways and cells maintain [NADPH] > [$NADP^+$]. The situation for [NADH] is just the opposite.

5. The ratio [NADH] to [NAD^+] is likely to be different in the cytosol and the mitochondria. In addition, in order to use oxidation of NADH to NAD^+ to drive synthesis of ATP, cytosolic NADH will have to donate electrons to the electron transport chain against the high [NADH] gradient in mitochondria.

6. In the Cori cycle, muscle produces lactic acid as a consequence of intense activity. Lactic acid is sent via the blood to the liver where it is used in gluconeogenesis to produce glucose. Liver cells are capable of releasing glucose back to the blood because they have glucose-6-phosphatase.

Summary

The metabolism of a typical heterotrophic cell can be represented as three interconnected functional blocks composing the metabolic pathways of: 1) catabolism, 2) anabolism and 3) macromolecular synthesis and growth. An energy cycle exists whose agents are ATP and NADPH. Energy and reducing power is delivered from catabolism to anabolism in the form of ATP and NADPH, and ATP and NADPH are regenerated from ADP and $NADP^+$ via catabolic reactions. Phototrophic cells contain a fourth metabolic system consisting of the photochemical apparatus for transforming light into chemical energy. In autotrophic cells, a fifth system occurs, the carbon dioxide fixation pathway for carbohydrate synthesis.

Three levels of stoichiometry can be recognized in metabolism: 1) simple reaction stoichiometry, 2) obligate coupling stoichiometry, and 3) evolved coupling stoichiometry, particularly as represented in ATP coupling. The net yield of 38 ATP/ glucose in cellular respiration is a biological adaptation, not the consequence of inviolable chemical laws. At 38 ATP/glucose, the K_{eq} for cellular respiration is 10^{170}. Considering the thermodynamics of a generally defined conversion, $A \rightarrow B$, the nature and magnitude of the ATP equivalent can be illustrated. Coupling the $A \rightarrow B$ conversion to ATP hydrolysis raises the equilibrium ratio, $[B]_{eq}/[A]_{eq}$, by 1.4×10^8-fold. The energetics of ATP is crucial to the solvent capacity of the cell: The thermodynamic favorability of phosphoryl transfer by ATP makes it possible for the cell to carry out reactions with great efficiency, even though [reactants] are low.

In the absence of regulation, competing substrate utilization and generation pathways could result in the net hydrolysis of ATP and wasteful dissipation of cellular energy. In addition to its role in metabolic thermodynamics, ATP also acts as an important allosteric effector in the kinetic regulation of metabolism. *Energy charge* is a concept that provides a measure of how fully charged the adenylate system is with high-energy phosphoryl groups. Enzymes can be classified as **R** or **U** in terms of their response to energy charge. **R**-type enzymes are members of ATP-regenerating metabolic pathways, while **U** enzymes are characteristically in biosynthetic, or ATP-utilizing sequences. Thus, **R** and **U** pathways are diametrically opposite with regard to ATP involvement. The reciprocal relationship between **R** and **U** systems means that energy charge reaches a point of metabolic steady-state; in healthy cells it is at an E.C. value of 0.85 - 0.88.

In multicellular organisms, organ systems have arisen to carry out specific physiological functions. Essentially all cells in these organisms have the same set of enzymes in the central pathways of intermediary metabolism, but the various organs do differ significantly in the metabolic fuels - glucose, glycogen, fatty acids, amino acids - that they prefer to use for energy production. Brain has a very high respiratory metabolism, essentially no fuel reserves, and is dependent on a supply of blood glucose as its principal fuel. Muscle is intermittently active as muscle contraction and relaxation takes place on demand. ATP is necessary to drive contraction in response to a increased [Ca^{2+}] pulse as the metabolic signal. The Ca^{2+}-ATPase operating during muscle relaxation uses almost as much ATP as the acto-myosin contractile system. Muscle is a major storage site of glycogen. During strenuous exercise, the rate of glycolysis in muscle may increase 2,000-fold. Muscle fatigue is the result of a drop in cytosolic pH due to the accumulation of H^+ released during glycolysis. Heart, a rhythmically active muscle system, has minimal reserves of fuel such as glycogen or triacylglycerols, and prefers free fatty acids as fuel. Adipose tissue is a metabolically active, amorphous tissue widely distributed throughout the body. A 70-kg person has enough triacylglycerols stored in adipose tissue to fuel 3 months' worth of modest activity. Glucose is the signal for release of fatty acids from adipocytes to the blood: When glucose levels are sufficient, adipocytes can form glycerol 3-phosphate allowing triacylglycerol synthesis from fatty acids generated by triacylglycerol turnover. Brown fat is a form of adipose tissue containing lots of mitochondria. In brown fat, oxidative phosphorylation is uncoupled by thermogenin and the energy of fatty acid oxidation is released as heat. The liver is a major metabolic processing organ. It serves an essential role in buffering [blood glucose]. It is also a center for fatty acid turnover, ketone body formation and conversion of amino acids into other metabolic fuels.

Chapter 23

DNA: Replication, Recombination, and Repair

• •

Chapter Outline

- Characteristics of DNA replication
  - Semiconservative
    - Meselson and Stahl (1958)
      - Cells grown in 15-N
      - Shifted to 14-N
      - DNA analyzed by CsCl gradient centrifugation after 1 or 2 generations
  - Semidiscontinuous
    - Leading strand: Continuous
    - Lagging strand: Discontinuous: Okazaki fragments
- Enzymology of DNA replication
  - Three DNA polymerases in *E. coli*
    - DNA polymerase I: First characterized by Kornberg (1959): Repair and removal of primers
      - 5' Polymerase activity
        - Low processivity
      - 3' Exonuclease
        - Proofreading
        - This and polymerase activity on large or Klenow fragment
      - 5' Exonuclease: Nick translation
    - Polymerase III: Replication
      - Core enzyme: Three subunits
        - α = polymerase
        - ε = 3' exonuclease: Proofreading
        - θ
      - DNA polymerase III
        - Two cores
        - γ = Clamp loader
      - DNA polymerase III holoenzyme
        - Two β dimers (one dimer per core): Sliding DNA clamp
  - DNA ligase: Joins 3' hydroxyl with 5' phosphate
- Termination
  - Ter locus contains oppositely oriented ter sequences
  - Tus protein: Contrahelicase
  - Catenated DNA disengaged by topoisomerase II (DNA gyrase)
- DNA polymerase is a replication factory
  - DNA polymerase is immobilized
  - DNA fed through
- Cell cycle

- M, G_1, S and G_2
- Checkpoints

- Eukaryotic DNA replication
  - Multiple replicators (ori's)
  - Pre-Replication Complex
    - ORC (origin recognition complex) binds replicators
    - Cdc6p binds to ORC
    - RLFs (replication licensing factors) bind to DNA: Two required
  - pre-RC substrate for two protein kinases
    - Cyclin-dependent protein kinase/cyclin B complex: Cyclin B-CDK
      - Phosphorylates ORC, Cdc6p, MCM, Cdc7p-Dbf4p
    - Cdc7p-Dbf4p kinase activated by cyclin B-CDK
      - Activity serves as replication switch
- Eukaryotic DNA polymerases
  - DNA polymerase α: Initiator of nuclear DNA replication
  - DNA polymerase δ: Principle enzyme
    - Forms complex with PCNA (DNA clamp)
    - High processivity
  - DNA polymerase ε: Role in replication
  - DNA polymerase β: DNA repair
  - DNA polymerase γ: Mitochondrial enzyme
- DNA ends: Telomeres
  - Tandem repeats of G-rich 5 to 8 bp segments
  - Telomerase maintains telomeres
    - RNA-dependent DNA polymerase
    - Ribonucleoprotein particle where RNA serves as template for telomere
- Reverse transcriptase: Three activities
  - RNA-dependent DNA polymerase
  - RNase H: Removes RNA of RNA/DNA hybrid
  - DNA-directed DNA polymerase
- Recombination
  - Homologous
  - Nonhomologous
  - Transposition
- General (homologous) recombination
  - Meselson and Weigle: Recombination in λ
    - Density label two strains: Recombinants had intermediate density
  - Holliday model
    - Chromosomes pair (synapse)
    - DNA unwinds
    - Strand invasion
    - Holliday junction
    - Branch migration
    - Resolution of junctions
      - Patch recombination
      - Splice recombination
- Enzymology of recombination
  - RecBCD
    - Helicase and nuclease activities
    - Chi site: GCTGGTGG: Induces RecBCD nuclease activity
    - Product is ssDNA tail with chi site at 3'-end
  - ssDNA binding protein: Coats DNA as RecBCD helicase unwinds dsDNA
  - RecA
    - ATP-dependent DNA strand exchange
    - Binds ssDNA and dsDNA: Forms extended filaments
    - High-affinity primary DNA-binding site binds ssDNA

        - Second dsDNA binds at secondary DNA-binding site
        - When homologous DNA encountered dsDNA converted to ssDNA
        - Strand exchange occurs
    - RuvA, RuvB, RuvC proteins
        - RuvA and RuvB: Holliday-junction specific helicase: Catalyzes branch migration
        - RuvC endonuclease that resolves Holliday junctions
- Transposons
    - McClintock identified activator gene in maize in 1950 (Nobel prize for work in 1983): Mobile genetic element
    - Insertion sequences: Simplest transposon
- DNA repair
    - Mismatch repair: Methyl-directed pathway determines template strand
    - Photoreaction of pyrimidine dimers: Photolyases use light energy to reverse pyrimidine dimers
    - Excision repair
        - Base excision: Removal of single altered base
            - DNA glycosidase removes damaged base
            - AP endonuclease cleaves at AP site
            - Exonuclease removes AP site
        - Nucleotide excision: Removes damaged base by removing short ssDNA fragment
- Molecular nature of mutations
    - Point mutations: Base substitutions
        - Transitions: Purine to purine or pyrimidine to pyrimidine
        - Transversions: Purine to pyrimidine or pyrimidine to purine
    - Insertions or deletions
        - Cause frame shifting
        - Produced by intercalating agents
    - Base analog mutations
        - 5-Bromouracil: Thymine analog whose tautomeric form pairs with G (not A)
        - 2-Aminopurine: Incorporates as A but pairs with C
    - Chemical mutagens
        - Nitrous acid: Oxidative deamination of bases: C changed to U gives G to A mutation
        - Hydroxylamine: Modifies C: Modified C pairs with A not G
        - Alkylating agents: Modified bases mispair

Chapter Objectives

DNA Replication

Review the structure of DNA before starting this chapter. You should already know that dsDNA is composed of two strands that run antiparallel to each other and that a single stand of DNA has a 5'-end and a 3'-end. Each strand is a polymer of deoxyribonucleoside monophosphates held together in phosphodiester linkage between the 5'-carbon of one nucleotide and the 3'-carbon of the next nucleotide. The strands are held to each other by complementary hydrogen bonds between pairs of bases AT, TA, GC, and CG.

DNA is replicated semiconservatively. Understand what this means and how the experiments of Meselson and Stahl proved it. These experiments employed the stable heavy isotope of nitrogen ^{15}N to grow a culture of *E. coli* to contain dense DNA. The culture was then shifted to ^{14}N-containing media, grown for a number of generations, and after each generation sampled to determine the density of DNA.

Enzymology

All DNA polymerases require a ssDNA template to direct the synthesis of a complementary strand. The complementary strand is produced by addition of deoxyribonucleoside monophosphate groups derived from dNTP's to the 3'-end of the growing complementary strand. Thus, the polymerase moves 3'→5' along the template and extends the newly synthesized strand 5'→3'.

The Kornberg polymerase, DNA pol I -a 109-kD, single-chain polypeptide- was the first DNA polymerase to be characterized. It plays a role in replication and repair. You should know the various activities exhibited by this DNA-dependent DNA polymerase including polymerase, 3' exonuclease, and 5' exonuclease. The 3' exonuclease functions in proofreading, whereas the 5' exonuclease, acting concurrently with polymerase activity, can cause nick translation or movement of a DNA gap by removing DNA (or RNA) on the 5'-end of the gap and adding DNA to the 3'-end of the gap. DNA polymerase I can cause strand displacement and has low processivity.

The principal polymerase in DNA replication in *E. coli* is DNA polymerase III, a complex of several proteins with separate functions. The α subunit has polymerase activity, the ε subunit has 3'→5' exonuclease activity or proofreading, and several other protein components are active in assembly of polymerase III, binding of polymerase III to template DNA, and high processivity. The ssDNA used as template by DNA polymerase III is supplied by a combination of actions by DNA gyrase and DNA helicase. DNA gyrase introduces negative supercoils into dsDNA in an ATP-dependent process. Negative supercoils overcome torsional stress caused by helicase-catalyzed unwinding of dsDNA, a process also driven by ATP hydrolysis. This action at an origin of replication leads to a replication bubble consisting of two replication forks. At each fork there is leading strand synthesis and lagging strand synthesis. Leading strand synthesis occurs in the direction of replication fork movement whereas lagging strand replication is away from the replication fork (see Figures 23.6). At a replication fork, DNA is replicated semidiscontinuously. Synthesis is primed on both strands at a replication fork by primase, a DNA-dependent RNA polymerase. As the replication fork moves to expose unreplicated DNA, the lagging strand is periodically reprimed. Later the RNA primers are removed by RNase H, an RNase specific for RNA:DNA hybrids, and by nick translation by DNA polymerase I. Any gaps are filled in leaving nicks that are sealed by DNA ligase.

RNA-dependent DNA polymerase or reverse transcriptase can convert an RNA template into DNA. Reverse transcriptases have, in addition to DNA polymerase activity, RNase H activity, and DNA-dependent DNA polymerase activity.

Recombination

General recombination involves exchange of DNA between two molecules that have similar sequences. DNA's with similar sequences that derived from a common sequence are homologous. Thus, the process is known as homologous recombination. In homologous recombination the two DNA's pair, single-stranded nicks are produced, and the ssDNA's displace each other and ligate to form a branched structure. The branch migrates, causing strand exchange, and is finally resolved by strand cleavage into either patch recombinants or splice recombinants (see answer to problem 6). The enzymology of general recombination is best understood in *E. coli* in which Rec A protein, RecBCD protein complex, and ssDNA binding protein or SSB play key roles. The RecBCD protein complex initiates recombination by attaching to the end of DNA and unwinding it. As DNA is unwound it begins to rewind but at a slower rate giving rise to a bubble of ssDNA. When a specific sequence GCTGGTGG known as a Chi site is encountered, RecBCD causes a single strand nick producing a ssDNA tail which is coated with SSB. RecA binds to SSB-coated ssDNA to form a nucleoprotein filament that binds to dsDNA and causes strand displacement. Ligation leads to branch formation, branch migration, and resolution of the Holliday junction.

DNA Repair

Mismatch repair removes mispaired bases on DNA by scanning duplex DNA for mispairs, excising the incorrect member of the pair, and replacing it with the correct base. Mispairs in newly synthesized DNA are corrected because the template DNA is methylated while the nascent DNA strand is unmethylated. UV-induced pyrimidine dimers are corrected by a light-dependent cleavage catalyzed by photolyase. Excision repair is active against several kinds of damaged DNA, including apurinic and apyrimidinic sites created by base removal. The general approach to excision repair is to create flanking single-strand breaks on the same stand in order to remove the damaged DNA.

Alteration in DNA sequence

DNA sequences can be altered by mutation, recombination, and transposon insertion. Mutations include insertions and deletions of one or more base pairs and changes in a single base (point mutations) leading to either a transition (a purine-purine change or a pyrimidine-pyrimidine change) or a transversion (purine-pyrimidine of vice versa). Small insertions and deletions can be produced by intercalating agents such as acridine orange whereas large deletions and insertions can be produced by recombination or by transposon mutagenesis. Transposons, first described by McClintock, are genetic elements that can change their location within DNA.

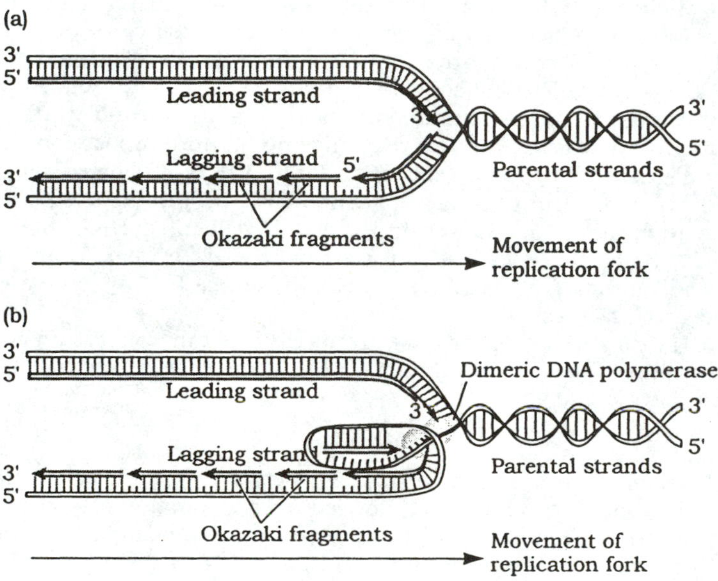

Figure 23.6 The semidiscontinuous model for DNA replication. Because DNA polymerases only polymerize nucleotides 5'→3', both strands must be synthesized in the 5'→3' direction. Thus, the copy of the parental 3'→5' strand is synthesized continuously; this newly made strand is designated the **leading strand**. (a) As the helix unwinds, the other parental strand (the 5'→3' strand) is copied in a discontinuous fashion through synthesis of a series of fragments 1000 to 2000 nucleotides in length, called the **Okazaki fragments**; the strand constructed from the Okazaki fragments is called the **lagging strand**. (b) Since both strands are synthesized in concert by a dimeric DNA polymerase situated at the replication fork, the 5'→3' parental strand must wrap around in *trombone fashion* so that the unit of the dimeric DNA polymerase replicating it can move it in the 3'→5' direction. This parental strand is copied in a discontinuous fashion because the DNA polymerase must occasionally dissociate from this strand and rejoin it further along. The Okazaki fragments are then covalently joined by DNA ligase to form an uninterrupted DNA strand.

Problems and Solutions

1. If ^{15}N-labeled E. coli DNA has a density of 1.724 g/mL, ^{14}N-labeled DNA has a density of 1.710 g/mL, and E. coli cells grown for many generations on $^{14}NH_4^+$ as nitrogen source are transferred to media containing $^{15}NH_4^+$ as sole N-source, what will be the density of

the DNA after one generation because replication is semiconservative? Assume the mode of replication were dispersive: what would be the density of DNA after one generation? Design an experiment to distinguish between semiconservative and dispersive modes of replication.

Answer: After one generation for both semiconservative and dispersive modes of replication, DNA of an intermediate density, (1.724 + 1.710)/2 = 1.717 g/mL), is expected. In order to distinguish between the two models, DNA is denatured into single-stranded DNA and subsequently analyzed by CsCl density gradient centrifugation to determine the density of the single strands. For the dispersive model, single-stranded DNA of only a single density (1.717 g/mL) is expected whereas for the semiconservative model, two DNA populations with slightly different densities (1.724 + 1.710 g/mL) will be observed.

2. What are the respective roles of the 5'-exonuclease and 3'-exonuclease activities of DNA polymerase I? What would be the phenotype of an E. coli strain that lacked DNA polymerase I 3'-exonuclease activity?

Answer: DNA polymerase I functions during replication to replace RNA primers with DNA and during DNA repair to fill in gaps produced by DNA repair endonucleases. The 5'-exonuclease activity is largely responsible for removing RNA primers and for the process known as nick translation. RNA primers are abundant in the lagging strand of newly synthesized DNA because this strand is reprimed frequently. Frequent repriming is necessary because the replication fork moves in a direction opposite to primer elongation on the lagging strand. During elongation, whenever DNA polymerase III encounters the 5'-end of a RNA primer, elongation stops, resulting in a gap between the 3'-end of newly synthesized DNA and the 5'-end of the previously used RNA primer. DNA polymerase I will extend these 3'-ends, using its DNA polymerase activity, while simultaneously removing the 5'-end of the RNA primer. The result is movement of the gap in a process known as nick translation and replacement of the RNA primer with DNA.

The 3'-exonuclease activity of DNA polymerase I is a proof-reading activity. During polymerization, when the 3'-end is being extended, if an incorrect base is incorporated into the 3'-end causing a mismatch, DNA polymerase I pauses at that site and removes the mismatched base on the 3'-end. Cells with DNA polymerase I lacking this 3'-exonuclease proof-reading capability will acquire random mutations at high frequency.

3. Assuming DNA replication proceeds at a rate of 750 base pairs per second, calculate how long it will take to replicate the entire E. coli genome. Under optimal conditions, E. coli cells divide every 20 minutes. What is the minimal number of replication forks per E. coli chromosome in order to sustain such a rate of cell division?

Answer: The *E. coli* genome is 4.64×10^6 bp. Assuming that DNA replication proceeds at a rate of 750 base pairs per second, the time required to replicate this amount of DNA assuming bidirectional replication (i.e., two replication forks starting at a single origin of replication and moving in opposite directions) is given by:

$$\frac{4.64 \times 10^6 \text{ bp}}{2 \times 750 \frac{\text{bp}}{\text{sec}}} = 3{,}093 \text{ sec} = 51.6 \text{ min} = 0.86 \text{ hr}$$

For the genome to be replicated within 20 min, the cell must be replicating DNA at the rate of:

$$\frac{4.64 \times 10^6 \text{ bp}}{20 \text{ min}} = 2.32 \times 10^5 \frac{\text{bp}}{\text{min}}$$

At a replication bubble, DNA is being synthesized at the rate of

$$2 \times 750 \frac{\text{bp}}{\text{sec}} = 1{,}500 \frac{\text{bp}}{\text{sec}} = 90{,}000 \frac{\text{bp}}{\text{min}}$$

There are $\dfrac{2.32 \times 10^5 \frac{\text{bp}}{\text{min}}}{90{,}000 \frac{\text{bp}}{\text{min}}}$ =2.58 replication bubbles or 5.16 replication forks

4. It is estimated that there are 10 molecules of DNA polymerase III per E. coli cell. It is likely that E. coli growth rate is limited by DNA polymerase III levels?

Answer: For maximum growth rate approximately 5 replication forks are active. If each replication fork has two copies of DNA polymerase III then 10 molecules of the enzyme should be sufficient to support maximum growth.

5. Approximately how many Okazaki fragments are synthesized in the course of replicating an E. coli chromosome? How many in replicating an "average" human chromosome?

Answer: Okazaki fragments are on the order of 1,000 to 2,000 nucleotides in length and are produced during lagging strand synthesis. The result of DNA replication is production of two duplex molecules. As a consequence of bidirectional replication, half of each newly synthesized strand is produced in lagging strand synthesis. Thus, the equivalent of a single strand of chromosomal DNA is produced as Okazaki fragments. The number of Okazaki fragments, assuming a length of 1000 nucleotides, is given by:

$$\frac{4.64\times10^6\,\text{nt}}{1{,}000\dfrac{\text{nt}}{\text{Okazaki fragment}}} = 4{,}640 \text{ Okazaki fragments, or}$$

2,320 Okazaki fragments 2,000 nucleotides long.

For a human chromosome:

$$\frac{3\times10^9\dfrac{\text{nt}}{\text{hapoid genome}}}{23\dfrac{\text{chromosomes}}{\text{genome}}} = 1.3\times10^8\frac{\text{nt}}{\text{chromosome}}$$

$$\frac{1.3\times10^8\,\text{nt}}{1{,}000\dfrac{\text{nt}}{\text{Okazaki fragment}}} = 1.3\times10^5 \text{ Okazaki fragments}$$

Note: This calculation assumes that Okazaki fragments in eukaryotes are the same size as those in prokaryotes. However, eukaryotic Okazaki fragments are typically smaller.

6. How do DNA gyrases and helicases differ in their respective functions and modes of action?

Answer: DNA gyrases introduce negative supercoils into DNA. These enzymes act by binding to DNA at crossover configurations caused by positive supercoils, catalyzing phosphodiester bond breakage of one of the duplex DNAs at the crossover, passage of the uncleaved DNA through the gap, and reformation of the phosphodiester bonds resulting in conversion of positive supercoil to negative supercoil. The process is driven by a conformational change in gyrase and ATP hydrolysis is required to re-establish the original conformation. Gyrases move ahead of replication forks, introducing negative supercoils thus facilitating strand unwinding at the replication fork.

Helicases are enzymes that separate DNA strands by unwinding DNA using ATP hydrolysis. Helicases bind to regions of single-stranded DNA, move along a DNA strand, and disrupt base pairs. Translocation and strand unwinding are coupled to ATP hydrolysis.

7. Assuming DNA replication proceeds at a rate of 100 base pairs per second in human cells, and origins of replication occur every 300 kbp, how long would it take to replicate the entire diploid human genome? How many molecules of DNA polymerase does each cell need to carry out this task?

Answer: First let us calculate how many replication forks would be active in a typical diploid cell replicating approximately 6×10^9 bp of DNA.

$$\frac{6\times 10^{9}\,\frac{\text{nt}}{\text{genome}}}{300\times 10^{3}\,\frac{\text{bp}}{\text{origin}}} = 2\times 10^{4}\text{ origins or }4\times 10^{4}\text{ replication forks.}$$

If each replication fork is producing 100 base pairs per sec it would take:

$$\frac{6\times 10^{9}\,\text{bp}}{4\times 10^{4}\,\text{replication forks}\times 100\,\frac{\text{bp}}{\text{sec}}} = 1,500\text{ sec or }25\text{ min}$$

If there are two molecules of DNA polymerase per replication fork the cell would require $8\text{x}10^4$ DNA polymerase molecules.

8. From the information in Figure 23.19, diagram the recombinational event leading to the formation of a heteroduplex DNA region with a bacteriophage chromosome.

Answer: Assume that the dsDNA illustrated below represents the same gene from two bacteriophage. The genes are not identical but are homologous. To make heteroduplex DNA, single strand nicking, followed by strand invasion, ligation, branch migration, and resolution of the Holliday junction produces two heteroduplex molecules. Subsequent DNA mismatch repair or replication will produce recombinant bacteriophage.

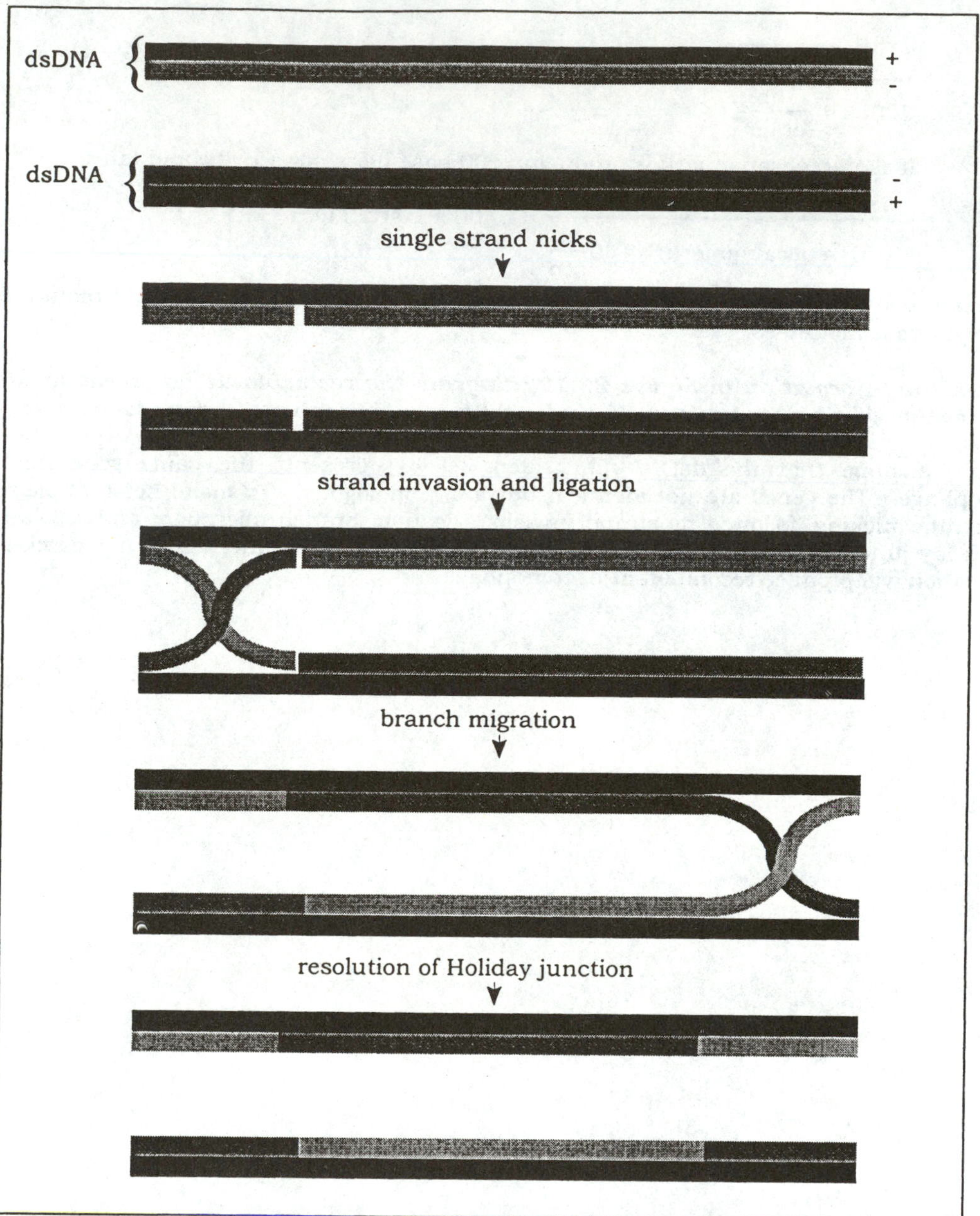

9. Homologous recombination in E. coli leads to the formation of regions of heteroduplex DNA. By definition, such regions contain mismatched bases. Why doesn't the mismatch repair system of E. coli eliminate these mismatches?

Answer: Mismatch repair systems in *E. coli* rely on DNA methylation in order to distinguish between two strands to determine which will serve as a template. In homologous recombination, methylated DNA duplexes may be joined. In this case, both strands are methylated and mismatch repair will not be able to distinguish between them.

10. If RecA protein unwinds duplex DNA so that there are about 18.6 bp per turn, what is the change in Δφ , the helical twist of DNA, compared to its value in B-DNA?

Answer: B-DNA has 10 bp per turn. The Δφ is calculated as follows:

$$\Delta\phi = \frac{360°}{10\text{ bp}} = 36°\text{per bp}$$

For RecA-unwound DNA, $\Delta\phi$ is:

$$\Delta\phi = \frac{360°}{18.6\ \text{bp}} = 19.4°\text{per bp}$$

The change in $\Delta\phi = 36° - 19.4° = 16.6°$

11. Diagram a Holliday junction between two duplex DNA molecules and show how the action of resolvase might give rise to either patch or splice recombinant DNA molecules.

Answer: The formation of a patch recombinant is shown below. The gaps (*) indicate the original nicks. Strand cleavage at the points indicated followed by religation to opposite cleavage points will produce a patch recombinant.

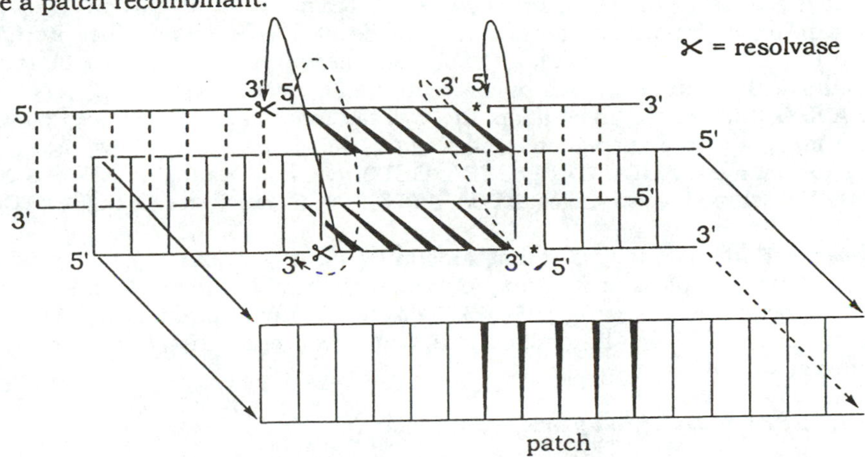

patch

The formation of a splice recombinant is shown below. The gaps (*) indicate the original nicks. Strand cleavage at the points indicated followed by religation to opposite cleavage points will produce a splice recombinant.

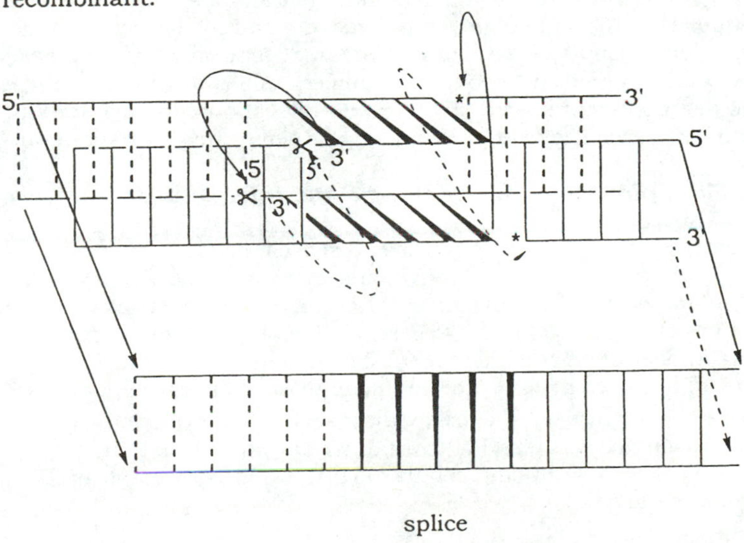

splice

12. Show the nucleotide sequence changes that might arise in a dsDNA (coding strand segment GCTA) upon mutagenesis with (a) HNO_2 (b) bromouracil, and (c) 2-aminopurine.

Answer: (a.) Nitrous acid (HNO_2) converts cytosine to uracil and adenine to hypoxanthine by oxidative deamination. Uracil base pairs with adenine, which subsequently pairs with thymine. Thus, cytosine to uracil results in C to T transitions. Hypoxanthine, produced by oxidative deamination of adenine, can pair with cytosine, which in turn pairs with G. The net result is an

A to G transition. For the sequence GCTA, transitions can occur at C and A on the coding strand to give the following mutant sequences: single base changes, G**T**TA and GCT**G**; two base changes (much less frequently), G**T**T**G**. Although only the coding strand is shown we must consider its complementary strand as a target for mutagenesis as well. The original sequence GCTA will be susceptible to mutagenesis at G and T by virtue of C and A on the complementary strand. So, we expect mutagenesis on the noncoding strand to produce the following mutations (shown as coding strand changes): single base changes, **A**CTA and GC**C**A; two base changes, **A**C**C**A. Multiple mutations could arise by mutagenesis of both strand to give the following set: **AT**TA, GC**CG**, G**TC**A, and **A**CT**G**; however, these would occur infrequently. Finally, it is possible, although highly improbable that three-base changes, **ATC**A, **AT**T**G**, G**TCG**, and four base changes, **ATCG**, could occur.

(b.) Bromouracil is a base analog of uracil with a bromine atom at carbon 5. It resembles thymine and can be incorporated into DNA in place of T. However, the presence of the electronegative Br influences the keto to enol tautomerization. In the enol tautomer, 5-bromouracil pairs with guanine. Thus, incorporation of 5-bromouracil for T produces T to C transitions or A to G transitions. In addition, the enol tautomer may be incorporated into DNA in place of C, resulting in C to T transitions and G to A transitions. The following sequences are possible: one base changes, **A**CTA, G**T**TA, GC**C**A, GCT**G**; two base changes, **AT**TA, **A**C**C**A, **A**CT**G**, G**TC**A, G**T**T**G**, GC**CG**; three base changes, **ATC**A, **AT**T**G**, G**TCG**; four base changes **ATCG**.

(c.) 2-Aminopurine is an adenine analog that substitutes for A but can pair with cytosine causing A to G and T to C transitions. The following mutant sequences are possible: one base changes, GC**C**A, GCT**G**; two base changes (rarely), GC**CG**. Less frequently, 2-aminopurine will pair with C leading to **A**CTA and G**T**TA. (All combinations could be considered, however, they will be infrequent mutations.)

13. Transposons are mutagenic agents. Why?

Answers: Transposons are capable of relocating within genomes and are capable of moving DNA sequences to new locations. This has several consequences. If a transposon moves into a gene it may inactivate the gene by insertional inactivation. If transposons, upon relocating, take with them genomic sequences this can cause problems depending on the nature of the genomic sequences moved. For example, if a poorly expressed gene is relocated nearby an enhancer element, expression of the gene may be greatly amplified. Alternatively if a gene is separated from an enhancer it will decrease expression. Transposon relocation can therefore inactivate genes or alter the level of gene expression by either increasing or decreasing transcription.

14. Give a plausible explanation for the genetic and infectious properties of PrP^{sc}.

Answer: Prion diseases are neurodegenerative diseases caused by proteinaceous infectious particles or prions. To date there is no convincing evidence that the infectious agent of prion diseases carries genetic material of any kind. Rather, the disease can be transmitted by an aberrant form of the prion protein PrP. PrP is a normal cellular protein found in uninfected individuals. (Using knockout mice it has been shown that PrP is an nonessential gene but its presence is required for prion disease development.) Normal (or cellular) PrP has a different conformation than PrP^{sc} (**sc**rapies is a prion disease first identified in sheep): cellular PrP is largely alpha helical whereas scrapies PrP contains beta strands. It has been postulated that scrapies PrP causes disease by inducing cellular PrP to undergo a conformational change to the beta structure-dominated form.

Questions for Self Study

1. The experiments of Meselson and Stahl showed that DNA is replicated semiconservatively. The experiments ruled out conservative and dispersive models. Describe the experiments and explain how they favored semiconservative replication over the other two models.

2. DNA polymerase I and DNA polymerase III in *E. coli* play key roles in DNA replication. Determine which polymerase best fits each of the statements below.

    a. Proof-reading on a separate subunit.

b. Highly processive enzyme.
c. Used to generate Klenow fragment.
d. Multisubunit enzyme.
e. Active in DNA repair.
f. Repairs gaps left after RNA primers removed.
g. Single polypeptide chain.
h. Functions at a replication fork.
i. Moderately processive enzyme
j. Carries out nick translation.

3. Match the columns.

a. DNA gyrase	1. Rep protein.
b. Helicase	2. Joins 3' hydroxyl to 5' phosphate.
c. Leading strand	3. ATP-dependent dsDNA unwinder.
d. Lagging strand	4. RNA polymerase.
e. Primase	5. Okazaki fragments.
f. RNase H	6. Attaches holoenzyme to DNA.
g. DNA ligase	7. 3' end toward replication fork.
h. SSB	8. Type II topoisomerase.
i. Localizes helicase to replication fork	9. Blocks DNA secondary structure formation.
j. β-subunit of DNA pol III	10. Removes primers from Okazaki fragments.

4. What are the three enzymatic activities of reverse transcriptase?

5. General recombination in *E. coli* depends on at least three proteins, RecA protein, RecBCD protein complex, and single-stranded DNA-binding protein (SSB). For each of the statements below indicate which protein fits best.
a. Promotes RecA binding to ssDNA.
b. Forms nucleoprotein filaments that unwind DNA to 18.6 bp per turn.
c. Recognizes the sequence GCTGGTGG (Chi site).
d. Catalyzes DNA strand exchange.
e. Binds to single-stranded DNA only.
f. Unwinds DNA in an ATP-dependent reaction.
g. Binds to the ends of a DNA duplex.

6. If an error occurs during replication resulting in a mismatched base pair, how does the repair mechanism distinguish between the correct, template base and the incorrect base?

7. UV-damaged DNA can be repaired by excision repair or by dimer reversal. Explain.

8. In 1983, Barbara McClintock was awarded the Nobel Prize in physiology or medicine for discoveries she made in the 1950s. What important class of genetic elements did McClintock discover?

9. Match.

a. Deletion	1. GGGACCC → GGGGCCC
b. Insertion	2. GGGGCCC → GGGCCCC
c. Transversion	3. GGGACCC → GGGAACCC
d. Transition	4. GGGACCC → GGGCCC

10. The base analog 5-bromouracil (5-BU) is a thymine analog that causes transitions because it tautomerizes readily from the keto form to the enol form. In the keto form, 5-BU pairs with A but in the enol form it pairs with G. The structure of 5-BU in the keto form is shown below. Draw the enol form and show how this form pairs with G and how the keto form pairs with A.

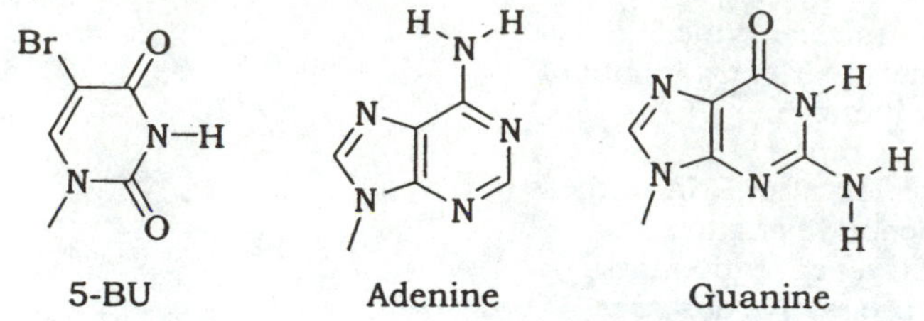

5-BU Adenine Guanine

Answers

1. *E. coli* was grown in ^{15}N for several generations to uniformly label DNA with this heavy nitrogen isotope. Cells were then shifted to ^{14}N for a few generations and DNA was isolated after each generation. The densities of DNA were then compared by density centrifugation. It was found that after one generation DNA of a density intermediate to that of ^{15}N-labeled DNA and ^{14}N-labeled DNA was produced. Subsequent generations produced two bands of density, one corresponding to the band after one generation and a second corresponding to ^{14}N-labeled DNA. The intermediate density DNA, produced after one generation, was further analyzed under denaturing conditions and it was found to be composed of single-stranded ^{15}N DNA and single-stranded ^{14}N DNA.

2. a. III; b. III; c. I; d. III; e. I; f. I; g. I; h. III; i. I; j. I.

3. a. 8; b. 3; c. 7; d. 5; e. 4; f. 10; g. 2; h. 9; i. 1; j. 6.

4. RNA-directed DNA polymerase, RNase H (degrades RNA of RNA/DNA duplex), DNA-directed DNA polymerase.

5. a. SSB; b. RecA; c. RecBCD; d. RecA; e. SSB; f. RecBCD; g. RecBCD.

6. The DNA is scanned for methyl groups located on the parent, template strand.

7. In excision repair, the damaged base is removed along with a few residues flanking the damage site. This action leaves a gap that is subsequently filled. In dimer reversal, photolyase binds to the dimer and, using the energy of visible light, disconnects the bases from each other.

8. Mobile elements (transposable elements or transposons).

9. a. 4; b. 3; c. 2; d. 1.

10.

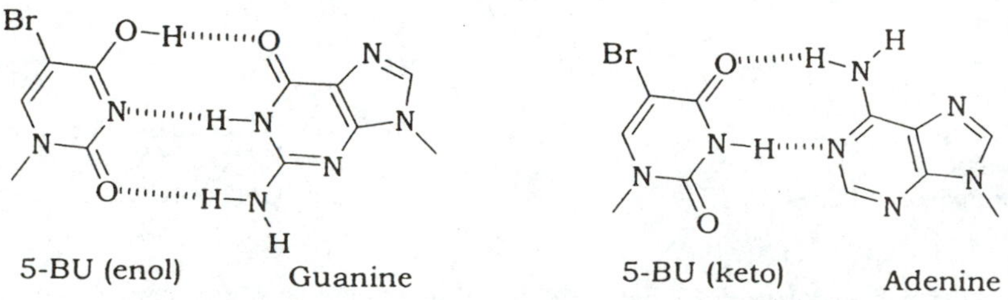

5-BU (enol) Guanine

5-BU (keto) Adenine

Additional Problems

1. The enzyme uracil-N-glycosylase is part of a DNA repair system that removes deoxyuracil from DNA. The enzyme scans DNA for uracil and hydrolyzes the glycosidic bond between sugar and base to leave an apyrimidinic site. How is this apyrimidinic site dealt with subsequently?

2. Suggest two mechanisms by which deoxyuracil gets into DNA. (Hint: one has to do with deoxynucleotide biosynthesis, the other has to do with instability of a certain base.)

3. Suggest a method of incorporating labeled dNTPs into DNA using DNA polymerase I.

4. Okazaki fragments arise from lagging strand synthesis. Yet in wild-type cells, no long fragments are initially observed. Can you explain why both long and short fragments are expected at early stages of replication? Bonus: Can you explain why long fragments are not observed in wild type cells but are observed in *ung* cells (*ung* codes for uracil-N-glycosylase).

5. The genetic locus of DNA Q mutations was first identified as one of a number of so-called *mut* (mutator) loci. What function of DNA metabolism might this locus be responsible for?

6. How was bidirectional replication first shown?

7. Explain why the base analog 5-bromouracil can give rise to T-A to C-G transitions.

8. The chemical mutagens and base analogs described in this chapter are very effective at inducing transitions. Transversions occur less frequently. Provide an explanation.

9. Based on your knowledge of enzymes that have been encountered in your studies on intermediary metabolism, can you correctly restate the "one-gene, one-enzyme hypothesis"?

Abbreviated Answers

1. An apurinic endonuclease recognizes the site and causes strand cleavage to remove the deoxyribose and several flanking positions, creating a gap that is repaired by DNA polymerase and ligase.

2. Deoxynucleotide biosynthesis involves formation of dNDPs from NDP without regard to base. Thus, dUDP is formed and converted to dUTP. Normally dUTPase hydrolyzes dUTP to dUMP and PP_i but with less than perfect efficiency. Occasionally dUTP is incorporated into DNA in place of dTTP. The second mechanism for formation of uracil in DNA is by nonenzymatic oxidative deamination of C's.

3. DNA can be labeled by taking advantage of the nick translation activity of DNA polymerase I. DNA is first nicked with limited digestion by DNase I, which creates randomly positioned single-strand breaks. These breaks or nicks are positions at which DNA polymerase I will initiate nick translation. If radioactively labeled dNTPs are employed, the newly synthesized stand will be appropriately labeled.

4. Because the lagging strand is periodically reprimed, it is synthesized as short DNA fragments or Okazaki fragments. With time, Okazaki fragments are converted into long DNA fragments by removal of the RNA primers and ligation. The leading strand is synthesized continuously because its elongation is in the same direction as replication fork migration. No long fragments are observed in wild type cells because of the uracil repair mechanism. Uracil is randomly incorporated into both the leading and lagging strands. Removal of the base followed by chain cleavage excises the uracil residues, leaving gaps to be repaired. The leading strand, although produced continuously, is fragmented by this repair system. Mutant *ung*⁻ cells do not repair uracil-containing DNA and produce long and short fragments

5. DNA Q codes for the proof reading subunit of DNA polymerase III. Certain mutations at this locus cause cells to undergo frequent mutagenesis because of error-prone DNA replication.

6. Cells were grown for a number of generations in low amounts ^{3}H-thymidine to uniformly label DNA with a small amount of radioactivity. Cells were then shifted for a brief period of time to high levels of ^{3}H-thymidine. Cells were subsequently fixed and autoradiographed to determine the location of heavily labeled DNA.

7. The base analog 5-bromouracil will more readily undergo keto-enol tautomerization because of the influence of the bromine group. In the enol tautomeric form, 5-bromouracil will pair with G. Thus, a T-A pair will be converted to a 5-bromouracil-G pair then to a C-G pair.

8. Mutagens are effective only because the damage they cause is not efficiently repaired by DNA repair mechanisms. To produce a transition requires a purine-pyrimidine mismatch during mutagenesis and these are less likely to be repaired than purine-purine or pyrimidine-pyrimidine mismatches produced during transversion mutagenesis.

9. On several occasions we have encountered single polypeptide chains that have two enzymatic activities. For example, phosphofructokinase 2 and fructose bisphosphatase 2 activities, responsible for fructose-2,6-bisphosphate metabolism, are found on a single protein. A more correct statement would be "one-gene, one-polypeptide".

Summary

The maintenance of the genetic information encoded in the sequence of bases in DNA is accomplished via replication; the expression of this information is achieved through transcription. DNA replication is semi-conservative: The two strands of the DNA double helix separate and each serves as a template for the synthesis of a new complementary strand. Thus, two daughter DNA double helices identical in every respect are reproduced from a single parental double helix. Each daughter double helix consists of one parental strand and one new strand. Using the techniques of ^{15}N-density labeling of DNA and CsC1 density gradient ultracentrifugation, Meselson and Stahl provided the experimental proof establishing the semi-conservative mechanism of replication.

In 1957, Arthur Kornberg and his colleagues reported the discovery of an enzyme from *E. coli* capable of DNA synthesis. This enzyme, DNA polymerase I, catalyzed the incorporation of the deoxynucleoside 5'-triphosphates dATP, dGTP, dCTP and dTTP into DNA in the presence of a template strand to copy and a primer strand that provided a free 3'-OH end for nucleotide addition. Successive addition of nucleotides at the 3'-end of the primer drove chain elongation in the 5'→3' direction. *E. coli* DNA pol I is a 109 kD polypeptide of 928 amino acid residues possessing three catalytic sites: one responsible for the 5'→3' DNA polymerase activity and two others that catalyzed distinct exonuclease reactions, one in the 3'→5' direction (the "3'-exonuclease") and one in the 5'→3' direction (the "5'-exonuclease"). The 3'-exonuclease activity improves the fidelity of replication by checking the accuracy of base-pairing between the base just incorporated and its complementary base in the template. If this pairing is inappropriate, the 3'-exonuclease excises the offending base, giving the polymerase another chance to insert the correct base. The 5'-exonuclease activity acts only on dsDNA and serves a repair function: It "edits out" sections of damaged DNA or any ribonucleotides incorporated during the initiation of DNA replication (see below).

All DNA polymerases discovered thus far catalyze chain growth in the 5'→3' direction, adding each new base as specified by a template DNA strand according to the A:T/G:C base-pairing rules of Watson and Crick. The template is read in the antiparallel 3'→5' direction and the incoming nucleotide is added to the 3'-end of a primer chain. It turns out that DNA pol III, not DNA pol I, is the principal DNA replicating enzyme of *E. coli*. DNA pol III consists of a 165 kD "core" polymerase of α, ε and θ subunits, where the 120 kD α subunit provides the polymerase active site and the 27.5 kD ε subunit contributes the proofreading 3'- exonuclease function. In vivo, this "core" polymerase is part of an 800 kD complex, the DNA polymerase holoenzyme. The processivity of DNA polymerase III holoenzyme exceeds 5,000; that is, it can associate with a template strand and read along it to synthesize a complementary DNA strand greater than 5,000 nucleotides long without once dissociating.

DNA replication is a complex process. First, the DNA helix must be unwound to expose single-stranded template regions. Unwinding can impose torsional stress and ATP-dependent DNA gyrases act to introduce negative supercoils to counteract this stress. ATP-dependent helicases catalyze the actual disruption of the double helix, breaking the H-bonds between the base pairs as they move along a strand of the DNA duplex. Replication is bidirectional: It begins at unique sites on chromosomes ("origins of replication"). The strands are separated here to form a "bubble", and so-called "replication forks" proceed away from the origin in both directions along the parental dsDNA, growing two daughter DNA duplexes in their wake. Replication is semi-discontinuous because DNA polymerases only work in the 5'→3' direction, reading a template in the 3'→5 sense. This polarity means that DNA polymerases can continuously copy the emerging

3'→5' strand, but the 5'→3' parental strand must be copied discontinuously: Only when a single-stranded stretch of this 5'→3' strand has been exposed can the DNA polymerase move long it in the 3'→5' direction to synthesize its complement. Reiji Okazaki provided the experimental verification for this semi-discontinuous mode of replication when he discovered that the radioactivity immediately incorporated into newly synthesized DNA occurred in short fragments, 1,000-2,000 nucleotides in length. With time, this radioactivity was associated with progressively longer DNA strands as the Okazaki fragments were ligated together to form a covalently contiguous DNA chain. DNA ligase catalyzes this reaction. The continuously synthesized DNA strand is called the "leading strand"; the discontinuously synthesized strand is called the "lagging strand". The primers for DNA synthesis in vivo are RNA oligonucleotides complementary to the DNA template. These RNA primers are synthesized by primase, an RNA polymerase. No RNA is found in mature DNA duplexes because DNA pol I binds at the 3'-OH nicks of Okazaki fragments. Its 5'→3' exonuclease activity then cuts out the RNA and the gaps this creates are filled in with DNA by the 5'→3' polymerase activity. DNA ligase then seals the junctions.

In eukaryotes, DNA replication occurs during the S phase of the cell cycle. A number of multimeric DNA polymerases are found in eukaryotic cells. DNA polymerase α and DNA polymerase δ are believed to constitute an "asymmetric dimer" replicase located at each replication fork, with DNA polymerase δ synthesizing the leading strand and DNA polymerase α forming the lagging strand.

RNA-directed synthesis of DNA occurs in certain RNA viruses known as retroviruses. The enzyme responsible is reverse transcriptase. Like all DNA polymerases, it incorporates nucleotides in the 5'→3' direction, reading its template in the 3'→5' direction. A tRNA H-bonded to the RNA template serves as primer. Reverse transcriptases display two other enzymatic activities in addition to their RNA-dependent DNA polymerase activity: an RNase H activity and a DNA-dependent DNA polymerase activity. RNase H degrades RNA chains in RNA:DNA hybrid duplexes. Its role in retroviral replication is to digest the genomic RNA chain so that the DNA-dependent DNA polymerase activity of reverse transcriptase can then copy the just-made DNA strand to give a duplex DNA. This DNA duplex then either mediates the subsequent course of the viral infection or becomes inserted into the host genome where it can lie dormant for many years. The reverse transcriptase of human immunodeficiency virus, the etiological agent in AIDS, is inhibited by the triphosphate derivative of AZT.

Genetic recombination involves the breakage and reunion of DNA strands, so that a physical exchange of parts takes place. The Holliday model for general recombination postulates a sequence of events including alignment of homologous sequences on two different duplex DNA molecules, introduction of single-strand nicks at analogous sites on both DNAs, invasion of each of the single strands into the other DNA duplex (strand invasion), and ligation of the free ends from different duplexes to create a cross-stranded intermediate, the Holliday junction. This junction can migrate along the DNA molecules so that considerable lengths of nucleotide sequence from one duplex become associated via base pairing with the other DNA duplex. If strands of the duplexes become nicked and then re-ligated with strands of the other, recombinant duplexes can be formed. The enzymology of general recombination is understood, at least in outline. The *Rec*BCD complex attaches to the end of a DNA duplex and progresses along it, unwinding the helix in an ATP-dependent reaction. *Rec*BCD enzyme also nicks the duplex at Chi "hotspots" of recombination to produce a ssDNA tail, a necessary prelude to the entry of *Rec*A protein into the process. The protein *Rec*A catalyzes the DNA strand exchange reaction, the reaction by which one strand of a duplex is displaced and replaced by an invading single strand from another duplex in an ATP-driven process. SSB (single-stranded DNA-binding protein) facilitates the action of RecA by binding to ssDNA and keeping it from assuming any secondary structure, which might impede strand exchange. The Holliday junction ensuing from RecBCD, RecA action upon two adjacent and homologous DNA duplexes is resolved in *E. coli* by the *ruv*C endonuclease/resolvase into one of the two isomeric recombinant forms, patch or splice. Cleavage of exchanged strands yields patch recombinants; cleavage of parental strands gives splice recombinants.

Chapter 24

Transcription and the Regulation of Gene Expression

• •

Chapter Outline

- Central Dogma: Crick: DNA → RNA → protein
  - Transcription: DNA to RNA
  - Translation: RNA to protein
  - Jacob and Monod: mRNA serves as intermediate between DNA and protein
- Transcription
  - Enzymology (*E. coli*)
    - DNA-dependent RNA polymerase (RNA polymerase)
      - 3' Polymerase activity
      - NTPs as substrates
    - Subunit structure
      - $\alpha_2\beta'\beta$: Core enzyme: Elongation
      - $\alpha_2\beta'\beta\sigma$: Holoenzyme: Initiation
      - β' = DNA binding subunit
      - β = NTP binding sites
      - σ = promoter recognition
  - Initiation
    - Closed promoter complex
      - σ of holoenzyme binds to promoter
      - Promoter: Two conserved sequences
        - Pribnow Box (-10): TATAAT
        - -35: TTGAGA
        - Separation between elements: 17 bp
    - Open promoter complex
      - dsDNA unwound
      - RNA polymerase at start of initiation (+1)
        - Negative numbers: Upstream of start site
        - Positive numbers: Downstream of start site
    - Several phosphodiester bonds made
      - Initiation site binds ATP or GTP (complementary to template start site)
      - Elongation site binds NTP complementary to template
      - Phosphodiester bond formation
      - After 5 to 9 bonds formed σ dissociates
  - Termination: Two types
    - Rho independent termination: Three terminator elements
      - G/C rich inverted repeat forms stem
      - Nonrepeated sequence separates inverted repeats and forms loop
      - 6 to 8 U's on end

        - RNA polymerase pauses and terminates after element is transcribed
    - Rho dependent termination
        - Rho: ATP-dependent helicase
        - Rho binding site: C-rich regions of transcript

- Transcriptional regulation in prokaryotes
  - Gene organization
    - Genes coding for enzymes with common goals often transcribed together
    - Genes organized into operon
    - Transcription produces polycistronic message
    - Operator near operon promoter regulates initiation of transcription
      - Induction: Increased synthesis in response to specific substrate
        - Inducer: Small molecule binds to protein
        - Gratuitous inducer: Not metabolized (e.g., *lac* operon's IPTG)
      - Repression: Decreased synthesis in response to specific metabolite
      - Constitutive expression: Expression not regulated
  - *lac* Operon: Jacob and Monod: operon hypothesis
    - Genes for lactose metabolism expressed when lactose present and glucose absent
    - Operator: Palindromic sequence near promoter recognized by *lac* repressor protein
    - *Lac* repressor protein: *lacI* gene product
      - In absence of inducer *lac* repressor binds to operator and blocks promoter
      - Inducer binding to *lac* repressor causes release of repressor from operator: Promoter available for transcription
    - CAP: Positive regulator
      - Catabolite activator protein: Binds near promoter increases promoter strength
        - Binding is cAMP dependent
        - cAMP levels determined by adenylyl cyclase levels
        - Glucose leads to low cAMP levels
  - *trp* Operon
    - Trp repressor: *trpR* Gene product
      - When complexed to tryptophan: Binds to operator and blocks transcription
      - Low tryptophan cause trp repressor to release from operator
      - Trp repressor regulates two other operons
        - *aroH*: Aromatic amino acid synthesis: Related metabolism
        - *trpR*: Example of autogenous regulation (autoregulation)
  - Transcriptional regulation through DNA:protein and protein:protein interactions
    - Transcriptional activator proteins interact with *both* DNA and RNA polymerase
- Eukaryotic transcription: Three RNA polymerases: Require transcription factors to bind to promoter
  - RNA polymerase I
    - Transcribes rRNA genes
    - α–Amanitin resistant
  - RNA polymerase II
    - Transcribes mRNA genes
    - α–Amanitin sensitive
  - RNA polymerase III
    - Transcribes tRNA and 5S rRNA genes
    - α–Amanitin sensitive but less than RNA polymerase II
- RNA polymerase II
  - Ten subunits: RPD1 to RPD10
    - RPD1: DNA binding (β'-like)
      - C-terminal domain (CDT): PTSPSYS repeats
        - Unphosphorylated CDT: Initiation competent
        - Phosphorylated CDT: Elongation competent
- Transcriptional regulation in eukaryotes
  - Promoter elements

        - TATA box (-25)
        - Proximal elements
            - CAAT box: Around -80: Strong promoter
            - GC box: Housekeeping genes
        - Enhancers (upstream activation sequences)
            - Distance and orientation insensitive
    - Response elements: Promoter modules that make genes responsible to common regulation
    - Initiation: Core promoter bound by basal apparatus: Components
        - RNA polymerase II
        - General transcription factors (GTP): TFIIB, TFIID, TFIIE, TFIIF, TFIIH
            - TFIIB
                - TATA-binding protein
                - TAFs: TBP-associated factors
    - Nucleosome structure
        - Swi/Snf complex disrupts nucleosomal arrays in ATP-dependent manner
        - Histone acetyltransferase: Acetylates lysines causing histone release
        - Histone deacetylases removes acetyl groups
- General model of eukaryotic gene activation
    - Requires two steps
        - Alteration of chromatin structure to allow access
        - Interaction of RNA polymerase II and GTFs with promoter
    - Mediator interacts with both transcriptional activators and the C-terminal domain of RNA polymerase II
    - Once transcription begins Elongator replaces Mediator
- DNA-binding motifs
    - Helix-turn-helix
        - C-terminal helix recognition helix
            - Binds to major groove
    - Zinc finger motifs
        - C_2H_2 class: Zn coordinated by 2 C and 2 H
        - C_x class: Zn coordinated by cysteine
    - bZIP motif: Basic region with leucine zipper
        - Leucine zipper: Protein dimerization domain: Coiled-coils formed between two zippers
        - Basic region: Rich in basic amino acids: DNA binding domain
- Post-transcriptional processing in eukaryotes
    - Capping: 5'-end activity
        - Guanylyl transferase adds guanylyl in 5'-5' linkage
        - 7 Position of G methylated
    - Polyadenylation: 3'-end activity
        - Consensus AAUAAA in mRNA signal for endonuclease and polyA polymerase activity
            - mRNA cut 10 to 30 nt downstream
            - 100 to 200 adenine residues added to 3'-end
    - Splicing
        - hnRNA (heterogeneous nuclear RNA) converted into mRNA
        - Introns removed
            - 5' Splice site
            - 3' Splice site
            - Branch site: Lariat formed
        - Exons joined
        - Process catalyzed by spliceosome
            - Small nuclear ribonucleoprotein particles (snRNPs)
            - Constitutive splicing: Formation of single mRNA
            - Alternative splicing: Formation of a variety of RNAs

Chapter Objectives

This chapter covers several important topics including RNA production or transcription, RNA processing, regulation of transcription and gene expression, and DNA-binding proteins.

RNA

RNAs are linear polymers of ribonucleotide monophosphates held together by phosphodiester bonds between the 5'-carbon of one nucleotide and the 3'-carbon of another nucleotide. The polymers have two distinct ends, a 5'-end and a 3'-end. Single-stranded RNAs often can form intrachain hydrogen-bonded structures by complementary base pairing. There are three general classes of RNAs: mRNAs, the agents of gene expression that bring genetic information from DNA to the ribosome to direct assembly of proteins; ribosomal RNAs (rRNA), structural and functional components of ribosomes; and, small stable RNAs like tRNAs, adaptor molecules that participate in protein synthesis.

RNA Polymerase

The principal RNA polymerase in *E. coli* is RNA polymerase I, a DNA-dependent RNA polymerase with $\alpha_2\beta\beta'$ subunit composition. In order to initiate transcription, the core polymerase requires the sigma factor σ which functions to identify specific DNA sequences termed promoters. A typical *E. coli* promoter is composed of a -35 region, with a consensus sequence of TCTTGACAT and a -10 region or Pribnow box with consensus sequence TATAAT. Initiation begins with the RNA polymerase holoenzyme, $\alpha_2\beta\beta'\sigma$, binding to the -35 region, and migrating to the -10 region where strand separation is initiated to expose the start-of-transcription site. Initiation of transcription usually begins with ATP or GTP in part because the initiation site, a nucleotide binding site on the RNA polymerase, preferentially binds these nucleotides. After a short oligonucleotide has been synthesized, the σ factor dissociates from the RNA polymerase leaving the task of elongation to the core complex. Elongation continues until termination signals are encountered of which there are two types, rho-independent and rho-dependent. In rho-independent termination, the 3'-end of the RNA is synthesized to include the termination signal, a sequence that forms a short G:C-rich stem and small loop with a run of unpaired Us. Termination sequence folding causes RNA polymerase to pause and terminate transcription. Rho-dependent termination, a less common mechanism, involves ρ factor, an ATP-dependent helicase that unwinds RNA:DNA duplexes in response to C-rich regions of the mRNA unoccupied by ribosomes.

Eukaryotes contain three classes of nuclear RNA polymerases: RNA polymerase I responsible for rRNA transcription, RNA polymerase II which transcribes protein-encoding genes, and RNA polymerase III responsible for 5S rRNA, tRNA, and other small stable RNAs. The three classes of RNA polymerases are distinguishable by their sensitivity to α-amanitin, a mushroom poison. RNA polymerase I is resistant to the compound, RNA polymerase II is sensitive to inhibition of chain elongation by α-amanitin, and RNA polymerase III is less sensitive. All three classes of RNA polymerases interact with specific promoter sequences with the aid of transcription factors.

RNA polymerase II promoters commonly have two sequence elements, a TATA box and an initiator element. The TATA box has a TATAAA consensus sequence and is located 25 base upstream of the start of transcription, which is within the initiator element.

Regulation of Transcription

Lactose is metabolized by a set of gene products encoded by the *lac* operon. Transcription of the *lac* operon results in a polycistronic mRNA, a mRNA with coding regions for three separate proteins. Regulation of the *lac* operon is achieved by repression and activation. RNA polymerase requires an additional protein to transcribe the *lac* operon because the promoter is a rather weak, inefficient promoter. The catabolite activator protein (CAP) acts as a positive regulator in this regard by binding to the *lac* promoter. However, binding occurs only in the presence of cAMP, which binds to CAP. *E. coli* prefers glucose and will metabolize it in preference to lactose when both sugars are present. Glucose preference results because glucose inhibits cAMP production leading to loss of promoter binding by CAP. Repression of the *lac* operon is achieved by *lac* repressor protein binding to an element termed the operator located near the promoter. Repressor binding blocks RNA polymerase from gaining access to the transcription start site. Release of repression occurs when lactose is present. Lactose is converted to allolactose, which functions as a inducer by binding to the *lac* repressor protein, causing it to dissociate from the operon. Derivatives of lactose that are nonmetabolizable yet function as inducers are termed gratuitous inducers. The *lac* operon is under both positive and negative control and is both inducible and repressible.

The *trp* operon is regulated by repression. Repression is achieved by *trp* repressor protein binding to the promoter in a trp-dependent manner.

Transcription in eukaryotes is regulated at initiation by promoter strength, by enhancers or upstream activation sequences (UAS), and by responsive elements. The function of enhancers is independent of proximity to the promoter and orientation of the enhancer with respect to the promoter. Responsive elements are located near the transcription start site and are recognized by specific proteins that bind to the responsive element.

DNA Binding Motifs

There are three prominent classes of DNA-binding motifs, helix-turn-helix, zinc finger, and leucine zippers. The helix-turn-helix motif is typically 20 residues arranged into a 7-residue helix, a small β-turn, and a second helix of about 7 residues that functions as the recognition helix. The recognition helix interacts with a specific DNA sequence via the major groove. The zinc finger motif contains a zinc-binding site defined by either two cysteines and two histidines (C_2H_2) or by several cysteines (C_x). The domain is at the base of a small loop that interacts with DNA via the major groove. The leucine zipper motif is a periodic repetition of leucine zippers along a helix such that leucines line one face of the helix. Proteins containing leucine zippers do not actually bind DNA via the leucine zipper. Rather DNA binding is a property of a sequence of predominantly basic amino acids N-terminal to the leucine zipper. The zipper functions as a protein dimerization site.

Post-Transcriptional Processing

In eukaryotes, transcription produces a class of primary transcripts in the nucleus known as heterogeneous nuclear RNA. This RNA is processed in three ways to produce mature mRNA. The primary transcripts are modified on the 5'-end by a cap structure, on the 3'-end by a polyA tail, and internally by splicing. The 5' cap consists of a guanine residue joined in 5'-5' phosphodiester linkage that is methylated at position N-7 on the base and may contain additional ribose methylations on the first and second nucleotide of the primary transcript. 3'-Polyadenylation occurs after transcription of a poly (A) processing signal (consensus sequence AAUAAA) has been transcribed in the primary transcript. Transcripts are cleaved downstream of this signal and 100 to 200 adenine residues are attached to the end. Splicing (Figure 24.36) removes introns or intervening sequences from the primary transcript and joins the remaining exons to produce mRNA. Splicing is carried out by nucleoprotein complexes in the nucleus. The process involves recognition of the two ends of an intron, cleavage and removal of the intron, and ligation of the exon. The splicing reaction passes through an intermediate structure known as a lariat structure or RNA branch with the 5'-end of the intron joined to a specific base within the intron in 5'-2' phosphodiester linkage.

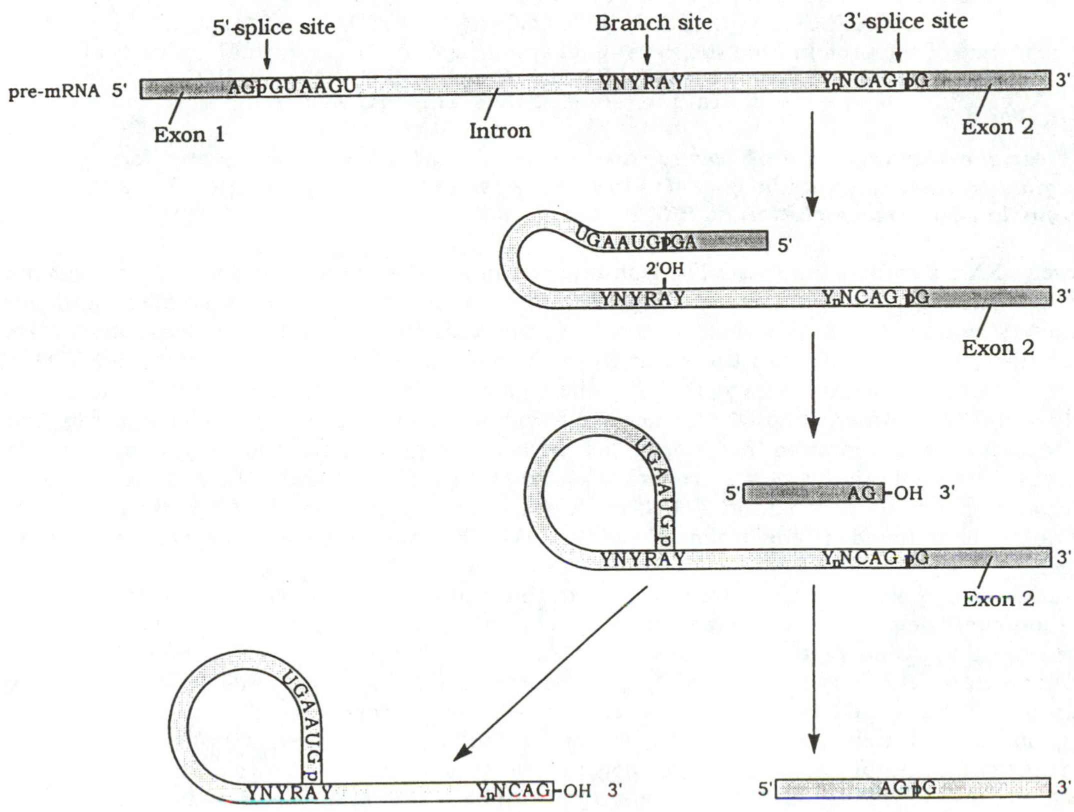

Figure 24.36 Splicing of mRNA precursors. A representative precursor mRNA is depicted. Exon 1 and Exon 2 indicate two exons separated by an intervening sequence (or intron) with consensus 5', 3', and branch sites. The fate of the phosphates at the 5' and 3' splice sites can be followed by tracing the fate of the respective *ps*. The products of the splicing reaction, the lariat form of the excised intron and the united exons, are shown at the bottom of the figure. The lariat intermediate is generated when the invariant G at the 5'-end of the intron attaches via its 5' phosphate to the 2'-OH of the invariant A within the branch site. The consensus guanosine residue at the 3' end of Exon 1 (the 5'-splice site) then reacts with the 5'-phosphate at the 3'-splice site (the 5'-end of Exon 2), ligating the two exons and releasing the lariat structure. Although the reaction is shown here in a stepwise fashion, 5' cleavage, lariat formation, and exon ligation/lariat excision are believed to occur in a concerted fashion.

Problems and Solutions

1. The 5'-end of an mRNA has the sequence:

...AGAUCCGUAUGGCGAUCUCGACGAAGACUCCUAGGGAAUCC...

What is the nucleotide sequence of the DNA template strand from which it was transcribed? If this mRNA is translated beginning with the first AUG codon in its sequence, what is the N-terminal amino acid sequence of the protein it encodes? (See Table 25.1 for the genetic code.)

Answer: The nucleotide sequence of the DNA sense strand has the following sequence, written 5' to 3'

...GGATTCCCTAGGAGTCTTCGTCGAGATCGCCATACGGATCT...

The first AUG codon, underlined, is shown below

...AGAUCCGUAUGGCGAUCUCGACGAAGACUCCUAGGGAAUCC...

The sequence of the protein from the N-terminal amino acid to the C-terminal amino acid is

```
...AGAUCCGUAUG GCG AUC UCG ACG AAG ACU CCU AGG GAA UCC...
           Met Ala Ile Ser Thr Lys Thr Pro Arg Glu Ser...
```

2. Describe the sequence of events involved in the initiation of transcription by E. coli RNA polymerase. Include in your description those features a gene must have for proper recognition and transcription by RNA polymerase.

Answer: RNA polymerase in *E. coli* is a multimeric protein of the form $\alpha_2\beta\beta'\sigma$. Sigma factor ($\sigma$) functions during initiation of transcription to recognize specific DNA regions called promoters. Promoters contain two DNA sequence elements important for initiation. Located approximately 35 bp upstream of the start of transcription is an element with a consensus sequence TTGACA called the -35 region that is recognized by the sigma factor and represents the initial point of binding of RNA polymerase to DNA. Once the RNA polymerase contacts this element it migrates to a second region, known as the Pribnow box or the -10 region where it forms an open complex. Whereas the initial binding of RNA polymerase occurs on double-stranded DNA, RNA polymerase eventually requires single-stranded DNA to serve as a template for RNA synthesis. Strand separation is initiated at the Pribnow box, an AT-rich region with a consensus sequence of TATAAT.

Initiation of polymerization starts with a purine nucleotide triphosphate that binds to the initiation site, one of two nucleotide binding sites on RNA polymerase. This nucleotide is destined to become the 5'-end of the RNA. The next nucleotide binds to the elongation site on RNA polymerase and the enzyme catalyzes phosphodiester bond formation between the 3' hydroxyl group of the nucleotide in the initiation site and the 5'-α-phosphate of the nucleotide in the elongation site with elimination of inorganic pyrophosphate (PP_i). Next, translocation along the antisense strand occurs moving the RNA polymerase to the next base. The elongation site is filled with a nucleotide triphosphate complementary to this base and a phosphodiester bond is formed. This process, catalyzed by RNA polymerase holoenzyme (i.e., $\alpha_2\beta\beta'\sigma$), is repeated until an oligonucleotide 6 to 10 nucleotide residues long has been formed. The sigma factor dissociates from the holoenzyme leaving the core enzyme to complete elongation, marking the end of initiation.

3. RNA polymerase has two binding sites for ribonucleoside triphosphates, the initiation site and the elongation site. The initiation site has a greater K_m for NTPs than the elongation site. Suggest what possible significance this fact might have for the control of transcription in cells.

Answer: The initiation site has a preference for purine ribonucleoside triphosphates in addition to having a greater K_m for NTPs than the elongation site. Thus, initiation occurs only when there is sufficient NTPs to support elongation. Having a preference for ATP or GTP ties initiation of polymerization intimately to energy levels in cells.

4. How does transcription in eukaryotes differ from transcription in prokaryotes?

Answer: One fundamental difference is the number of RNA polymerases involved in transcription. In prokaryotes, a single RNA polymerase is responsible for all classes of RNAs including mRNA, rRNA, and small, stable RNAs such as tRNAs. (The only other RNA polymerase activity, primase, functions during replication to produce RNAs used as primers.) In contrast, eukaryotes have several RNA polymerases each dedicated to production of one class of RNAs. RNA polymerase I is localized in the nucleolus and is responsible for transcription of ribosomal RNA genes except 5S rRNA. RNA polymerase II is active in the nucleus and is responsible for synthesis of mRNAs. RNA polymerase III also acts in the nucleus to produce small, stable RNAs such as tRNAs, 5S rRNA, and the small RNAs involved in mRNA processing and in protein targeting. Finally, mitochondria and chloroplasts have RNA polymerases responsible for transcription of organellar genes.

Eukaryotic RNA polymerases are themselves generally incapable of initiation of transcription and require additional protein factors or transcriptional factors to initiate protein synthesis. The enzyme complexes are larger than those found in prokaryotes.

5. DNA-binding proteins may recognize specific DNA regions by either reading the base sequence or by "indirect readout." How do these two modes of protein:DNA recognition differ? (Consult Chapter 8 to review the properties of nucleosomes.)

Answer: For "direct readout" interactions, DNA-binding proteins recognize base sequences directly. This is most easily accomplished by determinants in the major groove. Sufficient information is present in the major groove to distinguish specific DNA sequences. For "indirect readout" interactions, local conformational variations in the DNA structure are used in recognition. To a first approximation, double-stranded DNA is a rod-like structure. However, local regions may assume conformations that show subtle variations due to differences in nucleotide sequence. The type of conformation and the frequency of conformational changes are both dependent on base sequence.

6. The metallothionine promoter is illustrated in Figure 24.28. How long is this promoter, in nanometers? How many turns of B-DNA are found in this length of DNA? How many nucleosomes (approximately) would be bound to this much DNA?

Answer: Figure 24.28 shows the metallothionine promoter to be 260 basepairs in length. Since B-DNA is characterized by a distance of 0.34 nm/basepair, the promoter is 260 bp × 0.34 nm/bp = 88.4 nm in length. B-DNA has a pitch of one turn per 10 bp. Thus, the promoter has 260 bp × 1 turn/10 bp = 26 turns.

In Chapter 8 we are reminded that each nucleosomes are wrapped by 146 bp of DNA, and that each nucleosome is joined by 40 to 60 intervening bp. Choosing 50 bp of intervening DNA as an approximation, we find that each nucleosome is repeated every 196 bp. Accordingly, the metallothionine promoter would bind an average of approximately 260 bp × 1 nucleosome/196 bp = 1.3 nucleosomes.

7. Describe why the ability of bZIP proteins to form heterodimers increases the repertoire of genes whose transcription might be responsive to regulation by these proteins.

Answer: bZIP polypeptides form dimers by interacting through leucine zippers: amphipathic α-helices containing a periodic repetition of leucine residues spaced every seven residues such that they line up on one side of the helix. These proteins are DNA-binding proteins as well, interacting with DNA through a short stretch of basic residues located N-terminal to the leucine zipper motif. The specificity of DNA binding is determined by the sequence of basic residues. Because bZIP proteins function as dimers, two sequence elements on DNA are recognized. In the case of homodimer formation, the sequence elements form a dyad. However, heterodimers can bind to DNA lacking dyad symmetry. In fact, the bZIP-binding site may be composed of two distinct regions, each recognized separately by different bZIP proteins that function together as a heterodimer. The ability to form heterodimers allows for composite bZIP binding sites, lacking dyad symmetry, to be recognized by bZIP proteins.

8. Suppose exon 17 was deleted from the fast skeletal muscle troponin T gene (Figure 24.39). How many different mRNAs could now be generated by alternative splicing? Suppose that exon 7 in a wild-type troponin T gene was duplicated. How many different mRNAs might be generated from a transcript of this new gene by alternative splicing?

Answer: The fast skeletal troponin T gene consists of 18 exons. Exons 1 through 3, 9 through 14, and 18 are constitutive and are thus always found in mature mRNAs. Exons 4 through 8 are combinatorial and may be individually included or excluded in any combination. Finally, exons 16 and 17 are mutually exclusive, either one or the other if found. How many different mRNAs are possible? Let us first determine the number of different combinations of exons 4, 5, 6, 7, and 8. For each of these exons there are only two states or choices: present or not present. The number of combinations is given by 2 × 2 × 2 × 2 × 2 = 32. So, considering exons 4 through 8, 32 different combinations are possible. There is also a choice between exons 16 and 17. Thus, 2 × 32 or 64 different mRNAs may be produced.

If exon 17 is deleted, the number of different mRNAs possible is only 32. In the event that exon 7 is duplicated the number of combinations of exons 4, 5, 6, 7, duplicated 7, and 8 is given by the number of combinations of exons 4, 5, 6, and 8 ($2 \times 2 \times 2 \times 2 = 16$) multiplied by the number of different combinations of 7 and 7 duplicate which is 3 (either they are both present, both absent, or one or the other is present). This gives $3 \times 16 = 48$ different combinations of exons 4, 5, 6, 7, 7 duplicate, and 8. When we factor in the two possibilities for exon 16 or 17 we have $2 \times 48 = 96$ different mature mRNAs.

Questions for Self Study

1. Describe the events leading up to initiation of transcription by *E. coli* RNA polymerase. Be sure to include in your discussion the importance of promoter elements in a typical promoter.

2. Name the two types of transcriptional termination mechanisms found in bacteria.

3. The last two columns of the table shown below are incorrect. Correct them.

RNA polymerase Type	Genes Transcribed	α–Amanitin Sensitivity
RNA polymerase I	protein-encoding genes	intermediate sensitivity
RNA polymerase II	tRNAs, 5S rRNA and others	resistant
RNA polymerase II	ribosomal RNA genes	very sensitive

4. Eukaryotic RNA polymerase II requires what additional class of proteins to function?

5. The *lac* operon is under both positive and negative control. Explain how both the presence of lactose and the absence of glucose are required for initiation of transcription of the *lac* operon.

6. How does the availability of tryptophan regulate gene expression in the *trp* operon?

7. Eukaryotic genes. Match the terms.

a. A common element found in most promoters.	1. GC box
b. A promoter element found in "housekeeping " genes.	2. Subject promoters to regulation by some common signal.
c. Orientation and distance independent regulatory sequences	3. TATA box
d. Responsive elements	4. Enhancers (or upstream activation sequences)

8. What are the three major classes of DNA-binding motifs?

9. How are the 5'- and 3'- ends of eukaryotic transcripts typically modified?

10. The primary transcript of an eukaryotic gene is typically longer than its mRNA. Explain.

Answers

1. Initiation of transcription begins when RNA polymerase holoenzyme, through its σ subunit, binds to the -35 region of a promoter to form a closed complex. The polymerase moves along the promoter until it reaches the -10 region or Pribnow box where it unwinds about 12 base pairs of DNA exposing the transcription start site. This arrangement is known as the open promoter complex. Initiation begins with a nucleoside triphosphate, usually ATP or GTP, binding to the initiation site on RNA polymerase. The elongation site binds the next nucleotide and phosphodiester bond formation occurs. The holoenzyme produces a short oligonucleotide chain before its σ factor dissociates leaving the core complex to complete elongation.

2. ρ-Dependent (rho-dependent) and ρ-independent.

3.

RNA polymerase Type	Genes Transcribed	α–Amanitin Sensitivity
RNA polymerase I	ribosomal RNA genes	resistant
RNA polymerase II	protein-encoding genes	very sensitive
RNA polymerase II	tRNAs, 5S rRNA and others	intermediate sensitivity

4. RNA polymerase II requires general transcription factors.

5. Initiation of transcription is blocked by *lac* repressor protein binding to an element in the promoter of the *lac* operon known as the operator. Repressor protein binding occurs only in the absence of an inducer, a galactose derivative. The *lac* promoter is a weak promoter and requires an additional factor for transcription. This factor is a complex of catabolite activator protein (CAP) and cyclic AMP (cAMP). This complex binds to a region upstream of the RNA polymerase binding site in the promoter. cAMP is necessary for CAP binding and the levels of cAMP are sensitive to glucose such that glucose decreases cAMP levels. Thus, in the absence of glucose, cAMP levels are high enough to support CAP binding to the promoter.

6. Initiation of transcription is blocked by the Trp repressor protein. This protein binds tryptophan and associates with the promoter, at a site termed the operator, to block initiation of transcription. (The operon is also controlled by transcriptional attenuation. The operon codes for a small leader peptide. The rate at which this leader sequence is translated affects transcriptional termination.)

7. a. 3; b. 1; c. 4; d. 2.

8. Helix-turn-helix, zinc finger, and leucine zipper.

9. The 5' end of a transcript is capped by a 7-methylguanylyl residue in 5' to 5' linkage. The 3' end contains a polyA tail 100 to 200 adenine residues in length.

10. The primary transcript contains exons and introns. The exons are sequences destined to be present in the mature mRNA but they are not contiguous in the primary transcript. Rather, they are separated by introns. The process of splicing removes introns and joins exons to form the mature mRNA.

Additional Problems

1. What characteristics are associated with the -10 region (Pribnow box) of RNA polymerase promoters in *E. coli*?

2. What is a consensus sequence?

3. Describe the differences and similarities between RNA polymerase I and DNA polymerase I in *E. coli.*

4. Describe the structure of a rho-independent transcriptional terminator.

5. The operator of the *lac* operon was first identified by *cis-trans* complementation tests. In these tests a partial diploid is produced from two cell lines showing defects in lactose metabolism. The idea behind *cis-trans* complementation is to determine if the defects in lactose metabolism are due to mutations in the same gene or in different genes. If mutations are in different genes then a partial diploid will be capable of metabolizing lactose. For example, the *lac* operon contains two genes necessary for lactose metabolism that code for β-galactosidase and lactose permease. A partial diploid between a cell defective in β-galactosidase but containing a normal permease and a cell defective in permease but containing a normal β-galactosidase will result in a diploid capable of lactose metabolism. How might operator mutations behave in *cis-trans* complementation tests?

6. Why does IPTG cause induction of the *lac* operon? How does induction by IPTG differ from induction by lactose via allolactose?

7. How might splicing increase the versatility of eukaryotic genes?

8. How might the fact that mature mRNAs have 3'-polyA tails be used to isolate mature mRNAs from eukaryotic cells?

9. Modified nucleosides can be identified by reverse phase HPLC. How might you expect ribose-methylated and unmethylated nucleosides to differ on reverse phase HPLC? What about 7-methyl G *versus* unmodified G?

Abbreviated Answers

1. The -10 region or Pribnow box is an A-T rich sequence of about six bases located approximately 10 bases upstream of the start of transcription. The high A-T content allows this region to "melt" so that the strands separate. The RNA polymerase-promoter complex shifts from the closed form to the open form at the -10 region.

2. A consensus sequence is a sequence representing the most commonly found base in each position within a DNA element.

3. Both enzymes utilize nucleotide triphosphates to produce nucleic acid polymers, both require a template, read the template 3'→5' direction, and produce the nucleic acid is the 5'→3' direction. RNA polymerase is capable of initiating RNA synthesis by incorporating the very first 5' nucleotide whereas DNA polymerase requires a primer to function. RNA polymerase does not proofread nor does it have exonuclease activity. DNA polymerase has 3'- and 5'-exonuclease activity for proofreading and nick translation.

4. Rho-dependent terminators are DNA elements that when transcribed produce a short sequence of mRNA capable of forming a G:C-rich stem and loop followed by a stretch of Us.

5. Operator mutations may give rise to constitutive expression of the *lac* operon if for example the *lac* repressor protein binding site is altered. The phenotype is constitutive expression of the *lac* operon even in a partial diploid. Operator mutations are examples of *cis* acting mutations because they affect or influence genes only on the same DNA.

6. IPTG (isopropylthiogalactoside) is a galactose derivative capable of binding to the *lac* repressor protein, thereby altering its ability to bind to the operator of the *lac* promoter. Induction by allolactose is transient because allolactose is converted to lactose, which is metabolized to galactose and glucose by β-galactosidase. IPTG is not degraded by β-galactosidase and induction is long-lasting.

7. For certain genes, alternative splicing patterns give rise to different gene products that code for proteins with different properties.

8. A polyT derivatized matrix can be used to separate polyA mRNA from total cellular RNA.

9. In reverse phase HPLC, hydrophobic interactions are responsible for adsorption. Therefore, ribose methylated nucleosides are expected to bind with higher affinity and require higher concentrations of organic solvent to be removed from the reverse phase matrix. The same is true for modified bases, with the exception of 7-methyl modification of G (and A). This modification gives rise to a positive charge on N-7 leading to a decrease in hydrophobicity. 7-Methyl G will elute from a matrix at lower concentrations of organic solvent than unmodified G.

Summary

Transcription is the process whereby RNA chains are synthesized according to a DNA template. The enzymes responsible are DNA-dependent RNA polymerases. Prokaryotes have a single RNA polymerase, which synthesizes rRNA and tRNA as well as mRNA. RNA polymerases, like DNA polymerases, polymerize nucleotides in the 5' 3' direction, reading along templates in the 3' 5' direction. In contrast to DNA polymerases, RNA polymerases can catalyze the *de novo* synthesis of nucleotide chains (i.e., they do not require oligonucleotide primers).

E. coli RNA polymerase consists of a catalytic "core" of subunit composition $\alpha_2\beta\beta'$. This core requires a σ subunit in order to recognize promoter sequences, which signal the sites where genes begin. Addition of a σ subunit to the "core" RNA polymerase yields the RNA polymerase holoenzyme. This holoenzyme binds at promoter sites, which are characterized by two features: the Pribnow box, a TATAAT sequence element located at position -10, and a 9 bp consensus sequence TCTTGACAT near position -35. RNA polymerase holoenzyme then "melts" the DNA at the A:T-rich Pribnow box to form an "open complex" adjacent to the +1 site where transcription is initiated. Chain elongation proceeds at the rate of 20 to 50 nucleotides per sec, moving slower in G:C-rich regions where DNA strand separation is more difficult. Chain termination occurs when a G:C-rich inverted repeat sequence near the end of the transcript forms a stable stem-loop structure which halts further translocation of the RNA polymerase. This stem-loop is followed by a series of A residues in the template, which are transcribed into Us in the RNA. The hydrogen bonds holding these A:U base pairs are exchanged for A:T pairs between the "sense" and "anti-sense" DNA strands to re-anneal the DNA duplex and release the nascent transcript.

Eukaryotes have three classes of DNA-dependent RNA polymerases; all are found in the nucleus. RNA pol I transcribes the major rRNA genes, RNA pol II transcribes the protein-encoding genes and thus leads to mRNA synthesis, and RNA pol III transcribes tRNA genes as well as those genes encoding a number of other small cellular RNAs. All three RNA polymerases are large complex multimeric proteins. Each class of RNA polymerase recognizes a distinct set of promoter elements appropriate to the genes it transcribes. Distinct sets of DNA-binding proteins called transcription factors mediate these interactions, much as the σ subunit of the *E. coli* RNA polymerase directs it to its proper "transcription start sites".

Many genes in bacteria are organized into clusters known as operons. Expression of these genes to produce a single polycistronic mRNA is under the control of promoter and operator regulatory elements lying upstream. Transcription is controlled by induction and repression, processes in which the interaction of regulatory proteins with small metabolites determines whether these regulatory proteins will bind to DNA and facilitate or limit transcription. For example, the *lac* operon is expressed provided the regulatory protein, *lac* repressor, is not bound at the operator site. The presence of the *lac* repressor on the operator blocks transcription, an example of negative control. β-Galactosides induce expression of the *lac* operon by binding to *lac* repressor and promoting its dissociation from the operator. The affinity of *lac* repressor binding to operator DNA binding is 3 orders of magnitude greater than the affinity of the inducer(IPTG):*lac* repressor complex for operator DNA.

The *lac* operon is also under positive control by CAP, the catabolite activator protein. In the presence of cAMP, CAP binds near the *lac* promoter and assists transcription initiation by RNA polymerase. When [cAMP] is low, free CAP cannot associate with RNA polymerase and promote transcription, so the *lac* operon is not expressed. The availability of glucose, *E. coli*'s preferred carbon and energy source, leads to lower cAMP levels and repression of the *lac* operon as well as other operons involved in the metabolism of alternative carbon and energy sources. This phenomenon is called catabolite repression.

Transcriptional regulation in eukaryotes is more complicated. The promoters of eukaryotic genes are defined by modules of short conserved sequences, such as the TATA, CAAT and GC boxes, which bind transcription factors essential to RNA polymerase II function. In addition to these promoter elements, eukaryotic genes have enhancers, sequence elements whose presence is necessary for full expression of the gene. Enhancers may be located in either orientation with respect to the gene and may lie either upstream or downstream from it. Enhancers act nonspecifically to enhance transcription from any nearby promoter, but enhancer function is dependent on binding a specific transcription factor. Many genes respond to a common regulation and are characterized by the presence of common sequence elements in their promoters known as response elements. The heat shock element (HSE) and glucocorticoid response element (GRE) are examples. These response elements are recognized by specific transcription factors that coordinate expression of all genes containing the particular response element. DNA looping provides a mechanism for convening a variety of specific regulatory proteins at the transcription initiation site.

Certain structural motifs recur in DNA-binding proteins. In particular, DNA-binding proteins often employ an α-helix as their DNA recognition element. The diameter of an α-helix matches closely the width of the major groove in B-form DNA, and these recognition helices fit within the major groove, identifying specific nucleotide sequences through a variety of atomic interactions, including hydrogen bonding patterns with the edges of the bases, ionic and H-bonding with the sugar-phosphate backbone, as well as van der Waals fits and hydrophobic interactions with projecting nonpolar -CH_3 groups on thymine residues. DNA-binding proteins erect a variety of structural scaffolds serving to deliver the α-helix or other recognition element to the DNA site, and it is really these scaffolds that have lent their names to the structural motifs: the helix-turn helix, the Zn-finger and the leucine zipper. The great majority of DNA-binding proteins possess one of these three basic motifs

Chapter 25

Protein Synthesis and Degradation

● ●

Chapter Outline

- Characteristics of the Genetic Code
  - Triplet code: Three bases decoded as one amino acid
    - All 64 triplets used: 61 amino acid codons: 3 stop codons
  - Nonoverlapping, no commas
  - Unambiguous
  - Degenerate: Synonymous codons decode same amino acid
    - Third base degenerate (Except in two cases)
    - Second base
      - Pyrimidines code for hydrophobic amino acids
      - Purines code for polar or charged amino acids
  - Universal
- Aminoacyl-tRNA synthesis
  - Aminoacyl-tRNA synthetases: At least 20 charge the tRNAs that decode 61 codons
    - Cognate tRNA: All tRNAs charged by particular synthetase
    - Two classes of synthetases, both approximated as two-domain structures
      - Class I
        - Adds amino acid to 2'OH (but it moves to 3'OH)
      - Class II
        - Adds amino acid to 3'OH
    - Recognition
      - Amino acid binding site
      - tRNA binding site: Determinants on tRNA
        - At least one of anticodon bases
        - One or more of three base pairs in acceptor stem
        - Discriminator base N of NCCA at 3'-end
        - Anticodon
- Ribosomes: Cellular translational machinery
  - Prokaryotic ribosomes (*E. coli*): 70S particle: Two subunits
    - Small subunit: 30S
      - 21 proteins: S-proteins
      - 16S rRNA: 1542 nt
    - Large subunit: 50S
      - 31 distinct proteins: L-proteins
      - 23S rRNA: 2904 nt
      - 5S rRNA: 120 nt
    - rRNAs derive from one precursor transcript
  - Eukaryotic cytoplasmic ribosomes: 80S particle: Two subunits

- Small subunit: 40S
  - 33 proteins
  - 18S rRNA: Around 1800 nt
- Large subunit: 60S
  - 49 proteins
  - 28S rRNA: Around 4700 nt
  - 5S rRNA: 120 nt
  - 5.8S rRNA: 160 nt: Homologous to 5'-end of prokaryotic 23S rRNA

- Stages of protein synthesis in prokaryotes
  - Initiation: Formylmet-tRNA, mRNA, small subunit, large subunit
    - Formylmet-tRNA
      - Initiator tRNA
      - 5'-Base not paired
      - Three G:C bp in anticodon stem
      - Formyl group added to charged tRNA by methionyl-tRNA formyl transferase
    - mRNA
      - Ribosome binding site (Shine-Dalgarno sequence) near 5'-end interacts with 16S rRNA
      - Complementary sequence on 16S rRNA near 3'-end interacts with ribosome binding site
    - Initiation factors
      - IF-1, IF-3: Dissociate 30S subunits
      - IF-2: Binds GTP and fmet-tRNA and deposits fmet-tRNA into P site
        - GTP hydrolysis accompanied by IF release and 50S binding
        - 70S initiation complex formed
  - Elongation
    - Elongation factor Tu (EF-Tu)
      - Binds GTP and charged-tRNA
      - Deposits charged-tRNA into A-site
      - GTP hydrolysis releases EF-Tu·GDP
        - GDP exchanged for GTP by EF-Ts (Guanylyl exchange protein)
    - Peptide bond formation
      - Peptidyl transferase
      - 50S activity
      - 23S rRNA is the peptidyl transferase enzyme
    - Translocation
      - Deacylated tRNA removed from P site
      - Peptidyl-tRNA moved to P site
      - mRNA moved so that new codon in A site
      - EF-G: GTPase catalyzes translocation
      - Large and small subunits move relative to each other
  - Termination: Stop codon in A site
    - RF-1 recognizes UAA and UAG
    - RF-2 recognizes UAA and UGA
    - RF-3: GTPase
- Eukaryotic protein synthesis
  - Eukaryotic initiator $tRNA_i$: Charged with Met: Not formylated
  - Three stages of initiation
    - 43S preinitiation complex formation: eIF2, eIF4A, eIF3 and $Met\text{-}tRNA_i$
    - mRNA binding to 43S preinitiation complex
      - eIF4E: mRNA cap binding protein
    - 80S initiation complex formed upon large subunit binding
    - Initiation serves as a point of post-transcriptional regulation
  - Elongation

        - EF-1: Two components
            - EF1A: EF-Tu counterpart
            - EF1B: EF-Ts counterpart
    - Termination: Single RF recognizes all stop codons
- Inhibitors of protein synthesis
    - Useful scientifically in the elucidation of the mechanism of protein synthesis
    - Some effect prokaryotic and not eukaryotic protein synthesis, and vice versa
        - Useful as antibiotics
- Protein folding
    - Hsp70: Binds to nascent polypeptide chain on ribosome
    - Hsp60: Chaperonins
- Post-translational processing
    - Proteolytic cleavage of pro-enzymes
    - Protein translocation
        - Characteristics of translocation systems
            - Proteins made as preproteins with signal peptides
            - Specific protein receptors exist on target membrane
            - Movement catalyzed by complex structures: Translocons: ATP (or GTP) driven
            - Proteins generally maintained in loosely folded conformations for translocation competence
    - Prokaryotic translocation
        - N-terminal leader sequence
            - N-terminus of leader sequence: Basic amino acids
            - Central domain hydrophobic
            - C-terminus: Nonhelical structure
        - Leader peptidase: Removes leader sequence
    - Eukaryotic translocation and protein sorting
        - Secreted and membrane proteins synthesized on ER-localized ribosomes
            - Cytoplasmic ribosome initiates translation
            - N-terminal signal sequence detected by signal recognition particle (SRP)
            - SRP/ribosome complex binds to docking protein: ER membrane protein
            - Ribosome delivers peptide to translocon
            - Signal peptidase cleaves leader sequence
            - Membrane proteins carry 20-residue stop transfer sequence
        - Mitochondrial protein import
            - N-terminal sequence 10 to 70 residues long
                - Form amphiphilic α-helix
- Protein degradation: Ubiquitination most common pathway in eukaryotes
    - Ubiquitin: Conserved 76-residue protein
        - E1: Ubiquitin-activation protein: Attaches to C-terminal Gly of ubiquitin
        - E2: Ubiquitin-carrier protein: Accepts ubiquitin from activator protein: Carried on cysteine residue
        - E3: Ubiquitin-protein ligase: Binds target protein: Protein ubiquitinated on amino groups
    - Proteins with acidic N-termini
        - N-termini altered by Arg-tRNA
    - PEST sequences: Target proteins for degradation
    - Proteasomes
        - 20S proteasomes
        - 26S proteasomes

Chapter Objectives

Genetic Code

The genetic code is a triplet, non-overlapping code lacking commas. Understand the reason why a triplet code is necessary to code for 20 amino acids and why the code is degenerate. The 64 triplets all have meaning, with 61 triplets (codons) coding for amino acids and three functioning as stop-translation signals. The code is unambiguous and degenerate. The most common start signal is AUG, which codes for Met. AUG also codes for internal Met residues. The three stop codons also, called nonsense codons, are UGA, UAG, and UAA. (A convenient way to remember them is to start with AUG, move A to the end giving the stop codon UGA, by switching G and A, UAG, another stop codon is produced. Finally, if you remember that the three stop codons are URR (R = purine), the only choices left are UAA, a stop codon, and UGG the Trp codon. One other possibility is to remember that the three stop codons are the first three URR codes in alphabetical order.) Only two amino acids are coded by one codon each, namely Met (AUG) and Trp (UGG). The degeneracy in the code is not random. Codons NNY (Y is for pyrimidine) are always degenerate. Codons NNR (R, purine) are usually degenerate. The exceptions being AUG and AUA coding for Met and Ile and UGA and UGG coding for stop and Trp. The third position is completely degenerate for codons with C or G in the first and second positions. A pyrimidine in the second position codes for an uncharged amino acid whereas a purine in the second position codes for a polar amino acid. The amino acids Arg, Leu, and Ser have six codons each. The aromatic amino acids except Trp, and all the charged amino acids except Arg are coded for by two codons each. It may not be important to memorize the code; however, it would not be a difficult task to do so.

Aminoacyl-tRNAs

The aminoacyl-tRNA synthetase are key enzymes in information flow as they bridge the information gap between nucleotides and nucleic acids on the one hand and amino acids on the other. You should appreciate how specificity is achieved and how the incredible accuracy of aminoacylation is achieved. Know the following terms: codon usage, nonsense suppression, missense suppression, nonsense mutations, missense mutations, wobble.

Ribosomes

The site of cellular protein synthesis is the ribosome, a two-subunit ribonucleoprotein complex. You should be familiar with the ribosome vocabulary including the following terms: 16S rRNA, 18S rRNA, 23S rRNA, 28S rRNA, 5S rRNA, 5.8S rRNA, small subunit, 30S subunit, 40S subunit, large subunit, 50S subunit, 60S subunit, 70S ribosome, 80S ribosome, S-proteins, and L-proteins. The rRNAs derive from operons containing structural genes for the small- and large-subunit rRNAs and depending on cell type, the 5S rRNA. Ribosomes self assemble around the various rRNAs, which act as a scaffold for ordered protein binding.

Protein Synthesis

Although the ribosome plays the leading role in protein synthesis, it is assisted at each stage by nonribosomal protein factors specific for stages of protein synthesis including initiation, elongation, and termination. Initiation is a property of the small subunit and various initiation factors. A special tRNA charged with Met is used in initiation of protein synthesis. (In prokaryotes it is termed f-Met-$tRNA_f^{Met}$ where f indicates that Met is formylated. In eukaryotes, Met-$tRNA_i^{Met}$ is used.) Initiation begins when a ribosome, aided by initiation factors, dissociates into subunits in order to produce free small subunits. The small subunit binds mRNA and a complex of f-Met-$tRNA_f^{Met}$ and another initiation factor to form a preinitiation complex that is joined by the large subunit to form the initiation complex. Initiation complex formation is accompanied by GTP hydrolysis and release of ribosome-bound initiation factors. The outcome of initiation is an aminoacylated-tRNA bound in the so-called P-site on the ribosome. In peptide chain elongation (Figure 25.18), a second aminoacyl-tRNA binding site, the A-site, is filled, in an elongation factor-dependent manner, with an aminoacylated-tRNA appropriate for the codon on the mRNA. In *E. coli*, the elongation factor EF-Tu binds to aminoacyl-tRNA and GTP to form a complex. This complex binds to the A-site on the ribosome and deposits the charged tRNA at that location with subsequent hydrolysis of GDP and release of P_i and EF-Tu:GDP complex. For EF-Tu to participate in another round of elongation, it must be recharged by binding to another elongation factor, EF-Ts, which causes GDP to be released in forming a EF-Tu:EF-Ts complex that is resolved to EF-Tu:GTP and EF-Ts by GTP binding. Peptide bond formation is a property of the ribosome alone. The large subunit catalyzes the transpeptidation or peptidyl transfer of the P-site amino acid to the A-site aminoacylated-tRNA. The uncharged tRNA in the P-site is released and the A-site tRNA, now carrying a dipeptide and hence called a peptidyl-tRNA, is translocated along with the mRNA from the A- to the P-site. Translocation is dependent of elongation factors EF-G in the case of *E. coli*. The vacated A-site is now free to accept another aminoacylated tRNA to continue elongation. In the event that a stop codon enters the A-site, chain termination ensues. Protein release factors bind to the A-site and cause hydrolysis of the peptide from peptidyl-tRNA.

Protein Folding

Protein folding starts before the polypeptide is released from the ribosome. In many cases, folding is assisted by molecular chaperones. You should be familiar with the names, order of action and mechanism of a few of them: TF, NAC, HSP70, HSP60, GroES-GroEL and CCT (or TriC). Both GroES-GroEL and CCT are large structures into which proteins are sequestered for ATP-dependent folding.

Post-Translational Processing

The most common form of protein modification is proteolytic cleavage to activate a protein, or to release mature protein products from a larger primary translational product. Proteins are also targeted to specific locations within the cell or are destined for transport out of the cell. In prokaryotes, a signal sequence, located on the N-terminus of a protein, or at a more internal location, may direct the protein to be exported from the cell or to become a membrane-bound protein. Signal recognition particles play a role in halting protein synthesis shortly after a signal peptide is produced and, in conjunction with docking protein, directing the nascent protein to the endoplasmic reticulum membrane where protein synthesis resumes.

Protein Degradation

Ubiquitin plays a role in degradation of proteins in eukaryotes. This pathway is specific and efficient in removing defective proteins from the cell.

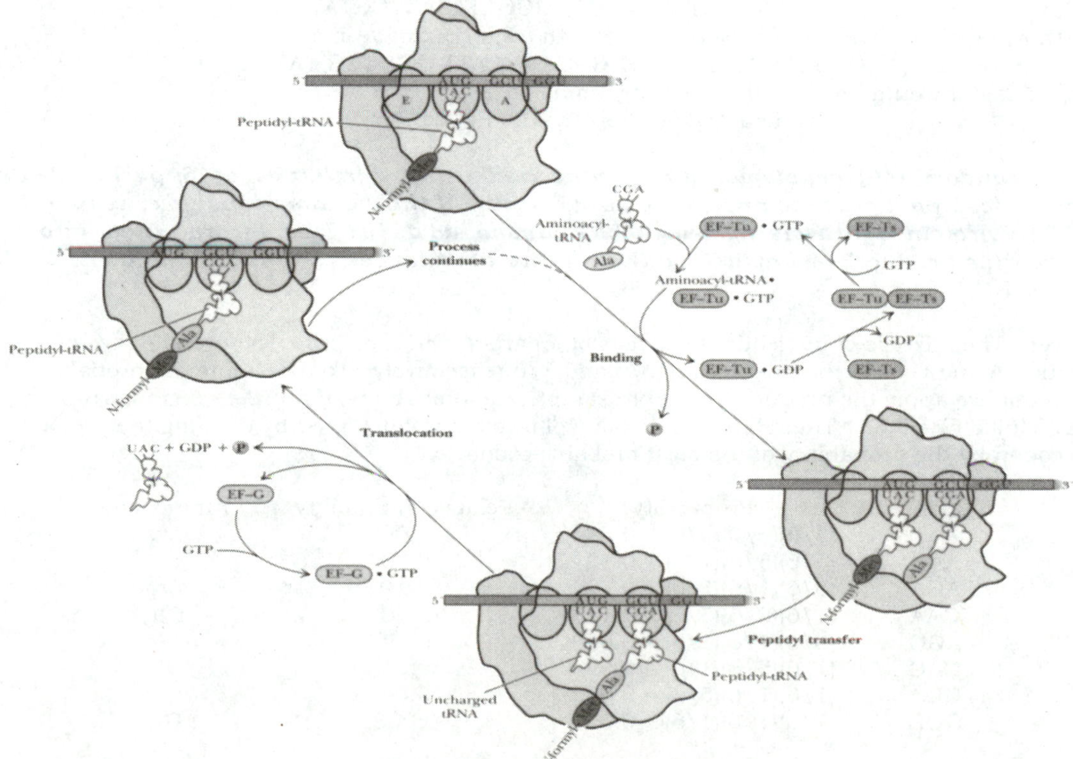

Figure 25.18 The cycle of events in peptide chain elongation on *E. coli* ribosomes.

Problems and Solutions

1. The following sequence represents part of the nucleotide sequence of a cloned cDNA:
. . . CAATACGAAGCAATCCCGCGACTAGACCTTAAC. . .
Can you reach an unambiguous conclusion from these data about the partial amino acid sequence of the protein encoded by this cDNA?

Answer: We are faced with the problem of deciding on a reading frame. Assuming that we have the noncoding strand sequence, we can inspect it for stop codons, UGA, UAG, UAA. (It makes no sense to look for AUG start codons because the sequence might have come from any location within the cDNA.) By inspection we find two stop codons as shown below.

. . . CAATACGAAGCAATCCCGCGAC**TAG**ACCT**TAA**C. . .

These two stop codons are not in the same reading frame (because they are not separated by an integral multiple of three nucleotides). Therefore, either of the two different reading frames that include TAG and TAA may represent the region coding for the C-terminal portion of the protein. The third possible reading frame, (the one that starts with CAA) does not contain a stop codon. More information is required in order to decide which is the correct reading frame. One possibility is to compare codon usage in the three frames with codon usage in other genes from the same organism. Alternatively, we might turn to amino acid composition analysis. The amino acid sequence is

Gln-Tyr-Glu-Ala-Ile-Pro-Arg-Leu-Asp-Leu-Asn

Certain amino acids are often found in low frequencies in proteins as for example Met (AUG), Trp (UGG), His (CAU/C), Cys (UGU/C), Tyr (UAU/C), and Phe (UUU/C). However, of these relatively rarely used amino acids, we find only the tyrosine codon shown below.

. . . CAA**TAC**GAAGCAATCCCGCGACTAGACCTTAAC. . .

It may be possible to check the protein for chymotrypsin cleavage (chymotrypsin cleaves after tyrosine) and this may help us decide if the tyrosine codon is in the appropriate reading frame.

A similar analysis may be applied to the complementary strand whose sequence, written 5' to 3', is

GTTAAGGTCTAGTCGCGGGATTGCTTCGTATTG

This sequence has two stop codons in different reading frames as shown below.

GT**TAA**GGTC**TAG**TCGCGGGATTGCTTCGTATTG

The third reading frame, shown below, may be an open reading frame.

G **TTA A**GG TC**T AG**T CGC GGG ATT GCT TCG TAT TG

If translated it would produce the following amino acid sequence

Leu-Arg-Ser-Ser-Arg-Gly-Ile-Ala-Ser-Tyr-

2. A random (AG) copolymer was synthesized using a mixture of 5 parts adenine nucleotide, 1 part guanine nucleotide as substrate. If this random copolymer is used in a cell-free protein synthesis system, which amino acids will be incorporated into the polypeptide product? What will be the relative abundances of these amino acids in the product?

Answer: The mixture of nucleotides consists of 5 parts A and 1 part G. The probabilities of selecting A and G from the mixture are 5/6 and 1/6 respectively. To determine the probability of any codon we apply the product rule. For example, the probability of an AAA codon is given by (5/6)(5/6)(5/6) = 0.58. The relative probability of a codon is calculated by dividing the probability of a codon by the probability of the most probable codon.

Codon	Probability	Relative Probability	Amino acid
AAA	(5/6)(5/6)(5/6) = 0.58	1.00	Lys
AAG	(5/6)(5/6)(1/6) = 0.12	0.20	Lys
AGA	(5/6)(1/6)(5/6) = 0.12	0.20	Arg
GAA	(1/6)(5/6)(5/6) = 0.12	0.20	Glu
AGG	(5/6)(1/6)(1/6) = 0.023	0.04	Arg
GAG	(1/6)(5/6)(1/6) = 0.023	0.04	Glu
GGA	(1/6)(1/6)(5/6) = 0.023	0.04	Gly
GGG	(1/6)(1/6)(1/6) = 0.005	0.008	Gly

To determine the abundance of amino acids we can apply the sum rule of probability. For example, the probability of a lysine codon is the sum of the probabilities of AAA and AAG. The relative probability is calculated by dividing the probability of an amino acid by the probability of the most probable amino acid.

Amino acid	Probability	Relative Probability	Normalized Probability
Lys	1.00 + 0.20 = 1.20	1.20	100
Arg	0.20 + 0.04 = 0.24	0.24	20
Glu	0.20 + 0.04 = 0.24	0.24	20
Gly	0.04 + 0.008 = 0.048	0.048	4

3. *Review the evidence establishing that aminoacyl-tRNA synthetases bridge the information gap between amino acids and codons. Indicate the various levels of specificity possessed by aminoacyl-tRNA synthetases that are essential for high-fidelity translation of messenger RNA molecules.*

Answer: Information flow in biological systems is described in the central dogma first postulated by Francis Crick. Genetic information is stored and transmitted from generation to generation in the form of a sequences of bases in nucleic acids. Nucleic acids are faithfully copied in the process of replication, which is directed by base-pair interactions involving complementary hydrogen bonds. The information content of a nucleic acid is its sequence of bases. A portion of this sequence is used to direct the synthesis of proteins, linear chains of amino acids, in a process known as gene expression. For most genes, gene expression involves at least two steps, transcription and translation. In transcription, a mRNA is produced by a base-pair directed polymerization. In translation, mRNAs are used to direct the synthesis of proteins by ribosomes. Even in this process, complementary base pair interactions between mRNAs and tRNAs are paramount in importance. Yet, translation results in the production of linear chains of amino acids. Thus, during gene expression a change in information content, from nucleotide sequences to amino acid sequences, occurs. The point at which this information gap is bridged is the production of aminoacyl-tRNAs by aminoacyl-tRNA synthetases. These enzymes must speak two languages: one based on nucleotides, the other based on amino acids.

In order for translation to occur with high fidelity, aminoacyl-tRNA synthetases must accurately charge tRNAs. This is accomplished by a two-step reaction. As the name implies, synthetases produce aminoacylated-tRNAs at the expense of ATP hydrolysis. In the first step in catalysis, an amino acid is adenlyated to form an aminoacyl-adenylate that is tightly bound to the enzyme. This reaction is accompanied by elimination of PP_i, which is subsequently hydrolyzed by pyrophosphatases rendering the reaction irreversible. In the second stage of catalysis, the amino acid is transferred to either the 2'-OH or the 3'-OH of a tRNA.

Specificity exists at each of these two stages. The aminoacyl adenylation reaction depends on the amino acid specificity of the amino acid binding site. This specificity is not absolute and activation of inappropriate amino acids occur with low frequency. This may lead to misacylation. Aminoacyl-tRNA synthetases carefully edit this step to insure that misacylated tRNAs are not released. Misacylated tRNAs trigger a deacylase that hydrolyzes inappropriate products.

4. *Draw base-pair structures for (a) a G:C base pair, (b) a C:G base pair, (c) a G:U base pair, and (d) a U:G base pair. Note how these various base pairs differ in the potential hydrogen-bonding patterns they present within the major groove and minor groove of a double-helical nucleic acid.*

Answer: For the G:C and C:G pairs, the pattern of hydrogen bond donor and acceptor groups in the major and in the minor groove are opposite to each other. For example, in the major groove of the G:C pair we find, on G, two acceptors, N-7, and the carbonyl oxygen of C-6, and on C, a donor, the amino group at C-4. In the minor groove, G presents an acceptor, N-3, and a donor, the amino group at C-2, whereas C presents an acceptor, the carbonyl oxygen of C-2. For the C:G pair the pattern is in the reverse order.

For the G:U and U:G pair the pattern of hydrogen-bonding groups is symmetrical. This is shown below where, reading from left to right in the major and minor groove, the pattern of hydrogen bond donors and acceptors is presented.

Base Pair	Groove	Hydrogen-Bonding Groups
G:C	major	A-A-D
G:C	minor	A-D-A
C:G	major	D-A-A
C:G	minor	A-D-D
G:U	major	A-A-A
G:U	minor	A-D-A
U:G	major	A-A-A
U:G	minor	A-D-A

G:C

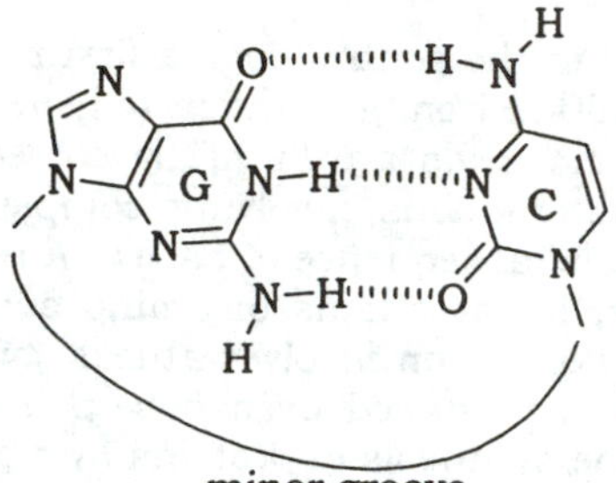

minor groove

C:G

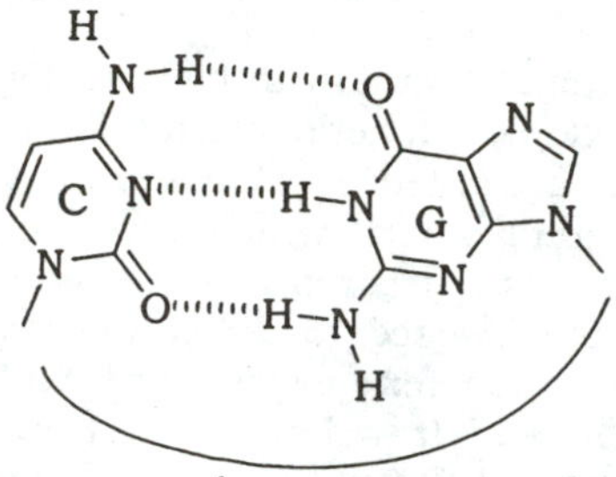

minor groove

G:U

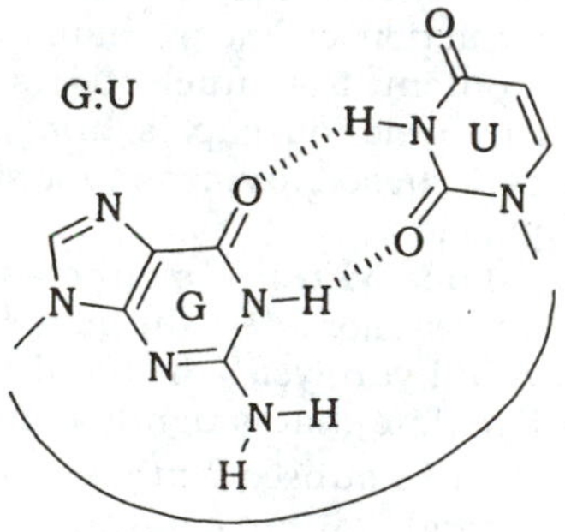

minor groove

U:G

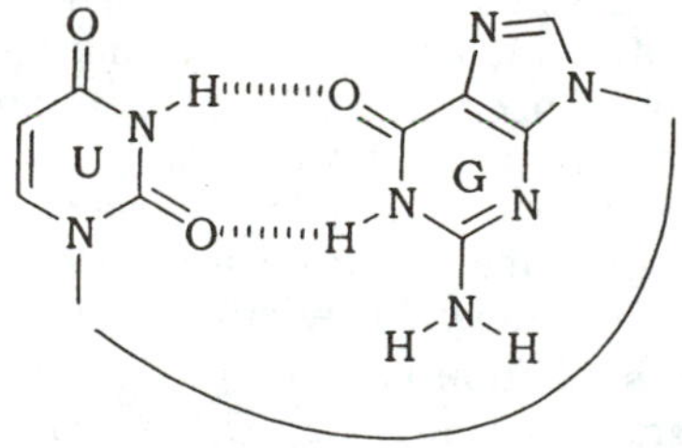

minor groove

5. *Point out why Crick's wobble hypothesis would allow fewer than 61 anticodons to be used to translate the 61 sense codons. How does "wobble" tend to accelerate the rate of translation?*

Answer: The wobble hypothesis allows base pairs to form between the 5' anticodon position and the 3' codon position according to the following rules:

5' Anticodon	3' Codon
C	G (canonical pair)
A	U (canonical pair)
U	A (canonical pair) or G
G	C (canonical pair) or U
I	U, C, or A

To decode a set of four codons that differ only in their 3' base, a minimum of two anticodons is necessary. For example to decode the four codons of threonine, ACU, ACC, ACA, and ACG, the anticodons IGU and CGU (written 5' to 3') are sufficient. Because inosine (I) can base pair with U, C, or A, the IGU anticodon will decode AGU, AGC, and AGA. The anticodon CGU decodes ACG only. In this example we can see that four codons can be decoded by only two anticodons.

The wobble rules explain the degeneracy of the genetic code. In all cases the codons NNU/C (or NNY) are always degenerate. To decode NNU requires a 5' anticodon base that pairs with U; the choices are A, G, or I. However, A is normally converted to I whenever it occurs in the 5' anticodon position. Therefore, only G or I are possible, both of which base pair with C as well. If G is used, then the anticodon will recognize both NNU and NNC. If I is used, this anticodon will recognize NNU, NNC, and NNA.

The codons NNA/G (NNR) are often degenerate because the 5' position of anticodons may contain U, which is capable of pairing with both A or G. Anticodons with C for 5' base will only decode a single codon and in fact there are two instances where this occurs in the genetic code. Both methionine and tryptophan are coded for by a single codon each, AUG and UGG respectively. Their anticodons must be CAU and CCA.

Wobble allows non-canonical base pairs to form. This makes codon-anticodon interactions less stable, allowing for an increase in the rate of translation.

6. How many codons can mutate to become nonsense codons through a single base change? Which amino acids do they encode?

Answer:

	Termination Codons		
	UAA	UAG	UGA
Third Base Change	UA**U** (Tyr) UA**C** (Tyr)	UA**U** (Tyr) UA**C** (Tyr)	UG**U** (Cys) UG**C** (Cys) UG**G** (Trp)
Second Base Change	U**U**A (Leu) U**C**A (Ser)	U**U**G (Leu) U**C**G (Ser) U**G**G (Trp)	U**U**A (Leu) U**C**A (Ser)
First Base Change	**C**AA (Gln) **A**AA (Lys) **G**AA (Glu)	**C**AG (Gln) **A**AG (Lys) **G**AG (Glu)	**C**GA (Arg) **A**GA (Arg) **G**GA (Gly)

7. Nonsense suppression occurs when a suppressor mutant arises that reads a nonsense codon and inserts an amino acid, as if the nonsense codon was actually a sense codon. Which amino acids do you think are most likely to be incorporated by nonsense suppressor mutants?

Answer: The most likely candidates for nonsense suppressors are tRNAs that decode codons that differ from stop codons by one base. For UAA and UAG termination codons, tRNAs that decode the following amino acid codons are candidates: Leu, Ser, Tyr, Trp, Gln, Lys, and Glu. For the UGA termination codon, tRNAs that decode Trp, Cys, Arg, Gly, Ser, Leu are candidates. However, there are certain constraints on nonsense suppression. The tRNA gene to be mutated must not be an essential gene to the cell. Ideally tRNA genes whose function is covered by another tRNA are ideally suited to become nonsense suppressors. In the above list Leu, Ser, and Arg tRNAs are good candidates because each of these amino acids have six codons. An amino acid coded by four codons, such as glycine, is also a good candidate. The remaining amino acids are coded by only one codon (e.g., Trp) or two codons each. These are the least likely candidates to serve as suppressor. The tRNA that decodes the UGG Trp codon would have to be first duplicated to provide an additional gene copy to mutate to a nonsense suppressor.

8. Why do you suppose eukaryotic protein synthesis is only 10% as fast as prokaryotic protein synthesis?

Answer: Protein synthesis, a fundamental reaction to all cell types, is carried out by ribosomes. Ribosomes from prokaryotes, eukaryotes, and organelles such as mitochondria and chloroplasts are all very similar in structure and they carrying out the process of protein synthesis in similar manners. At each stage of protein synthesis, initiation, elongation, and termination, the ribosome is dependent of protein factors, which together with the ribosome are responsible for a specific phase of protein synthesis.

One striking difference between prokaryotic ribosomes and eukaryotic ribosomes is the larger size and hence complexity of eukaryotic ribosomes. Not only are eukaryotic ribosomes considerably larger, they contain more ribosomal proteins and larger rRNAs. In addition, eukaryotic stage-specific protein factors are more complex. Because of this, eukaryotic ribosomes function at slower rates.

The advantage to slower ribosomal activity is an increase in fidelity. Eukaryotic cells produce a wider range of proteins whose proper functions are critical to cell survival. Thus, there is a greater demand for higher fidelity of protein synthesis in eukaryotes.

9. If the tunnel through the large ribosomal subunit is 10 nm long, how many amino acid residues might be contained within it? Assume that the growing polypeptide chain is in an extended (ε) (β-sheet-like) conformation.

Answer: There are actually two distance per residue values for β–sheet conformations, one for parallel and a different one for antiparallel structures. The more extended conformation is that of the antiparallel β–sheet, at 0.347 nm per residue, and since we might expect a single chain within the ribosome to be in its most extended form, we'll use that one for the calculation. Then, $\frac{10\text{nm}}{0.347\text{nm}/\text{residue}} = 28.8 \approx 29$ residues within the ribosomal large subunit. This corresponds nicely with an experimental determination of approximately 30 residues buried within the ribosome during translation.

10. Eukaryotic ribosomes are larger and more complex than prokaryotic ribosomes. What advantages and disadvantages might this greater ribosomal complexity bring to a eukaryotic cell?

Answer: More complex ribosomes allow for a greater repertoire of control mechanisms to function to regulate protein synthesis. In addition, the ability to produce proteins at a low error rate is improved. The costs of complexity include a lower rate of protein synthesis and a protein synthesis system that requires more components in order to function.

11. What ideas can you suggest to explain why ribosomes invariably exist as two-subunit structures, instead of a larger, single-subunit entity?

Answer: Ribosomes catalyze peptide bond formation using two aminoacyl-tRNAs as substrates that bind to two distinct tRNA binding sites on the ribosome known as the P-site and the A-site. During each round of elongation, the P-site is transiently occupied by a peptidyl-tRNA containing the newly synthesizes N-terminal portion of the polypeptide attached via aminoacyl linkage to the 3'-end of a tRNA. The A-site is filled with an aminoacyl-tRNA appropriate to the codon displayed at the A-site. The ribosome catalyzes peptide bond formation by nucleophilic attack by the amino group of the aminoacyl-tRNA in the A-site on the carbonyl carbon of the acyl bond on the peptidyl-tRNA in the P-site. The result is polypeptide chain transfer from the P-site to the A-site. As a result the P-site is now occupied by an uncharged tRNA whereas the A-site contains a peptidyl-tRNA one amino acid longer. In order for elongation to continue, the P-site must be vacated, the peptidyl-tRNA, now in the A-site, must be moved into the P-site, and the A-site must be refilled. This process, known as translocation, occurs in at least two steps or stages. In the first step, during peptide bond formation, the acceptor end of the A-site aminoacyl-tRNA moves into the P-site to form the peptide bond. The result of this movement is formation of a hybrid tRNA binding site: the 3'-end of the A-site tRNA is now in the P-site, a P/A hybrid state. What is the fate of the now deacylated tRNA? Interestingly, this tRNA is not immediately released from the ribosome. Rather, it is transferred to a third tRNA binding site, the E-site or exit site. However, the transfer is accomplished in two steps. As the 3'-end of the A-site tRNA moves into the P-site and peptide bond formation occurs, the 3'-end of the now deacylated tRNA moves into the E-site forming an E/P hybrid state. To occupy these two hybrid tRNA binding states, E/P and P/A, the tRNAs must move relative to the large subunit, an event that may be accomplished by movement of the small subunit. In order for elongation to continue, the A-site must be vacated. This is accomplished by resolving the P/A hybrid state into P-site binding and the E/P hybrid state into E-site binding. The A-site is now free to bind a new aminoacyl-tRNA, which initially interacts with the ribosome through a third hybrid binding site, the T/A. The transition of aminoacyl-tRNAs from T/A binding to the A-site may be a proof-reading step, and this transition may be influenced by E-site binding such that A-site binding results in release of an uncharged tRNA from the E-site.

The tRNA binding sites involve aspects of both the large and the small subunits. To produce hybrid sites, the subunits may be required to move relative to each other. The functional significance of this movement may be responsible for the high fidelity of protein synthesis.

12. How do prokaryotic cells determine whether a particular methionyl-tRNA$_{Met}$ is intended to initiate protein synthesis or to deliver a Met residue for internal

incorporation into a polypeptide chain? How do the Met codons for these two different purposes differ? How do eukaryotic cells handle these problems?

Answer: Initiator tRNA is a specific tRNA capable of decoding AUG codons responsible for the N-termini of proteins. In prokaryotes, this tRNA, $tRNA_f^{Met}$, is charged with methionine, which is subsequently formylated by methionyl-$tRNA_f^{Met}$ formyl transferase. Compared to tRNAs exclusively involved in elongation, $tRNA_f^{Met}$ has distinguishing features including an unpaired 5'-terminal base on the acceptor end, a unique CCU sequence it its D loop, and a set or three G:C base pairs in its anticodon stem. These features are essential to $tRNA_f^{Met}$ functioning during initiation.

There is only one codon for methionine, AUG, however it has two meanings, initiation or elongation, depending on context. Specifically, the context for initiation in prokaryotes is having a purine-rich sequence of from 4 to 8 bases located on the mRNA approximately 10 bases upstream of the AUG which will serve to initiate protein synthesis. This purine-rich sequence is known as the ribosome-binding site or the Shine-Dalgarno sequence. It interacts with a complementary, pyrimidine-rich sequence near the 3'-end of the small subunit rRNA. This interaction is responsible for determining the site of initiation of protein synthesis: initiation occurs, in most cases, at an AUG codon appropriately spaced from the ribosome-binding site. Any other AUG codon is decoded as methionine by methionyl-$tRNA^{Met}$.

In eukaryotes, a special initiator tRNA charged with methionine is used to initiate protein synthesis. However, $tRNA_i^{Met}$ is not formylated as is the case in prokaryotes but it is a special tRNA with features that distinguish it from tRNAs used in elongation. It is clear that initiation in eukaryotes does not involve an interaction between the 3'-end of the small subunit rRNA and the mRNA because eukaryotic small subunit rRNAs lack Shine-Dalgarno sequences. Initiation is dependent on formation of a 40S initiation complex that involves a cap binding protein responsible for recognizing the 5'-end cap modification.

13. What is the Shine-Dalgarno sequence? What does it do? The efficiency of protein synthesis initiation may vary by as much as 100-fold for different mRNAs. How might the Shine-Dalgarno sequence be responsible for this difference?

Answer: The Shine-Dalgarno sequence is a purine-rich sequence of 4 to 8 nucleotides located on prokaryotic mRNAs around 10 nucleotides upstream of an AUG codon used as an initiation codon. The Shine-Dalgarno sequence is often called the ribosome-binding site because mRNAs interact with ribosomes by interacting with a complementary, pyrimidine-rich sequence located near the 3'-end of the small subunit rRNA. This rRNA sequence is fixed for a given prokaryote; there is only one population of ribosomes. Shine-Dalgarno sequences on mRNAs show a range of variability from gene to gene. The degree to which a Shine-Dalgarno sequence is complementary to the pyrimidine-rich sequence on rRNA determines the efficiency with which a particular mRNA is used in initiation complex formation. mRNAs showing a high degree of complementarity are capable of interacting strongly with ribosomal small subunits and are therefore initiated frequently.

14. In the protein synthesis elongation events described under the section entitled "Translocation," which of the following seems the most apt account of the peptidyl transfer reaction: (a) The peptidyl-tRNA delivers its peptide chain to the newly arrived aminoacyl-tRNA situated in the A site, or (b) the aminoacyl end of the aminoacyl-tRNA moves toward the P site to accept the peptidyl chain? Which of these two scenarios makes most sense to you? Why?

Answer: A description of elongation is given in the answer to question 3 above and choice (b) agrees with this description. Using various conditions that stop or freeze the ribosome at various stages of protein synthesis, the state of occupancy of the P-site, A-site, and E-site have been determined. During elongation, a peptidyl-tRNA appears to bind to a hybrid P/A-site, thus exposing a portion of the A-site. In choice (a), a portion of the P site might be expected to be exposed whereas in choice (b), a portion of the A is exposed. The evidence favors (b).

For economy of effort it would be easier to move the aminoacylated end than the peptidyl group due to size. Finally, because the nascent protein appears to be threaded through the large subunit of the ribosome there may be fewer degrees of freedom for this group.

15. Why might you suspect that the elongation factors EF-Tu and EF-Ts are evolutionarily related to the G proteins of membrane signal transduction pathways described in Chapter 10?

Answer: EF-Tu functions during elongation as a ternary complex with GTP and an aminoacyl-tRNA. This ternary complex delivers charged tRNAs to an empty A-site on a ribosome. Upon binding to the ribosome, GTP hydrolysis occurs resulting in deposition of the aminoacyl-tRNA into the A-site and release of an EF-Tu:GDP complex and inorganic phosphate. EF-Tu functions catalytically and in order to participate in another round of elongation, GDP must be exchanged for GTP. This is accomplished indirectly by action of EF-Ts. (Ts and Tu refer to the temperature-stability of these two proteins. EF-Ts activity is stable to high temperatures whereas EF-Tu activity is unstable.) EF-Ts binds to the EF-Tu:GDP complex, displaces GDP, and forms a heterodimeric EF-Tu:EF-Ts protein complex. This complex dissociates upon GTP binding to EF-Tu, resulting in free EF-Ts and EF-Tu:GTP.

Many signal transduction pathways involve G proteins, GTP-binding proteins capable of stimulating or inhibiting adenylate cyclase. G proteins are capable of these activities only when they have bound GTP, and, these activities are transient because G proteins have intrinsic GTPase activity, which is responsible for hydrolyzing GTP to inactivate the protein. G proteins consist of three subunits α, β, and γ. The α subunit is a guanine nucleotide-binding protein that binds either GTP or GDP and exhibits GTPase activity. There are two types of α subunits, α_s and α_i, that stimulate or inhibit adenylate cyclase respectively. Following GTP hydrolysis, the α subunit contains bound GDP, which must be replaced with GTP. This is accomplished by the α subunit associating with the β and γ subunits to form a membrane-bound complex. This complex is capable of interacting with membrane-bound receptor proteins but only when the receptor protein is activated by hormone-binding. This interaction of G protein complex and hormone-loaded receptor protein results in dissociation of GDP, binding of GTP, and dissociation of an α:GTP complex that can now act on adenylate cyclase. Thus, G proteins participate in a GTP-GDP cycle reminiscent of EF-Tu:EF-Ts and it comes as no great surprise that the proteins are related.

16. Human rhodanese (33kD) consists of 296 amino acid residues. Approximately how many ATP equivalents are consumed in the synthesis of the rhodanese polypeptide chain from its constituent amino acids and the folding of this chain into an active tertiary structure?

Answer: The energetics of protein synthesis require at least 4 ATP equivalents per amino acid residue incorporated into protein. During elongation, the A-site is filled at the expense of GTP that occurs during recycling of EF-1, an eukaryotic elongation factor. A second GTP hydrolysis drives translocation catalyzed by EF-2, the eukaryotic translocation factor. Peptide bond formation is driven by the high-energy aminoacyl bond on the aminoacyl-tRNA substrate. Formation of aminoacyl-tRNAs is catalyzed by aminoacyl-tRNA synthetases, enzymes that consume ATP and produce AMP and PP_i. Pyrophosphatase hydrolysis of PP_i accounts for an additional high energy phosphate bond. For synthesis of human rhodanese with 296 amino acid residues, the A-site will have to be filled 295 times, accounting for ($295 \times 4 =$) 1,180 ATP equivalents. Met-$tRNA_i$ formation consumes two ATPs. Initiation requires two ATP equivalents, one in the form of GTP during eIF-2 mediated Met-$tRNA_i$ binding to form the 40S pre-initiation complex, and one in the form of ATP during 40S initiation complex formation. Peptide chain termination in eukaryotes requires GTP hydrolysis thus accounting for an additional ATP equivalent. Thus, $1{,}180 + 2 + 1 = 1{,}183$ ATP equivalents are consumed to synthesize rhodanese. In order to properly fold the polypeptide chain, an additional 130 equivalents of ATP are consumed during the folding cycle catalyzed by molecular chaperones. Active rhodanese production requires approximately 1,310 equivalents of ATP.

17. A single proteolytic break in a polypeptide chain of a native protein is often sufficient to initiate its total degradation. What does this fact suggest to you regarding the structural consequences of proteolytic nicks in proteins?

Answer: Although peptide bonds in proteins are quite stable, proteins are not made to last forever. During their lifetime they are subjected to a number of insults including oxidation of side chains, chemical modifications, and cleavage of peptide bonds. The consequences of these events may be inactivation of a protein or alteration of its activity to a new, undesirable form. To avoid accumulating inactive protein, cells must either correct the defect or degrade the protein and replace it through gene expression. In the case of peptide bond breakage, cells have evolved the ability to recognize and degrade nicked protein. In eukaryotic cells this process involves the protein ubiquitin, a highly conserved, 76 amino acid polypeptide. Ubiquitin is ligated to free amino groups on proteins and serves as a molecular tag, directing the protein's degradation by proteolysis. The ubiquitin pathway is apparently rapid and efficient and, because of this, cells never accumulate breakdown products of protein degradation.

Questions for Self Study

1. Match.

a. Nonsense codon	1. XXX → UGA, UAG, or UUA
b. Synonymous codons	2. AUG
c. Typical start codon	3. CCU, CCC, CCA, CCG = Pro
d. Third-base degeneracy	4. Stop translation
e. Nonsense mutation	5. AGA, AGG, CGA, CGG = Arg

2. Ribosomes played a key role in elucidating the genetic code. Explain.

3. Fill in the blanks. The enzymes responsible for attaching amino acids to tRNAs are the _________. There are at least twenty different enzymes for the 20 _____. Because of codon degeneracy, there are often more than one tRNA charged by the same amino acid. The different tRNA molecules recognized by the same charging enzyme are known as _____ tRNAs. All members of a set of tRNAs specific for a particular amino acid and served by one charging enzyme are known as _____.

There are two classes of charging enzymes. One important distinction between the two classes is _____. Independent of class, the charging enzymes catalyze the reaction in a two-step sequence. In the first step, the amino acid is activated by reaction with ATP and release of _____ to form _____. The enzyme-bound, activated amino acid is then transferred to the _____ end of the tRNA.

4. Briefly explain what is meant by the concept of wobble in codon base pairing.

5. Answer True or False.
a. Codon usage refers to the number of different tRNAs used to decode a message. _____
b. An example of codon usage is, of the two lysine codons AAA and AAG, AAA is used more frequently. _____
c. A nonsense suppressor is a tRNA obeying different wobble rules. _____
d. Typically, the most commonly used codon for an amino acid is decoded by the most abundant isoacceptor tRNAs. _____
e. Nonsense suppressor tRNAs decode stop codons as an amino acid. _____

6. Match

a. 30S	1. Large subunit rRNA of eukaryotic ribosome.
b. 50S	2. Small rRNA in large subunit of eukaryotes only.
c. 70S	3. Eukaryotic ribosome.
d. 40S	4. Small subunit rRNA of prokaryotic ribosome.
e. 60S	5. Small subunit rRNA of eukaryotic ribosome.
f. 80S	6. Small subunit of prokaryotic ribosome.
g. 23S	7. Large subunit rRNA of prokaryotic ribosome.
h. 16S	8. Large subunit rRNA in both prokaryotes and eukaryotes.
i. 28S	9. Prokaryotic ribosome.
j. 18S	10. Large subunit of eukaryotic ribosome.
k. 5S	11. Large subunit of prokaryotic ribosome.
l. 5.8S	12. Small subunit of eukaryotic ribosome.

7. Translation can be divided into three phases. What are they? During each phase, two tRNA binding sites on the ribosome, the A site and the P site, are occupied by various modified tRNAs

or release factors. For each phase of protein synthesis, state the occupancy of the A site and the P site.

8. What is the role of the Shine-Dalgarno sequence (ribosome-binding site), a purine rich region of prokaryotic mRNAs located upstream of the initiator codon?

9. Identify the following as associated with either initiation (I), elongation (E), or termination (T) of translation.

a. IF-2
b. EF-Tu
c. Cap binding protein
d. EF-Ts
e. EF-G
f. RF-1
g. IF-3
h. 40S preinitiation complex
i. eIF-2
j. polysomes

10. Referring to Table 25.10, match a protein synthesis inhibitor with its mode of action.

Inhibitor	Mode of Action
a. Streptomycin	1. Causes release of peptide in covalent linkage with antibiotic
b. Puromycin	2. ADP-ribosylated eEF-2.
c. Diptheria Toxin	3. Blocks peptidyl transferase activity of 50S subunit.
d. Ricin	4. Cleaves prokaryotic small subunit rRNA.
e. Cholicin E3	5. Inactivates eukaryotic large subunit rRNA by base removal.
f. Chloramphenicol	6. Codon misreading.

11. Fill in the blanks. The endoplasmic reticulum is an internal compartment typically composed of two regions, the smooth ER and the _____. The later region is so named because _____ stud its surface. These macromolecular complexes are translating proteins destined to be processed in the endoplasmic reticulum. Translation of these proteins actually starts in the cytosol. When a short N-terminal sequence of the nascent polypeptide, termed the _____, emerges from the ribosome it is recognized by the _____. This ribonucleoprotein complex binds to the ribosome and blocks further protein synthesis until the ribosome is escorted to the cytosolic surface of the endoplasmic reticulum. The _____ is responsible for docking the ribosome to the membrane. Protein synthesis then continues with the newly synthesized proteins being deposited into the lumen of the endoplasmic reticulum. There the N-terminal sequence is cleaved by _____.

12. What are the roles of ubiquitin and the large 26S ATP-dependent protease complex in degrading proteins?

Answers

1. a. 4; b. 5; c. 2; d. 3; e. 1.

2. *In vitro* translation systems were primed with synthetic polyribonucleotides to determine which amino acids were incorporated into protein in response to known nucleotide sequences. Trinucleotides were also used to attach aminoacylated tRNAs to the ribosome.

3. Aminoacyl-tRNA synthetases; amino acids; cognate; isoacceptor tRNAs; the position on the ribose moiety on the 3' end to which the amino acid is attached; PP_i; aminoacyl-adenylate; 3' end.

4. This concept, known as the wobble hypothesis, states that the first base (5' base) of an anticodon may be capable of pairing with more than one base in the 3' position (wobble position) of codons. Thus, a single tRNA may recognize more than one triplet codon.

5. a. F; b. T; c. F; d. T; e. T.

6. a. 6; b. 11; c. 9; d. 12; e. 10; f. 3; g. 7; h. 4; i. 1; j. 5; k. 8; l. 2.

7. During initiation of protein synthesis the P site is occupied by charged initiator tRNA and the A site is occupied by an aminoacyl-tRNA. During elongation the P site is occupied by a peptidyl tRNA or transiently by an uncharged tRNA. The A site is occupied by an aminoacyl-tRNA and transiently by a peptidyl tRNA. During termination, the P site is occupied by a peptidyl-tRNA and the A site is recognized by a release factor.

8. The ribosome-binding site interacts with a pyrimidine rich region near the 3' end of the small subunit ribosome to properly align the translational start site on the ribosome.

9. a. I; b. E; c. I; d. E; e. E; f. T; g. I; h. I; i. I; j. E.

10. a. 6; b. 1; c. 2; d. 5; e. 4; f. 3.

11. Rough endoplasmic reticulum or RER; ribosomes; signal sequence; signal recognition particle; docking protein or SRP receptor; signal peptidase.

12. Proteins destined for degradation are covalently modified by conjugation to ubiquitin. Ubiquitinated proteins are degraded by the protease complex.

Additional Problems

1. Using the genetic code, give an example showing that mutations in the second position of a codon do not give rise to drastic changes in the amino acid.

2. Using the wobble rules, predict all the possible anticodons used to decode Ile and Met.

3. Give two reasons why tRNAs responsible for decoding serine are good candidates for nonsense suppressors?

4. Mutations producing stop codons are generally referred to as nonsense mutations but specifically include amber mutations (mutations to UAG), ochre mutations (mutations to UAA), and opal mutations (mutations to UGA). Will ochre suppressors suppress amber mutations and vice versa?

5. Using the genetic code, indicate which sense codons can be converted by a single mutation to an ochre (UAA) nonsense codon. Are any problems anticipated in producing ochre suppressor tRNAs by mutating tRNAs used to decode these codons?

6. For mutations that do not result in amino acid changes, by what mechanism might they none the less effect gene expression?

7. In addition to ribosomes, *in vitro* protein synthesis systems from *E. coli* require a high-speed supernatant that supplies protein factors necessary for all of the side reactions necessary for protein synthesis. The reaction must be supplied with ATP, GTP, amino acids and a compound known as citrovoren. Citrovoren is an old name for folinic acid, N5-formyl-tetrahydrofolate. Explain why ATP, GTP and citrovoren are added. What role do they play in protein synthesis?

8. How are the 5S rRNA and 5.8S rRNA related? How is the 5.8S rRNA related to the large-subunit rRNA?

9. The substrates for protein synthesis are not free amino acids but aminoacylated tRNAs. Explain what an aminoacyl tRNA is and how is it formed.

10. During the elongation phase of protein synthesis, two tRNA binding sites orchestrate the elongation of a protein. Describe this process.

11. The small subunit of the ribosome is responsible for initiation of protein synthesis. It interacts with several proteins, and various RNAs to form a preinitiation complex. Later, the

preinitiation complex is joined by the large subunit to form the initiation complex, a complete ribosome lacking an aminoacyl-tRNA at the A site. Describe how a preinitiation complex forms.

12. In eukaryotic ribosomes, initiation of protein synthesis is regulated by phosphorylation of eIF-2. Explain.

13. The elongation factor EF-Tu is a GTPase required for binding of aminoacylated tRNA to the A site. This protein is recycled during protein synthesis by interacting with another protein, EF-Ts. Describe this interaction.

14. In terms of codon recognition, how is termination fundamentally different than elongation in protein synthesis?

Abbreviated Answers

1. The best examples of this are for codons with U or C in position two. With U in position two, any base in position one will code for a somewhat bulky, hydrophobic amino acid. With C in position two, the choices of amino acids include Ser, Pro, Thr, and Ala. Ser, Thr, and Ala are all quite similar in size.

2. AUU, AUC, and AUA code for Ile and AUG codes for Met. All three Ile codons can be decoded by IAU. GAU will decode Ile codons AUU and AUC. An anticodon with UAU will pair with the AUA code for Ile but it will also decode AUG so it can not be used. The Met code AUG must be decoded by CAU.

3. Two serine codons differ by one base from the stop codons making serine anticodons ideal candidates for nonsense suppressors. Further, serine is coded by 6 codons, indicating that multiple $tRNA^{Ser}$ genes exist.

4. An ochre suppressor must decode UAA and can thus have one of the following anticodon sequences, IUA or UUA. Using the wobble rules we see that IUA can decode UAU and UAC in addition to UAA but not UAG. However, the anticodon UUA can decode UAA and UAG. Thus, certain UAA suppressors will also suppress UAG nonsense mutations.

5. UAU (Tyr), UAC (Tyr), GAA (Gln), AAA (Lys), GAA (Glu), UUA (Leu) and UCA (Ser) are all candidates. With the exception of Leu and Ser, all the other amino acids are coded for by only two codons. If the cell has only single tRNAs for each of these amino acids, it will not be possible to mutate one of the tRNAs to read an ochre codon and still be able to decode the amino acid. There must be several tRNA genes for Leu and Ser and it may be possible to mutate one of them to read ochre codons while retaining the others for amino acid codons.

6. Even though a mutation might not change the amino acid encoded it may have an affect on gene expression if the change is from a rarely used codon to a frequently used codon or vice versa. Codon usage is not uniform and frequently used codons are decoded by tRNAs found in abundance in cells. Altering codons may influence the rate of translation, leading to abnormally high or low levels of gene product.

7. ATP is required for aminoacylation of tRNAs by aminoacyl-tRNA synthetases. Hydrolysis of GTP is used at various stages of protein synthesis to drive reactions. Formyl-tetrahydropteroly-glutamic acid serves as a formyl-group donor for methionyl-$tRNA_f^{Met}$ formyl transferase to formylate the charged initiator tRNA.

8. The 5S rRNA and the 5.8S rRNA are both components of the large subunit of the ribosome. The 5.8S is specific for eukaryotic cytoplasmic ribosomes whereas the 5S rRNA is found in eukaryotes and prokaryotes. The 5.8S rRNA of eukaryotes has a sequence similar to the 5'-end of the large subunit rRNA of prokaryotes.

9. Aminoacyl-tRNAs represent activated forms of amino acids used by the ribosome as substrates for peptide bond formation during protein synthesis. Aminoacyl-tRNAs are formed by aminoacyl-tRNA synthetases in a two-step reaction driven by hydrolysis of ATP.

10. During a cycle of elongation, the P-site is occupied by a peptidyl-tRNA produced by preceding rounds of elongation and the A-site is occupied by an aminoacyl-tRNA appropriate for the codon in the A-site. Peptide bond formation occurs by nucleophilic attack of the amino group of the A-site aminoacyl-tRNA on the carbonyl carbon of the ester bond in the peptidyl-tRNA resulting in peptide bond formation, deacylation of the P-site tRNA, and formation of a peptidyl-tRNA in the A-site. The P-site tRNA is released and the peptidyl-tRNA in the A-site is translocated to the P-site in a factor-dependent manner. Translocation of the tRNA is accompanied by movement of the mRNA as well, to bring a new codon into the A-site. The A-site is filled with the appropriate aminoacylated-tRNA by a factor-dependent process.

11. A ribosome is dissociated by initiation factors into a large subunit and a small subunit. The small subunit binds mRNA and a complex of a specific initiation factor (IF-2 in prokaryotes), GTP, and charged initiator-tRNA to form a preinitiation complex. The complex is joined by the large subunit with release of initiation factors to produce the initiation complex.

12. In eukaryotes, initiation of protein synthesis begins with formation of a ternary complex with eIF-2, GTP, and Met-$tRNA_i^{Met}$. The complex associates with a small subunit to produce a 40S preinitiation complex. The α-subunit of eIF-2 can be phosphorylated by a specific kinase. Phosphorylated eIF-2 binds eIF-2B tightly and effectively eliminates eIF-2B-dependent recycling of eIF-2.

13. EF-Tu forms a complex with GTP and an aminoacyl-tRNA before binding to the ribosome. Ribosome binding stimulates GTPase activity resulting in release of P_i, A-site binding of the aminoacyl-tRNA, and release of EF-Tu:GDP complex. EF-Ts binds to this complex and displaces GDP to produce an EF-Tu:EF-Ts complex. GTP can dissociate EF-Ts from this complex, thus allowing EF-Tu to participate in another round of elongation.

14. For termination codons, codon recognition is achieved by protein release factors instead of tRNAs.

Summary

The four-letter language of nucleic acid sequences is translated into the 20-letter language of proteins by a genetic code in which triplets of bases code for particular amino acids. Nucleotide sequences in mRNAs are colinear with amino acid sequences of proteins. The sequence of bases in a mRNA is read 5' 3' and encodes a polypeptide in the N C direction. Nucleotides and amino acids share no stereochemical complementarity and have no particular chemical affinity, which would be useful in translating nucleotide sequence information into amino acid sequence information. Transfer RNAs serve as adapter molecules that bridge this informational gap. Amino acids are covalently attached to the 3'-OH end of the tRNA and borne as aminoacyl-tRNA derivatives. The tRNAs possess anticodons, unpaired triplets of bases that are complementary to the "code words for amino acids", called codons, in mRNA. An aminoacyl-tRNA associates with the mRNA via specific base pairing between the anticodon and codon. As a consequence of such interaction, amino acids are aligned according to the sequence of codons in the messenger RNA and can be linked into polypeptide chains.

Since polynucleotides are composed of 4 different bases, there are $4^3 = 64$ possible triplets. The code is non-overlapping, thus any base is part of only one codon. The base sequence is read continuously from a fixed starting point without punctuation. Each successive triplet is a codon dictating an amino acid. A triplet genetic code has 64 possible codons for just 20 amino acids; the code is degenerate in that usually more than one codon exists for each of the amino acids. Of the 64 possible codons, 61 specify amino acids. The remaining three codons, the so-called nonsense, or "stop" codons, specify no amino acid, but instead signal the end of protein-coding regions within the mRNA. Thus, all 64 codons have meaning, though only 61 are "sense" codons. Further, the genetic code is unambiguous: each codon has only one meaning.

Amino acids of similar chemistry usually are represented by codons of similar sequence, e.g., codons with a pyrimidine in the second position encode amino acids with hydrophobic side chains, while second-base purine codons specify amino acids with polar or ionic R groups. Codon assignments are not random and synonymous codons (codons coding for the same amino acid) almost always have identical first and second bases, differing only in the third. (The only exceptions are for Arg, Leu and Ser; each of these amino acids is represented by six different codons). This third base degeneracy means that the consequences of single base change mutations are minimized. A striking feature of the genetic code is its universality: codons specify the same amino acid regardless of the organism.

The synthesis of aminoacyl-tRNA derivatives serves two ends: 1) It activates the amino acid for peptide bond formation by placing the carboxyl group in ester linkage with the 3'-OH end of a tRNA, and 2) It creates aminoacyl-tRNAs which bridge the informational gap between amino acids and codons. Aminoacyl-tRNA synthetases mediate the ATP-dependent synthesis of aminoacyl-tRNAs. One aminoacyl-tRNA synthetase exists for each of the twenty amino acids and it is capable of recognizing all of the cognate isoacceptor tRNAs specific for its amino acid. The various aminoacyl-tRNA synthetases are a remarkably eclectic set of proteins given the commonality of their function, and they recognize a diverse assortment of determinants in order to identify their cognate tRNA molecules. Some distinguish their tRNAs through interaction with the anticodon, an obvious discriminatory feature, but others rely only partially or not at all on the anticodon.

Francis Crick considered the nature of anticodon:codon pairing and third base degeneracy and hypothesized that a single anticodon might interact with several different codons if a certain amount of play, or wobble, was allowed at the third codon position. The wobble hypothesis is based on the permissibility of alternative, non-Watson:Crick-type base pairs. The wobble rules indicate that a first-base anticodon U could recognize either an A or G in the codon third base position, first-base anticodon G might recognize either U or C in the third base position of the codon, and first-base anticodon I might interact with U, C or A in the codon third position.

Codon usage is not random; some codons are used more frequently in mRNAs than their synonyms, and the more commonly used codons are represented by a greater abundance of the corresponding isoacceptor tRNAs. Nonsense mutations lead to premature termination of protein synthesis and the formation of truncated polypeptides. Intergenic suppression occurs if a second mutation arises in a tRNA gene and produces a tRNA whose anticodon can read the nonsense codon and insert its amino acid

Translation is the process whereby genetic information is expressed through synthesis of a protein. Ribosomes are the agents of protein synthesis. Ribosomes are self-assembling ribonucleoprotein particles ubiquitous to all cells, invariably consisting of two unequal subunits, each containing more than twenty different proteins and at least one characteristic ribosomal RNA (rRNA) molecule.

Like chemical polymerization processes, protein biosynthesis has three phases: initiation, elongation and termination. In initiation, the small ribosomal subunit joins with an mRNA molecule and a particular methionine-bearing tRNA that is used only in peptide chain initiation, never in peptide chain elongation. The large ribosomal subunit then adds to the small subunit:mRNA:Met-tRNA complex and the elongation phase commences.

All the peptide bonds in the protein, from the first to the last, are formed in the elongation phase. Elongation proceeds via the codon-directed association of aminoacyl-tRNAs with the ribosome:mRNA complex, as the ribosome moves along the mRNA and encounters successive codons. The ribosome:mRNA complex has two aminoacyl-tRNA binding sites: the A site where an incoming aminoacyl-tRNA binds and the P site where peptidyl-tRNA (the tRNA carrying the growing polypeptide chain) is held. Peptide bond formation is achieved when the α-carboxyl end of the polypeptide chain is transferred from its tRNA to the free α-NH_2 group of the aminoacyl-tRNA in the A site. The deacylated tRNA (the tRNA left after transfer of the polypeptide) leaves the P site. The tRNA in the A site now bears the polypeptidyl chain and it is translocated from the A site to the P site as the ribosome moves one codon further along the mRNA. This codon now directs the binding of the next incoming aminoacyl-tRNA in the A site, and the process repeats itself until all the codons have been translated. Termination of polypeptide synthesis is triggered by the appearance of a nonsense codon (UAA, UAG or UGA) in the A site. Release factor proteins bind at the A site, transforming the peptidyl transferase activity of the ribosome into a peptidyl hydrolase, and the free polypeptide is released. The ribosome:mRNA complex then dissociates.

The reactions of peptide chain initiation, elongation and termination are orchestrated at each step by specific soluble proteins - initiation factors, elongation factors and release factors. GTP hydrolysis provides the energy to drive conformational changes in the protein-synthesizing apparatus. These conformational changes represent the workings of the protein synthesis machinery in its cycle of successive amino acid addition to the growing peptidyl chain. Two GTPs are expended for each amino acid added to the chain: one in aminoacyl-tRNA binding at the A site and one in the coupled movement of the ribosome along the mRNA which translocates the peptidyl-tRNA from the A site to the P site. Polysomes, consisting of an mRNA molecule with several ribosomes simultaneously translating it, serve as the active protein-synthesizing units.

While the translation is similar in prokaryotes and eukaryotes, both the process and the accompanying apparatus are considerably more complex in eukaryotes. Protein synthesis inhibitors have been very useful in probing the mechanism of protein synthesis; some have clinical value as antibiotics.

A great variety of modifications are introduced post-translationally into proteins, proteolytic cleavage being a particularly prevalent change. Proteolytic cleavage can serve a number of ends: Generating diversity in the protein, activating its biological function, or facilitating the sorting and dispatching of proteins to their proper destinations in the cell. In the latter role, a leader peptide located at the N-terminus of the protein acts as a signal sequence, which tags the protein as belonging in a particular compartment or organelle. These signals are recognized by protein translocation systems.

Some of these systems identify the protein before its synthesis is complete. For example, the eukaryotic "signal recognition particle" (SRP) binds to the leader sequence of proteins destined for processing in the endoplasmic reticulum, just as this signal emerges from the ribosome. The SRP then chauffeurs the ribosome to the cytosolic face of the ER and docks it there so that the growing polypeptide chain is threaded through the ER membrane into the ER lumen. Within the lumen, the leader sequence is removed by signal peptidase. Additional modifications (such as glycosylation) may be introduced before the nascent polypeptide becomes a mature protein and gains its final destination.

Other protein translocation systems, such as those serving in mitochondrial protein import or prokaryotic protein export, employ molecular chaperones to keep the completed polypeptide in a partially unfolded state as it is shepherded through the cytoplasm from the ribosome to the membrane-embedded translocation apparatus. Once there, this apparatus catalyzes the energy-dependent translocation of the protein across the membrane. Typically, signal sequences are "cleavable presequences" clipped from the polypeptide once they have reached the other side of the membrane.

Proteins are in a dynamic state of turnover, with protein degradation serving an important role in determining the half-life of a protein in the cell. Protein degradation may be non-selective, as in lysosomal degradation, or selective, as in the ubiquitin-mediated pathway. The latter system targets proteins for degradation by a large multi-catalytic ATP-dependent protease complex by labeling them with ubiquitin. Ubiquitin is a 76-residue polypeptide that becomes covalently attached to proteins via linking its C-terminal glycyl α-carboxyl to free $-NH_2$ groups in the protein. Ubiquitinylation of a protein is determined by the nature of the amino acid residue at the protein's N-terminus.

Chapter 26

The Reception and Transmission of Extracellular Information

Chapter Outline

- Hormones: Three types
  - Steroid hormones: Cholesterol derivatives
    - Regulate metabolism, salt and water balance, inflammatory process, sexual function
  - Amino acid derivatives
    - Epinephrine and norepinephrine
      - Regulate smooth muscle contraction, cardiac rate, lipolysis, glycogenolysis
    - Thyroid hormone
      - Stimulates metabolism
  - Peptide hormones
- Transmembrane signal processing hormone receptors: Three families
  - 7-Transmembrane segment
    - Extracellular hormone-binding domain
    - 7 Transmembrane segments
    - Intracellular G protein-binding site
  - Single transmembrane segment
    - Extracellular ligand recognition site
    - Intracellular catalytic domain: Either of two types
      - Tyrosine kinase
      - Guanylyl cyclase
  - Oligomeric ion channels: Multi subunit proteins
    - Ligand-gated channels: Ligands are neurotransmitters
- Intracellular second messengers: cAMP, cGMP, Ca^{2+}, IP_3, DAG, phosphatidic acid, ceramide, nitric oxide, cyclic ADP-ribose
- cAMP produced by stimulation of adenylyl cyclase: Stimulation cascade
  - Hormone binding to receptor induces GTP for GDP exchange on G protein heterotrimer
  - G_α-GTP dissociates from trimer: Binds to and stimulates adenylyl cyclase
  - GTPase activity of G_α hydrolyzes GTP: Adenylyl cyclase stimulation stops
  - G_α-GDP reassociates with $G_\beta G_\gamma$ to reform trimer
  - Two kinds of G_α
    - Stimulatory
      - Cholera toxin: ADP-ribosylation of $G_{s\alpha}$-protein using NAD^+ as substrate: Inhibits GTPase activity
    - Inhibitory
      - Pertussis toxin: ADP-ribosylation of $G_{i\alpha}$-protein
  - *ras* GTP-binding protein

    - Protooncogene: Normal *ras* has low GTPase activity: Stimulated by GTPase-activating protein
    - Oncogene: Impaired GTPase activity due to protein mutation
- 7 TMS receptors
  - α_1 Adrenergic receptors: Stimulate inositol phosphate pathway
  - α_2 Adrenergic receptors: Inhibit adenylyl cyclase
  - β Adrenergic receptors: Stimulate adenylyl cyclase
- Phospholipase activation produces second messengers
  - Hydrolysis of phosphatidylinositol produces several products
    - DAG (diacylglycerol)
      - Remains in membrane: Activates protein kinase C
    - Inositol phosphate (IP) or IP_2 (which is converted to IP_3)
      - IP_3: Water soluble stimulates intracellular calcium release
  - Activation of phospholipase C
    - Phospholipase C-β, -γ, -δ calcium stimulated
    - Phospholipase C-β also stimulated by G protein
    - Phospholipase C-γ also stimulated by receptor tyrosine kinase
  - Hydrolysis of phosphatidylcholine produce diacylglycerol or phosphatidic acid
  - Hydrolysis of sphingomyelin produces ceramide: Stimulates ceramide-activated protein kinase
- Calcium as second messenger
  - Two sources
    - Extracellular: Entry via opening of plasma membrane channels in response to cAMP
    - Intracellular: Endoplasmic reticulum and calciosomes release calcium in response to IP_3
      - Calcium released from ER opens plasma membrane channels and together with DAG stimulates protein kinase C
      - Calcium oscillations often induced by excitation of phosphoinositide pathway
  - Calcium-binding proteins: Three groups
    - Calcium-modulated proteins: Calmodulin, parvalbumin, troponin-C: Contain EF-hands
      - Calcium binding allows protein to bind to target
      - Target is basic amphiphilic alpha (Baa) helix
    - Annexin proteins: Peripheral membrane proteins
    - Protein kinase C
      - Soluble protein inactive
      - Target-bound protein active
        - DAG binding to soluble protein opens substrate binding domain
          - Phorbol esters mimic DAG binding
        - Calcium binding activates kinase
- Protein phosphatases reverse action of kinases
- Hormone receptors with enzymatic activity: Two classes
  - Tyrosine kinases
    - Nonreceptor tyrosine kinase: $pp60^{c\text{-}src}$
    - Receptor tyrosine kinases
      - Extracellular glycosylated, receptor-binding domain
      - Transmembrane alpha helix
      - Intracellular tyrosine kinase domain
      - Three classes of tyrosine kinases
        - Class I: Extracellular domain contains 2 Cys-rich repeats; EGF
        - Class II: $\alpha_2\beta_2$: β transmembrane subunit, α Cys-rich extracellular domain; insulin receptor
        - Class III: Immunoglobulin-like domain; PGDF
- Guanylyl cyclases
  - Membrane-bound cyclases: Single-TMS receptors

- GTP to cGMP
- Stimulated by peptide hormones
    - Peptide hormones: Largest class of hormones
        - Synthesized as prepropeptide that contains signal sequence
        - Signal sequence cleavage produces propeptide
        - Propeptide cleavage and modification leads to active hormone (or hormones)
- Soluble guanylyl cyclases: Stimulated by nitric oxide (NO·)
    - Roles of NO· in cells
        - Neurotransmitter
        - Second messenger: Stimulation of cyclase: cGMP
        - Diffuses rapidly across cell membranes
        - Short half-life

- Protein modules in signal transduction
    - SH2 domain: Binds to phosphotyrosine-containing target
    - SH3 and WW: Binds to proline-rich target
    - PDZ: Binds to terminus of target protein
    - PH: Associates with phosphoinositides, directs protein to plasma membrane
- Steroid hormones
    - Glucocorticoids
    - Mineralocorticoids
    - Vitamin D
    - Sex hormones
- Modes of action
    - Transcriptional regulators in nucleus
        - Two functions
            - Bind DNA directly
            - Bind to transcriptional factors (Jun, Fos)
    - Regulation of ligand-gated ion channels
        - Modulates ion channels
- Nervous system cells
    - Neurons: Transmit nerve impulses
        - Axons: Cell processes that transmit impulse away from cell body
            - Synaptic terminal: Specialized ends of axon
        - Dendrite: Receives input
    - Glial cells: Protect and support neurons
- Electrical signals: Change in electrical potential: Potential generated by ion gradients
    - Resting neuron exhibits a –60 mV membrane potential
    - Action potential
        - Activated by depolarization
        - Voltage-gated sodium channels open allow Na^+ in
        - Sodium causes depolarization until channels close
        - Potassium channels open to allow K^+ out
        - Potassium causes repolarization (hyperpolarization)
        - Propagates at speeds up to 100 m/s
        - Not attenuated with distance
- Synapses: Junctions
    - Electrical synapse: Small gaps: Pass depolarization electrically
    - Chemical synapse: Axon releases neurotransmitter into synaptic cleft
        - Acetylcholine
            - Cholinergic synapse receptors
                - Nicotinic receptors: Ligand-gated cation channel
                - Muscarinic receptors: Interact with G proteins
                    - 7-TMS proteins: Stimulation causes
                        - Inhibition of adenylyl cyclase
                        - Stimulation of phospholipase C

        - Potassium channel opening
- Amino acid neurotransmitters
    - Glu, Asp: Excitatory
    - GABA, Gly: Inhibitory
- Catecholamines: Epinephrine, norepinephrine, dopamine, L-dopa
- Peptides: Endorphins, enkephalins

Chapter Objectives

Hormones

There are three major classes of hormones: steroid hormones, which are derived from cholesterol, amino acid derived hormones, and peptide hormones. The steroid hormones either interact with surface receptors or bind to intracellular proteins. Amino acid-derived hormones and peptide hormones bind to surface receptors and lead to stimulation of a signal transduction pathway leading to second messenger production. There are three families of membrane receptors. The 7-transmembrane segment (7-TMS) receptors have seven helical segments that span the membrane, an intracellular recognition site for G-proteins, and an extracellular domain for hormone binding. The single-transmembrane segment catalytic receptors have a single transmembrane segment, an extracellular domain containing the hormone binding site, and an internal domain with catalytic activity, either tyrosine kinase or guanylyl cyclase. The third family is the oligomeric ion channels. These are ligand-gated ion channels that open in response to neurotransmitters.

Second Messages

cAMP is produced by adenylyl cyclase and degraded by phosphodiesterase. cGMP is produced by guanylyl cyclase. Ca^{2+} is released from internal stores or moves across the membrane via a calcium channel from the extracellular environment. Inositol trisphosphate, diacylglycerol, and phosphatidic acid are released by metabolism of membrane components. Nitric oxide is produced by action of NO synthase.

G-Proteins

G-proteins function to transduce a cell surface signal into an internal signal. Hormone binding to a hormone receptor protein on the cell surface results in release of G-protein. The G-proteins are heterotrimers with the α-subunit functioning as the messenger between hormone binding to the receptor and production of an intracellular signal. This subunit is a guanine nucleotide binding protein. GTP bound to G_α leads to release of a G_α-GTP complex that activates adenylyl cyclases. G_α has intrinsic GTPase activity and hydrolysis of GTP leads to inactivation of G_α. There are many G-proteins exhibiting a range of activities from stimulatory to inhibitory. ADP-ribosylation by cholera toxin inhibits the $G_{s\alpha}$ GTPase activity, locking it into the active messenger state. ADP-ribosylation of $G_{i\alpha}$ inhibits GDP/GTP exchange, thus preventing its inhibition of adenylyl cyclase.

Membrane-Derived Second Messages

Specific phospholipases function to produce a host of second messages including inositol trisphosphate (IP_3), diacylglycerol, free arachidonic acid, which serves as the precursor of eicosanoids, and ceramide. IP_3 stimulates release of calcium from internal stores and calcium in concert with diacylglycerol stimulates protein kinase C.

Calcium

The importance of calcium as a second message has already been covered in muscle contraction, where troponin C functions as the key calcium binding protein. Calcium can be released from internal stores by IP_3 or from the plasma membrane by other signals. Calcium binds to calcium-modulated proteins, including calmodulin, parvalbumin and troponin C or to annexin proteins. The first three proteins all have virtually identical calcium binding domains, an EF-hand, consisting of a two helices separated by a calcium binding loop. Calmodulin can bind to a number of other proteins and modulate their activity in a calcium-dependent manner.

Tyrosine Kinases

The first tyrosine kinase to be discovered was an oncogene carried by Rous sarcoma virus. The gene product, $pp60^{v\text{-}src}$, is capable of phosphorylating a number of cellular target proteins. This activity is responsible for transformation. Epidermal growth factor, insulin, and platelet-derived growth factor all function by binding to surface receptor proteins that have tyrosine kinase activity that is dependent on hormone binding.

Peptide Hormones, NO·, and Steroids

The peptide hormones, a diverse group of hormones, derive from a primary translation product by post-translational processing events including: proteolytic cleavage of a hydrophobic N-terminal signal peptide, proteolytic cleavage internally, and post-translational covalent modifications. NO· is a small, gaseous hormone produced from arginine. It functions as a hormone by stimulating guanylyl cyclase. The steroid hormones bind to internal receptor proteins that function as transcriptional activators leading to synthesis of new proteins.

Neurons

Understand the basic anatomy of a neuron and the role of dendrites, axons, nodes of Ranvier, and synaptic knobs. Sensory neurons are the input gateways to the nervous system, interneurons are the connections between neurons, and motor neurons pass information to muscle cells. All cells have an electrical potential across their membrane. The sequence of events responsible for an action potential includes opening and closing of voltage-sensitive sodium channels followed by opening and closing of potassium channels. The anatomy of synaptic clefts should be understood. Be familiar with these terms: presynaptic cell, postsynaptic cell, synaptic cleft, synaptic vesicle. Migration and fusion of synaptic vesicles is a calcium-mediated event. Voltage-gated calcium channels open in response to the arrival of an action potential, allowing calcium to enter the presynaptic cell followed by synaptic membrane fusion and release of neurotransmitter into the synaptic cleft. Neurotransmitter receptors on the postsynaptic cell are often voltage-gated ion channels. Acetylcholine is the neuro-transmitter of cholinergic synapses. There are muscarinic receptors and nicotinic receptors for acetylcholine. Nicotinic receptors are ligand-gated ion channels. Antagonists of nicotinic receptors include tubocurarine, cobratoxin and α-bungarotoxin. Muscarinic receptors cause the opening of potassium channels in a G-protein mediated process. Agents active against muscarinic receptors include atropine and, of course, muscarine. DIFP, malathion, parathion, sarin, and tabun all inhibit acetylcholinesterase. Other important neurotransmitters include the amino acids glutamate, aspartate, and glycine; γ-aminobutyric acid; the catecholamines, epinephrine, norepinephrine, dopamine, and L-dopa; and various peptide neurotransmitters.

Problems and Solutions

1. Compare and contrast the features and physiological advantages of each of the major classes of hormones, including the steroid hormones, polypeptide hormones, and the amino acid-derived hormones.

Answer: The steroid hormones have two modes of action. They can bind to surface receptors leading to regulation of ion channels. Alternately, steroid hormones can diffuse into the cell and interact with specific internal receptor proteins leading to stimulation of transcription and gene expression. The time course and duration of stimulation is quite different for these two modes of action. Ion channel stimulation is a rapid though brief response. Alteration in gene expression is slower but long-lasting. The polypeptide hormones function by receptor binding at the cell surface resulting in production or release of a second messenger. Production of the peptide hormones requires gene expression and several post-translational modifications including proteolytic cleavages and covalent modifications. Peptide hormones are the most diverse group of hormones and as a result can participate in a wide variety of specific hormone-receptor interactions. The amino-acid derived hormones include two groups. The catecholamine hormones, including epinephrine and norepinephrine, are derived from the amino acid tyrosine and function by surface receptor binding and second message production. The second group of amino-acid derived hormones are the thyroid hormones thyroxine and triiodothyronine, which bind to intracellular receptors that in turn regulate gene expression.

2. Compare and contrast the features and physiological advantages of each of the known classes of second messengers.

Answer: Second messengers include Ca^{2+}, the cyclic nucleotides cAMP and cGMP, diacylglycerol (DAG), inositol trisphosphate (IP_3), phosphatidic acid, ceramide, nitric oxide (NO·), and cyclic ADP-ribose. At rest, cells maintain low concentrations of intracellular calcium but maintain calcium stores in internal compartments, such as the endoplasmic reticulum and other vesicles, and cells are typically bathed in an extracellular solution of high calcium concentration. The internal calcium concentration can be rapidly changed by opening calcium channels either in the plasma membrane or in an intracellular calcium storage compartment, or both. Calcium is a small molecule that can rapidly diffuse to intracellular targets, which include calcium binding proteins. Calcium-dependent signals are fast-acting but require the action of calcium pumps and sequestering agents to be switched off.

The cyclic nucleotides derive from ATP and GTP by adenylyl and guanylyl cyclase. They bind to target proteins in a highly specific manner and the levels of cyclic nucleotides are regulated by controlling both activity of the cyclases and the phosphodiesterases (that degrade cyclic nucleotides into nucleoside monophosphates by hydrolysis). The cyclic nucleotides and inositol trisphosphate and its derivatives are strictly intracellular agents. They are impermeable to membranes and remain within the cytosol. Inositol trisphosphate can be converted to a large number of phosphorylated intermediates that can serve as second messengers. Diacylglycerol, phosphatidic acid, and ceramide are second messages that act at membrane surfaces. Nitric oxide is the smallest second message. (Calcium is smaller but its hydrate is larger). It can diffuse rapidly to target sites and requires no special uptake mechanism because it is freely permeable across the membrane.

3. Nitric oxide may be merely the first of a new class of gaseous second messenger/neurotransmitter molecules. Based on your knowledge of the molecular action of nitric oxide, suggest another gaseous molecule that might act as a second messenger and propose a molecular function for it.

Answer: Nitric oxide is synthesized from arginine by NO synthase. This gaseous compound is capable of rapidly diffusing across cell membranes. The target of NO· is the heme prosthetic group of soluble guanylyl cyclase. NO· binding results in an increase in enzyme activity leading to increased concentrations of cGMP. In studying hemoglobin we discovered that carbon monoxide has a high affinity for heme groups. Carbon monoxide is produced during heme catabolism when one of the methylene carbons of heme is converted to CO while the remainder of the heme group is converted to bilirubin. Researchers at John Hopkins University have reported that CO can attenuate the NO·-induced stimulation of guanylyl cyclase.

4. Herbimycin A is an antibiotic that inhibits tyrosine kinase activity by binding to SH groups of cysteine in the src gene tyrosine kinase and other similar tyrosine kinases. What effect might you expect it to have on normal rat kidney cells that have been transformed by Rous sarcoma virus? Can you think of other effects you might expect for this interesting antibiotic?

Answer: Transformed cells exhibit a number of properties distinct from nontransformed cells including loss of contact inhibition, continuous division, increased glucose uptake and glycolysis, loss of cytoskeletal elements leading to cell shape changes, alteration in surface antigens, loss of differentiation, and growth in liquid culture (anchorage-independent growth). Rous sarcoma virus carries the src oncogene, which codes for a tyrosine kinase that is responsible for transformation by the sarcoma virus. By inhibiting the tyrosine kinase with herbimycin A, the characteristics of transformed cells are all reversed.

5. Monoclonal antibodies that recognize phosphotyrosine are commercially available. How could such a monoclonal antibody be used in studies of cell signaling pathways and mechanisms?

Answer: The ability to measure changes of phosphotyrosine levels in response to a hormonal signal is necessary to establish a role for tyrosine kinases in a particular signaling pathway. It is an easy task to use the radioactive isotope of phosphorous, ^{32}P, to establish that a protein is phosphorylated but it is difficult to distinguish between phosphotyrosine, phosphoserine and phosphothreonine. The protein would have to be converted to amino acids but this presents a problem because phosphotyrosine is not particularly stable. Further, phosphoproteins are not abundant and therefore difficult to isolate. Monoclonal antibodies against protein phosphotyrosine would allow rapid quantitation of phosphotyrosine levels, and using Western blots, identification of phosphotyrosine-containing proteins.

6. Explain and comment upon this statement: The principal function of hormone receptors is that of signal amplification.

Answer: In the case of a hormone receptor that functions enzymatically, the statement is true and the receptor tyrosine kinases are good examples. Hormone binding leads to simulation of protein kinase activity, which leads to phosphorylation of multiple protein targets. In G-protein-mediated pathways, hormone receptors serve the purpose of transmitting information from the surface to the cell interior. Amplification occurs at two stages. A hormone receptor with bound hormone can stimulate several G-proteins. The G-proteins stimulate adenylyl cyclase, leading to an increase in cAMP, stimulation of cAMP-dependent protein kinase, and phosphorylation of multiple targets. When hormone receptors function to stimulate transcription, the hormone signal is converted to a hormone-hormone receptor complex with a fixed stoichiometry. The complex then stimulates transcription and only at this stage is the signal amplified.

7. Synaptic vesicles are approximately 40 nm in outside diameter, and each vesicle contains about 10,000 acetylcholine molecules. Calculate the concentration of acetylcholine in a synaptic vesicle.

Answer: A synaptic vesicle with a 40 nm outside diameter has an inside diameter of approximately 36 nm. The volume occupied by one synaptic vesicle is given by:

$$V = \frac{4}{3}\pi r^3 = \frac{4}{3} \times 3.1416 \times \left(\frac{36 \times 10^{-9}\ \text{m}}{2}\right)^3 = 2.44 \times 10^{-23}\ \text{m}^3$$

$$V = 2.44 \times 10^{-23}\ \text{m}^3 \times \left(\frac{100\ \text{cm}}{1\ \text{m}}\right)^3 \times \left(\frac{1\ \text{L}}{1000\ \text{cm}^3}\right)$$

$$V = 2.44 \times 10^{-20}\ \text{L}$$

The concentration of acetylcholine in a vesicle is given by:

$$[\text{Acetylcholine}] = \frac{\dfrac{10{,}000\ \text{molecules}}{6.02 \times 10^{23}\ \text{molecules/mol}}}{2.44 \times 10^{-20}\ \text{L}} = 0.68\ \text{M}$$

8. GTPγS is a nonhydrolyzable analog of GTP. Experiments with giant squid synapses reveal that injection of GTPγS into the presynaptic end (terminal) of the neuron inhibits neurotransmitter release (slowly and irreversibly). The calcium signals produced by presynaptic action potentials and the number of synaptic vesicles docking on the presynaptic membrane are unchanged by GTPγS. Propose a model for neurotransmitter release that accounts for all of these observations.

Answer: The fact that the nonhydrolyzable analog of GTP leads to inhibition of neurotransmitter release implicates a G-protein. Since calcium release occurs normally and the number of synaptic vesicles that dock on the presynaptic membrane is unchanged by GTPγS, the defect may be in fusion of the synaptic vesicle to the membrane. A cAMP-dependent vesicle fusion step that is blocked by an inhibitory G-protein would account for the observations.

Questions for Self Study

1. Name three classes of chemical species that act as hormones.

2. List the three receptor superfamilies that mediate transmembrane signal processing and give a brief description how signal processing occurs.

3. What two enzymatic activities are responsible for regulating cAMP levels?

4. cAMP is an example of a second messenger. Name five other second messengers.

5. A large family of second messengers can be derived from phospholipase C activity on phosphatidylinositol and its derivatives. Explain.

6. Protein kinase C is sensitive to what two intercellular signals?

7. Membrane-bound guanylyl cyclases and soluble guanylyl cyclases are stimulated by very different signals. What are they?

8. Match a term with its definition.

a. Node of Ranvier	1. Connects to sensory receptor.
b. Synapse	2. Carries nerve impulses away from cell body.
c. Schwann cell	3. Moves nerve impulses to the cell body.
d. Interneuron	4. Insulating layer around axons.
e. Sensory neuron	5. Gap between an axon and a dentrite.
f. Dendrite	6. Moves signals from one neuron to another.
g. Axon	7. Gap between Schwann cells along an axon's length.

9. Answer True of False.

a. Typically, action potentials are induced by a hyperpolarization of the membrane voltage. _____

b. The resting potential of an axon is determined by the concentration gradients of all impermeable ions. _____

c. During an action potential, the membrane voltage changes from approximately -60 mV to +40 mV because of changes in sodium permeability. _____

d. Potassium permeability changes alone are responsible for returning the action potential back to resting values. _____

e. Action potentials attenuate with distance. _____

10. Fill in the blanks. In cholinergic synapses, small vesicles termed _____ are localized on the inside of the synaptic knob and contain large amounts of the neurotransmitter _____. When an action potential arrives at the synaptic knob, voltage-gated _____ channels open. This is followed by fusion of vesicles with the plasma membrane and release of neurotransmitter into the synaptic cleft. Neurotransmitter induces action potentials in postsynaptic cells by binding to receptors. There are two kinds of receptors _____ and _____ distinguished by there responses to a toxic alkaloid in toadstools or to nicotine. Neurotransmitter action is usually short lived because it is rapidly hydrolyzed by the enzyme _____.

11. Match

a. Endorphin	1. Binds to nicotinic receptors and blocks their opening.
b. Catecholamine	2. Insecticide that blocks muscarinic receptor.
c. Glycine	3. Excitatory amino acid transmitter.
d. Glutamate	4. Epinephrine.
e. Malathion	5. Peptide neurotransmitter.
f. d-tubocurarine	6. Inhibitory amino acid transmitter.

Answers

1. Steroids, amino acid derivatives, and peptides.

2. 7-Transmembrane segmented receptors are integral membrane proteins composed of seven helical segments, an extracellular hormone-binding domain, and an intracellular G-protein binding domain. GTP-binding proteins or G-proteins bind GTP and are released from the transmembrane receptor upon hormone binding. Depending on the type of G-protein, the G protein-GTP complex activates or inhibits adenylyl cyclase leading to changes in cyclic AMP levels. Single-transmembrane segment catalytic receptors are integral membrane proteins with an extracellular hormone receptor domain, a transmembrane segment, and an intracellular domain with either tyrosine kinase or guanylyl cyclase activity. Ligand gated oligomeric ion

channels open or close in response to ligand binding. Many such ion channels are regulated by neurotransmitters.

3. Adenylyl cyclase and phosphodiesterase.

4. cGMP, inositol phosphates, diacylglycerol (DAG), calcium, phosphatidic acid, ceramide, nitric oxide, cyclic ADP-ribose.

5. Phospholipase C activity produces diacylglycerol (DAG) and phosphoinositol. Depending on the phosphorylation state of phosphatidylinositol, inositol-1-P, inositol-1,4-P, or inositol-1,4,5-P may be released. Inositol-1,4,5-P leads to inositol-1,3,4,5-P, inositol-1,3,4-P, inositol-pentaphosphate, and inositol hexaphosphate. DAG may also be converted to phosphatidic acid.

6. Calcium and diacylglycerol.

7. Membrane-bound guanylyl cyclases are stimulated by a variety of peptide hormones. Soluble guanylyl cyclases are stimulated by nitric oxide.

8. a. 7; b. 5; c. 4; d. 6; e. 1; f. 3; g. 2.

9. a. F; b. F; c. T; d. F; e. F.

10. Synaptic vesicles; acetylcholine; calcium; muscarinic; nicotinic; acetylcholinesterase.

11. a. 5; b. 4; c. 6; d. 3; e. 2; f. 1.

Additional Problems

1. Caffeine (1,3,7-trimethylxanthine), theophylline (1,3-dimethylxanthine found in teas), and theobromine (3,7-dimethylxanthine found in chocolate) mimic the action of hormones that cause an increase in cAMP. Why?

2. Anabolic steroids have been used to increase athletic performance because these hormones lead to increase in mass and strength of muscle. However, their use for this purpose is banned in many competitions and random tests are often conducted to determine if an athlete has taken anabolic steroids. (Notable examples of athletes who have tested positive for drug use include Ben Johnson, Diego Maradona, and Hulk Hogan.) The tests are conducted on urine samples and steroid use can be detected even if active steroid use had been discontinued at some time prior to testing. From the properties of steroid hormones, can you suggest why synthetic steroids might have long biological half-lives?

3. Explain why the GTP analogs GMP-PNP and GMP-PCP are activators of adenylyl cyclase.

4. Draw the structure of inositol and explain why it can be metabolized into a large number of active compounds.

5. Based on the description of calmodulin target proteins and the Baa helix and referring to the wheel plot of Figure 6.53, suggest a possible function for melittin.

6. The cellular membrane potential is established by ion pumping mechanisms that are electrogenic. How can an electrical potential across a synthetic membrane be established in the absence of electrogenic pumps?

7. How does hyperpolarization lead to inhibition of action potentials?

8. Electrical recordings from an unstimulated neuromuscular junction reveal small, transient fluctuations in the membrane voltage termed miniature end-plate potentials. The magnitude of these end-plate potentials is reduced in size with curare. Further, their frequency is altered with changes in the membrane potential with hyperpolarization of the presynaptic membrane decreasing frequency and depolarization increasing frequency. Based on these observations, explain how miniature end-plate potentials arise.

9. In myelinated axons, sodium channels are clustered in the nodes of Ranvier where gaps in the myelin sheath leave the axon exposed. How does a nerve impulse propagate along a myelinated axon and is the spacing of nodes of Ranvier at all important for this process?

10. What mechanism insures that nerve stimuli move in one direction along a neuron?

Abbreviated Answers

1. These agents inhibit phosphodiesterase, the enzyme responsible for cAMP metabolism to AMP. With phosphodiesterase inhibited, levels of cAMP increase and remain high for prolonged periods of time.

2. Steroids are fat soluble substances that can partition into hydrophobic substances like fats and be retained there for long periods of time. Further, the steroid ring is very stable and only slowly metabolized by reduction to more water soluble forms that are excreted by the kidney.

3. These GTP analogs bind to the GTP binding site on G-proteins; however, because they lack the normal phosphate ester linkage they are not hydrolyzed, resulting in prolonged stimulation of adenylyl cyclase.

4. Inositol can be converted to a large number of phosphorylated derivatives that have different biological functions. The structure of inositol is shown below.

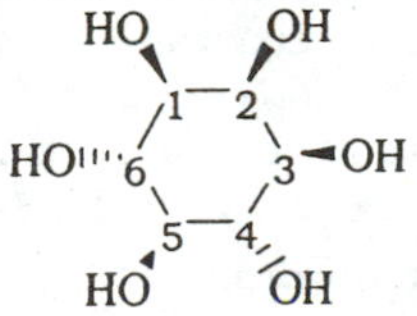

HO OH
1 2
HO 6 3 OH
5 4
HO OH

5. The α-helical wheel plot of melittin shows a helix with hydrophobic residues on one face and a collection of hydrophilic residues on another face with several basic amino acids at one end of the peptide. Melittin is a component of bee venom and shows hemolytic activity. In addition, it binds to calmodulin and can block calmodulin stimulation of calmodulin-binding proteins.

6. An electrical potential across a semipermeable membrane can be formed by placing a salt, composed or a permeable cation or anion and an impermeable counterion, on one side of the membrane. For example, consider a salt composed of a negatively-charged, large, impermeable protein and a small cation like sodium. Upon dissolving the salt in solution on one side of the membrane, the salt dissociates into the large protein anion and Na^+. Sodium freely diffuses across the membrane because it moves down its concentration gradient. But because the impermeable anion cannot follow, sodium exiting from the membrane produces a charge imbalance across the membrane creating a voltage gradient.

7. An action potential is initiated above a certain threshold voltage for sodium channels. This threshold is slightly less negative than the normal resting potential of the cell. Hyperpolarization drives the membrane to a more negative value and a greater voltage change is required to reach the threshold voltage.

8. Miniature end-plate potentials arise from random release of packets of acetylcholine as synaptic vesicles fuse with the presynaptic membrane and release their contents into the synaptic cleft. Curare blocks end-plate potentials by blocking acetylcholine receptors. Hyperpolarization lowers the frequency of their occurrence whereas depolarization increases it by altering the presynaptic calcium flux and thus reducing or increasing spontaneous synaptic vesicle release.

9. For an action potential to be propagated along an axon, a change in the membrane potential must be large enough to exceed the threshold voltage for sodium-channel opening. Once a voltage change due to an action potential occurs in a local region of the axon, it must be conducted to nearby regions of the axon to cause a depolarization of the membrane to a value above the threshold value. The myelin sheath greatly increases the transverse resistance,

causing current, induced by an action potential, to be carried down the axon. As the current moves down the axon it is attenuated; however, the myelin sheath reduces attenuation and the nodes of Ranvier are spaced such that the attenuated current is still large enough to cause a local depolarization of the membrane exceeding the threshold voltage. In effect, action potentials jump from node to node. Between nodes the stimulus is carried as a passive electrical current.

10. If an axon is stimulated electrically to initiate an action potential, the action potential will be conducted in both directions along an axon. However, the transmission of information across a synaptic cleft occurs only in one direction.

Summary

Hormones are chemical signal molecules that control and coordinate the many and diverse processes that occur in different parts of an organism. Steroid hormones regulate metabolism, salt and water balances, inflammatory processes and sexual function. Epinephrine and norepinephrine regulate smooth muscle contraction and relaxation, blood pressure, cardiac rate and other processes. Peptide hormones likewise regulate processes in all body tissues, as well as the release of other hormones. Hormones bind with very high affinity to their receptors.

Non-steroid hormones bind exclusively to outward-facing membrane receptors and activate signal transduction pathways which mobilize second messengers, including cyclic AMP, cyclic GMP, Ca^{2+}, inositol-1,4,5-trisphosphate, diacylglycerol and nitric oxide. The receptors that mediate transmembrane signaling include three receptor superfamilies, including 1) the 7-transmembrane segment receptors, which transmit their signals via GTP-binding proteins, 2) the single transmembrane segment catalytic receptors, which possess tyrosine kinase or guanylyl cyclase activity, and 3) oligomeric ion channels.

The first second messenger to be discovered was cAMP, produced by adenylyl cyclase and hydrolyzed by phosphodiesterase. The activity of cAMP is mediated by G proteins, which exchange GDP for GTP upon binding of hormone to the associated receptor protein. G_α-GTP dissociates from $G_{\beta\gamma}$ and binds to adenylyl cyclase to activate the synthesis of cAMP. Inhibitory G proteins act in a similar manner to inhibit adenylyl cyclase. The toxic effects of cholera toxin are due to ADP-ribosylation and consequently activation of G_s, whereas pertussis toxin ADP-ribosylates and inactivates G_i. The small G proteins, typified by the protein product of the *ras* gene, function in a manner similar to the heterotrimeric G proteins. Mutant forms of the *ras* protein have been implicated in the tumorigenic activity of certain tumor viruses.

The 7-transmembrane segment receptors, typified by the α- and β-adrenergic receptors, are postulated to possess 7 transmembrane α-helical segments, with a glycosylated, extracellular N-terminal segment and a cytoplasmic C-terminal domain that can be phosphorylated by receptor kinases such as β-adrenergic receptor kinase (BARK) and protein kinase A (a cAMP-dependent protein kinase).

A diverse array of second messengers is generated by breakdown of membrane phospholipids. Phospholipase C hydrolyzes phosphatidylinositol-4,5-bisphosphate to produce inositol-1,4,5-trisphosphate (IP_3) and diacylglycerol (DAG). IP_3 stimulates an increase in intracellular [Ca^{2+}], whereas DAG activates protein kinase C. Activation of phospholipase C enzymes is mediated by G proteins or by receptor tyrosine kinases. IP_3 has a half-life of only a few seconds. It is catabolized by phosphatases and/or kinases to produce a series of other inositol-phosphate metabolites which may themselves be second messengers in certain cellular processes. Second messengers may also be produced by breakdown of phosphatidylcholine. Action of phospholipase A_2 on PC produces arachidonic acid, which can be metabolized to eicosanoid compounds, and phospholipases C and D can produce DAG and phosphatidic acid, respectively. Sphingomyelinase action on sphingomyelin produces ceramide, which stimulates a protein kinase, and gangliosides and their breakdown products modulate the activity of protein kinases and G protein-coupled receptors.

Calcium ion is an important second messenger, which is released by action of IP_3 into the cytoplasm from intracellular stores including calciosomes and ER. Extracellular calcium can enter the cell via ligand gated, voltage gated, or second messenger regulated channels. Ca^{2+} binds to a series of calcium-binding proteins, such as calmodulin, which act to modulate enzyme and ion channel activity within the cell. Calmodulin target proteins possess a basic amphiphilic α-helix, to which calmodulin binds specifically and with high affinity. Ca^{2+} is also an activator of protein kinase C, which elicits a variety of cellular responses by phosphorylation of target proteins at Ser and Thr residues. Protein kinase C contains a psuedosubstrate sequence that masks the active site to inhibit PKC activity in the absence of DAG and Ca^{2+}. Many protein kinases and protein phosphatases are regulated by such intrasteric control. Protein phosphatases may either be specific for Ser- and Thr-phosphates or for Tyr-phosphates. The tumor promoting activity of okadaic acid, the major cause of diarrhetic shellfish poisoning, arises from its potent inhibition of protein phosphatases PP1 and PP2A.

The single transmembrane segment receptor proteins display either a tyrosine kinase or guanylyl cyclase activity. Additionally, soluble tyrosine kinases and guanylyl cyclases also participate in cellular signaling processes. Receptor tyrosine kinases are integral transmembrane proteins, whereas non-receptor tyrosine kinases, which are related to retroviral transforming proteins, are peripheral, lipid-anchored proteins. Receptor tyrosine kinases are membrane-associated allosteric enzymes, since hormone binding induces oligomerization of receptors in the membrane, activating phosphorylation of the cytoplasmic domains and stimulating tyrosine kinase activity by the receptors.

Membrane-bound guanylyl cyclases are single-TMS receptors, which synthesize cGMP in response to hormone binding. Peptides, which activate membrane-bound guanylyl cyclases, include atrial natriuretic peptides, enterotoxins and peptides such as speract and resact secreted by mammalian eggs, which stimulate sperm motility and act as sperm chemoattractants. Soluble guanylyl cyclases are receptors for nitric oxide, which is synthesized by NO· synthase. NO· acts both as a neurotransmitter and as a second messenger. NO· is unique among second messengers, because, as a dissolved gas, it is capable of rapid diffusion across cell membranes without the assistance of carriers. NO· has a half-life of 1-5 seconds and is degraded by nonenzymatic pathways. NO· activates guanylyl cyclase by binding to the heme prosthetic group to form a nitrosoheme.

Steroid hormones exert their effects in two ways: 1) in the nucleus, steroids act as transcription regulators, modulating gene expression. Steroids also can act at the cell membrane, directly regulating ligand-gated ion channels and other processes. Specialized receptor proteins carry steroids to the nucleus.

The nervous system is composed of neurons and neuroglia or glial cells. Schwann cells envelope neurons and form an insulating layer. Neurons contain three distinct regions: the cell body, which contains the nucleus, mitochondria, ribosomes, endoplasmic reticulum and other organelles; the axon, a long extension of the cell body whose primary function is to carry nerve impulses from the cell body to the cell periphery; and, dendrites, short, branched structures that carry impulses to the cell body from the periphery. Certain axons have myelin sheath derived from Schwann cells. The Schwann cells wrap around the axon to produce an insulating layer with periodic gaps called the nodes of Ranvier where the axon is exposed. The distal end of the axon ends in a synaptic knob or synaptic bulb.

There are three kinds of neurons: sensory neurons that transduce sensory signals into nervous signals; interneurons that pass signals from one neuron to another; and, motor neurons that pass signals to muscles. The signals carried by neurons are electrical signals that arise from changes in the electrical potential across the neuron membrane as a result of changes in permeability of the membrane to various ions. Nerve impulses are called action potentials and they arise when the membrane becomes permeable to specific ions. A resting cell maintains a membrane potential of about -60 mV (with the inside negative relative to the outside), due in large part to the Na^+-K^+ pump. When a cell is depolarized, the membrane potential is made more positive (the opposite, to hyperpolarize, is to change the membrane potential to a more negative value) by about 20 mV, voltage-gated ion channels specific for sodium are opened. Because sodium is in high concentration external to the cell and the inside of the cell is at a negative potential, sodium rushes into the cell causing the cell to depolarize. At this point, the sodium channels close and potassium specific voltage-gated channels open allowing potassium to rush out of the cell down its electrochemical gradient to hyperpolarize the cell to a value below the resting membrane potential. Eventually, potassium channels close and the voltage returns to the resting potential. Action potentials are propagated very rapidly, are not attenuated with distance, and show a fixed amplitude. Because the amplitude is fixed, the information content is the number and frequency of action potentials. The voltage-gated sodium channels are integral membrane proteins.

Transmission from one neuron to another involves information transfer across the synaptic cleft. There are two kinds of synapses, electrical synapses with a very small gap, approximately 2 nm, separating the presynaptic cell from the postsynaptic cell, and chemical synapses, with 20 nm to 50 nm gaps. Chemical substances called neurotransmitters are released by the presynaptic cell, diffuse across the synaptic cleft, and bind to receptors in the postsynaptic cell. Neurotransmitters include acetylcholine, various amino acids, catecholamines, peptides, and gaseous compounds. The cholinergic synapse employs acetylcholine as a neurotransmitter. The presynaptic cell contains synaptic vesicles filled with acetylcholine. A voltage-gated calcium channel in the presynaptic membrane opens in response to the arrival of an action potential and allows calcium to enter the cell. Acetylcholine release can be specifically blocked by several toxins produced by *Clostridium botulinum* the bacteria responsible for botulism poisoning. α-Latrotoxin, a black widow venom protein, stimulates acetylcholine release.

Acetylcholine released into the synaptic cleft binds to acetylcholine receptors in the postsynaptic membrane, leading to membrane depolarization, and initiation of an action potential in the postsynaptic cell. There are two classes of acetylcholine receptors that are distinguished by their responses to muscarine, a toxic toadstool alkaloid, and nicotine. Nicotinic receptors are K^+-Na^+ channels located in motor endplates of skeletal muscles. Nicotine binding causes the receptor to lock into the open conformation. The nicotinic acetylcholine receptors are ligand-gated ion channels that undergo a conformational change upon acetylcholine binding that results in the opening of a channel equally permeable to both K^+ and Na^+. The effects of sodium on the membrane voltage are greater than those of potassium and as a result the postsynaptic cell depolarizes and an action potential is initiated. The channel remains open for only a few milliseconds, then closes and as long as acetylcholine remains bound to the receptor it will not reopen. To resensitize the receptor, acetylcholine must be removed. This occurs when the concentration of acetylcholine in the synaptic cleft drops to below 10 nM in response to acetylcholinesterase-dependent degradation of acetylcholine. Repackaging of acetylcholine into synaptic vesicles occurs in the presynaptic cell. Empty synaptic vesicles are produced by endocytosis of the presynaptic membrane. A proton pump, a V-type ATPase, produces a proton gradient across the vesicle membrane by pumping protons into the vesicle and acetylcholine transport protein uses this gradient to concentrate acetylcholine in the vesicles. Agents that prevent nicotinic acetylcholine receptors from opening include d-tubocurarine, cobratoxin, and α-bungarotoxin.

The muscarinic receptors, located in smooth muscle and glands, are stimulated by muscarine. Binding of acetylcholine to muscarinic receptors results in 1) inhibition of adenylyl cyclase, 2) stimulation of phospholipase C, and 3) opening of potassium channels. All of these effects are mediated by G-proteins. As mentioned, muscarine stimulates these receptors. A potent antagonist is atropine from the deadly nightshade plant.

Organophosphorus compounds are potent inhibitors of acetylcholinesterase, a serine esterase with an active site serine that is a target for DIFP, malathion and parathion (commonly used insecticides), and sarin and tabun (nerve gases). Physostigmine and neostigmine are mild inhibitors of acetylcholinesterase that have been used in the treatment of myasthenia gravis.

The amino acids glutamate and aspartate are also excitatory neurotransmitters that are stored in presynaptic vesicles and released into the presynaptic cleft by calcium-dependent exocytosis. Synaptic vesicles are filled with glutamate in a manner similar to acetylcholine-filled vesicles. A proton pump creates a proton gradient that provides the driving force for glutamate accumulation. To remove glutamate from the synaptic cleft, glutamate is taken into glial cells and converted to glutamine. Glutamine is transported to the presynaptic neuron and recycled by the mitochondria back to glutamate. Glutamate receptors include N-methyl-D-aspartate (NMDA), kainate, and AMPA receptors, examples of ligand-gated ion channels, and metabotropic receptors, which are G-protein mediated receptors that are coupled to phosphotidylinositol metabolism. The NMDA receptor is the best characterized glutamate receptor. It functions as a Ca^{2+}, Na^+, and K^+ channel that is closed by Mg^{2+} in a voltage-dependent manner. The hallucinogenic drug, phencyclidine (PCP, a.k.a angel dust) is a specific antagonist of the NMDA receptor.

γ-Aminobutyric acid (GABA) and glycine are inhibitory neurotransmitters that function by hyperpolarizing the postsynaptic neuron. The receptors are ligand-gated chloride channels. GABA receptors are localized in the brain and opened by ethanol. Glycine receptors are found in the spinal cord and are blocked specifically by strychnine.

The catecholamine neurotransmitters include L-dopa, dopamine, norepinephrine, and epinephrine, all derived from the amino acid tyrosine. Norepinephrine is the neurotransmitter in junctions between sympathetic nerves and smooth muscle. Excessive production of dopamine or hypersensitivity of dopamine receptors is responsible for schizophrenia while lowered production of dopamine is found in Parkinson's disease.

Peptide neurotransmitters include endorphins and enkephalins, natural opioid substances that act as pain relievers, endothelins, regulatory peptides that act on smooth muscle and connective tissue, vasoactive intestinal peptide, and many more. Vasoactive intestinal peptide causes an increase in cAMP via a G-protein-mediated pathway.

Glossary

abzymes Catalytic antibodies elicited in an organism in response to immunological challenge by a foreign substance

accessory pigments Visible light-absorbing pigments in plants and photosynthetic bacteria that, along with chlorophylls, absorb sunlight.

action potential (nerve impulse) Transient changes in membrane potential that move rapidly along nerve cells.

active site The region of the enzyme that binds the substrate and catalytically transforms it.

active transport Energy-dependent transport of a substance across a membrane resulting in its accumulation against a concentration gradient.

acyl-carrier protein A protein that binds activated intermediates of fatty acid synthesis through a thioester linkage.

adipocytes (adipose cells) Animal cells where fats (triacylglycerols) are stored.

aerobes Organisms that use oxygen as an electron acceptor in energy-producing pathways.

alcaptonuria A genetic disease in which homogentisate collects in the urine as a result of a deficiency in homogentisate dioxygenase. The only malady suffered by carriers of this disease is a tendency toward arthritis in later life.

aldose A monosaccharide containing an aldehyde group.

alginates Polysaccharides that bind metal ions, particularly calcium.

alleles Alternative forms of a gene at a particular location or locus.

allosteric regulation The activation or inhibition of enzymatic activity through noncovalent interaction of the enzyme with small molecules other than the substrate.

α-helix A common secondary structure in proteins, which is almost always right handed.

amino acids Building blocks of proteins with a tetrahedral α-carbon covalently linked to an amino group and a carboxyl group.

ammonotelic organisms Aquatic animals that release free ammonia to surrounding water.

amphibolic pathway A metabolic pathway that can be both catabolic and anabolic.

amphipathic Possessing both polar and nonpolar groups.

amphiphilic Having both strongly polar and strongly nonpolar groups on a single molecule. Also known as amphipathic.

anabolism Metabolism in which complex biomolecules, such as carbohydrates, lipids and proteins, are generated from simpler precursors.

anaerobes Organisms that can subsist without O_2.

anaplerotic reaction A reaction that replenishes metabolites which are removed from a central metabolic pathway.

androgens Male sex hormone steroids.

antibody A substance capable of recognizing and binding a "foreign substance" known as an antigen.

anticodon The three-nucleotide unit in tRNA that recognizes and base pairs with a particular mRNA codon.

antigen A molecule specifically recognized and bound by an antibody.

antiparallel β-sheet A tertiary protein conformation in which the adjacent strands of a β-strand run in opposite directions. In these structures, hydrophobic residues are usually found on just one side of the sheet.

antiparallel α-helix proteins Proteins whose structures are heavily dominated by α-helices.

apoenzyme The protein portion of an enzyme devoid of its cofactors or prosthetic groups.

apoptosis Programmed cell death.

archea Along with eukaryotes and eubacteria, one of the three major groups into which living cells are classified.

ATP equivalent A metabolic unit of exchange; the conversion of ATP to ADP.

autogenous regulation of a gene (autoregulation of a gene) The regulation of expression of a gene by the product of the gene.

autotroph An organism that uses carbon dioxide as its sole source of carbon.

axoneme A complex bundle of microtubule fibers that include two central, separated microtubules surrounded by nine pairs of joined microtubules.

• •

bacteriophage (phage) Viruses that infect bacteria.

ballistic method A technique of transformation in which microprojectiles coated with DNA are fired into recipient cells by a gas-powered gun.

base pair Two nucleotides in nucleic acid chains that are paired because of hydrogen bonding between their bases, such as A with T and C with G.

β-bulge A structure in which one residue of a β-strand is not H-bonded but residues on either side of it are H-bonded to contiguous residues in the adjacent strand.

β-oxidation The oxidative degradation of a fatty acid into acetyl-CoA by repeated oxidations and cleavages at the β-carbon of a fatty acid.

β-pleated sheet A common secondary structure in proteins in which the polypeptide chain is arranged in a pleated fashion. Several such strands side-by-side are typically joined by H-bonds to form a sheet. Adjacent strands may be either parallel or antiparallel.

β-turn A simple structure observed in many proteins in which the peptide chain forms a tight loop with the carbonyl oxygen of one residue hydrogen-bonded with the amide proton of the residue three positions down the chain.

bifunctional enzyme (tandem enzyme) A protein possessing two distinct enzymatic functions.

bile acids Polar carboxylic acid derivatives of cholesterol that are important in the digestion of food, especially the solubilization of ingested fats.

binuclear center A closely-associated pair of metal ions.

Bohr effect When the oxygen affinity of hemoglobin in red blood cells decreases due to a lowering of the pH.

branch migration The procession of base-pairing between an invading DNA strand and one strand of a DNA duplex during recombination.

brown fat A specialized type of adipose tissue found in newborns and animals involved in heat production.

buffer A solution component that tends to resist changes in pH as acid or base is added. Typically, a buffer is composed of a weak acid and its conjugate base.

• •

caloric homeostasis The constant availability of fuels in the blood.

Calvin-Benson cycle (Calvin cycle) The set of reactions that transform 3-phosphoglycerate into hexose during photosynthesis.

carbohydrate A class of molecules with molecular formula $(CH_2O)_n$, where n is three or more.

carbon dioxide fixation The synthesis of carbohydrates from carbon dioxide and water.

catabolism Energy-yielding metabolism, which involves the oxidative degradation of complex nutrient molecules such as carbohydrates, lipids, and proteins.

catalytic power The ability of enzymes to accelerate reaction rates. The enhancement may be as much as 10^{20} over uncatalyzed levels.

cDNA library A DNA library constructed by synthesizing cDNA from purified cellular mRNA.

cellulose A structural polysaccharide and linear homopolymer of D-glucose units linked by β(1∏4) glycosidic bonds.

cerebroside A sphingolipid containing one sugar residue as the head group.

chemical mutagens Agents that chemically modify bases so that their base-pairing characteristics are altered.

chemotrophic organism An organism that obtains energy by metabolizing organic compounds.

chiral compound A compound with an asymmetric center, such that it occurs in two nonsuperimposible forms.

chitin A structural polysaccharide with repeating units of N-acetyl-D-glucosamine in β(1∏4) linkage.

cholesterol The most common steroid in animals and the precursor to all other animal steroids.

chloroplasts Chlorophyll-containing photosynthetic organelles, found in plants and algae, that harvest light and transform it into metabolically useful chemical forms.

chromatin A complex of DNA and protein in a eukaryotic cell nucleus.

chromatography Experimental techniques for the separation of mixtures. Methods include ion exchange chromatography, high-performance liquid chromatography and gas chromatography.

chromosome walking A technique for ordering DNA fragments in a genomic library. It involves hybridization, restriction mapping, and isolation of overlapping DNA molecules.

chromosome A single large DNA molecule, and its associated proteins, containing many genes. It stores and transmits genetic information.

cloning vector A plasmid used to carry a segment of a foreign DNA and to introduce it into a cell where it can be replicated.

cloning The amplification of identical DNA molecules or cells from a single parental DNA molecule or cell.

codon A sequence of three adjacent nucleotides in mRNA that codes for a specific amino acid.

coenzyme An organic molecule that helps an enzyme to carry out its catalytic function.

cofactor An inorganic ion or organic molecule required by an enzyme for catalytic activity.

coiled coil A bundle of α-helices wound into a superhelix, characterized by a repeat distance in the individual helices of 3.5 residues per turn.

collagen A fibrous protein that is a principal constituent of connective tissue in animals.

colligative properties Properties of solutions that depend only on the concentration of solute species and not on their nature.

competitive inhibitor A reversible inhibitor of an enzyme that competes with the substrate for binding at the active site of the enzyme.

complementarity-determining region (CDR) In an immunoglobulin, regions that form the structural site that is complementary to some part of an antigen's structure, providing the basis for the antibody:antigen recognition.

complementary DNA (cDNA) A DNA copy of a mRNA molecule.

configuration A spatial arrangement of atoms, which cannot be altered without breaking and reforming covalent bonds.

conformation A three-dimensional structure or spatial arrangement of a molecule that results from free rotation of substituent groups around single bonds.

conjugate acid/conjugate base The acidic and basic forms, respectively, of a dissociable compound.

constitutive expression The expression of genes independently of regulation.

constitutive splicing A type of splicing in which every intron is removed and every exon is incorporated into the mature RNA without exception.

converter enzyme (modifying enzyme) An enzyme that catalyzes the covalent modification of another enzyme, thereby changing its catalytic activity.

cooperativity A behavior in which the binding of one ligand to a protein influences the affinity of the protein for additional molecules of the same ligand. Cooperativity can be positive or negative.

corepressors A metabolite that depresses the synthesis of its own biosynthetic enzymes.

corticosteroids Steroid hormones formed by the adrenal cortex.

coupled reactions Two chemical reactions that have a common intermediate to transfer energy to one another.

covalent modification The reversible covalent attachment of a chemical group to an enzyme in order to regulate the activity of the enzyme.

cristae The folds of the inner mitochondrial membrane, which provide the inner membrane with a large surface area in a small volume.

critical micelle concentration A lipid-specific minimal lipid concentration at which micelles will form.

cruciform A cross-shaped tertiary structure of DNA that arises when inverted repeats form a structure that involves intrastrand base pairing.

cytochromes Proteins containing heme prosthetic groups; they serve as electron carriers in oxidation-reduction reactions such as respiration and photosynthesis.

• •

dark reactions of photosynthesis The reactions of photosynthesis that do not require light, notably carbon dioxide fixation.

debranching enzyme An enzyme that can degrade a branched polymer, such as a limit dextrin, by repositioning or removing branches.

degenerate codons (synonymous codons) Codons that specify the same amino acid.

denaturation The loss of structural order in a macromolecule; it is accompanied by loss of function.

denitrifying bacteria Bacteria capable of using NO_3^- and similar forms of oxidized inorganic nitrogen as electron acceptors in place of oxygen in energy-producing pathways.

density gradient ultracentrifugation A technique for separating macromolecules on the basis of their densities.

deoxyribonucleotide Nucleotides containing 2-deoxy-D-ribose as the pentose.

desensitization The loss of regulatory properties in an enzyme.

designer enzymes Synthetic enzymes tailored to carry out specific catalytic processes.

dextrorotatory isomer A stereoisomer that rotates the plane of plane-polarized light clockwise.

diabetes mellitus A common endocrine disease characterized by an abnormally high level of glucose in the blood.

dialysis Removal of small molecules and ions from macromolecules in solution by allowing them to pass through a semipermeable membrane.

diastereomer One of a pair of non-mirror image isomers that differ in configuration at only one of the asymmetric centers.

dielectric constant An index of the polarity of solvents, increasing with increasing polarity.

diploid cells Cells containing two sets of chromosomes.

docking protein (SRP receptor) A heterodimeric transmembrane protein that binds to the SRP-ribosome complex to stimulate the ribosome to resume polypeptide synthesis.

double-reciprocal plot (Lineweaver-Burk plot) A graph of the reciprocal of the initial velocity versus the reciprocal of the substrate concentration for an enzyme-catalyzed reaction. It allows an accurate determination of the K_m and the V_{max}.

• •

eicosanoids (prostaglandins, thromboxanes, and leukotrienes) Breakdown products of phospholipids. Eicosanoids are derived from 20-carbon polyunsaturated fatty acids used to synthesize prostaglandins, thromboxanes, and leukotrienes.

electrolytes Substances capable of generating ions in solution, thereby causing an increase in the electrical conductivity of the solution.

electron transport pathway A metabolic pathway in which electrons from the tricarboxylic acid cycle are passed through membrane-associated proteins to reduce O_2 to H_2O.

electroporation A transformation procedure in which the membranes of cells exposed to pulses of high voltage are rendered momentarily permeable to DNA molecules.

elongation factors Proteins required for the ribosome-catalyzed growth of a polypeptide chain.

elongation In protein synthesis, the successive addition of amino acids to a polypeptide chain by ribosomes.

Embden-Meyerhof pathway The steps of glycolysis.

enantiomer Mirror image, non-superimposible stereoisomers.

endergonic A process that has a positive ΔG value (energy absorbing).

energy charge An index of the capacity of the adenylate system (ATP/ADP/AMP system) to provide high-energy phosphoryl groups in order to drive thermodynamically unfavorable reactions.

energy transduction The transformation of energy from one form to another.

enhancer (upstream activation sequence; UAS) Nucleotide sequence involved in regulating the transcription of a gene.

enthalpy (H) A thermodynamic quantity, defined as the heat of a reaction at constant pressure.

entropy A measure of disorder and randomness in a system or its surroundings.

enzyme A protein (or RNA, ribozyme) that catalyzes a specific chemical reaction.

enzyme cascade A series of reactions in a signal transduction pathway in which one enzyme responds to a signal by altering the activity of another, which in turn alters the activity of another, etc., resulting in a vast amplification of a original signal.
enzyme kinetics The study of rates of enzyme-catalyzed reactions.

enzyme-substrate complex (ES) A complex formed when substrate molecule(s) bind noncovalently to the active site of an enzyme.

episomes Plasmids capable of chromosomal integration.

essential amino acids Amino acids that cannot be synthesized by the body and must be obtained from dietary sources.

estrogens Female sex hormone steroids.

eukaryote A single-celled or multi-celled organism whose cells have a membrane-bound nucleus, organelles and many chromosomes.

excision repair The replacement of damaged or modified bases in DNA by endonucleolytic cleavage.

exergonic A process with a negative ΔG value (energy releasing).

exon The region of a eukaryotic gene that codes for sequences joined during the process of splicing.

exon insertion The process whereby exons encoding particular functional or structural domains of proteins from one gene are inserted into another.

exon shuffling The exchange of exons between genes via recombination events taking place in introns.

expression vector A cloning vector that allows a cloned inserted DNA to be transcribed into RNA and, often translated into protein.

• •

facilitated diffusion Spontaneous diffusion of a substance across a membrane mediated by a carrier.

facultative anaerobes Organisms that can adapt to anaerobic conditions by substituting other electron acceptors for O_2 in their energy-producing pathways.

familial hypercholesterolemia A variety of inherited metabolic defects that lead to greatly elevated levels of serum cholesterol, much of it in the form of LDL particles.

fasting state A situation in which food intake by an organism is zero.

fermentation The anaerobic catabolism of organic molecules for the production of energy. In alcoholic fermentation, pyruvate is converted to ethanol and carbon dioxide.

filamentous phage particles Single-stranded DNA molecules packaged into long, thin tubes constructed from thousands of protein monomers.

first law of thermodynamics A principle stating that the total energy of an isolated system is conserved.

fitness The ability of an organism to survive and reproduce.

flavoprotein Protein conjugated with a flavin coenzyme.

flippase Protein that can flip phospholipids from one side of a lipid bilayer to the other.

fluid mosaic model A model proposed for the structure of biological membranes. In this model, membranes are dynamic structures composed of protein and phospholipid molecules which can rotate and move laterally.

freeze fracture electron microscopy A technique used to visualize the structure of biological membranes.

furanose A five-membered ring monosaccharide formed by intramolecular hemiacetal formation.

• •

galactosemia A disease in which toxic levels of galactose accumulate in the body, causing cataracts and permanent neurological disorders.

gene conversion A recombination process whereby two different sequences in a genome interact in such a way that one is converted to the other.

gene pool The sum of all the alleles of a particular gene.

gene sharing When the product of a single gene gains a second function without losing its primary function.

gene therapy The repair of a genetic deficiency by introduction of a functional version of the gene.

gene-replacement therapy A technique in which cells are removed from a patient, genetically manipulated and returned to the patient in order to correct a human genetic disorder.

general genetic recombination The process whereby DNA sequences are exchanged between homologous chromosomes, resulting in the arrangement of genes into new combinations.

general transcription factors Proteins required for transcription by RNA polymerase II.

genes The elements or units carrying and transferring inherited characteristics from parent to offspring. Genes are contained in chromosomes.

genetic recombination The formation of chromosomes with combinations of gene types different from those found in the parental chromosomes.

genomic library A DNA library prepared by isolating the total DNA from an organism, digesting it into fragments and cloning the fragments into an appropriate vector.

genotype The genetic make-up of an organism, as opposed to its phenotype or outward characteristics.

germ cells The sperm and the eggs of multicellular organisms.

Gibbs free energy (G) A thermodynamic state function relating enthalpy, entropy and temperature, $G = H - TS$.

globular protein A water-soluble protein that is roughly spherical in shape. It consists of one or more compactly-folded polypeptide chains.

glucogenic compound A compound that can be used for gluconeogenesis in animals. Most amino acids are glucogenic.

gluconeogenesis An anabolic pathway for the production of glucose from noncarbohydrate precursors.

glycerophosphate shuttle A process whereby two different glycerophosphate dehydrogenases, one in the cytoplasm and one in the outer face of the mitochondrial inner membrane, work together to deliver electrons to the mitochondrial electron transport system.

glycoconjugate A carbohydrate in which one or more sugar residues is covalently linked to a peptide chain, a lipid or a protein.

glycogen The major form of storage polysaccharide in animals.

glycolipid A carbohydrate covalently linked to a lipid molecule.

glycolysis The metabolic pathway responsible for the stepwise degradation of glucose and other simple sugars to pyruvate during which energy is released from the sugar and captured in the form of ATP under anaerobic conditions.

glycoprotein A protein containing a covalently linked carbohydrate.

glycosaminoglycans A class of polysaccharides consisting of linear chains of repeating disaccharides in which one of the monosaccharide units is an amino sugar and one (or both) of the monosaccharides contains at least one negatively charged sulfate or carboxylate group.

glyoxylate cycle A modification of the tricarboxylic acid cycle, used by plants and bacteria to produce four-carbon dicarboxylic acids (and eventually even sugars) from two-carbon acetate units.

glyoxysomes A plant peroxisome which contains the enzymes for the glyoxylate cycle.

grana Stacks of flattened vesicles formed from the thylakoid membrane in chloroplasts.

gratuitous inducers Nonmetabolizable substrate analogs capable of activating the synthesis of enzymes.

group transfer potential The measure of the ability of a compound to donate a functional group to a specific receptor molecule or to water.

growth medium The nutrients needed for a microorganism to grow in a laboratory.

• •

haploid cells Cells containing only one set of chromosomes.

helix-turn-helix A description of a structural motif found in DNA-binding proteins.

heme The iron-porphyrin prosthetic group of heme proteins.

hemiacetal The product formed when an alcohol reacts with an aldehyde.

hemiketal The product formed when an alcohol reacts with a ketone.

hemoproteins Heme-containing proteins.

Henderson-Hasselbach equation An equation relating the pH, the pKa and the concentrations of the proton-acceptor [A^-] and proton-donor species [HA] in a solution.

heredity Loosely defined as the tendency of an organism to possess the characteristics of its parent(s).

heterogeneous nuclear RNA (hnRNA) A large mRNA precursor with stretches of nucleotide sequence that have no protein-coding capacity.

heterotroph An organism that requires an organic form of carbon, such as glucose, in order to generate metabolic energy and to synthesize other essential carbon compounds.

high performance liquid chromatography (HPLC) A chromatographic technique for separating components of a mixture by forcing the liquid mixture though a column under high pressure.

high-density lipoprotein (HDL) A plasma lipoprotein that contains a significant amount of protein and that transports cholesterol and cholesterol esters from tissues to the liver.

histones A class of arginine- and lysine-rich basic proteins that interacts ionically with anionic phosphate groups in the DNA backbone to form nucleosomes

HIV-1 Human immunodeficiency retrovirus that causes AIDS.

Holliday junction A cross-strand intermediate in genetic recombination in which two double-stranded DNA molecules are joined by the reciprocal crossover involving one strand of each molecule.

holoenzyme A catalytically-active complex of protein and cofactor.

homeobox domain A DNA motif that encodes a related 60-amino acid sequence found in a number of proteins that act as a sequence specific transcriptional factor.

homeostasis The maintenance by cells of a relatively constant internal environment.

homologous genetic recombination The recombination of two DNA molecules of similar sequences.

homologous proteins Proteins from different species that have similar sequences and functions.

hydration A state in which a molecule or ion is surrounded by water molecules.

hydrogen bonds Weak chemical forces between a hydrogen atom covalently bonded to an electronegative atom such as oxygen or nitrogen and a second electronegative atom that serves as the hydrogen bond acceptor.

hydrophobic interactions The association of nonpolar groups or molecules with each other in aqueous environments.

hydroxyurea A free-radical quenching agent and inhibitor of ribonucleotide reductase.

hyperuricemia Chronic elevation of blood uric acid levels, which occurs in about 3% of the population as a consequence of impaired excretion of uric acid or overproduction of purines.

• •

i

illegitimate recombination A rare recombination which occurs between nonhomologous DNA independently of any unique sequence element.

immunoglobulin A major class of antibody molecules found circulating in the bloodstream.

indirect readout The ability of a protein to indirectly recognize a particular nucleotide sequence by recognizing local conformational variations in double-helical DNA resulting from the effects that base sequences have on DNA structure.

inducers Substrates capable of activating the synthesis of enzymes that metabolize them.

induction of enzyme synthesis An increase in enzyme synthesis by the activation of transcription of the gene encoding the enzyme.

inhibitor A compound that decreases the velocity of an enzymatic reaction.

initiation complex A complex consisting of 70S ribosome with GTP, N-formyl-Met-$tRNA_f^{Met}$, and mRNA ready for the elongation steps in protein synthesis.

initiation factors (IFs) Proteins necessary for the initiation of protein synthesis.

initiation In protein synthesis, the first phase that primes the ribosome for polypeptide formation.

insulin Polypeptide hormone of the β-cells of the islets of Langerhans in the pancreas of all vertebrates that regulates carbohydrate metabolism.

integral membrane protein (intrinsic membrane protein) A membrane protein with hydrophobic surfaces that penetrate or extend all the way across the lipid bilayer.

intercalation The insertion of a molecule between stacked structural elements such as the insertion of a planar molecule between two successive bases in a nucleic acid.

intermembrane space The space between an inner and outer membrane.

international unit (IU) One IU is the amount of enzyme that catalyzes the formation of one micromole of product in one minute at a specified pH, temperature and ionic strength.

intron (intervening sequence) A sequence of nucleotides in a gene that is transcribed but is not found in the mature mRNA.

ion-exchange chromatography A method for separating substances on the basis on their charge.

ionic bonds Weak bonds that are the result of attractive forces between oppositely charged polar functions, such as negative carboxyl groups and positive amino groups.

iron-sulfur proteins Proteins that contain one of various iron-sulfur clusters as prosthetic groups; they participate in oxidation-reduction reactions.

islets of Langerhans Cells in the pancreas that secrete insulin.

isoacceptor tRNAs The set of tRNAs specific for a particular amino acid.

isoelectric focusing An electrophoretic technique for separating proteins according to their isoelectric point or pI.

isozyme An enzyme that exists in more than one quaternary form, differing in their catalytic activities.

• •

katal One katal is the amount of enzyme catalyzing the conversion of one mole of substrate to product per second. One katal equals 6 x 10^7 international units.

keratins Insoluble structural proteins with parallel polypeptide chains in the α-helical or β-sheet conformations.

ketogenesis The synthesis of acetone, acetoacetate and β-hydroxybutyrate generated from acetyl-CoA in the oxidation of fatty acids.

ketogenic compounds A compound such as an amino acid that can be degraded to yield acetyl-CoA, thereby contributing to the synthesis of fatty acids or ketone bodies.

ketone bodies Fuel molecules (acetone, acetoacetate and β-hydroxybutyrate), synthesized in the liver from acetyl-CoA. During starvation and in untreated diabetes mellitus, ketone bodies become a major source of fuel.

ketose A monosaccharide containing a ketone group.

kinase A class of enzymes that phosphorylate target molecules.

Klenow fragment The larger fragment of DNA polymerase I which contains the polymerase and 3'-exonuclease activity. A Klenow fragment is formed when DNA polymerase I is cleaved by limited proteolysis.

Krebs cycle The tricarboxylic acid cycle.

• •

L

lactose intolerance A metabolic disorder caused by the inability to digest lactose due to the absence of the enzyme lactase in the intestine.

lamellae The paired folds of the thylakoid membrane of chloroplasts.

lariat A branched, covalently closed loop of RNA formed as an intermediate during RNA splicing.

leader peptide An N-terminal sequence of amino acids in a polypeptide.

Leloir pathway The catabolism of galactose.

lethal synthesis The metabolic transformation of an otherwise innocuous compound into a poisonous derivative.

leucine zipper A dimerization structural motif found in DNA-binding proteins.

leukotrienes A class of molecules derived from arachidonic acid; they stimulate contractions in vascular, respiratory, and intestinal smooth muscle.

levorotatory isomer A stereoisomer that rotates the plane of plane-polarized light counterclockwise.

ligand A molecule that is bound to another, usually larger molecule.

light reactions of photosynthesis The reactions of photosynthesis that require light and cannot occur in the dark.

limit dextrans Highly-branched oligosaccharides that are left over after extensive removal of glucose units from starch or glycogen by α-amylase, glycogen phosphorylase or starch phosphorylase.

Lineweaver-Burke plot A linear transformation of the Michaelis-Menten equation in which $1/v_0$ is plotted as a function of $1/[S]$.

linking number (L) The basic parameter characterizing supercoiled DNA. The number of times the two strands of DNA are intertwined.

lipid bilayer A back-to-back arrangement of lipid monolayers in which the hydrophobic tails aggregate in the interior of the bilayer and the polar head groups face outward into the aqueous environment.

lipid A class of biological molecules defined by low solubility in water and high solubility in nonpolar solvents.

lipopolysaccharide A lipid group joined to a polysaccharide made up of long chains with many different and characteristic repeating structures.

lipoproteins Proteins conjugated with lipids.

liposome A closed, spherical structure formed from a single phospholipid bilayer, which can be used as a drug and enzyme delivery system in therapeutic applications.

locus The chromosomal location of a gene.

logic A system of reasoning, using principles of valid inference.

low-density lipoproteins (LDLs) Plasma lipoproteins, with a low protein-to-lipid ratio.

lymphocytes White blood cells involved in the immune response. B lymphocytes, from bone marrow, synthesize and secrete antibodies. T lymphocytes, from the thymus gland, play a regulatory role in the immune response or act as a killer of foreign and virus-infected cells.

• •

macromolecules Proteins, polysaccharides, polynucleotides (DNA and RNA), and certain lipids.

major groove The larger of the two grooves created on the surface when DNA forms a double helix.

malate-aspartate shuttle An electron-transport shuttle system in which malate and aspartate are carried across the inner mitochondrial membrane.

maple syrup urine disease A heredity defect in the oxidative decarboxylation of branched chain α-keto acids leading to elevated levels of valine, leucine and isoleucine in blood and urine. The urine of these individuals smells like maple syrup; the disease is fatal unless dietary intake of these amino acids is restricted in early life.

matrix The space inside the inner mitochondrial membrane.

messenger RNA (mRNA) A ribonucleic acid, which serves to carry the information or "message" that is encoded in genes to the sites of protein synthesis in the cell where this information is translated into a polypeptide.

metabolic channeling A process whereby the product of an enzymatic reaction in a pathway is delivered directly to the next enzyme in the pathway for which it serves as the substrate.

metabolism The sum of the chemical changes that convert nutrients into energy and finished products in an organism.

metabolites Simple organic compounds that are substrates, intermediates or products in cellular energy transformation and in the biosynthesis or degradation of biological molecules, such as amino acids, sugars, fatty acids, and nucleotides.

metalloenzyme An enzyme that binds a metal tightly or that requires a metal ion to maintain activity or stability.

metalloproteins Proteins conjugated with metals.

micelle A structure formed by amphipathic molecules in aqueous solution in which the hydrophobic portions aggregate in the interior of the structure and the hydrophilic portions of the molecules project into the aqueous environment.

Michaelis constant (K_m) The concentration of substrate at which an enzyme-catalyzed reaction proceeds at one-half its maximum velocity.

Michaelis-Menten equation A rate equation relating the initial velocity (v_o) of an enzymatic reaction to the substrate concentration [S]. $v_o = V_{max}[S]/(Km + [S])$ where V_{max} is the maximum velocity and K_m is the Michaelis constant.

minor groove The smaller of the two grooves created on the surface when DNA forms a double helix.

Mitchell's chemiosmotic hypothesis A proposal stating that the energy stored in a proton gradient across a membrane by electron transport drives the synthesis of ATP in cells.

mitochondria The power plants of cells, which carry out the energy-releasing aerobic metabolism of carbohydrates and fatty acids with the concomitant capture of energy in metabolically useful forms such as ATP.

modulator proteins Proteins that bind to enzymes and by binding influence the activity of the enzyme.

molecular chaperone A family of proteins which, among other things, function to guide protein folding.

monolayer An arrangement of lipids at an air/water interface in which the hydrophilic lipid head groups are in the water phase and the fatty acids point up into the air.

monosaccharide A simple sugar of three or more carbon atoms with the formula $(CH_2O)_n$.

mosaic proteins Proteins composed of different structural motifs. Each such motif is found in other proteins.

multifunctional polypeptides Single polypeptide chains having two or more enzymatic centers.

mutant An organism with a change in its genetic information.

• •

natural selection The differential reproduction of genetically distinct individuals within a population.

neo-Darwinian theory of evolution The view that natural selection determines the frequency of alleles in a population and ultimately, the genetic makeup of populations.

neurotransmitters Compounds that are released into the synaptic cleft, thereby propagating the transmission of an action from one neuron to another.

neutral theory of molecular evolution A postulate stating that the majority of molecular changes in evolution are due to the random fixation of neutral mutations within the population.

nitrate assimilation A two-step metabolic pathway, where NO_3^- is reduced to NH_4^+ in green plants, various fungi and certain bacteria.

nitrifying bacteria A group of chemoautotrophs, which oxidize NH_4^+ to NO_3^-.

nitrogen fixation The formation of NH_4^+ from N_2 gas; this reduction occurs only in certain prokaryotic cells.

nondegenerate site A nucleotide site in which all changes at the site are nonsynonymous.

nonessential amino acids Amino acids that can be synthesized by an organism and thus are not required in the diet.

nonsense codons Codons that do not specify any amino acids. They serve as termination codons.

nonsense suppressors Mutations in tRNA genes that alter the tRNA so that the mutant tRNA can now read a particular stop codon and insert an amino acid.

nuclear magnetic resonance (NMR) A spectroscopic technique used to study the structures of molecules in solution. In NMR, the absorption of electromagnetic energy by molecules in magnetic fields provides information about molecular structure and dynamics.

nucleases Enzymes that hydrolyze nucleic acids.

nucleic acid Linear polymers of nucleotides linked in a 3' to 5' fashion by phosphodiester bridges.

nucleoproteins Proteins conjugated with nucleic acids.

nucleoside A compound formed by the linkage of pentose sugar to a purine or pyrimidine base.

nucleosomes Structures in which the DNA double helix is wound around a protein core composed of pairs of four different histone polypeptides.

nucleotide A nucleoside with a phosphoric acid esterified to a sugar hydroxyl group.

nucleus The organelle that is the repository of genetic information in the form of linear sequences of nucleotides in the DNA of eukaryotes.

• •

Obligate aerobes Organisms, like humans, for which O_2 is essential to sustain life

obligate anaerobes Organisms that cannot use O_2 at all and are even poisoned by it.

Okazaki fragments Short ssDNA chains of about 1000 residues in length, which are formed during the discontinuous synthesis of the lagging strand of DNA.

oligopeptides Peptide chains of more than 12 and less than 20 amino acid residues.

oligosaccharide A polymer of approximately 2 to 10 monosaccharide units linked by glycosidic bonds.

oncogenes Genes implicated in tumor growth.

operon hypothesis A theory which accounts for the coordinate regulation of related metabolic enzymes.

operons A cluster of genes encoding the enzymes of a particular metabolic pathway, along with the regulatory sequences that control their transcription. Inducible operons are expressed only in the absence of their corepressors or only in the presence of small molecule inducers.

optical activity The ability of a substance to rotate the plane of plane-polarized light.

organelles Membrane-bound structures found in eukaryotic cells; they perform specialized cell functions. Examples include nucleus, mitochondria, chloroplasts, endoplasmic reticulum, Golgi apparatus, vacuoles, peroxisomes, lysosomes, and chromoplasts.

origin(s) of replication A DNA sequence site for the beginning of DNA replication.

oxidative phosphorylation The enzymatic phosphorylation of ADP to ATP coupled to the transfer of electrons from a substrate to O_2.

oxygen-evolving complex The PSII complex of photosynthesis which is responsible for the photolysis of water.

• •

P/O ratio The ratio of the number of molecules of ATP synthesized to the number of atoms of oxygen reduced in oxidative phosphorylation.

palindromes (inverted repeats) A segment of duplex DNA in which the base sequences of the two strands exhibit two-fold rotational symmetry.

passive diffusion Unassisted spontaneous diffusion of a substance across a membrane.

patch recombinants Recombinant heteroduplexes that are formed by (-) strand cleavage at a Holliday junction.

pentose phosphate pathway (hexose monophosphate shunt; phosphogluconate pathway) A metabolic pathway that interconverts pentoses and hexoses and is a source of NADPH.

peptide bond An amide bond between amino acids.

peptidoglycan A strong peptide-polysaccharide layer contributing to the cell wall surrounding a bacterium.

peripheral membrane protein (extrinsic membrane protein) A membrane protein which does not penetrate the bilayer to any significant extent but which is associated with the membrane by ionic interactions and hydrogen bonds.

peroxisomes Eukaryotic, cytoplasmic, membrane-bound organelles which carry out a variety of flavin-dependent oxidation reactions, regenerating oxidized flavins by reaction with oxygen to produce hydrogen peroxide, H_2O_2.

phenotype The observable characteristics of an organism.

phenylketonuria A genetic disease in which phenylpyruvate collects in the urine, causing severe mental retardation unless a newborn is placed on a diet low in phenylalanine.

phospholipids A class of lipids, each member of which contains a phosphate moiety.

phosphoproteins Proteins that have phosphate groups esterified to the hydroxyls of serine, threonine, or tyrosine residues.

photophosphorylation Light-driven ATP synthesis from ADP and P_i catalyzed by ATP synthase.

photorespiration The light-dependent uptake of O_2 accompanied by the release of CO_2 and metabolism of phosphoglycolate that occurs primarily in C3 photosynthetic plants.

photosynthesis The use of light energy to produce ATP and NADPH. These energy-rich substances can drive carbohydrate synthesis from carbon dioxide and water.

photosystem (photosynthetic unit) In photosynthetic cells, a membrane-bound reaction center with an antenna of several hundred light-harvesting chlorophyll molecules.

phototrophic organisms Organisms that grow by transforming light energy into chemical energy

phylogeny The origin and evolution of the many types and species of organisms.

pitch The axial distance required to complete one turn of a helix.

plasma membrane The outer membrane surrounding cells and delineating their cytoplasm from the external environment.

plasmalogens Ether glycerophospholipids in which the alkyl moiety is *cis*-α,β-unsaturated.

plasmids Circular, extrachromosomal DNA.

plastids Self-replicating organelles in plants.

platelet activating factor The compound, 1-alkyl-2-acetylglycerophosphocholine, which has the ability to dilate blood vessels in order to reduce blood pressure in hypertensive animals and to aggregate platelets.

point mutations A class of mutations in which one base pair is substituted for another. Two possible types are transitions (a purine-purine or pyrimidine-pyrimidine replacement) and transversions (a purine-pyrimidine or pyrimidine-purine replacement).

polylinker A short region of DNA sequence bearing numerous restriction sites.

polymerase chain reaction (PCR) A repetitive polymerization technique for dramatically amplifying the amount of a specific DNA segment.

polypeptide A long chain of amino acids linked by amide bonds.

polyribosomes (polysomes) Multiple ribosomes attached to mRNA.

polysaccharide A polymer of many monosaccharide units linked by glycosidic bonds. These polymers can be either linear or branched.

post-translational modification Enzymatic processing of a polypeptide chain after genetic information from DNA has been translated into newly formed protein.

Pribnow box A nucleotide sequence involved in the initiation of transcription of prokaryotic genes.

primary structure The amino acid sequence of a protein.

primer A short nucleotide oligomer to which a polymerase adds monomers.

processing Alterations that convert a newly synthesized RNA into mature messenger RNA.

processivity In DNA synthesis, the degree to which an enzyme remains associated with the template through successive cycles of nucleotide addition.

prokaryotes Single-celled organisms that lack nuclei and other organelles.

promoter A nucleotide sequence at which transcription initiation is regulated and initiated.

proproteins Larger inactive protein precursors that are activated through proteolysis.

prostaglandins Cyclopentanoic acids derived from arachidonic acid and other polyunsaturated fatty acids.

prosthetic group A molecule that is tightly bound to an enzyme.

protein isoforms A set of related polypeptides derived from a common gene by differential RNA splicing.

protein module A tertiary structural motif or domain that may exist in several different proteins or that may occur two or more times in the same protein.

protein translocation The targeting of proteins to their proper destinations in cells.

protein A molecule composed of one or more polypeptide chains, each with a characteristic sequence of amino acids linked by peptide bonds.

proteoglycans A family of glycoproteins whose carbohydrate moieties are predominately glycosaminoglycans.

proteosome A large oligomeric structure inclosing a central cavity in which degradation of proteins takes place.

proto-oncogene Non-cancerous genes involved in cell proliferation, mutations in which often convert them into oncogenes.

purines A structure containing a pyrimidine ring fused to a five-membered imidazole ring.

pyranose A six-membered ring monosaccharide formed by intramolecular hemiacetal formation.

pyrimidine A six-membered heterocyclic aromatic ring containing two nitrogen atoms.

• •

Quaternary structure The three-dimensional organization of two or more polypeptide chains in a multisubunit protein.

• •

R state The active conformation of an allosteric enzyme.

Ramachandran plot A representation of the possible secondary structural elements obtained by plotting the degrees of rotation of the φ and ϕ bonds linking amino acids α carbons joined into a polypeptide.

random genetic drift A random change in allelic frequency.

rate of nucleotide substitution The number of nucleotide substitutions per site per year.

rate of product formation In an enzyme-catalyzed reaction, the rate of the formation of product with time: $v = d[P]/dt$.

rate-limiting step The slowest step in a chemical reaction. This step has the highest activation energy among the steps leading to the formation of a product from a reactant.

reaction center A complex of a pair of photochemically-reactive chlorophyll a molecules that forms the core of a photosystem. The reaction center is the site of conversion of photochemical energy into electrochemical energy during photosynthesis.

recombinant DNA technology (genetic engineering) The process of isolating, manipulating and cloning a sequence of DNA.

recombinant DNA A DNA molecule that includes DNA from different sources.

recombinant plasmids Hybrid DNA molecules consisting of plasmid DNA sequences plus inserted DNA elements (called inserts). Also known as chimeric plasmids.

redox couple The oxidized and reduced forms of a substance.

reducing sugar A carbohydrate whose free anomeric OH group can reduce oxidizing agents.

Refsum's disease An inherited metabolic disorder that results in defective night vision, tremors, and other neurological abnormalities, caused by an accumulation of phytanic acid in the body.

regulatory proteins Proteins that do not perform any obvious chemical transformation, but can regulate the ability of other proteins to carry out their physiological functions.

release factors Proteins that promote polypeptide release from the ribosome.

replication forks During DNA replication, it is the Y-shaped junction where the double-stranded DNA template is unwound and the new DNA strands are synthesized.

replicons The units of DNA replication.

repression A decrease in protein synthesis arising from cessation of transcription of the gene encoding the protein.

response elements Promoter modules in genes that serve as binding sites for proteins that activate transcription.

restriction endonucleases Enzymes, isolated chiefly from bacteria, that have the ability to cleave double-stranded DNA.

retroviruses A class of eukaryotic viruses that has single-stranded RNA genomes that replicate through a double-stranded DNA intermediate.

reverse transcriptase An enzyme that synthesizes DNA using RNA as a template.

ribonucleotide A nucleotide containing D-ribose as its pentose.

ribosomal RNA (rRNA) Ribonucleic acid, which serves as a component of ribosomes.

ribosomes Compact ribonucleoprotein particles responsible for protein synthesis and found in the cytosol of all cells, as well as in the matrix of mitochondria and the stroma of chloroplasts.

ribozymes Catalytic RNAs.

• •

Sacromere The basic structural unit of muscle contraction.

saponification The process of hydrolysis of acylglycerols with alkali to yield glycerol and salts of free fatty acids.

saturated fatty acid A fatty acid that does not contain a carbon-carbon double bond.

second law of thermodynamics A law stating in part that the total entropy of the universe always increases in a spontaneous process.

second messenger An intracellular agent synthesized in response to an external signal or first messenger, such as a hormone.

secondary structure The arrangement in space of atoms in the backbone of a polypeptide chain or a nucleic acid.

semiconservative model for DNA replication The process for duplicating DNA in which the nucleotide sequence in one strand dictates the sequence in the other, complementary strand, resulting in two daughter molecules of double-stranded DNA, each of which contains one of the parent strands.

sequence alignments A comparison of two sequences by juxtaposing two nucleotide or amino acid sequences and analyzing their similarities. Two methods are the dot-matrix method and the sequence-distance method.

sequential model for allosteric behavior A model for the cooperative binding of identical ligands to an oligomeric protein.

serine proteases A class of proteolytic enzymes whose catalytic mechanisms are based on an active site serine residue.

severe combined immunodeficiency syndrome (SCID) A group of related inherited disorders characterized by the lack of an immune response to infectious disease.

shikimate pathway A metabolic pathway leading to the formation of chlorismate, a key intermediate in the synthesis of aromatic amino acids.

Shine-Dalgarno sequence A mRNA sequence, rich in purine, which is required for mRNA binding to prokaryotic ribosomes.

shuttle vectors Plasmids capable of propagating and transferring genes between two different organisms, one of which is typically a prokaryote and the other a eukaryote.

sickle-cell anemia A human disease characterized by crescent-shaped red blood cells.

signal recognition particle (SRP) A nucleoprotein assembly that binds to the signal sequence to halt further protein synthesis by the ribosome.

signal sequence (signal peptide) The N-terminal sequence of residues in a newly synthesized polypeptide that targets the protein for translocation across a membrane.

signal transduction pathway The sequence of reactions linking the perception of a signal with a cellular response to that signal.

single-stranded assimilation (single-stranded uptake) A process driven by branch migration, which displaces the homologous DNA strand from the DNA duplex and replaces it with the invading ssDNA strand.

site-specific genetic recombination Genetic recombination occurring only at specific sequences.

small nuclear RNA (snRNA) A class of RNA molecules found only in eukaryotic cells.

snRNPs ("snurps") Small nuclear ribonucleoprotein particles.

solvent capacity The capacity of the cell to keep all of its essential metabolites and macromolecules in an appropriate state of solvation.

somatic cells All body cells except for the germ cells (sperm and eggs).

SOS response A system that converts a lesion in DNA to an error-prone site and restores DNA replication.

specific activity The number of micromoles of a substrate transformed by an enzyme per minute per milligram of protein at 25°C; it is a measure of enzyme activity and purity.

specificity The ability of an enzyme to discriminate among competing substrates.

spectroscopic methods Techniques which measure the absorption and emission of energy of different frequencies by molecules and atoms.

sphingolipids A class of lipids that contains an 18-carbon amino alcohol backbone, called a sphingosine, joined to a fatty acid.

splice recombinants Recombinant heteroduplexes that are formed by (+) strand cleavage at a Holliday junction.

spliceosome A multicomponent complex formed from the association of various snRNPs with pre-RNA.

standard reduction potential A quantity that indicates the tendency of a chemical species to donate or accept electrons.

starch A storage polysaccharide in plants.

steady state A state in which the rate of disappearance of a compound is equal to its rate of synthesis. That is, the change in concentration of the compound with time is equal to zero.

steady state (cellular) A state of apparent constancy, which is actually very dynamic. In this state, energy and material are consumed by the cell and used to maintain the harmonious stability and order of the cell.

steroid A lipid containing 18 or more carbon atoms and a fused four-ring polyprenyl structure.

stoichiometry The measurement of the amounts of chemical elements involved in chemical reactions.

stomata Microscopic pores on a leaf through which carbon dioxide diffuses directly into photosynthetic cells.

stop-transfer sequence A hydrophobic 20-residue peptide that stops the passage of a growing polypeptide chain through the ER membrane.

storage proteins A class of proteins whose biological function is to provide a reservoir of an essential nutrient.

stroma The soluble portion of a chloroplast.

structural complementarity Having a molecular surface with chemical groups arranged to specifically interact with complementary chemical groups on another molecule.

substrate-level phosphorylation Phosphorylation of a nucleoside diphosphate to a nucleoside triphosphate by transfer of a phosphoryl group from a non-nucleoside substrate.

substrate The substance upon which an enzyme acts.

suicide substrate A substrate analog that is transformed by an enzyme into a substance that irreversibly inactivates the enzyme.

supercoil Underwound (negative supercoil) or overwound (positive supercoil) double-stranded DNA.

supramolecular complexes The combination of various members of one or more classes of macromolecules. Examples include multifunctional enzyme complexes, ribosomes, chromosomes, and cytoskeletal elements.

synapsis A process of chromosome pairing in which two homologous DNA duplexes are juxtaposed so that their sequences are aligned.

synonymous codons Different codons that specify the same amino acid.

system That portion of the universe with which we are concerned; the rest of the universe is called the surroundings. Systems may be closed, isolated, or open to the surroundings.

• •

T state The less active conformation of an allosteric enzyme.

template A strand of DNA or RNA whose sequence of nucleotide residues acts as a pattern for the synthesis of a complementary strand.

termination The final step in a process such as protein synthesis.

terpenes A class of lipids formed from combinations of two or more molecules of 2-methyl-1,3-butadiene, also known as isoprene.

tertiary structure The compact three-dimensional folded shape of a polymer.

thermodynamics A collection of laws and principles describing the flows and interchanges of heat, energy and matter in systems of interest.

third law of thermodynamics A law which states that entropy of any crystalline, perfectly ordered substance must approach zero as the temperature approaches 0 K, and at T = 0 K, entropy is exactly zero.

third-base degeneracy The irrelevance of the third base in a codon.

thromboxanes Molecules derived from arachidonic acid which are involved in platelet aggregation during blood clotting.

thylakoid membrane The inner-membrane system of chloroplasts, which is organized into paired folds that extend throughout the organelle. It is the site of light-dependent reactions of photosynthesis, leading to the formation of NADPH and ATP.

thylakoid space (thylakoid lumen) The interior of the thylakoid vesicles.

thylakoid vesicles Flattened sacs or disks arising from the paired folds (lamellae) in chloroplasts.

titration curve A graph of the pH versus the equivalents of base added during the titration of an acid.

Tobacco mosaic virus An RNA virus infecting plants.

topoisomerase An enzyme capable of changing the linking number of a DNA molecule by breaking one or both strands of DNA, and rejoining the ends.

transcription An enzymatic process in which an RNA copy is made of the sequence of bases along one strand of DNA.

transcription attenuation A regulatory mechanism that manipulates transcription termination or transcription pausing to regulate gene transcription downstream.

transduction The transfer of genetic information from one cell to another by means of a viral vector.

transfection The uptake of viral DNA by competent cells.

transfer RNA (tRNA) Ribonucleic acids that serve as carriers of amino acid residues for protein synthesis.

transformation The uptake, integration and expression of naked DNA by competent cells.

transgenic animals Animals that have acquired new genetic information as a consequence of the introduction of foreign genes.

transition state analogs Stable molecules that are chemically and structurally similar to the transition state and that bind more strongly than a substrate or a product to the active site of an enzyme.

transition state A high-energy, unstable arrangement of atoms, in which bonds are being broken and formed. It represents the transition between reactants and products in a chemical reaction.

translation The process that converts genetic information embodied in the base sequence of a messenger RNA molecule into the amino acid sequence of a polypeptide chain.

transpeptidation The peptide bond-forming reaction in protein synthesis.

transport proteins A class of proteins whose job is to transport specific substances from one place to another.

transposons Segments of DNA that are moved enzymatically from place to place in the genome, the smallest of which is called an insertion sequence or IS.

triacylglycerols (triglycerides) Molecules consisting of glycerol esterified with three fatty acids.

tricarboxylic acid cycle (TCA cycle; citric acid cycle; Krebs cycle) A metabolic pathway in which an acetyl group is oxidized to CO_2, and the energy released is captured as ATP, NADH, and $FADH_2$.

turnover number (k_{cat}) The measure of an enzyme's maximal catalytic activity, or the number of substrate molecules converted into product per unit of time when the enzyme is saturated with substrate.

• •

Ultrafiltration A technique for removing small molecules from solutions of macromolecules and/or for concentrating dilute solutions of macromolecules.

uncoupler A compound that disrupts the normally tight coupling between electron transport and the phosphorylation of ADP.

unsaturated fatty acid A fatty acid that contains one or more carbon-carbon double bonds.

ureotelic organisms Organisms, such as terrestrial vertebrates, which excrete excess nitrogen as urea.

uricotelic organisms Organisms, such as birds and reptiles, that excrete nitrogen as uric acid.

• •

Van der Waals forces Weak, chemical forces induced by electrical interactions between approaching atoms or molecules as their electron clouds fluctuate instantaneously in time.

viruses Supramolecular complexes of nucleic acid (DNA or RNA), encapsulated in a protein coat and in some instances, surrounded by a membrane envelope. Within a host cell, they become self-replicating.

• •

Waxes Esters of long-chain alcohols with long-chain fatty acids.

weak chemical forces Forces ranging from 4 to 30 kJ/mol, including hydrogen bonds, van der Waals forces, ionic bonds and hydrophobic interactions.

wild-type The reference phenotype.

wobble position The first base of the anticodon. It pairs with the third base of a codon.

• •

Z-scheme A representation of the photosynthetic electron transport chain in which the electron carriers are arranged according to their redox potentials.

zinc finger A description of a structural motif found in DNA-binding proteins.

zwitterion A neutral molecule with both a positive and a negative charge.

zymogen (proenzyme) An enzyme that exists as an inactive precursor until one or several of its peptide bonds are cleaved. At that point the enzyme acquires full enzymatic activity.